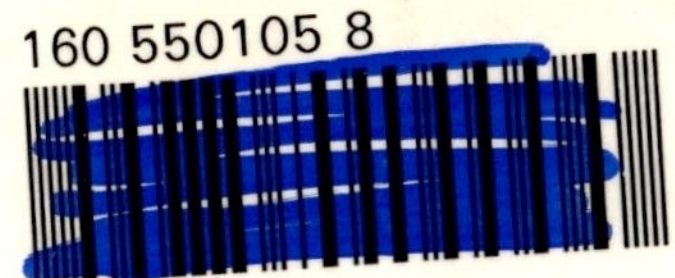
160 550105 8

AF616301

Prediction of Solar Radiation on Inclined Surfaces

Solar Energy R&D in the European Community

Series F:

Solar Radiation Data

Volume 3

Publication arrangements: T. C. Jones

Solar Energy R&D
in the European Community

Series F Volume 3
Solar Radiation Data

Prediction of Solar Radiation on Inclined Surfaces

edited by

J. K. PAGE
Department of Building Science,
University of Sheffield, U.K.

D. Reidel Publishing Company
A MEMBER OF THE KLUWER ACADEMIC PUBLISHERS GROUP

Dordrecht / Boston / Lancaster / Tokyo

for the Commission of the European Communities

Library of Congress Cataloging in Publication Data

CIP

Solar radiation data.

(Solar energy R & D in the European Community. Series F, Solar radiation data; v. 3)
Includes index.
1. Solar radiation–Europe–Observations. 2. Solar energy–Europe–Observations. I. Page, J.K. II. Series.
QC911.82.E85S67 1986 551.5'271'094 86-6552
ISBN 90-277-2260-9

Publication arrangements by
Commission of the European Communities
Directorate-General Information Market and Innovation, Luxembourg

EUR 10482

LEGAL NOTICE
Neither the Commission of the European Communities nor any person acting on behalf of the Commission is responsible for the use which might be made of the following information.

Published by D. Reidel Publishing Company
P.O. Box 17, 3300 AA Dordrecht, Holland

Sold and distributed in the U.S.A. and Canada
by Kluwer Academic Publishers,
190 Old Derby Street, Hingham, MA 02043, U.S.A.

In all other countries, sold and distributed
by Kluwer Academic Publishers Group,
P.O. Box 322, 3300 AH Dordrecht, Holland

Printed in The Netherlands

PREFACE

This book is aimed to help solar energy practitioners in all fields by making available well documented and tested meteorologically based calculation techniques for estimating the solar radiation falling on inclined planes, presented in a practical form that can be used by designers. It is the result of several years' work involving several institutions in the European Community associated with the first and second four year programmes in Project F - Solar Radiation Data, which has formed part of the Community's Solar Energy Research and Development programme. The publication deals with methods for the mathematical prediction of hourly and daily values of solar radiation falling on inclined planes under clear conditions in climates of differing atmospheric clarity, under overcast conditions, and also monthly mean hourly and daily values of solar radiation on inclined surfaces, using as inputs, monthly means of the daily duration of bright sunshine, a widely observed meteorological parameter and other data from the CEC European Solar Radiation Atlas, Vol. I, Global radiation on horizontal surfaces. This project has been based on the combination of theoretical studies and pragmatic studies. The work reported has been extensively checked against practical measurements of actual inclined surface radiation in Europe.

In the first four-year Community programme, a number of institutions participating in the programme worked on the development of methodologies for the estimation of inclined surface solar radiation. Each institution was asked to compare the results of their prediction method with actual observations of inclined surface irradiation made by reliable observing organisations. These observations were made available on a special CEC solar radiation magnetic tape prepared collaboratively under the first phase of CEC Project F. This tape included a considerable number of simultaneous observations of global and diffuse horizontal and inclined surface irradiation, made at the hourly level, with associated data on duration of bright sunshine usually at the daily level. At an inter-comparison of models at a CEC meeting held in Berlin towards the termination of the first four-year programme in November 1979, two models emerged as being significantly more accurate than the others at the hourly prediction level. One model had been developed in the Institut für Lichttechnik in the Technical University of Berlin by a team led by Professor Krochmann, the key research associates being Dr. Rattunde and Dr. Aydinli. The other successful model had been developed by Professor Page and his team in Sheffield, the key research associates being Dr. Rodgers and Dr. Souster. Both models drew on the two institutions' experience relating to daylighting design in order to derive models that took proper account of the actual radiance distribution of clear and overcast skies. It was shown that isotropic sky models produced unreliable results. The results of the calculations using these two methods showed good agreement between calculated and measured values of the monthly mean global slope irradiation with an uncertainty of $<\pm 10\%$ for most stations.

The Commission accordingly invited Professor Krochmann and Professor Page to liaise on the production of a final recommended computational method for Europe as part of the work of the second four-year programme under Project F. A considerable amount of additional work was carried out to refine and improve the original models before combining them to give a unified model. The second stage of the contract with Berlin was with the Lichtemesstechnik Berlin, but the same personnel were involved as before, Professor Krochmann, Dr. Rattunde and Dr. Aydinli. This publication refers to the Berlin team in many places, but does not attempt to distinguish which part of the work was done within which Berlin institution. The Berlin team had the lead responsibility for the overcast day model, which was extensively checked against observations made in the Community area. The Sheffield team consisting primarily of Professor Page and R.J. Flynn, had the lead responsibility for further development of the clear day model. A formal scientific liaison concerning the measured radiance distribution of the sky was set up with Dr. Valko of the Schweizerische Meteorologische Anstalt, Zurich, Switzerland. New clear day diffuse models for predicting horizontal and inclined surface diffuse sky irradiances have been developed and successfully checked against data sets of horizontal and slope observations for a number of European sites. The merging of the two models for monthly mean predictions was a joint task between the two institutions. The detailed assembly and description of the algorithms was carried out in Sheffield. The preparation of this book and also most of the tables for the manual computing process with its associated methodology was handled in the Department of Building Science, University of Sheffield.

Substantial contributions and important advice were received from other participants in the programme. The long experience of the team leader of Project F, Action 2.2, Inclined Surface Prediction, Monsieur Dogniaux of the Institut Royal Meteorologique de Belgique, was particularly valuable to the project. He provided sound meteorological advice throughout and many detailed comments on the draft reports as well as many practical contributions. Dr. Kasten of the Deutscher Wetterdienst, Hamburg, also made many important contributions to the success of this project. Liaison with him was particularly important in view of his lead role in the preparation of the Community's European Solar Radiation Atlas, Vol. I, 2nd Edition, Global radiation data - horizontal surfaces, the key source of inputs for the inclined plane prediction process.

Specially selected clear day data for checking the clear day data was made available for checking in Sheffield by Mr. Slob of the Koninklijk Nederlands Meteorologisch Institut, by Dr. Kasten of the Deutscher Wetterdienst, Hamburg and by Mr. Durbin of the UK Meteorological Office, Bracknell. One must also not forget to acknowledge the vital contribution of the many observing scientists whose reliable data mounted on the magnetic tape from careful measurements with well calibrated instruments over many years provided a sound observational basis for the detailed checking of these theoretical models against field measurements on inclined surfaces. In concluding this preface, it is again stressed that the primary aim throughout the project has been to produce a sound technical document of real practical value to engineers, architects and other system designers, providing full details of all algorithms used, so that better estimates can be made of solar energy available on slopes in the Community area in the future.

During the four year programme just completed, more observational data has been available for those areas north of the Alps than for the areas to the south of the Alps, so more checks have been carried out for this region of Europe than has been possible further south. However, there is no reason to believe why the computing methods proposed in this publication will not prove equally satisfactory in all areas of Europe. It is also believed that the techniques proposed will have a world wide validity, provided appropriate knowledge can be made available about the local clarity of the atmosphere.

The techniques described in this book were used to produce Tables of solar radiation on inclined surfaces for 102 sites in Europe of solar radiation for monthly mean conditions and for clear day conditions. These tables were also used to draw maps of the solar radiation on inclined surfaces across Europe. These maps and tables have already been published in the CEC, European Solar Radiation Atlas, Vol. II, Inclined Planes, TUV, Verlag, 1984.

The methodologies given in this Book are also now available as colloquial interactive programs for microcomputer and main frame computers. This will facilitate the use of the methodologies for practical engineering design.

Principle sources of contributions

The material provided came from many sources. It is only possible to acknowledge in detail the principal contributors. Chapter 1, which was prepared by J.K.Page, drew heavily on the work of a CEC Working Party chaired by J.W. Gruter, concerned with units and symbols. Chapter 2 was written by J.K. Page and his research collaborators, R.J. Flynn and P.R. Gill. Chapter 3, Parts I and III drew heavily on material submitted by P. Valko of the Schweizerische Meteorologische Anstalt, Zurich, Switzerland. Theoretical material was added by J.K. Page, especially in Parts II and IV. Chapter 4 was prepared by S. Aydinli of the Institut für Lichttechnik, Technical University of Berlin, F.R. Germany. Chapter 5 was prepared by J.K. Page. The practical instructions for computation in Chapter 6 were prepared by J.K. Page and his research assistant, R.J. Flynn. The chapter on other applications including daylighting was prepared by J.K. Page, who also prepared Appendices 1 to 3.

Appendix 1 drew heavily on the work of the Working Group on Symbols and Terminology chaired by J.W. Gruter. Appendix 2 was strongly based on the work of M. Dogniaux of the Institut Royal Meteorologique de Belgique, Brussels, Belgium. Appendices 3 and 5 were written by J.K. Page. Appendix 4 was written by P. Valko of the Schweizerische Meteorologische Anstalt, Zurich, Switzerland. Appendix 6 was an edited version of work by S. Aydinli, R. Rattunde and J. Krochmann of the Institut für Lichttechnik, Berlin, F.R. Germany. The remaining appendices, 7, 8, 9 and 10 were prepared by J.K. Page. The Print Unit of the University of Sheffield was responsible for the printed proformas, and the final preparation of the tables and diagrams for the book. The extensive detailed typing was carried out by Mrs. Lynn Mann of the Department of Building Science, University of Sheffield. Professor J.K. Page, formerly of the Department of Building Science, University of Sheffield, was responsible for editing all the material of this book for publication. The CEC would also like to thank other collaborators, not specifically mentioned for their various contributions, including both data, scientific information, and comments on drafts. It was not possible to list all the people concerned, but their contribution is gratefully acknowledged.

W. Palz, CEC, Brussels

December 1985

CONTENTS

Preface v

CHAPTER 1

SCIENTIFIC CONCEPTS, SYMBOLS AND TERMINOLOGY USED 1

CHAPTER 2

DEVELOPMENT OF A VALIDATED OBJECTIVE RADIATION MODEL FOR PREDICTING THE SOLAR RADIATION ON CLOUDLESS DAYS ON HORIZONTAL SURFACES IN THE EUROPEAN REGION

Abstract 13

Part I The prediction of solar geometry and daylength 14

Part II Transmission of the direct beam radiation through the cloudless atmosphere 17

Part III The estimation of diffuse solar radiation from cloudless skies 30

Part IV Construction of a daily irradiation model for horizontal surfaces for clear days and its application to the production of clear day tables for the European region using an objective method for estimating Linke Turbidity Factors for clear days 69

CHAPTER 3

THE RADIANCE OF CLOUDLESS SKIES AND THE GROUND IN EUROPE IN RELATION TO THE PREDICTION OF THE DIFFUSE RADIATION FROM THE CLOUDLESS SKY AND GROUND ON SLOPES AND THE CONSTRUCTION OF A CLOUDLESS DAY SLOPE RADIATION MODEL

Abstract 93

Part I The observed radiance distribution of cloudless skies and the prediction of the associated components of the diffuse irradiance on inclined planes 94

Part II The clear sky radiance model used as the basis for the CEC prediction process for estimating diffuse irradiance on inclined surfaces on clear days 147

Part III Estimation of reflected radiation from the ground 179

Part IV Clear day model adopted to produce inclined surface tables for CEC European Solar Radiation Atlas, Vol. II 207

CHAPTER 4

PREDICTION OF SOLAR RADIATION ON SLOPES - OVERCAST DAYS - WITH SPECIAL REFERENCE TO THE EEC REGION 213

CHAPTER 5

THE ESTIMATION OF MONTHLY MEAN HOURLY IRRADIANCE AND DAILY IRRADIATION ON INCLINED PLANES

Abstract 265

Part I Prediction of monthly mean hourly values of the direct beam irradiance 266

Part II Estimation of monthly mean diffuse irradiation on horizontal surfaces 280

Part III Modelling global, beam and diffuse irradiation on slopes 309

CHAPTER 6

INSTRUCTIONS FOR THE MANUAL CALCULATION OF HOURLY AND DAILY VALUES OF CLEAR DAY AND MONTHLY MEAN SOLAR IRRADIATION ON HORIZONTAL AND INCLINED SURFACES USING THE METHODOLOGY ADOPTED TO PRODUCE THE CEC INCLINED SURFACE RADIATION ATLAS 319

CHAPTER 7

OTHER APPLICATIONS OF THE EUROPEAN COMMUNITY SOLAR RADIATION MODEL 379

APPENDICES

APPENDIX 1

BASIC ALGORITHMS NEEDED TO PREDICT THE MOTION OF THE SUN, DAYLENGTH AND THE EXTRATERRESTRIAL IRRADIANCE 391

APPENDIX 2

A REVIEW OF METHODS FOR ASSESSING THE LINKE TURBIDITY FACTOR IN THE ABSENCE OF MEASUREMENTS OF SOLAR DIRECT BEAM INTENSITY 399

APPENDIX 3

THE RELATIONSHIP BETWEEN MONTHLY MEAN MAXIMUM DAILY SUNSHINE, S_{max} AND MONTHLY MEAN DAILY SUNSHINE 411

APPENDIX 4

THE USE OF SWISS RADIANCE SKY MEASUREMENTS TO ESTIMATE DIFFUSE IRRADIANCE FROM THE SKY 415

APPENDIX 5

DEVELOPMENT OF SUITABLE SIMPLIFIED ALGORITHMS FOR ESTIMATING THE CLEAR SKY DIFFUSE RADIATION ON SLOPES 421

APPENDIX 6

THE ESTIMATION OF MEAN MONTHLY HOURLY VALUES OF BRIGHT SUNSHINE FROM MONTHLY MEAN DAILY BRIGHT SUNSHINE. 433

APPENDIX 7

DETERMINATION OF THE RATIO OF THE CLOUDLESS SKY DIFFUSE SKY IRRADIANCE/ILLUMINANCE ON A HORIZONTAL SURFACE TO THE ZENITH RADIANCE/LUMINANCE OF THAT SKY 439

APPENDIX 8

ACCURACY OF PREDICTION OF DAILY GLOBAL RADIATION ON A HORIZONTAL SURFACE ACHIEVED BY THE EUROPEAN COMMUNITY SOLAR RADIATION MODEL 443

APPENDIX 9

COMPARISON OF OBSERVED RADIANCE RATIOS FOR LOW ELEVATIONS IN THE SKY WITH PREDICTED VALUES 451

APPENDIX 10

BLANK PROFORMAS FOR RADIATION CALCULATIONS USING EC METHODOLOGY, DESK TOP METHOD 455

CHAPTER 1

SCIENTIFIC CONCEPTS, SYMBOLS AND TERMINOLOGY USED

Introduction

1. Basic scientific concepts

This book is entirely dedicated to the prediction of solar radiation on inclined planes. This chapter deals only with basic concepts, terminology and symbols.

2. Solar radiation

The solar radiation, sometimes called short wave radiation, received at the surface of the earth, lies almost entirely in the wavelength band of 0.29 - 4 μm. The amount of solar energy beyond 2.5 μm is very small.

3. Thermal radiation

Thermal radiation from the atmosphere, usually called terrestrial radiation or long wave radiation has most of its energy at wavelengths greater than 4 μm. Thermal radiation is not discussed in this book. Its estimation has been discussed by Page (1) in another publication.

4. Pyrheliometric reference scale

This book presents most data on the World Radiometric Reference scale (WRR). Where possible any data measured on the International Pyrheliometric Scale (IPS), which was the international standard before the 1st January, 1981, were converted to WRR. The basic conversion factor is $I_{WRR} = 1.022\ I_{IPS}$ where I is the direct beam irradiance normal to the beam.

5. Basic scientific concepts concerning solar radiation

Three basic scientific concepts concerning short wave solar radiation fluxes are used in this book.

a) Irradiance, which is the flux of short wave solar radiation falling on unit area per unit time, i.e. power per unit area. The S.I. units for irradiance used here are watts per square metre, abbreviated as Wm^{-2}.

b) Irradiation, which is the amount of solar energy falling on unit area over a stated time. The irradiation is thus the integral with respect to time of the irradiance. The integration time has to be stated. The daily solar irradiation, often referred to as the solar radiation, is stated in the CEC Atlas in kilowatt hours per square metre, abbreviated as $kWhm^{-2}$. This unit is not an international standard unit. The S.I. unit normally used in meteorology is a megajoule per square metre, MJm^{-2}. The conversion factor is $3.6\ MJm^{-2} = 1.0\ kWhm^{-2}$. The unit of $kWhm^{-2}$ was adopted as the unit of daily irradiation in this book and in the two CEC Solar Radiation Atlases, in view of the familiarity of architects, engineers and householders with the unit kilowatt hour (kWh).

c) Radiance, sometimes loosely called the brightness, here normally refers to the short wave radiation received by scattering from the sky from a specific direction. The radiance is defined as the radiant flux per unit apparent area per unit solid angle emitted in a stated direction from a specific position in the sky. The units of radiance are watts per square metre per steradian (Wm^{-2} sr^{-1}).

6. Differentiation between irradiance and irradiation

In this book the same symbols are used for irradiance and for irradiation. However, the two concepts can be clearly differentiated both by context, and also by inspection of the attached units.

7. The components of solar radiation on inclined planes

There are three components of the solar irradiance/solar irradiation falling on an inclined plane of slope β and orientation α :

i) the direct beam component described by the symbol $I(\beta,\alpha)$

ii) the sky diffuse component described by the symbol $D_s(\beta,\alpha)$

iii) the ground reflected diffuse component described by the symbol $R(\beta,\alpha)$

The ground reflected component depends on the albedo of the ground in front of the plane. The albedo is the reflectance across the whole solar spectrum, visible and infra red. The term "ground albedo" thus describes the proportion of the incident solar radiation reflected from the ground. The sum of the sky diffuse component and the ground reflected diffuse component is described as the total diffuse component $D(\beta,\alpha)$.

8. Global radiation

The sum of the three components of solar irradiance/irradiation on any surface is known as the global irradiance/irradiation and is described by the symbol $G(\beta,\alpha)$. The global irradiation is often called the global radiation on a given plane.

9. Omission of reference to slope for horizontal planes

When reference is made to a horizontal plane, the convention normally used in this book is that the references to the slope tilt and orientation are omitted, i.e. G $kWhm^{-2}$ d^{-1}, refers to the daily global irradiation on a horizontal surface. This is usually called the global solar radiation. The only exception to the above rule concerns the direct beam irradiance, where the reference orientation is set perpendicular to the solar beam, i.e. I Wm^{-2} refers to the irradiance normal to the solar beam and $I(0,0)$ Wm^{-2} refers to the beam irradiance on the horizontal plane.

10. Solar constant

The irradiance normal to the solar beam outside the atmosphere at mean solar distance is known as the solar constant I_o. A value of 1367 Wm^{-2} has been used throughout this book for the solar constant.

11. Reference directions for angular measure

a) Latitudes are set positive in the northern hemisphere and negative in the southern hemisphere. Longitudes are set negative, if west of Greenwich and positive, if east of Greenwich.

b) Solar azimuth angles are measured from due south in northern hemisphere and from due north in southern hemisphere. Easterly directions are by convention negative, and westerly directions are positive.

c) The angular position of the sun is described with reference to the centre of solar disc. No angular allowance is made to allow for the effects of atmospheric refraction, except in the computation of optical air mass. Refraction effects are only significant at low solar altitudes and even then are small. The elevation of the sun is measured from the horizontal plane.

12. Choice of solar declination for computational purposes

The angle the rays of the sun make with the equatorial plane is known as the solar declination. The declination varies from a maximum value of 23°27' on June 21st to -23°27' on December 22nd. The geometrical movements of the sun on any particular day are determined by the declination. The tables of the monthly mean radiation values are computed using the monthly mean of the solar declination. Different declinations, chosen to be statistically associated with days of high radiation, are used to produce the mean monthly clear day radiation values. The specific dates for the declination chosen to compute clear day values are stated in Table 2.14 for estimates for clear sky conditions. The actual values of declinations used are given in Table 6.2. All computations were carried out in true local time (i.e. solar time).

13. Subscript conventions

The following subscript conventions apply throughout this book.

a) Subscript c refers to clear conditions. Such conditions have low mean daily cloud amounts, but not necessarily no cloud throughout the day.

b) Subscript m refers to monthly mean conditions.

c) Subscript b refers to overcast conditions (based on German bewolken = to cloud over).

d) Subscript s refers to diffuse solar radiation received by scattering from the sky.

e) Subscript o refers to an extraterrestrial or theoretical surface value, i.e. maximum possible daylength on a specific day S_0 hours.

14. Other basic features of terminology adopted

Roman letters are used to describe mainly quantities which are not dimensionless.

Greek letters are used to describe angles and other dimensionless quantities.

The letter f is reserved to describe correction and conversion functions. Each specific correction/conversion function is described by a specific numerical subscript, f_1 - f_7.

15. Specific symbols and terminology for solar radiation, and illuminance predictions

Roman letters - mainly physical quantities which are not dimensionless

Note: Conversion/correction functions with a symbol f_n are listed at the end.

a,b	Angstrom regression coefficients	dimensionless
B_s	Schuepp Turbidity Coefficient	dimensionless
c	Correction for summer time - hours	h
D_c	Clear sky horizontal surface diffuse irradiance	Wm^{-2}
$D_c(\beta,\alpha)$	Clear sky inclined surface total diffuse irradiance (sky + ground)	Wm^{-2}
$D_{cs}(\beta,\alpha)$	Clear sky inclined surface diffuse irradiance from the sky alone	Wm^{-2}
D_m	Monthly mean horizontal surface diffuse irradiance	Wm^{-2}
$D_m(\beta,\alpha)$	Monthly mean inclined surface total diffuse irradiance (sky + ground)	Wm^{-2}
$D_{ms}(\beta,\alpha)$	Monthly mean inclined surface diffuse irradiance from sky alone	Wm^{-2}

Special diffuse radiation symbols used in mean monthly diffuse computations only

${}_cD_m(\beta,\alpha)$	Clear sky contribution to the monthly mean inclined surface diffuse irradiance (If a u is added to the subscript on the left, it refers to an uncorrected value)	Wm^{-2}
${}_{pc}D_m$	Contribution of overcast/partially clouded sky to monthly mean diffuse irradiance on a horizontal surface	Wm^{-2}
${}_{pc}D_m(\beta,\alpha)$	Contribution of overcast/partially clouded sky to an inclined surface monthly mean diffuse sky irradiance (If a u is added in the subscript, it refers to an uncorrected value)	Wm^{-2}

Symbol	Description	Units
$D_{.25}$	Estimated monthly mean horizontal diffuse irradiance for a value of σ_{4m} = .25	Wm^{-2}
$_uD_m$	Uncorrected horizontal diffuse irradiance for actual relative sunshine duration σ_{4m}	Wm^{-2}
$_uD_{ms}(\beta,\alpha)$	Sum of clear and overcast partially clouded skies contributions to inclined surface radiation before correction	Wm^{-2}
D_b	Overcast sky horizontal surface diffuse irradiance	Wm^{-2}
$D_b(\beta,\alpha)$	Overcast sky inclined surface total diffuse irradiance (sky + ground)	Wm^{-2}
$D_{bs}(\beta,\alpha)$	Overcast sky inclined surface diffuse irradiance from the sky alone	Wm^{-2}

Illuminance symbols

Symbol	Description	Units
$(ED)_c$	Clear sky diffuse illuminance on a horizontal surface	lux
$(ED)_c(\beta,\alpha)$	Clear sky diffuse illuminance on an inclined plane	lux
$ED_m(\beta,\alpha)$	Monthly mean diffuse illuminance on a slope	lux
$ED_b(\beta,\alpha)$	Monthly mean diffuse illuminance on a slope on overcast days	lux
$ER(\beta,\alpha)$	Ground reflected illuminance on a slope (subscript c, clear, m, mean, o, overcast	lux
$(EI)_c$	Clear sky direct beam illuminance normal to beam	lux
$(EL)_{cz}$	Clear sky zenith luminance	$cdm^{-2}sr^{-1}$
E.T.	Equation of time	hours
G_o	Monthly mean daily extraterrestrial global irradiation on horizontal surface corrected to mean solar distance	$Whm^{-2}\ d^{-1}$
G_c	Clear sky horizontal surface global irradiance	Wm^{-2}
G_c in Tables in Atlas	Clear sky daily global irradiation on stated slope	$kWhm^{-2}\ d^{-1}$

$G_c(\beta,\alpha)$	Clear sky inclined surface global irradiance	Wm^{-2}
G_m	Monthly mean horizontal surface global irradiance	Wm^{-2}
G_m in Tables in Atlas	Monthly mean daily global irradiation on a stated slope	$kWhm^{-2}\ d^{-}$
$G_m(\beta,\alpha)$	Monthly mean inclined surface global irradiance	Wm^{-2}
G_{max}	Mean monthly maximum daily global irradiation on a horizontal surface	$Whm^{-2}\ d^{-1}$
G_b	Overcast sky horizontal surface global irradiance	Wm^{-2}
$G_b(\beta,\alpha)$	Overcast sky inclined surface global irradiance	Wm^{-2}
H	Scale height of the atmosphere 8 km	Km
h	Station height above sea level - metres	m
I_c	Clear sky direct beam irradiance normal to beam	Wm^{-2}
$I_c(0,0)$	Clear sky horizontal surface direct beam irradiance	Wm^{-2}
$I_c(\beta,\alpha)$	Clear sky inclined surface direct beam irradiance	Wm^{-2}
I_m	Monthly mean direct beam irradiance normal to beam	Wm^{-2}
$I_m(0,0)$	Monthly mean horizontal surface direct beam irradiance	Wm^{-2}
$I_m(\beta,\alpha)$	Monthly mean inclined surface direct beam irradiance	Wm^{-2}
$I_{max}(\beta,\alpha)$	Mean monthly maximum direct beam irradiance	Wm^{-2}
I_o	Solar constant (1367 Wm^{-2})	Wm^{-2}
I_{oj}	Extraterrestrial irradiance at normal incidence on day J	Wm^{-2}
I^*	Direct beam irradiance normal to beam for perfectly clear atmosphere - Monteith and Unsworth formulation	Wm^{-2}
J	Day number	D

J'	Day angle	degrees
K_b	Luminous efficiency of overcast sky diffuse radiation	$lm\ W^{-1}$
K_d	Correction to the solar constant for the annual variation of sun-earth distance	dimensionless
K_{dc}	Luminous efficiency of clear sky diffuse irradiation on horizontal surface	$lm\ W^{-1}$
K_s	Luminous efficiency of direct beam	$lm\ W^{-1}$
$L_c(\Theta,\alpha_s)$	Clear sky radiance at elevation degrees and orientation of α_s to sun's direction	$Wm^{-2}\ sr^{-1}$
L_{cz}	Clear sky zenith radiance	$Wm^{-2}\ sr^{-1}$
m	Relative optical air mass	dimensionless
p/p_o	Station height correction to the relative air mass	dimensionless
$q_a{}^m$	Atmospheric transmittance coefficient due to absorption alone at air mass m	dimensionless
$R_c(\beta,\alpha)$	Clear sky ground reflected diffuse irradiance on an inclined surface	Wm^{-2}
$R_m(\beta,\alpha)$	Monthly mean ground reflected irradiance on an inclined surface	Wm^{-2}
$R_b(\beta,\alpha)$	Overcast day ground reflected irradiance on an inclined surface	Wm^{-2}
S_m	Monthly mean daily hours of bright sunshine	h
S_{max}	Mean monthly maximum daily hours of bright sunshine	h
S_o	Astronomical daylength hours	h
S_{04}	Mean monthly daylength calculated on the basis of a 4 degree horizon for sunrise and sunset using monthly mean declination	h
S_{04max}	Daylength for calculating monthly maximum radiation on basis of 4 degree horizon for sunrise and sunset using declinations indicated in Table 6.2 in Chapter 6	h
T_L	Linke Turbidity Factor at air mass 2	dimensionless
$T_L(\gamma)$	Linke Turbidity Factor at solar altitude γ	dimensionless

t	Time of day in solar time	true local time
t_r	Time of sunrise in solar time	true local time
t_s	Time of sunset in solar time	true local time
w_c	Precipitable water vapour clear sky conditions	mm
w_m	Monthly mean precipitable water vapour	mm

Greek symbols - Angular measure and other dimensionless quantities*

α	Surface azimuth angle, easterly negative, westerly positive	degrees
α_S	Wall solar azimuth angle or horizontal shadow angle	degrees
β	Surface inclination of plane from horizontal plane (Also used for Angstrom Turbidity Coefficient)	degrees
γ	Solar altitude	degrees
γ_v	Vertical shadow angle	degrees
δ	Solar declination	degrees
δ_R	Rayleigh optical thickness	dimensionless
λ	Longitude, west of Greenwich negative east of Greenwich positive	degrees
λ_{ST}	Longitude of the time reference zone	degrees
ν	Angle of incidence of the solar beam on an inclined surface	degrees
ρ_s	Ground albedo (of the surrounding ground)	dimensionless
σ_h	Monthly mean hourly relative sunshine duration	dimensionless
$\sigma_m = S_m/S_o$	Monthly mean daily relative sunshine duration	dimensionless
$\sigma_{4m} = S_m/S_{04}$	Monthly mean daily relative sunshine duration based on 4 degree altitude of sunrise and sunset	dimensionless

* Some additional Greek symbols were used for radiance studies in Chapter 3. These are given in Section 17.

$\sigma_{4max} = S_{max}/S_{04max}$	Mean monthly maximum daily relative sunshine duration based on a 4 degree altitude of sunrise and sunset.	dimensionless
$cb^{\sigma}{}_{m}$	Monthly mean relative daily duration of sunshine on a card burn daylength	dimensionless
σ_{rel}	Relative bright sunshine probability	dimensionless
τ_{m}	Monthly mean Monteith and Unsworth pseudo-turbidity coefficient	dimensionless
Φ	Latitude, northern hemisphere positive southern hemisphere negative	degrees
ψ	Solar azimuth, east of N/S negative, west of N/S positive	degrees
ω	Solar hour angle, a.m. negative p.m. positive	degrees
ω_{o}	Solar hour angle at sunrise and sunset sunrise negative, sunset positive	degrees

Correction/conversion functions

f_1	Additional scattering correction for estimating cloudless sky horizontal diffuse irradiance as a function of solar altitude γ and air mass 2 Linke Turbidity Factor T_L	dimensionless
f_2	Conversion ratio to obtain inclined surface clear sky irradiance from the clear sky predicted horizontal surface irradiance as a function of slope tilt wall solar azimuth angle α_S, and air mass 2 Linke Turbidity Factor T_L	dimensionless
f_3	Conversion ratio to obtain ground reflected diffuse irradiance on an inclined plane from the horizontal surface irradiance as a function of plane inclination β	dimensionless
f_4	Conversion ratio to obtain the inclined surface overcast sky irradiance on an inclined plane β from the horizontal overcast horizontal surface irradiance	dimensionless
f_5	Additional correction function for estimating monthly mean direct irradiance as a function of σ_{4m} and solar altitude γ	dimensionless

f_6	Correction function to adjust uncorrected value of the monthly mean horizontal diffuse sky irradiance as a function of σ_{4m}, solar altitude γ, and annual mean Ångstrom (a + b)	dimensionless
f_7	Additional shading ring correction to allow for non-isotropic distribution of sky radiance	dimensionless

16. Symbols for other quantities

Water vapour in the atmosphere

e_s	Saturation vapour pressure of water in air at a given temperature	mb
e	Vapour pressure of water vapour	mb
h	Specific enthalpy of air	$kJkg^{-1}$
T_D	Dew point	°C
U	Relative humidity	%

Dry bulb temperature

T_c	Dry bulb temperature cloudless day	°C
T_m	Dry bulb temperature monthly mean	°C
T_b	Dry bulb temperature overcast day	°C

Wind speed

u_c	10m wind speed cloudless day	ms^{-1}
u_m	10m wind speed monthly mean	ms^{-1}
u_b	10m wind speed overcast day	ms^{-1}
h_{oc}	External surface heat transfer convection coefficient for forced convection	$Wm^{-2}C^{-1}$

17. Symbols used in the description of radiance patterns across sky and over ground

α	Horizontal angle between meridian direction of a patch and due south	degrees
α_S	Horizontal angle between meridian direction of a patch and the meridian direction of the sun	degrees
Θ	Angle of elevation of a patch viewed in the upper hemisphere, normally sky	degrees
$-\Theta$	Angle of depression of a patch viewed in the lower hemisphere	degrees
η	Angle between a specified patch of sky and the line joining the centre of the solar disc	degrees
ξ	Angle between the line joining the centre of the solar disc and the zenith direction	degrees
n (as a subscript)	In direction of the nadir i.e. straight down	
z (as a subscript)	In direction of zenith i.e. straight up	

CHAPTER 1 - References

1. Page, J.K. (1985), Estimation of long wave radiation exchanges between the external surfaces of buildings and sky and ground, in UK Data for Solar Energy Applications, Ed. J.K. Page & R. Lebens, to be published by H.M.S.O., UK Dept. of Energy, Energy Technology Support Unit.

CHAPTER 2

DEVELOPMENT OF A VALIDATED OBJECTIVE RADIATION MODEL FOR PREDICTING THE SOLAR RADIATION ON CLOUDLESS DAYS ON HORIZONTAL SURFACES IN THE EUROPEAN REGION

Abstract

Part I of this Chapter starts by describing the principles by which the solar geometry is computed for any date and time of day in the year. The detailed algorithms for predicting the solar geometry are given in Appendix 1.

Part II discusses the modelling of the transmission of the atmosphere, as a function of its clarity, to estimate the direct beam irradiance perpendicular to the solar beam, I_c and its subsequent conversion to the clear sky beam irradiance on a horizontal surface $I_c(0,0)$. The Linke Turbidity Factor is adopted to describe the clarity of the atmosphere. A detailed review of European based methods for the estimation of atmospheric turbidity is given in Appendix 2.

Part III develops the related model for estimating the cloudless day horizontal surface diffuse irradiance from the sky as a function of turbidity and solar altitude from observed diffuse radiation data. Thus combining the two models, provided a representative atmospheric turbidity for any month can be estimated, cloudless day estimates of horizontal global, beam and diffuse irradiance G_c, $I_c(0,0)$ and D_c can be made for any specific site. Detailed checks on the clear day diffuse irradiance model are presented.

Part IV of this Chapter discusses the construction of a mathematical model for predicting daily global and diffuse irradiation on horizontal planes, and its practical use to determine representative values of the Linke Turbidity Factor for clear days in Europe using observed values of monthly mean maximum global solar radiation on a horizontal surface and the monthly mean values of the sum of the Angstrom regression coefficients (a + b).

The modelling methods developed to determine the Linke Turbidity Factors are objective and require no judgements concerning atmospheric clarity at specific sites. The model used to construct the horizontal component of the CEC Inclined Surface radiation tables is explained in detail and recent improvements are described.

CHAPTER 2 - PART I

THE PREDICTION OF SOLAR GEOMETRY AND DAYLENGTH

1. Introduction

The first stage in systematic solar radiation modelling is to develop a satisfactory method for assessing the solar radiation falling on horizontal surfaces on cloudless days. Establishing the solar geometry is obviously the starting point. The solar energy transmitted through the atmosphere depends on a number of factors, but the path length through the atmosphere is critical. The amount of aerosols in the atmosphere at the site in question is obviously very important. The lower the sun, the greater the additional attenuation due to aerosols. The aerosols may be naturally generated, for example water droplets or dust from deserts, or manmade, for example smoke pollution.

The aerosols have two effects on solar radiation from a cloudless sky:

i) they decrease the intensity of direct irradiance normal to the beam I_c.

ii) they augment the intensity of diffuse energy reaching a horizontal surface from the sky, D_c, because over half the energy scattered from the beam by the aerosol tends to be scattered forwards and reaches the surface as additional diffuse energy.

The colour of the sky, in fact, is a useful indicator of the clarity of the atmosphere. The deeper the blue, the greater the clarity. In very humid tropical areas, there may be so much additional scattered energy, that the sky takes on a milky white appearance even under cloudless sky conditions.

2. Assessing the position of the sun

Before any irradiance predictions can be made, one must determine the position of the centre of the solar disc. This position is most conveniently expressed in relation to the horizontal plane. The position of the centre of the solar disc may be defined by two angles, the solar altitude which is the angle between a line joining the centre of the solar disc and the horizontal plane, and the solar azimuth, which is the angle, measured on the horizontal plane, between the solar meridian and, in the Northern Hemisphere, a line running due south. In the Southern Hemisphere, the azimuth angle is measured from a line running due North. The solar meridian is the vertical plane which runs through the centre of the solar disc. Its direction is defined by its intercept on the horizontal plane. These angles are illustrated in Figure 2.1.

The CEC computational method neglects the small effects of atmospheric refraction in determining the solar geometry. Such effects are normally negligibly small and only important at very low solar altitudes, when the energy available from the sun is too low to be of practical engineering significance. The solar altitude and azimuth angle for a given latitude are a function of the solar declination, which is the angle that the sun's rays make with the equatorial plane at any instant. The solar declination varies continuously throughout the year but the change across any one day is sufficiently small to make it possible to neglect the small differences occurring across the day. The representative solar altitude angle and azimuth angle for each month in this publication are determined from the noon values of the solar declination for two given dates in each month, using a standard algorithm to estimate the declination for any date, which is approximated as being invariant from year to year. The EC method uses long term mean values of the declination based on a year length of 365.25 days. For very accurate calculations on specific days, it is necessary to take account of leap years, etc.

3. Inclined surface solar geometry

When predictions are required for surfaces other than horizontal, the position of the sun must be stated in relation to that particular surface. This position in relation to the sun may be defined by two angles, i) the angle of incidence of the sun's rays on the surface and ii) the wall solar azimuth angle. The angle of incidence of the sun's rays is always measured from a line normal (i.e. perpendicular) to the surface. The wall solar azimuth angle is the angle measured on the horizontal plane, between the solar meridian and the vertical meridian containing the line normal to the surface. These two angles are also illustrated in Figure 2.1.

4. Computing the solar geometry

The detailed algorithms for computing the solar geometry are given in Appendix 1, which also contains details of how to compute the astronomical daylength, and also the daylength with the solar altitude above 4°, which is used in the methodology developed in this book to define the practical daylength for sunshine recorders. A system of time has also to be established. All calculations in this book have been carried out using local apparent (solar) time. This system of time offers the advantage of symmetry of the solar geometry about the north-south line. Appendix 1 contains the algorithms needed to convert from local standard time to local apparent (solar) time.

5. Horizontal surface global irradiance

The global irradiance on a horizontal surface under cloudless sky conditions may be estimated from the sum of its two components, the direct beam irradiance, $I_c(0,0)$, and the sky diffuse irradiance, D_c, on a horizontal surface. Thus, noting that the angle of incidence of the sun's rays on a horizontal surface is $\sin \gamma$, where γ is the solar altitude measured from the horizontal plane, G_c is given by:

$$G_c = I_c \sin \gamma + D_c \qquad Wm^{-2} \tag{2.1}$$

where I_c is the cloudless sky irradiance normal to the beam.

Part II of this Chapter deals with the estimation of I_c and Part III with the estimation of D_c.

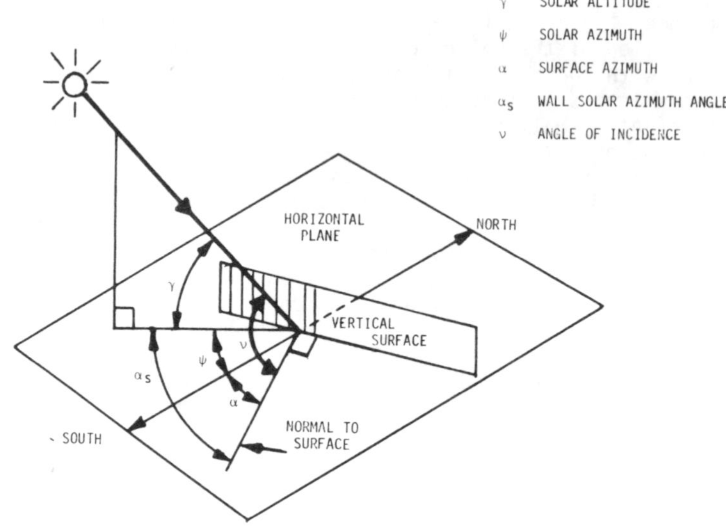

Figure 2.1 Solar and inclined surface geometry for a vertical surface ($\beta = 90°$). For other slopes, the normal would not be in the horizontal plane. The meridian direction of inclined planes is obtained by projection of the normal onto the horizontal.

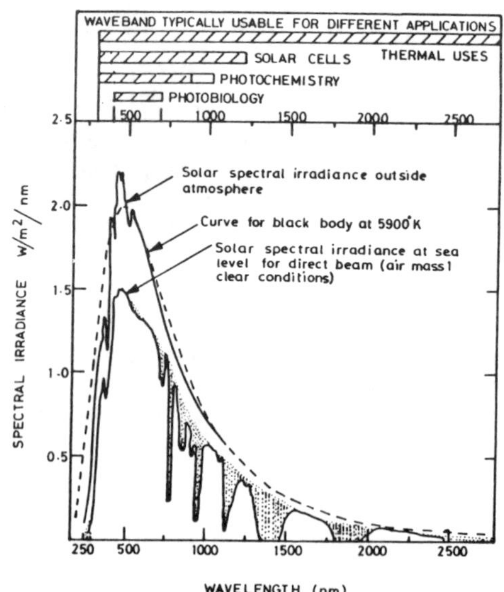

Figure 2.2 Spectral irradiance curves for direct sunlight extraterrestrially and at sea level; shaded areas indicate absorption due to atmospheric constituents, mainly H_2O, CO_2 and O_3. Wavelengths potentially utilised in different solar energy applications are indicated at the top. Air mass 1

CHAPTER 2 - PART II

TRANSMISSION OF THE DIRECT BEAM RADIATION THROUGH THE CLOUDLESS ATMOSPHERE

6. Introduction

This part of Chapter 2 deals with the transmission of direct solar beam radiation through the cloudless atmosphere. It starts with a theoretical review of the transmission of solar energy through the cloudless atmosphere. It then discusses the practical assessment of the transmission properties of the atmosphere for solar radiation using the Linke Turbidity Factor to describe the clarity of the atmosphere. The corrections adopted to allow for the effects of solar altitude on the Linke Turbidity Factor are explained. Appendix 2 reviews more specialised scientific information inter-relating the various approaches to estimating the Linke Turbidity Factor in Europe.

7. Extraterrestrial irradiance

The atmosphere attenuates the extraterrestrial irradiance. For practical applications the output of the sun may be considered to be invariant with time. However, account has to be taken of variations in the distance between the sun and the earth in determining the extraterrestrial irradiance at any given date. The extraterrestrial irradiance normal to the solar beam at mean solar distance is known as the solar constant. The internationally agreed value of 1367 W/m^2 has been adopted in the EC inclined surface calculation method. The standard algorithm used to correct the solar constant for solar distance to calculate the extraterrestrial irradiance on any date is given in paragraph 18.1.

8. Theoretical considerations - monochromatic transmission theory

Figure 2.2 shows the solar spectrum outside the atmosphere, and also the solar spectrum after the energy has traversed the atmosphere along a zenith path. The substantial modifications are obvious. Some energy is lost by scattering, and some by absorption. The absorption tends to be concentrated in specific bands, and the solar spectrum after transmission through the atmosphere is no longer a continuous smooth spectrum with a form approaching blackbody emission.

The effects at any specific wavelength λ can be described in terms of Beer's Law as

$$I(\lambda) = K_d \, I_o(\lambda) \, e^{-m\,\delta(\lambda)} \qquad (2.2)$$

where $I(\lambda)$ is the spectral irradiance received in a given waveband

$I_o(\lambda)$ is the spectral irradiance in that waveband outside the atmosphere at mean solar distance

K_d is the correction to allow for variations in solar distance during the course of the year m	is the optical path length through the atmosphere expressed as a ratio to the zenith path length. Refer Figure 2.3.

$\delta(\lambda)$	is the extinction coefficient at unit air mass at wavelength λ.

The fraction of energy at wavelength λ transmitted at unit air mass is given by $e^{-\delta(\lambda)}$ and this is known as the spectral transmittance $T(\lambda)$ per unit air mass.

It follows that

$$I(\lambda) = K_d \, I_o(\lambda)(T(\lambda))^m \tag{2.3}$$

9. Factors reducing the beam strength

One can identify six major factors reducing the direct solar beam strength at any given wavelength:

i) the relative optical path length through the atmosphere which is termed the relative air mass m, the zenith path length to sea level being defined by m = 1. Refer Figure 2.3. Allowances have to be made for site elevation above sea level.

ii) the Rayleigh scattering due to the air molecules. The scattering is inversely proportional to approximately the fourth power of the wavelength, so blue light is preferentially scattered out of the beam and appears as a dominant feature of the cloudless sky.

iii) Absorption by ozone, especially dominant in the ultra violet region, but also exerting some influence on the middle part of the visible spectrum.

iv) Absorption by water vapour, where the absorption bands lie principally in the infra red part of the spectrum. Refer Figure 2.2.

v) Absorption by other gases, like CO_2, oxides of nitrogen, mainly important again in the infra red part of the spectrum. Refer Figure 2.2.

vi) Absorption/scattering by the atmospheric aerosol, both natural and manmade, e.g. dust from deserts, smoke, etc.

The overall transmittance of the atmosphere $T(\lambda)$ per unit air mass at wavelength λ can thus be expressed as the product of the separate transmittance factors due to the above factors.

$$T(\lambda) = [T_R(\lambda) \times T_O(\lambda) \times T_G(\lambda) \times T_W(\lambda) \times T_D(\lambda)] \tag{2.4}$$

where $T_R(\lambda)$	is the Rayleigh transmittance per unit air mass at wavelength λ

$T_O(\lambda)$	is the ozone transmittance per unit air mass at wavelength λ

$T_G(\lambda)$	is the transmittance factor for gases other than ozone and water vapour, like CO_2, NO_2, etc., per unit air mass at wavelength λ.

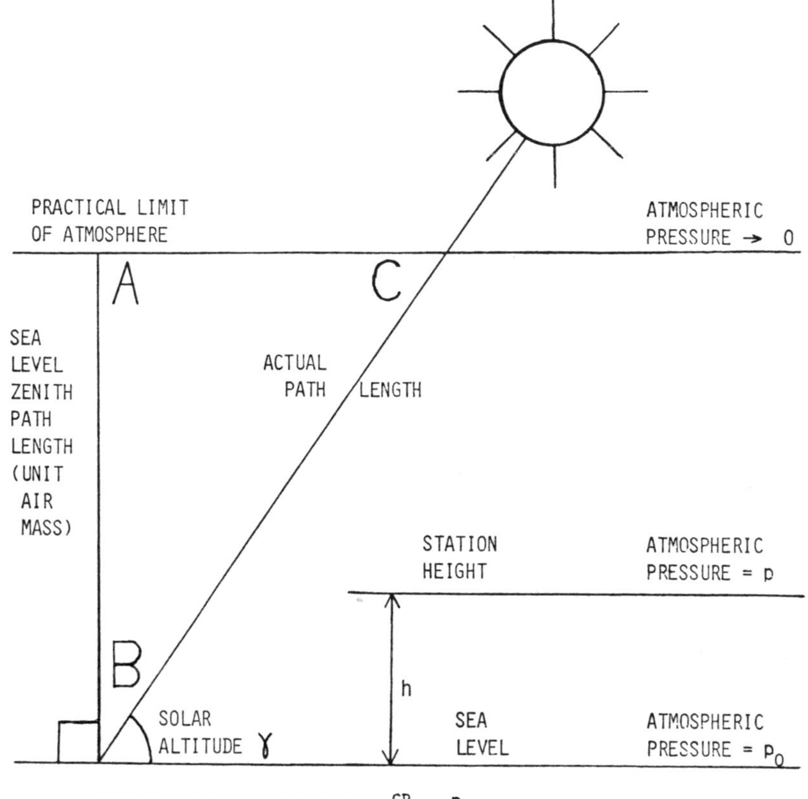

Figure 2.3 The concept of relative air mass (also known as optical air mass). At solar altitudes below 10°, allowances have to be made for atmospheric refraction and the curvature of the earth's surface. These allowances are covered by Kasten's formula (7).

$T_W(\lambda)$ is the water vapour transmittance per unit air mass at wavelength λ

$T_D(\lambda)$ is the aerosol transmittance per unit air mass at wavelength λ

The last two factors are very site-dependent; the other factors relate more to the general properties of the atmosphere, and are only site-dependent in so far as the values of $T_R(\lambda)$ and the optical air mass are influenced by station altitude above sea level.

10. Rayleigh scattering

The Rayleigh scattering can be predicted theoretically from the properties of the atmosphere, so enabling $T_R(\lambda)$ to be estimated with reasonable accuracy, for example the recent work of Frohlich (1) as corrected by Young (2).

11. Ozone absorption

Ozone absorbs very strongly in the region 290 - 350 nm and rather less strongly in the region 450 - 770 nm. The ozone content of the atmosphere varies with time of year. Representative values are tabulated in Robinson (3). At 50°N latitude, the typical minimum occurs in October at about 0.28 cm NTP and the maximum in April at 0.38 cm NTP. Most of the effects of variations are experienced in the ultra violet region, but there is some influence on the visible region. At 550 nm the transmittance factors due to absorption by ozone are, for O_3 content of 0.38 cm NTP, .967 at air mass 1, .935 at air mass 2 and .874 at air mass 4, while for O_3 content of 0.28 cm NTP, the corresponding values are .975 at air mass 1, .950 at air mass 2 and .904 at air mass 4. The annual percentage differences in the direct beam spectral irradiance at 550 nm due to changes in ozone content thus vary from about 1% at high solar altitudes to about 3.5% at low solar altitudes.

12. Water vapour absorption and the concept of precipitable water vapour

The water vapour bands are located in the infra red part of the spectrum. Water vapour removes considerable amounts of energy in specific wavebands of the solar spectrum, refer Figure 2.2. The amount of absorption depends on the amount of water vapour in the path length w_p. w_p may be derived from the amount of water vapour in the atmosphere in a vertical column above the site. If all the water vapour in the vertical column could be condensed to liquid, the resulting depth of liquid water is known as the precipitable water vapour content w. The water vapour in the path length m is given by:

$$w_p = w \cdot m$$

where m is the relative air mass.

w is normally quoted in centimetres. The amount of water vapour w in the atmosphere varies from area to area and from day to day. For low level stations, it may be estimated to a first approximation from Hann's formula.

$$w = 0.17\, e \quad \text{cm} \tag{2.5}$$

where e is the surface water vapour pressure in the atmosphere in mb or hPa.

Table 2.1 provides typical values of w for a wide range of latitudes.

Table 2.2 contains representative monthly values of precipitable water vapour for the UK for cloudless days and the average for all days. It will be noted that there is less water vapour present on cloudless days than on average days.

Once the precipitable water vapour is known, it is theoretically possible to use the water vapour absorption coefficient for any waveband to find the spectral transmittance, using the fundamental measured properties of water vapour at different wavelengths which are a function of pressure and the vertical pressure characteristics of the atmosphere.

In the humid tropics the amount of precipitable water vapour is very high, reaching sometimes 10 cm, compared with typical temperate country values of 1-2 cm. This reduces the strength of the direct beam appreciably but the reduction is not pronounced in the visible region, as the reduction lies mainly in the infra red region of the spectrum. In desert areas, the precipitable water vapour is normally low and very little of the infra red radiation is removed by water vapour absorption, though the beam may be substantially weakened by the high dust content.

13. Influence of other gases

The most important of these gases is CO_2 which has absorption bands mainly at the far end of the infra red solar spectrum. Accurate calculations of spectral irradiance have to consider absorption effects of CO_2, oxides of nitrogen and so on. It is normally assumed there are no significant variations from site to site due to these causes.

14. Absorption and scattering due to aerosol

There are strong variations in absorption and scattering due to aerosols from site to site due to differences in the amount of atmospheric aerosol present. The absorption and scattering due to aerosol is complicated to estimate. The size and number of the particles and their refractive index influence the scattering patterns. Under certain conditions there may be marked forward scattering. There are two widely used approaches to describe the absorption effects of aerosols.

a) Angstrom Turbidity Coefficient β_A

Angstrom suggested the transmittance due to aerosol could be expressed as a function of wavelength and air mass m in the exponential form:

$$T_D(\lambda) = e^{-f(\lambda)} \tag{2.6}$$

where $f(\lambda)$ is a wavelength-dependent function.

The practical function suggested by Angstrom is:

$$f(\lambda) = \beta_A/\lambda^{\alpha} \tag{2.7}$$

where λ is the wavelength in micrometres

β_A is the extinction coefficient corresponding to a wavelength of 1000 nm.

and α is an exponent closely correlated to the size of the scattering particles, and the pattern of their size distribution which determines the wavelength dependence of the attenuation. β_A is known as the Angstrom turbidity coefficient.

In Europe the value of the Angstrom turbidity coefficient β_A normally increases from winter to summer, and is greatest in urban areas, and lowest in rural areas. A commonly adopted representative value of the wavelength-dependent coefficient α is 1.3. Adoption of a value of α enables the aerosol transmittance at any wavelength λ, $T_D(\lambda)$, to be established from the value of β_A.

Typical European values of β_A are:

Rural sites	$\beta_A = 0.05$
Urban sites	$\beta_A = 0.10$
Industrial sites	$\beta_A = 0.20$

b) Schuepp Turbidity Coefficient B_S

In the Schuepp formulation, the overall Rayleigh scattering, water, aerosol and ozone absorption are also allowed for separately. However, when, considering aerosol effects, a reference wavelength of 500 nm is used for defining the Schuepp Turbidity Coefficient to allow for the atmospheric aerosols instead of a wavelength of 1000 nm as used by Angstrom. The wavelength of 500 nm is close to the waveband of maximum energy in the terrestrial solar spectrum. The additional transmittance factor per unit air mass due to aerosol, $T_D(\lambda)$, is defined in the Schuepp system as a power function of 10 rather than an exponential power function.

Thus

$$T_D(\lambda) = 10^{-B_S/(2\lambda)^{\alpha}} \tag{2.8}$$

where B_S is the Schuepp turbidity coefficient

and α is the wavelength exponent as before, the wavelength λ being expressed in micrometres.

If one adopts a value of $\alpha = 1.3$, then the Schuepp B_S coefficient can be related to the Angstrom β_A coefficient which is expressed in the exponential form by the expression:

$$B_S = \log_{10} e \times \beta_A \times 2^{\alpha} \tag{2.9}$$

$$= 0.434 \times 2^{1.3} \times \beta_A = 1.07\ \beta_A.$$

If a wavelength-dependent exponent of $\alpha = 2.0$ is used then

$$B_S = 1.74\ \beta_A \tag{2.10}$$

Table 2.1 Depth of precipitable water vapour in the atmosphere w, typical seasonal variations with site elevation and latitude. Units: centimetres.

Altitude expressed as atmospheric pressure		Warm or humid season Latitude 0°	30°	45°	60°	70°	Cold or dry season Latitude 0°	30°	45°	60°	70°
1000 mb	mean	5.0	4.0	2.5	2.0	1.8	3.0	1.5	0.8	0.5	0.3
	min	2.0	2.0	1.0	0.7	0.7	1.0	0.4	0.4	0.2	0.1
	max	10.0	7.0	4.0	4.0	4.0	7.0	4.0	2.0	1.0	1.0
900 mb	mean	3.0	1.9	1.6	1.25	1.1	2.0	1.0	0.5	0.35	0.2
	min	1.0	0.7	0.7	0.4	0.4	0.7	0.4	0.2	0.2	0.1
	max	7.0	4.0	4.0	2.0	2.0	4.0	2.0	1.0	1.0	0.7
800 mb	mean	2.0	1.5	1.0	0.8	0.7	1.0	0.6	0.3	0.2	0.1
	min	1.0	0.7	0.4	0.4	0.2	0.4	0.2	0.1	0.1	0.05
	max	4.0	4.0	2.0	2.0	2.0	2.0	2.0	1.0	0.7	0.2
700 mb	mean	1.0	0.8	0.5	0.4	0.35	0.6	0.3	0.15	0.1	0.05
	min	0.4	0.4	0.2	0.1	0.2	0.2	0.1	0.1	0.05	0.02
	max	2.0	2.0	1.0	1.0	1.0	1.0	1.0	0.4	0.2	0.1

Table 2.2 Precipitable water content above the surface for the UK from i) the UK Meteorological Office and ii) Bannon and Steele.

	Precipitable water cm		
	i) Meteorological Office Clear days		ii) Bannon and Steele
	Mean	Range	Average of all days
January	0.5	0.3-0.7	1.1
February	0.5	0.3-0.7	1.1
March	0.6	0.3-1.0	1.2
April	0.7	0.3-1.5	1.3
May	1.0	0.7-2.0	1.6
June	1.5	1.0-2.5	2.0
July	2.0	1.5-2.5	2.4
August	2.0	1.5-2.5	2.4
September	1.5	1.0-2.0	2.1
October	1.0	0.7-1.5	1.7
November	0.7	0.3-1.0	1.3
December	0.5	0.3-0.7	1.1

The Schuepp formulation is widely used in certain countries of Europe. For example, it is referred to quite often in Valko's contribution to Chapter 3 from Switzerland.

15. Overall transmittance for whole spectrum based on monochromatic theory

As the values of $T_R(\lambda)$, $T_O(\lambda)$, $T_W(\lambda)$ and $T_G(\lambda)$ can be established from the basic properties of the atmosphere and the absorbing characteristics of the different gases, a complete solution to find the direct beam irradiance is achievable from knowledge of the extraterrestrial solar spectrum and the application of the physical values of the various transmittance coefficients, but it is not normally practicable to do it for specific observations. Thus

$$I_c = \sum_{\text{waveband } n=1}^{\text{waveband } n=N} K_d \; I_o(\lambda)_n \; (T_R(\lambda) \times T_O(\lambda) \times T_G(\lambda) \times T_W(\lambda) \times T_A(\lambda))^m \qquad (2.11)$$

where I_c is the direct beam normal irradiance, Wm^{-2}.

$I_o(\lambda)_n$ is the spectral irradiance in waveband n outside the atmosphere at mean solar distance Wm^{-2} per defined band width

K_d is the correction to mean solar distance

m is the optical air mass.

The integration has to be carried out waveband by waveband over the whole solar spectrum from waveband 1 to waveband N using appropriate band widths. Such fundamental approaches have most value in solar energy studies in appraising the performance of systems whose output is very dependent on the spectral inputs, for example silicon solar cells. In practice, however, they are too complicated for day to day applications. It is, therefore, necessary to adopt a very much simplified approach which scientifically respects the above features of the transmission of the solar spectrum through the atmosphere, but which at the same time is practical in its applicability.

16. Practical relationships for assessing atmospheric turbidity

Practical approaches attempt to describe the overall transmission properties of the atmosphere in terms of one or at most two variables when examining the transmission of the overall solar spectrum through the atmosphere. Two basic approaches are used to achieve the necessary simplification:

i) the development of simplified models which require separate inputs to describe site atmospheric aerosol content and site atmospheric water vapour content.

ii) the development of simplified models which require a single input to describe the combined effects of atmospheric aerosol and atmospheric water vapour content at a specific site.

In the first CEC programme on Solar Radiation Data, Global Irradiation on Inclined Surfaces, Action 3.2, the University of Sheffield adopted the Monteith and Unsworth Turbidity Coefficient as the basis of their studies (4). This was a beam radiation model of the first type requiring two inputs, the Monteith and Unsworth Turbidity Coefficient (5), and the precipitable water vapour content. The University of Berlin in contrast (6) adopted the Linke Turbidity Factor which requires the user to provide only a single input, T_L.

In the second phase of the CEC programme, it was decided, as an issue of policy, to adopt the Linke Turbidity Factor approach, because it was far simpler for use by practical engineers and solar designers. It also did not demand an input, namely precipitable water vapour, which was often difficult to obtain for specific sites. The relative clarity of the site is thus expressed in terms of a single number.

Models of the first type will, therefore, not be discussed further in the main body of the text, but some discussion of the inter-relationships between the different approaches is included in Appendix 2 and in Chapter 3. The quantitative relationship between the Schuepp Turbidity Coefficient B_s and $T_L(\lambda)$ is illustrated in Chapter 3, Figure 3.8.

17. Linke Turbidity Factor

The Linke Turbidity Factor is based on the concept of comparing, as a ratio, the actual atmospheric extinction coefficient of the atmosphere measured over the whole solar spectrum with the theoretical atmospheric extinction coefficient of pure dry air over that same spectrum over the same defined path length.

For any given solar altitude γ, we may rearrange equation 2.4 for pure dry air and, by integrating numerically on a waveband basis, obtain the clear air values. Setting the bands to run from 1 to N at defined bandwidths over the whole solar spectrum and inputting the known physical properties of the atmosphere at each waveband, one may determine $T_R(\lambda)$, $T_O(\lambda)$ and $T_G(\lambda)$ for each waveband. Hence inputting the values of $I_o(\lambda)_n$, the spectral irradiance can be estimated for each waveband, and the result numerically integrated to get the beam irradiance for perfectly clear air.

$$I_{oc} = \sum_{\text{waveband } n=1}^{\text{waveband } n=N} K_d (I_o(\lambda)_n \times (T_R(\lambda) \times T_O(\lambda) \times T_G(\lambda))^m \quad Wm^{-2} \qquad (2.12)$$

i.e omitting the terms $T_D(\lambda)$ and $T_W(\lambda)$.

By carrying out an integration of equation 2.12 over the whole solar spectrum, and expressing the attenuation in terms of Beer's Law, using a solar altitude/air mass dependent function, namely the Rayleigh optical depth of the pure dry atmosphere δ_R per unit air mass, one may express the cloudless day direct beam irradiance for pure dry air I_{oc} as

$$I_{oc} = K_d \, I_o \, e^{-\delta_R m} \qquad (2.13)$$

where I_o is the solar constant, 1367 Wm^{-2}

K_d is the correction for solar distance on day J

δ_R is the Rayleigh optical depth per unit air mass at the stated solar altitude

m is the relative optical air mass.

This provides the reference condition. δ_R can be evaluated on theoretical considerations following the principles set out above.

If one then takes the real atmospheric situation, defining a multiplying factor $T_L(\gamma)$, termed the Linke Turbidity Factor at solar altitude γ, by which to multiply δ_R to obtain the actual overall optical depth for the site, one obtains

$$I_c = K_d \, I_o \, e^{-\delta_R m \, T_L(\gamma)} \tag{2.14}$$

Thus

$$T_L(\gamma) = \frac{\ln I_o - \ln I_c - \ln K_d}{m \;\delta_R} \tag{2.15}$$

It should be noted that $T_L(\gamma)$ by definition must be greater than 1, and that the value combines the influence of precipitable water vapour in the atmosphere and attenuation due to aerosol into a single index, $T_L(\gamma)$.

Therefore, given the value of K_d, which is a function of the day of year, the optical air mass which is a function of solar altitude and station height, and the value of δ_R which is a function of air mass (and therefore of site altitude and solar elevation), the expression can be easily evaluated for any value of $T_L(\gamma)$ to find I_c.

In the past the majority of workers in the field of solar radiation studies have assumed the derived values $T_L(\gamma)$ were independent of solar altitude γ. However, as explained in paragraph 19, this has not proved a satisfactory assumption and it has not been accepted in the studies for the CEC.

Standard algorithms were used in this study to compute the above quantities. These are now set out in detail.

18. Standard algorithms used in computation of the direct beam irradiance

18.1 Correction for solar distance K_d

The path of the earth round the sun is elliptical, so there are variations in the distance of the earth from the sun which have to be allowed for using the inverse square law.

The correction K_d may be calculated from the following expression:

$$K_d = 1.0 + 0.03344 \cos(J' - 2.80) \quad \text{dimensionless} \tag{2.16}$$

where J' is the day angle, which is equal to the day number J, (Jan 1st 1, Feb 1st 32, etc.) divided by 365.25, multiplied by 360 degrees

18.2 Computation of relative optical air mass

The methodology of Kasten (7) for computation of relative optical air mass was adopted in the CEC programme. The value is first estimated for sea level and then a correction is applied for site elevation above sea level. The basic concepts are illustrated by Figure 2.3. Increase in station height reduces the air mass for a given solar altitude.

For solar altitudes below 10°

$$m = (p/p_0)/(\sin\gamma + 0.15(\gamma + 3.885)^{-1.253}) \quad \text{dimensionless} \qquad (2.17)$$

where p/p_0 is the correction for site elevation, see below

γ is the solar altitude in degrees.

Above 10°, where effects of atmospheric refraction are small, a simplified formula is satisfactory

$$m = (p/p_0)/\sin\gamma \quad \text{dimensionless} \qquad (2.18)$$

The ratio p/p_0 may be evaluated from the actual site atmospheric pressure measurements p^0 and the standard sea level atmospheric pressure p_0 at standard temperature. Alternatively, and more simply, a suitable algorithmic correction formula may be used.

In the preparation of the CEC European Solar Radiation Atlas, Vol. II, Inclined Surfaces, the following formula was used:

For site elevations < 4000 metres

$$p/p_0 = 1.0 - z/10000 \quad \text{dimensionless} \qquad (2.19)$$

where z is the station elevation in metres.

A more accurate formulation according to Kasten is

$$p/p_0 = e^{-z/z_h} \quad \text{dimensionless} \qquad (2.20)$$

where z is again the station elevation in metres

z_h is the scale height of the atmosphere ≈ 8 km

This reduces to a first approximation to

$$p/p_0 = (1 - z/8000) \quad \text{dimensionless} \qquad (2.21)$$

18.3 Rayleigh optical thickness δ_R per unit air mass

The Rayleigh optical thickness is the optical thickness of a pure Rayleigh scattering atmosphere, per unit air mass along a specified path length. As solar radiation is not monochromatic radiation, Beer's law cannot be applied in its simple form. The Rayleigh optical thickness is dependent on the precise optical path and hence on the relative air mass. The following formula proposed by Kasten (8) has been used throughout the CEC solar radiation atlas programme to determine δ_R.

$$\delta_R = 1/(0.9\ m + 9.4) \text{ per unit air mass} \qquad (2.22)$$

19. The effects of solar altitude on the values of $T_L(\gamma)$

Ideally the values of the Linke Turbidity Factor should be independent of solar altitude. In practice, as several European studies have shown, the Linke Turbidity Factor derived from pyrheliometric measurements tends to increase with increase of solar altitude. Table 2.3 based on Dogniaux and Doyen (9) shows the type of effect observed in practice at a typical European urban site.

Analysis of the very much earlier UK data by Stagg (10) and the more recent data by Armstrong (11) reveals similar differences with air mass for the UK. A method had to be adopted, therefore, to adjust for the influence of solar altitude on the Linke Turbidity Factor. As Appendix 2 contains a general review of the observed European relationships, only the correction method actually adopted for the standard computations which follow in this book will be described here in detail. The corrections for solar altitude were based on the method proposed by WMO (12). The full details of the WMO method are given in Appendix 2. The key element of the WMO methodology, extracted for these systematic European studies, were the corrections applied to assess the impact of solar altitude on the Linke Turbidity Factor. The reference mean Linke Turbidity Factor for any month was standardised to an air mass of 2, i.e. a solar altitude of 30° at sea level. These air mass 2 values of T_L were then corrected for actual solar altitude. The WMO corrections proposed for this correction are discontinuous. A continuous algorithm was developed which was fitted to the discontinuous corrections proposed by WMO to adjust the Linke Turbidity Factor for solar elevation. This correction took the following form.

For $T_L > 2.5$

$$T_L(\gamma) = T_L - (0.85 - 2.25 \sin\gamma + 1.11 \sin^2\gamma) \quad \text{dimensionless} \qquad (2.23)$$

where $T_L(\gamma)$ is the Linke Turbidity Factor at solar altitude γ.

T_L is the air mass 2 Linke Turbidity Factor.

The WMO method is based on a reference turbidity of 3.3 before various corrections are applied. It was noted that, if T_L was very low, values of $T_L(\gamma)$ could result that were below 1, which is in conflict with the scientific definition of the Linke Turbidity Factor. An arbitrary correction factor therefore was adopted to phase out gradually the corrections to air mass 2 Linke Turbidity Factors below 2.5.

Table 2.3 Variation in mean monthly Linke Turbidity Factor in the morning and afternoon as a function of air mass, Uccle. Dogniaux & Doyen (9)

Air mass	Month	1	2	3	4	5	6	7	8	9	10	11	12
m = 2	a.m.	-	-	4.2	4.5	4.5	5.0	4.5	4.4	4.3	4.3	-	-
m = 2	p.m.	-	-	4.4	4.5	4.5	5.0	4.6	4.6	4.5	4.4	-	-
m = 4	a.m.	3.5	4.0	4.0	4.1	4.2	4.1	4.3	4.1	4.0	4.0	3.8	3.7
m = 4	p.m.	3.8	4.2	4.1	4.3	4.3	4.3	4.4	4.2	4.2	4.3	4.0	3.7

Thus, for $T_L < 2.5$, the following formula was used

$$T_L(\gamma) = T_L - (0.85 - 2.25 \sin\gamma + 1.11 \sin^2\gamma)(T_L - 1)/1.5$$

dimensionless (2.24)

These solar altitude dependent corrections were then applied to all subsequent computations involving the use of the Linke Turbidity Factor throughout the solar radiation modelling studies that are developed in this book. Appendix 2 should be consulted for further details of various European methods for estimating the Linke Turbidity Factor. A new methodology was developed in this programme to objectively estimate the Linke Turbidity Factor at air mass 2 from the Angstrom regression coefficients to enable the CEC Solar Radiation Atlas, Vol. II Inclined Surfaces to be prepared. This procedure was based on the use of a global irradiance model, so it is now necessary to consider the development of the diffuse irradiance model before returning to a fuller discussion of the practical values of T_L used to develop the CEC European Solar Radiation Atlas, Inclined Surface tables and maps.

CHAPTER 2 - PART III

THE ESTIMATION OF DIFFUSE SOLAR RADIATION FROM CLOUDLESS SKIES

20. Development of a validated solar radiation model to predict horizontal surface diffuse irradiance from cloudless skies in Europe

As a key part of the process of developing an inclined surface clear and average day irradiance prediction model for the EEC area, it was necessary to develop, and test against observation, a turbidity sensitive model for predicting the clear sky diffuse irradiance on horizontal surfaces for different conditions of atmospheric clarity. The development and validation of the new model was carried out using data selected from the CEC solar radiation tapes prepared during the first four year Solar Energy Programme of the Community, Project F - Solar Radiation Data. These tapes contained hourly observed data of global and diffuse irradiation on horizontal surfaces as well as hourly values of the observed bright sunshine for a number of stations in the Community area. A clear day for the purpose of validation was assumed to occur when the observed daily sunshine fraction was greater than 0.9. Where insufficient clear days were available, data were selected as individual clear hours with a sunshine fraction of 1.00.

The development of the final CEC model began with two existing models, one produced by Page and Rodgers (13) at the University of Sheffield, and the other by Krochmann (14) at the Technische Universität, Berlin. Figure 2.4 shows a comparison of the predictions made by these two models for sites at sea level at mean sun-earth distance as a function of the solar altitude and Linke Turbidity Factor. The Sheffield model was pragmatically derived from observations in several temperate climates having a typical range of turbidities. The Berlin model had a better theoretical basis which gave rise to a simpler algorithm than the Sheffield model. It was, therefore, decided to make the first predictive comparisons with the selected observed data using the Berlin model. The decision to adopt the Berlin model was confirmed at the meeting between the Lichtmesstechnik of Berlin and the University of Sheffield in Sheffield on the 10th April, 1981.

The Berlin diffuse model assumes that the molecules and particles of the atmosphere scatter solar radiation equally in all directions, i.e. isotropically. The radiation reaching a horizontal surface may be divided into three parts, 1) direct beam radiation which passes straight through the atmosphere, 2) radiation absorbed by the atmosphere and emitted again as long wave radiation (this radiation is not available for scattering) and 3) diffuse radiation scattered by the molecules and aerosols in the atmosphere. If one knows the extraterrestrial irradiance, it is possible, for any given Linke Turbidity Factor, to calculate levels of the direct beam and absorbed irradiance on any given date. The amount of scattered diffuse radiation may then be easily calculated if one makes the assumption that, with isotropic scattering, exactly one half of the scattered energy is scattered downwards to be received as diffuse radiation. This assumption allows the clear sky horizontal surface diffuse irradiance D_c to be estimated as:

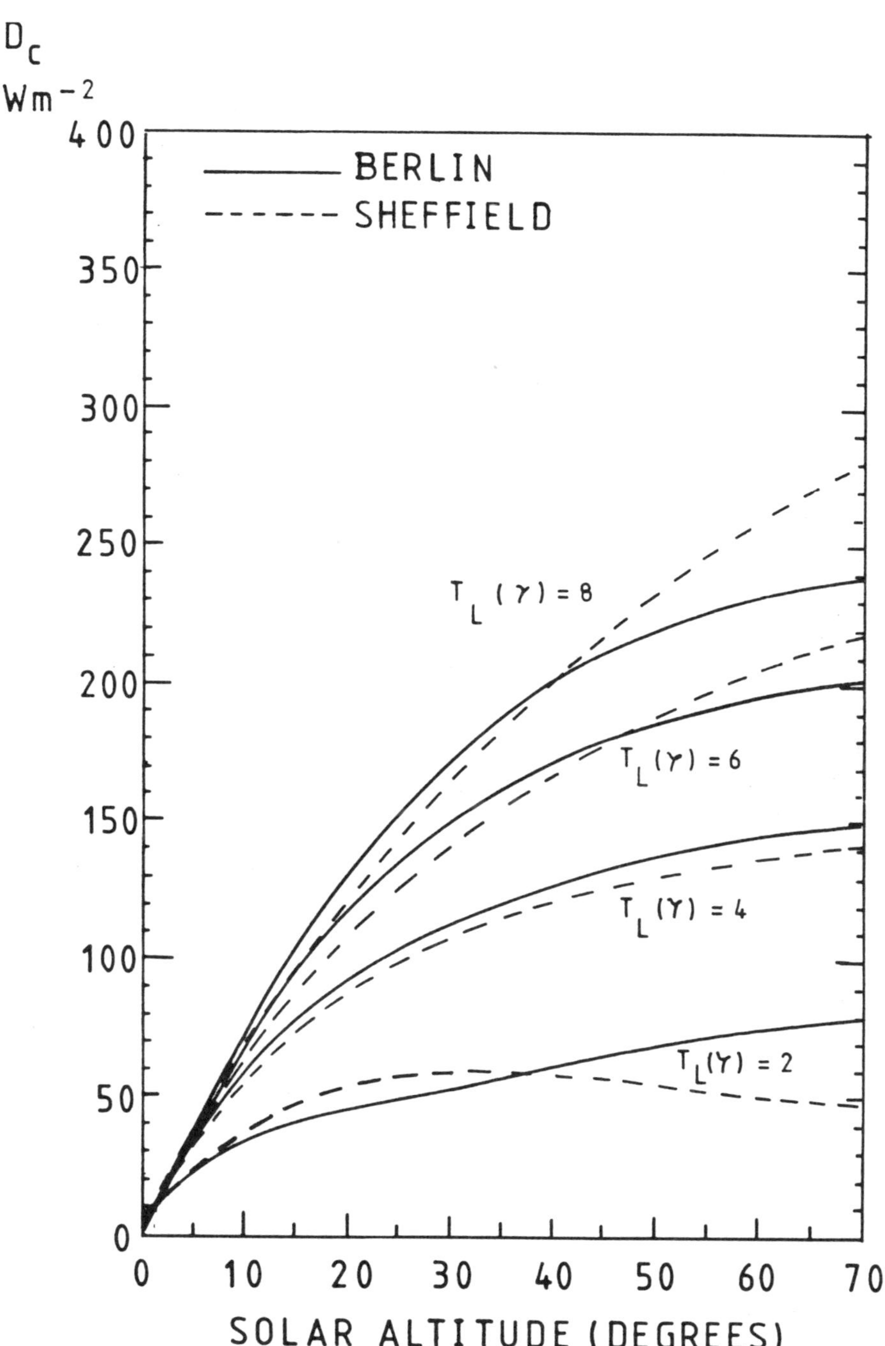

Figure 2.4 Comparison of Sheffield and Berlin model predictions of clear sky horizontal surface diffuse irradiance at mean solar distance at sea level.

$$D_c = 0.5(I_{oj} - I_c - I_{abs})\sin\gamma \qquad Wm^{-2} \tag{2.25}$$

where D_c is the clear sky horizontal surface diffuse irradiance, Wm^{-2}

I_{oj} is the extraterrestrial irradiance at normal incidence on day J, Wm^{-2}

I_c is the direct beam irradiance at normal incidence, Wm^{-2}

I_{abs} is the absorbed irradiance, not available for scattering, Wm^{-2}

γ is the solar altitude, degrees.

Aydinli (15) indicated that the absorbed irradiance I_{abs} could be calculated from values of extraterrestrial irradiance, solar altitude and Linke Turbidity Factor using equations 2.26 and 2.27 below:

$$q_a^{\ m} = \left(\sum_{i=0}^{5} a_i\, \gamma^i\right)(0.506 - 0.010788\ T_L(\gamma)) \tag{2.26}$$

and

$$I_{abs} = I_{oj}(1 - q_a^{\ m}) \qquad Wm^{-2} \tag{2.27}$$

where $q_a^{\ m}$ is the atmospheric transmittance coefficient for absorption by a path length of air mass m, dimensionless

$T_L(\gamma)$ is the Linke Turbidity Factor at solar altitude γ degrees

and where

$a_o = 1.294$ $\quad a_2 = -3.973 \times 10^{-4}$ $\quad a_4 = -2.2145 \times 10^{-8}$
$a_1 = 2.4417 \times 10^{-2}$ $\quad a_3 = 3.8034 \times 10^{-6}$ $\quad a_5 = 5.8332 \times 10^{-11}$

The direct irradiance at normal incidence I_c may be calculated from values of the Rayleigh optical thickness, the relative optical air mass, the Linke Turbidity Factor and extraterrestrial irradiance, using equations 2.22, 2.17, 2.14, 2.16 given earlier in Part II of this Chapter.

Thus substituting into equation 2.25, it follows that

$$D_c = 0.5(I_{oj}\ q_a^{\ m} - I_c)\sin\gamma \qquad Wm^{-2} \tag{2.28}$$

where, using equation 2.14, I_c is given by:

$$I_c = I_{oj} \exp(-\ m\ \delta_R\ T_L(\gamma)) \qquad Wm^{-2}$$

where m is relative optical air mass, dimensionless

δ_R is the Rayleigh optical thickness, dimensionless

Unfortunately, the assumption of isotropic scattering is not valid, as the larger particles and water droplets in the atmosphere make the scattering process more complex by introducing Mie scattering with a considerable forward scattered component. The present work aimed to provide an empirically derived correction to the Berlin model to allow for this additional forward scattering and any other systematic errors in the existing model.

21. The selection of observed EEC hourly cloudless sky irradiation data for checking the Berlin model, and developing improvements to it

The intention was, where possible, to select for each site, i) three cloudless days associated with conditions of low turbidity, ii) three cloudless days associated with conditions of medium turbidity, and iii) three cloudless days associated with conditions of high turbidity for the months of January, April, July and October for each station where appropriate data were available. A clear day data set for one site was thus intended to provide (9 x 4) daily sets of observations of diffuse and global solar radiation on the horizontal plane. A cloudless day was defined as a day when the relative daily duration of sunshine, assessed on the basis of the daylength during which the solar elevation was greater than 4°, was $>$ 0.9. The aim was not to derive representative monthly values of the turbidity factor, but to check the diffuse radiation model over as wide a range of clear day atmospheric conditions as possible at each station. However, insufficient clear day data were available for certain sites in the selected months and some changes in selection policy were necessary. In certain cases the data did not cover a full year and the criteria had to be adjusted accordingly.

Horizontal surface global and diffuse irradiation data are available for all the sites in the data set. Hourly sunshine data were not available from the two French sites. The selected data were as follows:

i) Bracknell, United Kingdom 51°23'N, 0°47'W, 73m

Source of data: Tape EEC 3.2

Period of data: 1967-76

Size of data set selected: 144 observations (17 days)

Inclined surface data: G_{VN} G_{VS} G_{VE} G_{VW}.

The data were selected as complete days where $S_d/S_{04} > 0.9$. Three days of high, medium and low turbidity were selected from each of January, April, July and October. A further five days were added on the suggestion of the UK Meteorological Office.

ii) Cabauw, Netherlands 51°58'N, 4°56'E, 2m

Source of data: K.N.M.I.(Slob)

Period of data: 1979

Size of data set selected: 141 observations (13 days)

Inclined surface data: G_{VN} G_{VS} G_{VE} G_{VW} $G_{45°E}$ $G_{45°S}$ $G_{45°SW}$ $G_{45°W}$. $G_{67°S}$ $G_{22°S}$.

Thirteen days were selected by the K.N.M.I. with a sunshine fraction greater than 0.85 during the months April-September.

iii) <u>Carpentras, France</u> 44°5'N, 5°3'E, 105m

Source of data: Tape EEC 3.2

Period of data: 1973-75

Size of data set selected: 280 observations (24 days)

Inclined surface data: G_{VS} (12 days) $G_{44°S}$ (12 days)

The data were selected as whole days where $S_d/S_{04} > 0.9$. As the two sets of inclined surface measurements covered different periods, two sets of twelve days (low, medium and high turbidity in January, April, July and September) were chosen. No hourly sunshine data were available so it was assumed that S = 1.0 in each hour of the day. Any apparently erroneous results caused by this assumption could be recognised in the subsequent editing process and were rejected later.

iv) <u>Hamburg, Germany</u> 53°38'N, 10°0'E, 14m

Source of data: Tape received from Dr. Kasten

Period of data: 1964 and 1974

Size of data set selected: 853 observations (499 from 1964 and 354 from 1974)

No inclined surface data.

The data were selected by Dr. Kasten as individual hours according to the following criteria:

a) Sunshine duration = 1.0 hour

b) Cloud cover $\leqslant$ 2 oktas at the end of the hour.

Two years were chosen arbitrarily from the 12 year record on the tape. Initial examination revealed that the turbidity factor tended to be higher during isolated hours than successive groups of hours which met the two selection criteria. It was assumed that this was a result of the presence of clouds and, therefore, only groups of three or more consecutive hours were selected from the tape.

v) <u>Locarno-Monti, Switzerland</u> 46°10'N, 8°47'E, 380m

Source of data: Swiss Meteorological Service (Valko)

Period of data: 1964

Size of data set selected: 41 observations (5 days)

Inclined surface data: G_{VN} G_{VE} G_{VS} G_{VW}.

The data were selected from two sets of two weeks of hourly data in June and December as complete days by examination of hourly sunshine records.

vi) Norrkopping, Sweden — 58°38'N, 16°7'E, 10m

Source of data: Tape EEC 3.2

Period of data: 1978

Size of data set selected: 119 observations

Inclined surface data: $G_{60°S}$ $D_{60°S}$.

Only a limited amount of data were available. Rather than selecting complete days, intervals of three or more hours where S = 1.0 hour were selected.

vii) Trappes, France — 48°46'N, 2°1'E, 168m

Source of data: Tape EEC 3.2

Period of data: 1973-75

Size of data set selected: 274 observations (23 days)

Inclined surface data: $G_{48°46'S}$ (12 days) $G_{50°SE}$ (8 days) G_{VN} (3 days)

The data were selected as complete days where $S_d/S_{04} > 0.9$. The inclined surface data sets covered different periods so days were selected separately from each set to give as good a coverage of the data as possible. No hourly sunshine data were available so it was assumed that S = 1.0 in each hour of the day. Any erroneous results caused by this assumption could be recognised and rejected later.

viii) Uccle, Belgium — 50°49'N, 4°21'E, 105m

Source of data: "Distribution spectrale du rayonnement solaire à Uccle" (I.R.M., 1981)

Period of data: 1980

Size of data set selected: 28 observations

No inclined surface data.

ix) Valentia, Republic of Ireland — 51°56'N, 10°15'W, 20m

Source of data: Tape EEC 3.2

Period of data: 1976-78

Size of data set selected: 133 observations (14 days)

Inclined surface data: G_{VS}.

There were insufficient cloudless days in the months January, April, July and October so cloudless days had to be selected from February, May, August and November instead. Some of the days selected did not meet the criterion $S_d/S_{04} > 0.9$, but were included to give a sufficiently large set of hourly observations. To make up for lack of cloudless day data on the tape EEC 3.2 in specific months two cloudless June days were also added to the data set.

22. Rejected data

Before further analysis was carried out the data were examined hour by hour to ensure that conditions were cloudless. The criteria used in this rejection process were:

i) Sunshine fraction below 90% in a given hour

The data had been selected on the basis of whole day data so there remained some individual hours when the hourly recorded sunshine was less than ninety percent which indicated some cloud must have been present and which, therefore, needed to be rejected. This process also eliminates potentially inaccurate measurements made when the angle of incidence is large around sunrise and sunset when the beam intensity may be too low to burn the sunshine card. Hourly sunshine data were available only for Bracknell, Cabauw, Locarno Monti and Valentia.

ii) Exceptionally high values of turbidity factor

Since the data set was selected on the basis of sunshine and not cloud cover, we must assume that some measurements could have been made when thin cloud was present. This would result in higher calculated values of the turbidity than would normally occur under true clear sky conditions. In order to eliminate such measurements all hourly data associated with a Linke Turbidity Factor greater than 12 were rejected at this stage. It is accepted that this is a rather high value for cloudless conditions, but the aim was to stretch the model over as wide a range of conditions as reasonable. Inter-comparison of the Linke Turbidity Factors with cloud cover observations has shown that the Linke Turbidity Factor can reach a value of 10 under a completely cloudless sky. Since less than one per cent of the data set was associated with a turbidity factor above 10, the distortion caused by such data must be negligible.

iii) Exceptionally high observed/predicted diffuse irradiation ratio

At low solar altitudes the predicted and observed values of diffuse irradiation tend to be only a few watts per square metre. It is thus possible for an error in either the prediction or observation to be small in absolute terms, but to have a drastic effect on the observed/predicted ratio. To eliminate these data, all measurements associated with an observed/predicted diffuse ratio greater than two were eliminated. With one exception, all the points eliminated this way were either the first or the last observation during a particular day.

23. Statistical analysis of errors using the Berlin model

The analysis of errors was carried out using the selected set of EEC clear sky global and diffuse irradiance data, the choice of which is described above. These data were first used to derive hourly values of the Linke Turbidity Factor which were then used in the unmodified Berlin model to predict the hourly clear sky diffuse irradiance on a horizontal surface. The predicted diffuse radiation values were compared with the observed diffuse values. The errors ε were expressed as the ratio of observed over predicted horizontal surface diffuse irradiance, which is calculated by equation 2.29.

$$\varepsilon = D_c(obs)/D_c(pred) \qquad (2.29)$$

where $D_c(obs)$ is the observed clear sky horizontal surface diffuse irradiance, Wm^{-2}

$D_c(pred)$ is the predicted clear sky horizontal surface diffuse irradiance, Wm^{-2}

The derived values of ε were first classified against solar altitude using ten degree intervals of solar altitude (0-10, 10-20, etc.) before further analysis was carried out. Above sixty degrees solar altitude, there were relatively few observations,so all observations taken at solar altitudes greater than fifty degrees were grouped together. A number of studies were then undertaken on this data, stratified by solar altitude range.

24. Statistical study of mean errors

For each site and for each solar altitude range the mean observed/predicted diffuse irradiance ratios were calculated using the unmodified Berlin model. These means for individual stations are shown in Table 2.4, together with the combined values for rural sites and urban sites considered separately as well as the overall mean values for all sites.

From Table 2.4, one can see that the unmodified Berlin model produced consistently high values of the observed/predicted ratio. In other words it consistently under-predicted the observed values of horizontal surface diffuse irradiation in the data set, a result one would expect from the neglect of forward scattering. The information in Figure 2.4 was first used to judge whether the original Berlin model or the original Sheffield model was the more accurate. Figure 2.5, which is derived from Figure 2.4, shows that, over most of the conditions commonly encountered in Western Europe, the unmodified Berlin model produces higher predictions than the Sheffield model. Thus, although the original Berlin model consistently under-estimates the observations, it is slightly more accurate than the Sheffield model at the hourly level. It also had the advantage of being simpler.

Table 2.4 also shows that there is quite a large variation in the performance of the model in relation to measurements made at different sites. For example the model was found to underpredict by an average of 1%

Table 2.4 Ratio of observed/predicted diffuse irradiation - mean values for each site for different solar altitude ranges.

Site	Solar altitude range Degrees						
	0-9.9	10.0-19-9	20.0-29.9	30.0-39.9	40.0-49.9	50.0-90.0	Mean
Bracknell	1.14*	1.27	1.21	1.22	1.23	1.24	1.23
Cabauw	-	0.98	1.07	1.10	1.14	1.21	1.10
Carpentras	1.21	1.06	1.09	1.13	1.11	1.06	1.10
Hamburg 1964	-	1.18	1.17	1.16	1.16	1.25	1.17
Hamburg 1974	-	1.09	1.11	1.16	1.13	1.08	1.12
Locarno-Monti	-	1.23	1.03*	0.92*	0.94*	0.87	1.07
Norrkopping	1.15*	1.06	1.05	1.11	1.27	-	1.13
Trappes	1.16	1.17	1.08	1.07	1.06	1.08	1.10
Uccle	-	-	1.07*	1.17*	1.16*	1.32*	1.18
Valentia	-	1.05	1.01	1.00	0.99	0.98	1.01
Urban Sites	1.16	1.16	1.13	1.14	1.14	1.18	1.15
Rural Sites	1.20	1.07	1.06	1.09	1.12	1.02	1.08
All Sites	1.18	1.13	1.11	1.13	1.14	1.12	1.13

* Values based on < 10 observations

Table 2.5 Statistical comparison of observed/predicted diffuse irradiance ratio for urban and rural sites in different solar altitude ranges.

	SOLAR ALTITUDE RANGE					
Urban sites	0-9.9	10-19.9	20-29.9	30-39.9	40-49.9	> 50
No. points	40	315	335	327	268	155
Mean ε	1.156	1.155	1.132	1.141	1.142	1.175
Standard deviation	0.253	0.172	0.122	0.131	0.152	0.208
Rural sites						
No. points	48	127	136	85	85	92
Mean ε	1.198	1.070	1.057	1.087	1.120	1.023
Standard deviation	0.260	0.171	0.164	0.184	0.224	0.212
t-statistic	0.764	4.709	5.447	3.092	1.027	5.513
Highest level of significance at which the two means are different	50%	0.1%	0.1%	1.0%	60%	0.1%

at Valentia, an exceptionally clear site, while at Bracknell it appeared to underpredict by as much as 20%. The Bracknell results appeared anomalous and enquiries of the UK Meteorological Service showed additional corrections had been applied to the UK data on the tape by Cowley which appeared to make the prediction anomalously high compared with other EEC sites, especially at low solar altitudes. The UK meteorological services have now reverted to lower corrections. Part of the variations may be due to the different shade ring dimensions and corrections applied in different countries. It was not possible to resolve this issue explicitly as it is the responsibility of those who make the observations to decide on the precise correction procedures.

25. Influence of Linke Turbidity Factor on the accuracy of the estimates

Table 2.4 deals with mean errors only. However, the major physical influence on the model's accuracy is likely to be the turbidity of the local atmosphere. The simplest way of characterising this, is to subjectively divide the sites into the categories of rural and urban. The mean values of the observed/predicted diffuse irradiance ratio for the two types of site are given at the bottom of Table 2.4. Table 2.5 shows a more useful comparison using the t-statistic. The differences are highly significant over a wide range of solar altitudes. However, there are two solar altitude divisions out of six where the differences are not statistically significant at the 5% level.

An alternative approach is to assess the accuracy of the model as a function of the Linke Turbidity Factor. For this analysis the data were divided into the same 10° ranges of solar altitude as before. Linear regression was used to examine the relationship between the ratio of the observed/predicted diffuse and the Linke Turbidity Factor for each site for each solar altitude range. These regressions for each station are shown graphically with their associated data sets in Figures 2.6 - 2.16. Though the graphs are marked T_L the value referred to is $T_L(\gamma)$. The derived regression coefficients are given in Tables 2.6a and 2.6b and the associated correlation coefficients in Table 2.6c. The graphs in Figures 2.6 - 2.16 show a relatively wide scatter of the data. This scatter reflects the relatively simple nature of the model compared with the range of complex factors and short term variations affecting clear sky diffuse irradiance levels. One also has to recognise that the accuracy of pyranometers is not high when measuring small signals with the radiation incident at oblique angles. There is considerable experimental variance present in the diffuse data set. Since this study could not deal with such small scale variations, any improvements to the clear sky diffuse model could only be based on relatively simple statistical treatments of the prediction errors of the basic model based on linear regression. The regression was based on equation 2.30

$$\varepsilon = a' + b' \; T_L(\gamma) \tag{2.30}$$

where γ is the mid point of the solar altitude interval in question, i.e. in the interval 10°-19.9, solar altitude $\gamma = 15°$

a' & b' are regression constants for solar altitude band $\gamma' - 5°$ to $\gamma + 5°$

and $T_L(\gamma)$ is the Linke Turbidity Factor at solar altitude γ.

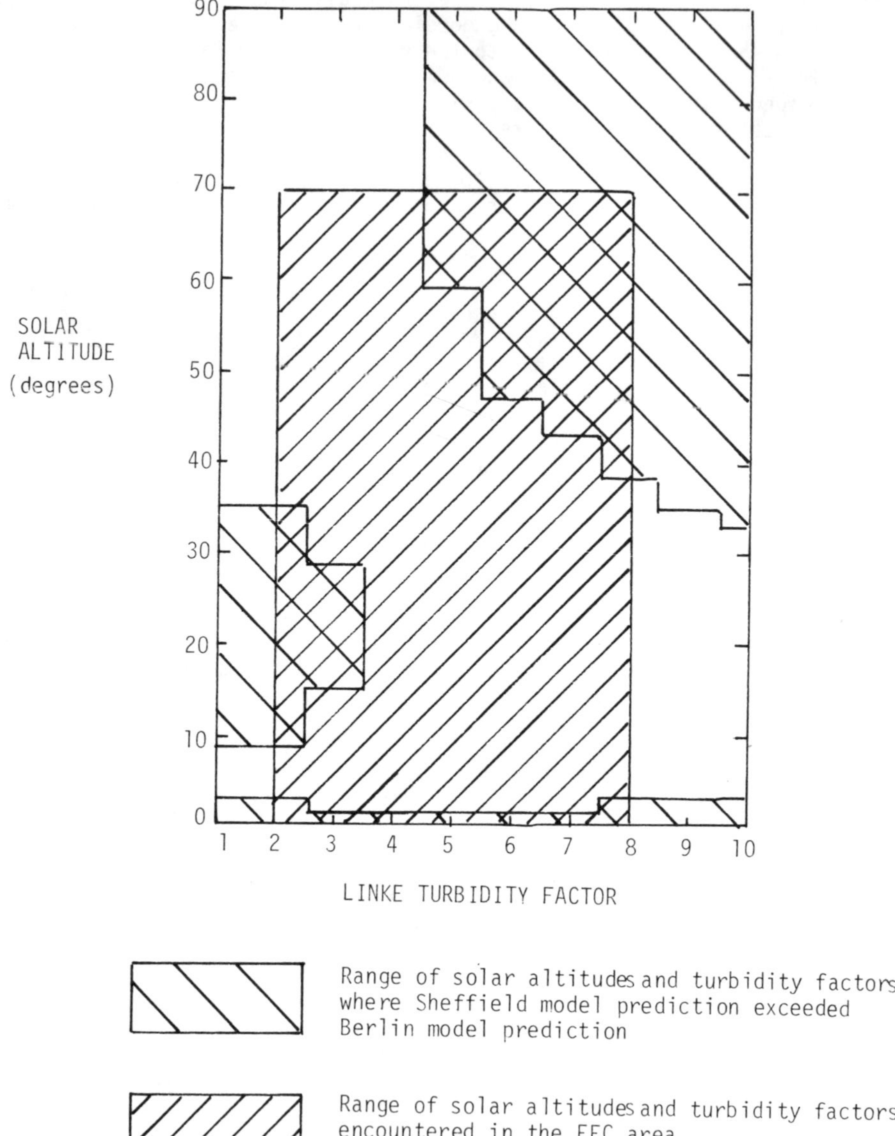

Figure 2.5 Comparison of Sheffield and Berlin clear sky horizontal surface diffuse irradiance model predictions.

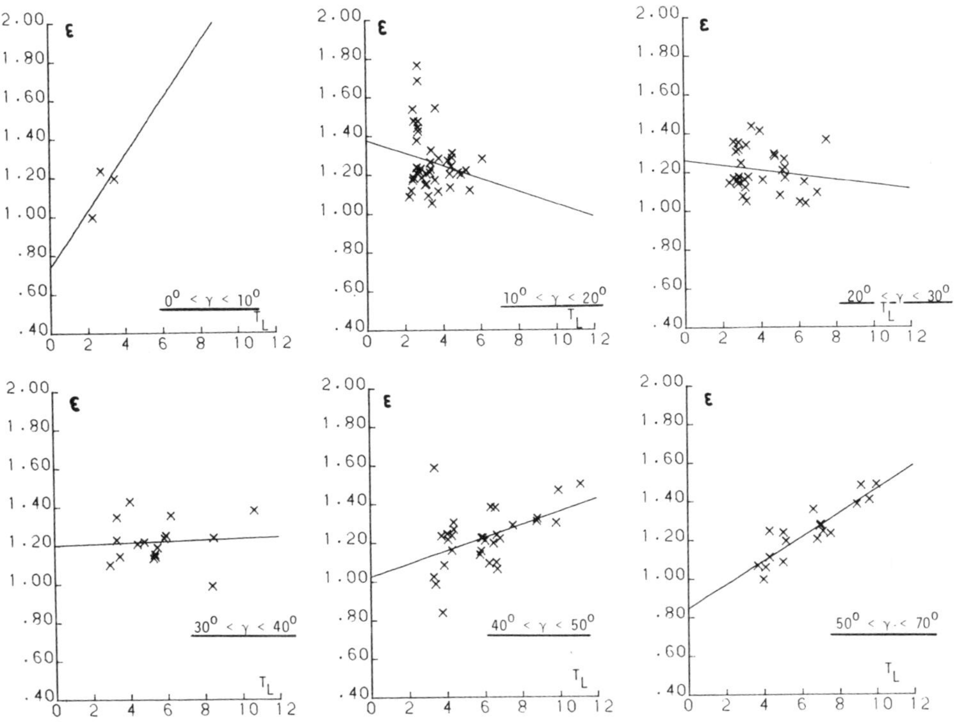

Figure 2.6 Prediction errors as a function of Linke Turbidity Factor $T_L(\gamma)$, classified by solar altitude band for Bracknell.

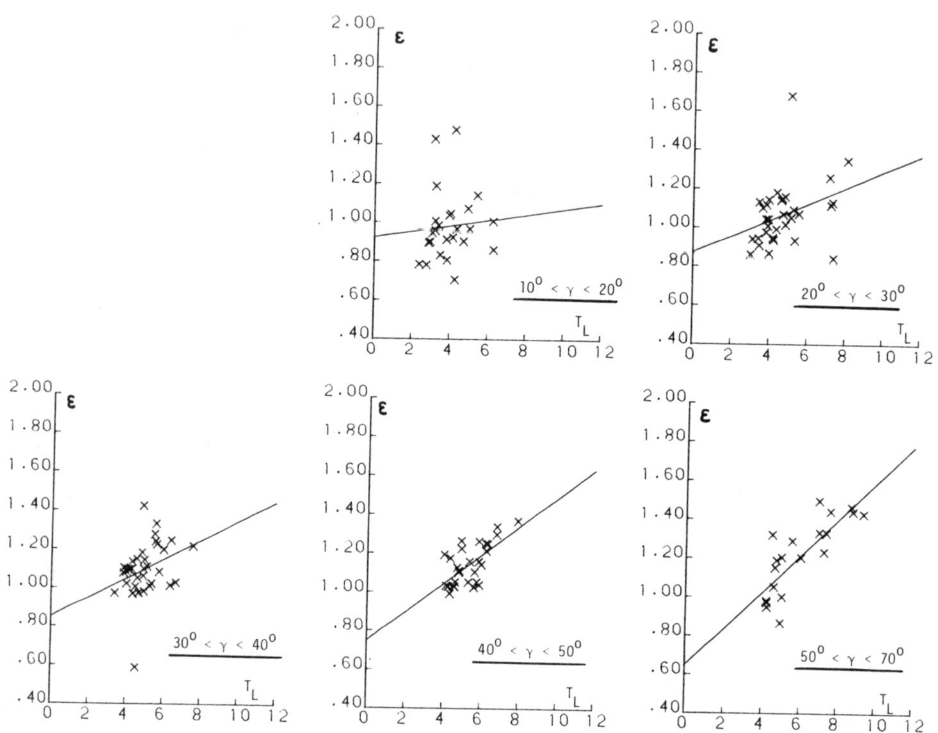

Figure 2.7 Prediction errors as a function of Linke Turbidity Factor $T_L(\gamma)$, classified by solar altitude band for Cabauw, Netherlands.

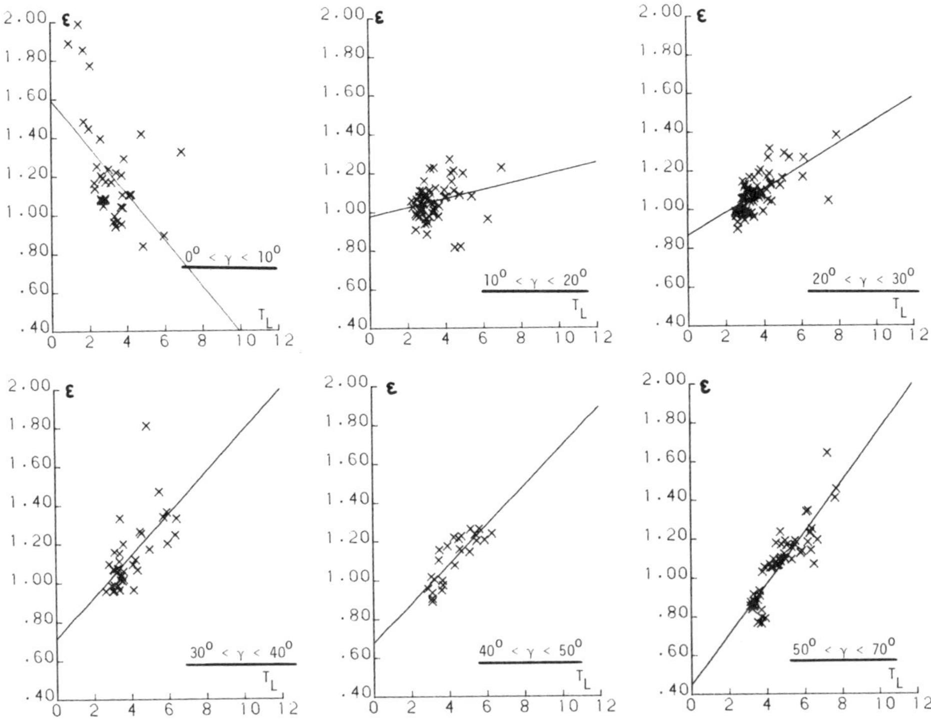

Figure 2.8 Prediction errors as a function of Linke Turbidity Factor $T_L(\gamma)$, classified by solar altitude band for Carpentras, France.

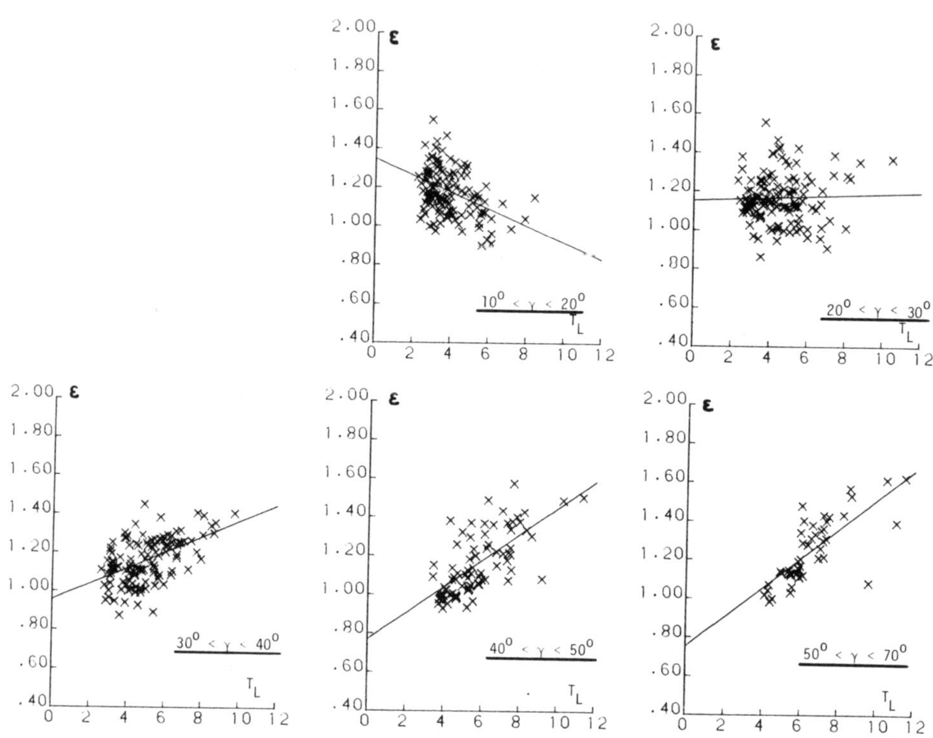

Figure 2.9 Prediction errors as a function of Linke Turbidity Factor $T_L(\gamma)$, classified by solar altitude band for Hamburg 1964, Germany.

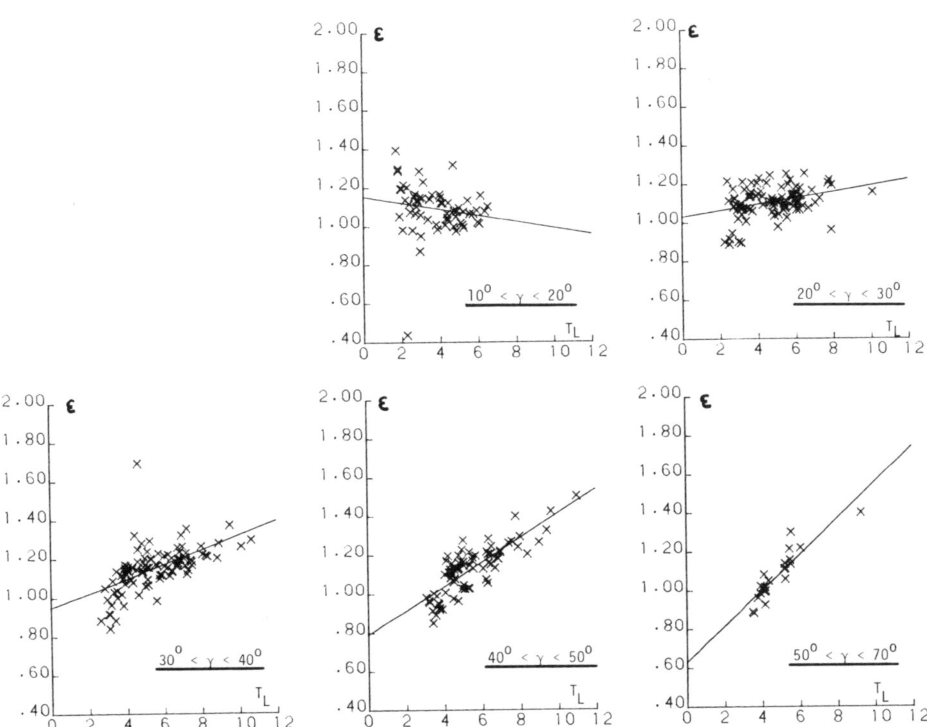

Figure 2.10 Prediction errors as a function of Linke Turbidity Factor $T_L(\gamma)$, classified by solar altitude band for Hamburg 1974, Germany.

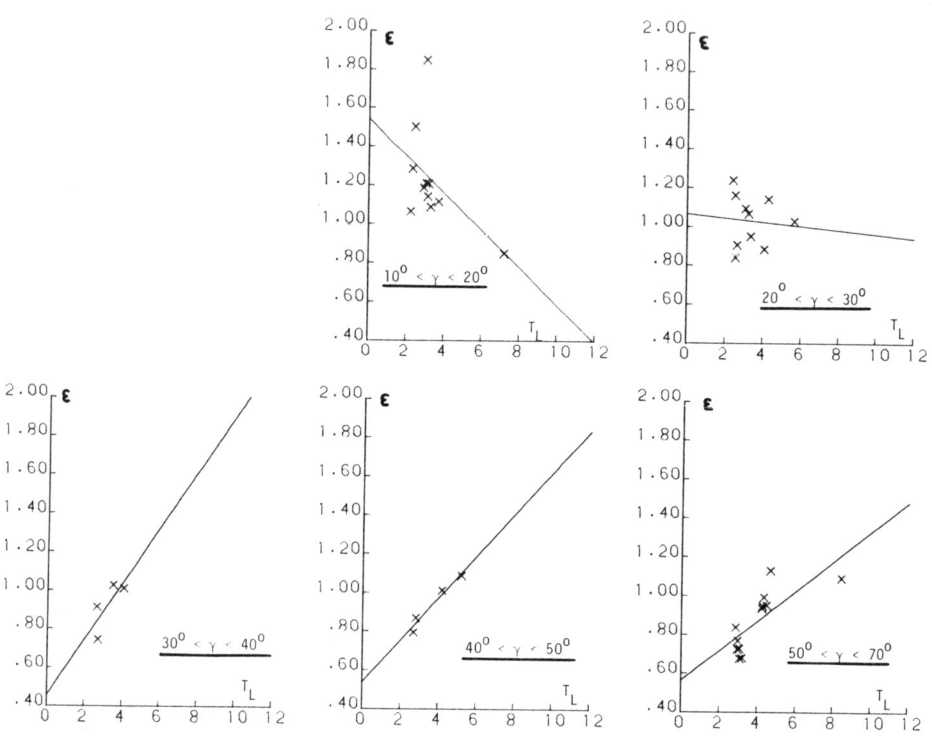

Figure 2.11 Prediction errors as a function of Linke Turbidity Factor $T_L(\gamma)$, classified by solar altitude band for Locarno Monti, Switzerland.

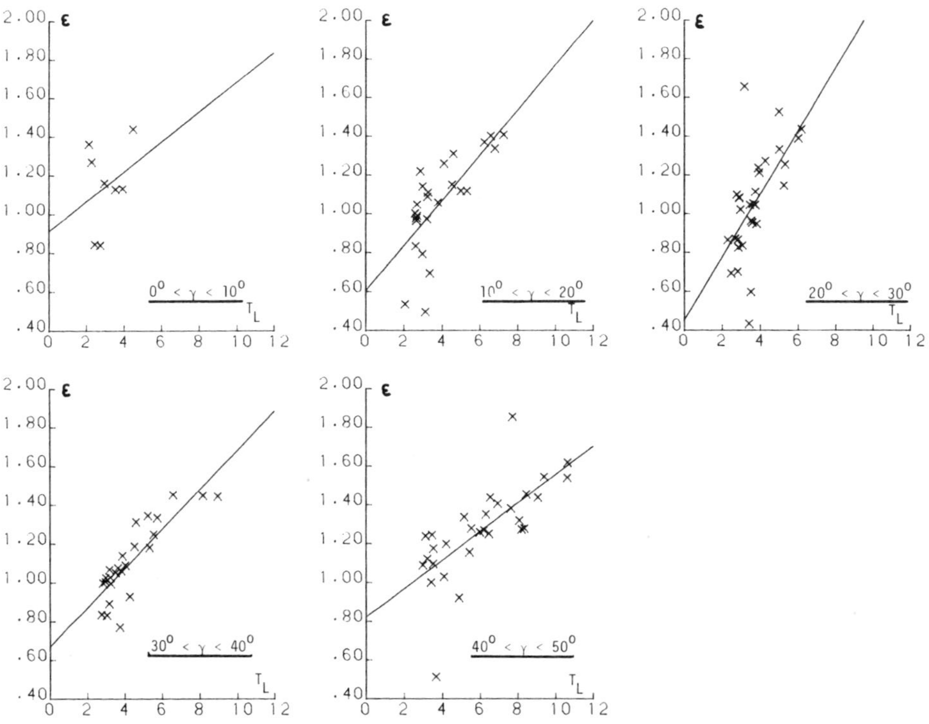

Figure 2.12 Prediction errors as a function of Linke Turbidity Factor $T_L(\gamma)$, classified by solar altitude band for Norrkopping, Sweden.

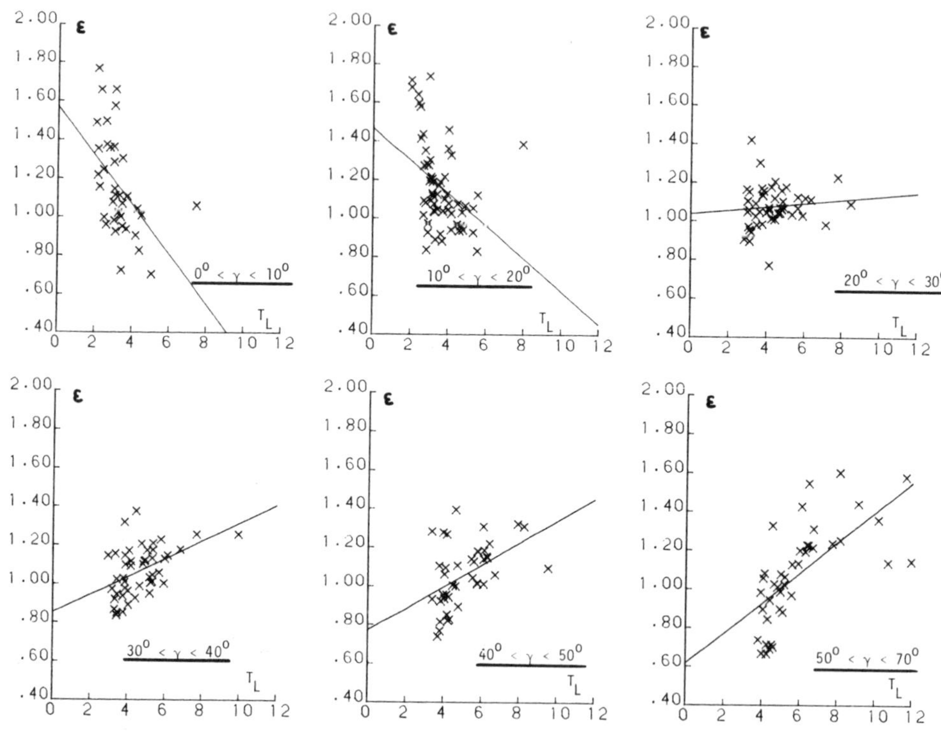

Figure 2.13 Prediction errors as a function of Linke Turbidity Factor $T_L(\gamma)$, classified by solar altitude band for Trappes, France.

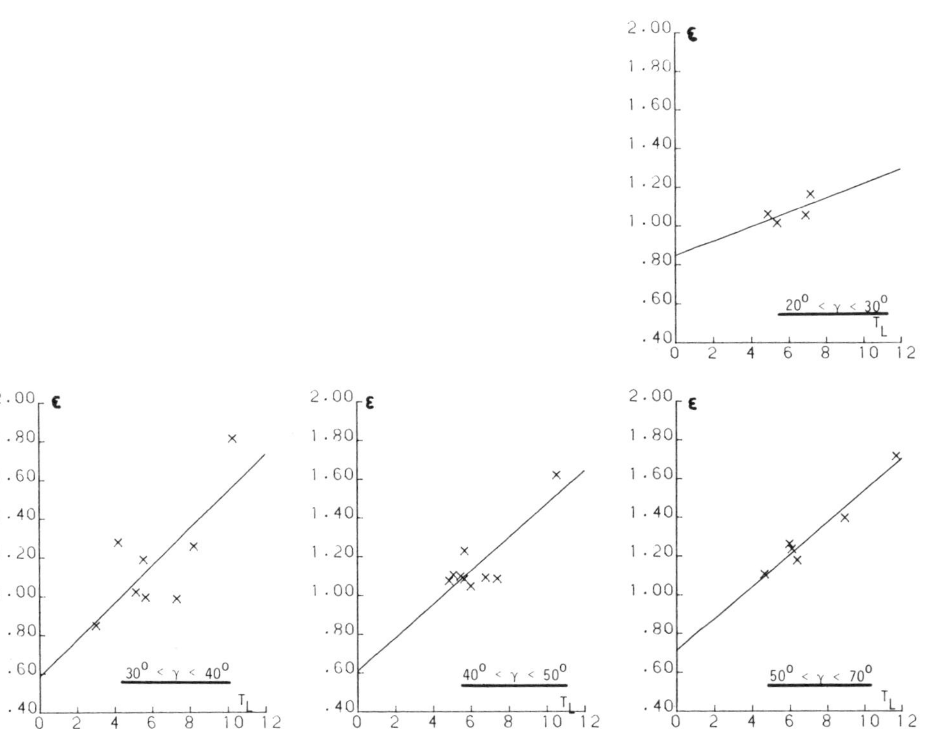

Figure 2.14 Prediction errors as a function of Linke Turbidity Factor $T_L(\gamma)$, classified by solar altitude band for Uccle, Belgium.

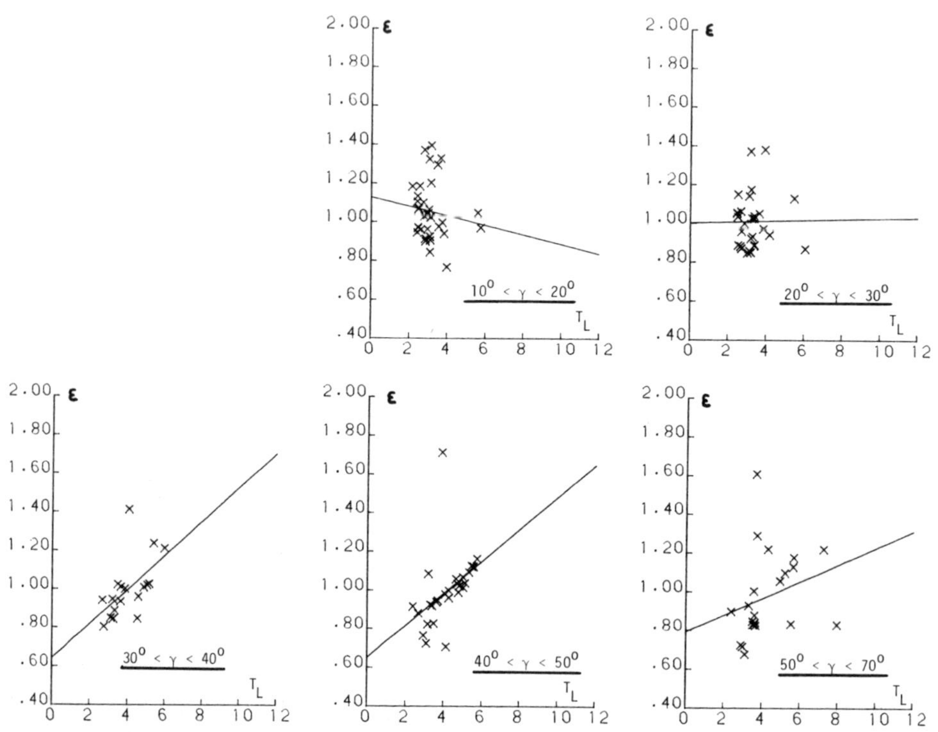

Figure 2.15 Prediction errors as a function of Linke Turbidity Factor $T_L(\gamma)$, classified by solar altitude band for Valentia, Ireland.

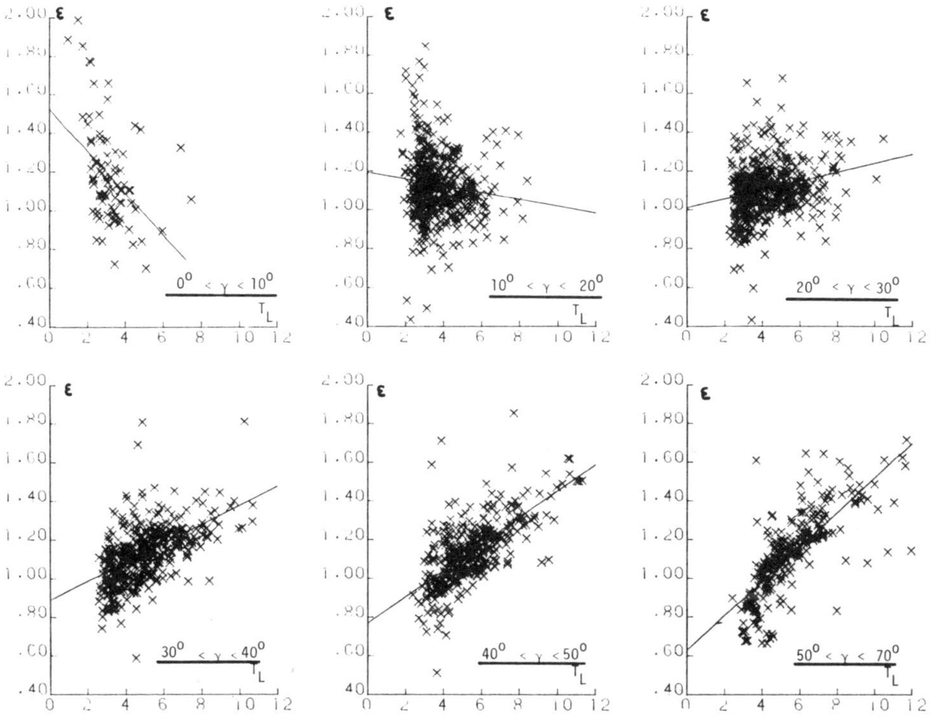

Figure 2.16 Prediction errors as a function of Linke Turbidity Factor $T_L(\gamma)$, classified by solar altitude band, combined data for all sites.

Table 2.6 Statistically derived corrections to Berlin horizontal surface diffuse model. Coefficients a'_γ and b'_γ for linear regression functions of turbidity factor for different ranges of solar altitude.

Correction to Berlin model = $a'_\gamma + b'_\gamma T_L$

where T_L = Linke Turbidity.

a) Regression coefficient a'_γ

Site	Solar altitude range Degrees					
	0-9.9	10-19.9	20-29.9	30-39.9	40-49.9	50+
Bracknell	0.735*	1.379	1.261	1.201	1.025	0.848
Cabauw	-	0.921	0.870	0.849	0.746	0.646
Carpentras	1.595	0.979	0.867	0.711	0.677	0.448
Hamburg 1964	-	1.347	1.152	0.956	0.768	0.753
1974	-	1.154	1.031	0.950	0.791	0.632
Mean	-	1.251	1.092	0.953	0.780	0.693
Locarno-Monti	-	1.543	1.065*	0.452*	0.536*	0.561
Norrkopping	0.911*	0.604	0.451	0.667	0.821	-
Trappes	1.576	1.470	1.038	0.853	0.770	0.614
Uccle	-	-	0.846*	0.584*	0.610*	0.713*
Valentia	-	1.128	1.002	0.639	0.649	0.791

b) Regression coefficient b'_γ

Site	Solar altitude range Degrees					
	0-9.9	10-19.9	20-29.9	30-39.9	40-49.9	50+
Bracknell	0.144*	-0.033	-0.012	0.003	0.034	0.062
Cabauw	-	0.015	0.042	0.050	0.074	0.094
Carpentras	-0.120	0.023	0.059	0.107	0.101	0.131
Hamburg 1964	-	-0.043	0.004	0.040	0.068	0.076
1974	-	-0.016	0.016	0.037	0.063	0.093
Mean	-	-0.029	0.009	0.039	0.066	0.085
Locarno-Monti	-	-0.096	-0.010*	0.144*	0.108*	0.077
Norrkopping	0.077*	0.116	0.162	0.102	0.074	-
Trappes	-0.128	-0.085	0.009	0.046	0.057	0.078
Uccle	-	-	0.037*	0.096*	0.086*	0.083*
Valentia	-	-0.024	0.002	0.089	0.083	0.044

* Values based on ⩽ 10 observations.

Table 2.6 continued — Statistically derived corrections to Berlin horizontal surface diffuse model. Coefficients a'_γ and b'_γ for linear regression functions of turbidity factor for different ranges of solar altitude.

c) Correlation coefficients associated with above regression functions (the number of observations in each case is given in brackets)

	Solar altitude Degrees					
	0-9.9	10-19.9	20-29.9	30-39.9	40-49.9	50-90
Bracknell	0.691 (3)	-0.203 (43)	-0.159 (31)	0.065 (17)	0.463 S1 (32)	0.895 S1 (18)
Cabauw	-	0.085 (25)	0.349 S5 (32)	0.321 (34)	0.663 S1 (29)	0.812 S1 (21)
Carpentras	-0.517 S1 (39)	0.236 (56)	0.654 S1 (66)	0.658 S1 (38)	0.829 S1 (24)	0.887 S1 (57)
Hamburg 1964	-	-0.412 S1 (124)	0.042 (127)	0.530 S1 (123)	0.686 S1 (79)	0.753 S1 (46)
Hamburg 1974	-	-0.175 (62)	0.328 S1 (93)	0.565 S1 (99)	0.814 S1 (77)	0.896 S1 (23)
Norrkoping	0.298 (8)	0.714 S1 (25)	0.608 S1 (30)	0.842 S1 (25)	0.710 S1 (30)	-
Trappes	-0.493 S1 (37)	-0.391 S1 (60)	0.111 (48)	0.473 S1 (46)	0.471 S1 (42)	0.659 S1 (41)
Valentia	-	-0.125 (35)	0.014 (30)	0.557 S5 (18)	0.421 S5 (27)	0.272 (22)
Sites with limited amounts of data						
Locarno-Monti	-	-0.501 (11)	-0.080 (10)	-0.758 (4)	0.974 S5 (4)	0.753 S1 (12)
Uccle	-	-	0.653 (4)	0.749 S5 (8)	0.839 S1 (9)	0.975 S1 (9)

S1 Correlation coefficient r is significant at the 1% level

S5 Correlation coefficient r is significant at the 5% level but not at the 1% level.

Tables 2.6a and 2.6b show a clear relationship between the linear regression coefficients a' and b' and solar altitude band for the individual stations. With the exception of Norrkopping, for which very few data were available, the regression coefficient a' (in Table 2.6a) decreased with increasing solar altitude. The slope of the regression equation, measured by coefficient b' (in Table 2.6b), tends to increase from a negative value at low solar altitude to a positive value at high solar altitude.

Table 2.6c shows that for most of these regressions the correlation coefficient r is significant at the 1% or 5% level. A number of the regressions do not give a good fit to the data for individual sites. However, the final analysis uses all the data and gives more satisfactory statistical results.

26. Improvements to the basic Berlin model

The improvement made to the horizontal surface clear day model for the EEC area were based on the analysis of the combined data set from all the available sites. As before, the values of observed/predicted diffuse irradiance ratio ε were divided according to solar altitude. For each range of solar altitude, a linear regression function relating ε to the derived Linke Turbidity Factor was fitted to the data. These regressions were in the form of equation 2.30.

The combined data and the regression lines are shown in Figure 2.16. Table 2.7 shows the associated regression coefficients and the associated correlation coefficients. The correlation coefficients are relatively low due to the high degree of variation in the data. However, they are all significant at a high level. Only in the solar altitude range 10° -19.9° is the level of significance of the correlation below 1%. However, since this regression line is almost horizontal in this range we are, in effect, taking an average value of the ratio.

The final stage of the modelling process was to collapse these two sets of regression coefficients, a_0' and a_1' into polynomial functions of Linke Turbidity Factor and solar altitude. It was decided to express the error correction function at a particular solar altitude in the form of equation 2.31 to emphasise the typical observed European Linke Turbidity Factor of 5.

$$f_1 = a' + b' \, (T_L(\gamma) - 5) \qquad (2.31)$$

where f_1 is the correction to the Berlin model at solar altitude γ

$T_L(\gamma)$ is the Linke Turbidity Factor at solar altitude γ

a' is the value of the correction factor at solar altitude γ for a Linke Turbidity Factor of 5

b' is the slope of the linear regression function at solar altitude γ.

Figure 2.17 shows graphs of the two sets of regression coefficients. A good approximation to each set of coefficients was achieved with fifth order polynomial functions. These have been used to modify equation 2.31 to give equation 2.32.

Figure 2.17 Regression coefficients a' and b' derived from European data set for use in Equation 2.31 (dotted lines are extrapolated). These curves are fitted as polynomials in Equation 2.32.

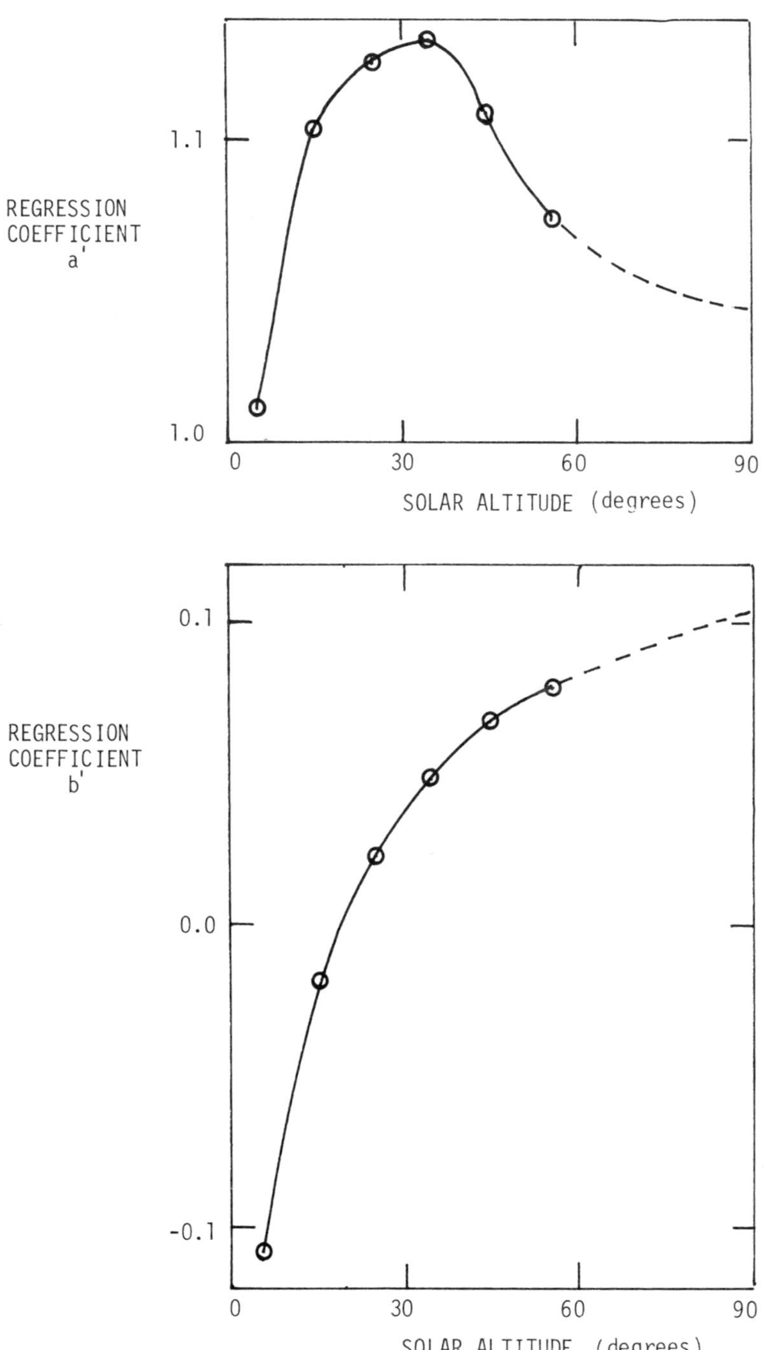

Table 2.7 Statistically derived corrections to Berlin horizontal surface diffuse model. Coefficients a'_γ and b'_γ for linear regression functions of turbidity factor for different ranges of solar altitude. Correction to Berlin model = $a'_\gamma + b'_\gamma T_L$.

a) Values derived using combined data from all sites

	Solar Altitude Range Degrees					
	0 - 9.9	10-19.9	20-29.9	30-39.9	40-49.9	50+
Regression Coefficient a'_γ	1.522	1.195	1.012	0.888	0.768	0.629
Regression Coefficient b'_γ	-0.108	-0.018	0.023	0.049	0.068	0.089
Correlation Coefficient r	-0.440	-0.118	0.238	0.538	0.673	0.757
Level of significance of r	0.2%	10%	0.2%	0.2%	0.2%	0.2%

b) Mean of values calculated for individual sites

	Solar Altitude Range Degrees					
	0 - 9.9	10- 19.9	20-29.9	30-39.9	40-49.9	50+
Regression Coefficient a'_γ	1.586	1.240	1.032	0.880	0.775	0.662
Regression Coefficient b'_γ	-0.124	-0.032	0.017	0.053	0.069	0.082

Table 2.8 Correction function to correct Berlin diffuse model. Function f_1. Values of a_i and b_i used in equation 2.32.

i	a_i	b_i
0	9.272×10^{-1}	-1.9043×10^{-1}.
1	1.850×10^{-2}	1.8226×10^{-2}.
2	-5.377×10^{-4}	-6.0133×10^{-4}.
3	5.512×10^{-6}	1.1015×10^{-5}.
4	-1.502×10^{-8}	-1.0043×10^{-7}.
5	-3.816×10^{-11}	3.5385×10^{-10}.

$$f_1 = (\sum_{i=0}^{5} a_i \gamma^i) + (\sum_{i=0}^{5} b_i \gamma^i)(T_L(\gamma) - 5)) \qquad (2.32)$$

The values of a_i and b_i are given in Table 2.8.

The correction factor f_1 obtained from equation 2.32 acts as a multiplier for the results produced by the original Berlin horizontal surface diffuse irradiance model and corrects for non-isotropic scattering effects which were not allowed for in the original model. Thus the improved diffuse model based on observation data for clear days is

$$D_c = 0.5\ f_1(I_{oj}\ q_a^{\ m} - I_c)\ \sin \gamma \qquad (2.33)$$

Figure 2.18 shows a graphical representation of f_1 as a function of solar altitude and Linke Turbidity Factor. Figure 2.19 gives the values of $q_a^{\ m}$.

The results produced by the final EC recommended model are presented graphically in Figure 2.20. Using equation 2.1 we can therefore now combine the results of Chapter 2, Part II and Part III, and estimate the global irradiance on a horizontal surface as

$$G_c = I_c \sin \gamma + D_c \qquad Wm^{-2}$$

It is then straightforward to construct a daily irradiation model from this irradiance model using numerical integration techniques.

27. Checking the accuracy of the cloudless sky diffuse irradiance model - horizontal surfaces

27.1 Checking at the individual hourly level

Checks were carried out on the prediction accuracy of the cloudless sky model at the hourly level for two sites, using data for specific selected cloudless hours on selected days of exceptionally low cloud amount. The two sites used were Bracknell, UK and Cabauw, Netherlands.

27.2 Bracknell checks

Cloudless day data for four selected months observed between 1967 and 1976 were used. The months were January, April, July and December. The hours used for testing were selected to meet the following criteria

i) Observed sunshine in hour = 1

ii) Observed cloud during hour 0 or 1 okta

The basic data set from which the selected hours were extracted was provided by the UK Meteorological Office. The extracted cloudless day data set provided 166 hours of data for checking: 29 hours from 7 days in January, 55 hours from 8 days in April, 56 hours from 8 days in July, and 26 hours from 5 days in October.

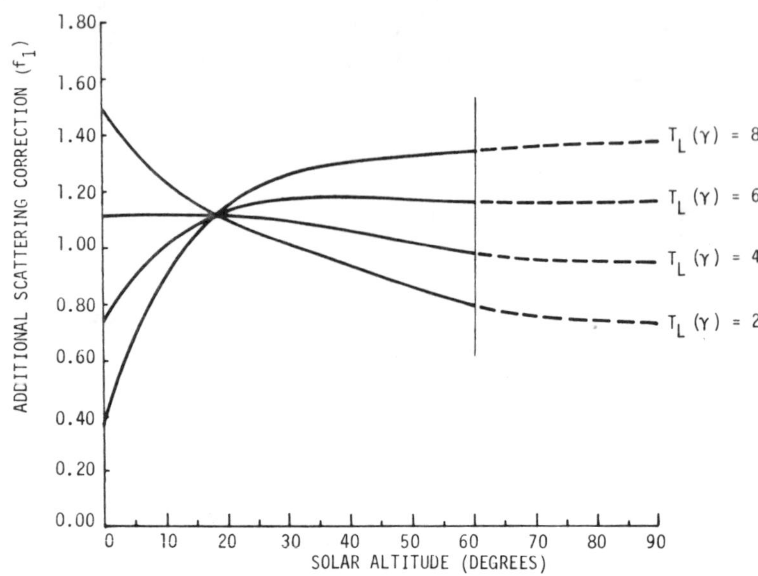

Figure 2.18 Pragmatically determined additional scattering correction f_1 to the basic Berlin model derived by comparison of the actual clear day observations in the EEC area.

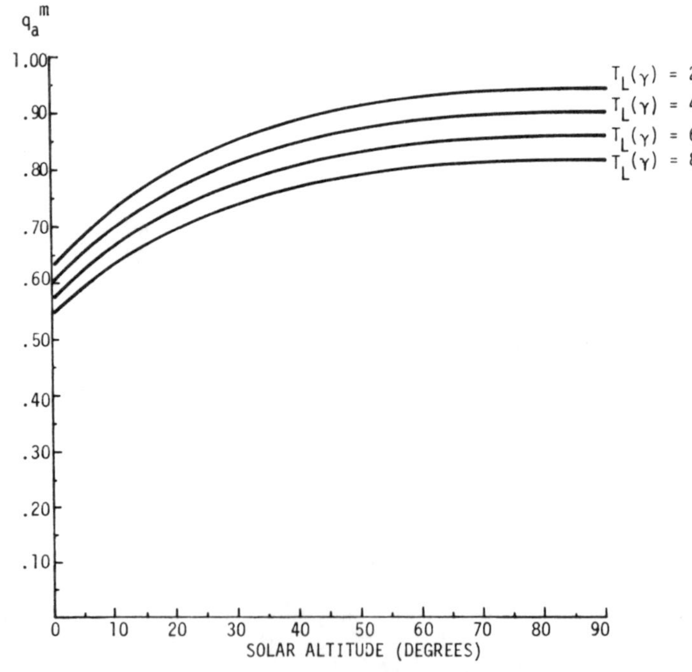

Figure 2.19 Values of q_a^m, the transmittance of the atmosphere due to absorption alone in the absence of scattering as a function of solar altitude and Linke Turbidity Factor.

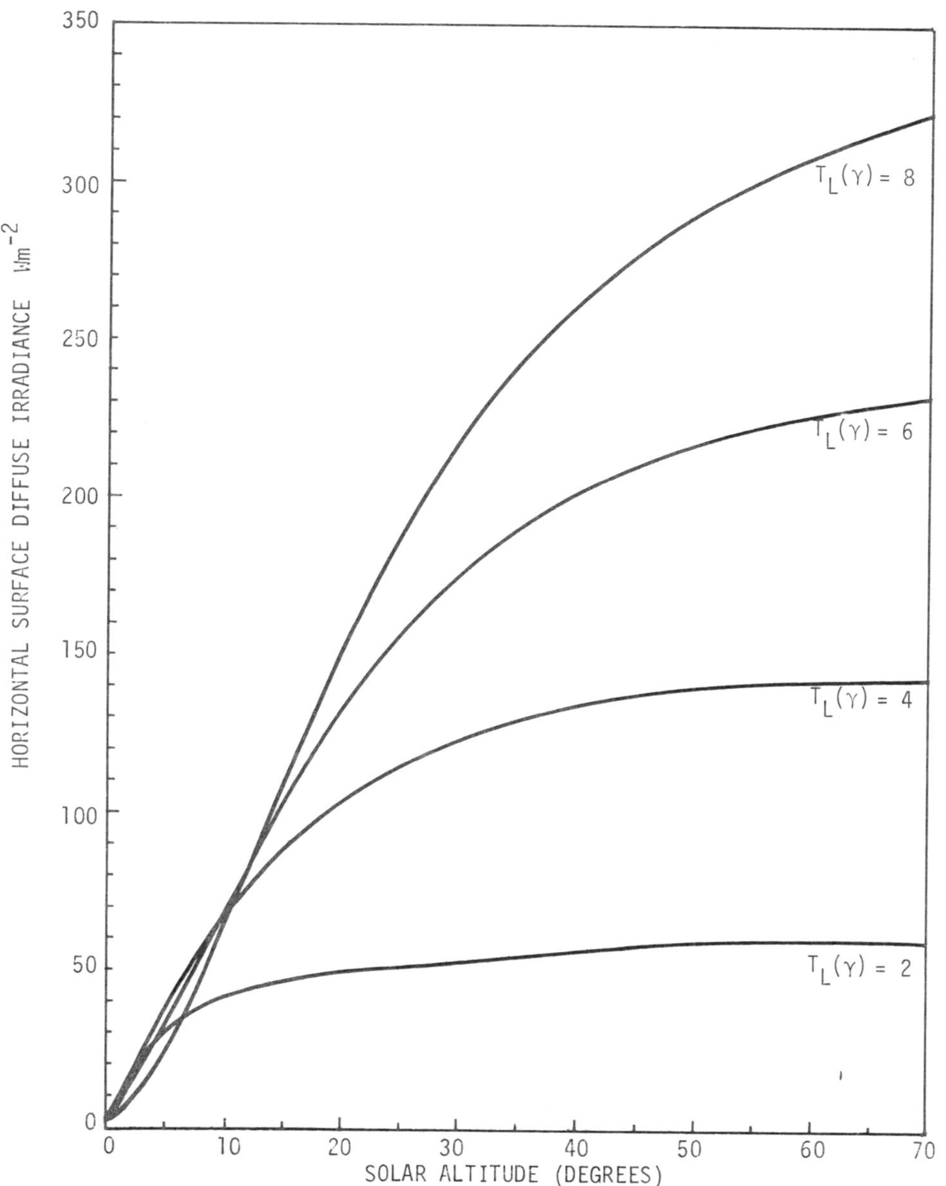

Figure 2.20 The relationship between horizontal surface diffuse irradiance at sea level on clear days with the sun at mean solar distance and solar altitude for different values of the Linke Turbidity Factor. EC clear day model.

The values of diffuse irradiation were first of all corrected using the corrections for non-isotropic sky effects suggested by Painter (16). It was assumed that the hourly irradiation was representative of the irradiance at the mid hour. The beam irradiance on the horizontal surface was then estimated from the hourly difference $G_C - D_C$ (corrected). The mean solar altitude for the hour was available in the data set and so the beam irradiance normal to the beam I_C could be estimated as $I_C(0,0)/\sin\gamma$. This value was then used to estimate the observed Linke Turbidity Factor. The observed Linke Turbidity Factor was then fed into the EC clear sky diffuse irradiance model, together with the solar altitude and the correction to mean solar distance to predict the diffuse horizontal irradiance. The ratio of the corrected observed value of the diffuse irradiance to the predicted value was then calculated. The process was repeated for each of the 166 hours. The data were then grouped into 10 degree bands of solar altitude 0-10, 10-20, etc., and the mean errors and standard deviations for each band were extracted. The results are given in Table 2.9. The mean errors computed over a very wide range of turbidities were relatively small and the associated standard deviations ranged from .032 to .085.

An additional study was made using the observed values of the dew point temperature at noon to see if the errors were correlated to the moisture content of the atmosphere. It was found there was a correlation between the noon dew point and the prediction error of the form

$$\varepsilon = 1.017 - 0.0075\ T_D \quad (r = -.4534)$$

where T_D was the noon dewpoint in degrees Celsius.

Such a correlation would be expected on theoretical grounds, as the precipitable water vapour is related to the dew point. If the water content of the atmosphere is low, the absorption due to water vapour is low. Therefore, if a certain Linke Turbidity Factor is observed, the contribution from the aerosol scattering must be correspondingly greater than average so the model will underestimate the scattered diffuse. When the water content of the atmosphere is very high, the scattered energy will be less than the average for that turbidity factor, so the model will overestimate the scattered diffuse. The Linke Turbidity Factor approach thus confers simplicity at the expense of some loss of accuracy. However, this refinement was not used in the studies that follow, as the necessary data werenot generally available for the project.

27.3 Cabauw checks

A corresponding study was carried out for Cabauw using a special set of data provided by Slob (17). The selection of the data set is discussed in Chapter 3. In this case the diffuse radiation was estimated as the difference between the global radiation and the resolved component of the pyrheliometer value on the horizontal surface. Using the same rejection criteria as for Bracknell, the results are given in Table 2.10. It should be noted this data set was for the summer period only. A lot of the data for the band 10-20 degrees did not meet the selection criteria. It is in this zone that angle of incidence errors begin to become large with pyranometers, so obtaining the diffuse as the difference of two relatively

Table 2.9 Accuracy in prediction of shadow band corrected diffuse solar radiation on a horizontal surface for cloudless hours at Bracknell, using CEC model, as a function of solar altitude.

Solar altitude band degrees	Number* of obs	Mean solar alt degrees	Mean $T_L(\gamma)$	Range of $T_L(\gamma)$	Horizontal diffuse prediction error Obs/pred		
					Mean	S.D.	Mean
0-10	4 (1)	9.0	2.83	2.43- 3.03	.93	.032	(All .90)
10-20	56 (6)	15.1	3.00	2.00- 5.15	1.01	.075	(All 1.02)
20-30	33 (5)	24.2	3.41	2.13- 5.34	1.02	.065	(All 1.03)
30-40	25 (2)	33.6	4.77	2.77- 9.78	1.02	.075	(All 1.03)
40-50	31 (2)	44.6	5.47	3.09-10.2	0.98	.085	(All 1.00)
50-60	10 (3)	54.6	6.09	3.45- 9.26	0.96	.040	(All .92)
60-70	7 (1)	60.7	6.50	3.85- 8.99	0.96	.061	(All .93)
Total obs used	166(20)						

*Data were rejected if diffuse error was greater than 1.15 or less than 0.85, because data set was to be used for subsequent slope predictions. The number of rejected values is given in brackets. The values (All .90) refer to the means from the whole data set.

Table 2.10 Accuracy of prediction of hourly diffuse solar radiation using EC model against observed horizontal diffuse solar radiation for individual selected cloudless hours derived from global pyranometer observations and associated pyrheliometer observations at Cabauw (Data source: Slob (17)).

Solar altitude band degrees	Number* of obs	Mean solar alt degrees	Mean $T_L(\gamma)$	Range of $T_L(\gamma)$	Horizontal diffuse prediction error Obs/pred		
					Mean	S.D	Mean of all
10-20	23 (15)	15.7	4.81	3.42- 6.68	0.91	.045	(All .81)
20-30	27 (4)	25.2	4.64	3.36- 7.08	0.93	.065	(All .89)
30-40	38 (4)	34.9	5.07	3.53- 8.33	0.98	.064	(All .95)
40-50	30 (1)	45.1	5.60	4.17- 8.67	1.02	.064	(All 1.02)
50-60	19 (2)	55.3	6.10	4.21- 9.21	1.04	.070	(All 1.06)
Total obs used	135						

*Data were rejected if diffuse error was greater than 1.15 or less than 0.85, because data set was to be used for subsequent slope predictions. The number of rejected values is given in brackets. The values (All .90) refer to the means from the whole data set.

large numbers, one of which contains systematic errors due to angle of incidence effects, can lead to relatively large errors in the estimated diffuse. The advantage of using a shading band with a pyranometer to determine diffuse is that both instruments should have the same cosine errors, so errors in the measurement in the beam are not transferred into the diffuse component by difference. It will be seen that in general the new EC model again produces reasonable estimates of the diffuse irradiance over a wide range of Linke Turbidity Factors, except at low solar altitudes.

Figure 2.21 shows a plot of the hourly values of the error ratio (observed diffuse horizontal irradiance/predicted diffuse horizontal irradiance) for the 13 selected days for Cabauw plotted against solar altitude. The morning values are marked with circles and the afternoon values with crosses. In general it will be noted that the errors are less than 10%. The hour by hour changes occur reasonably smoothly indicating that changes in atmospheric properties on specific days rather than random measurement errors account for the form of the curves. Attention has already been drawn to the solar altitude error effect latent in this figure.

It will be noted in both the Cabauw and the Bracknell data, the mean Linke Turbidity Factor is strongly associated with solar altitude. The Bracknell data at lower solar altitudes contain a considerable number of observations in January and October when the atmosphere is clearer, whereas the Cabauw data are confined to the summer period, which accounts for the difference between the two data sets.

28. Climatological studies of the accuracy of the EC cloudless day sky irradiance model at the hourly level

28.1 Belgium, Uccle

The longest series of published observations of the Linke Turbidity Factor in Europe are undoubtedly those published by Dogniaux and his various collaborators at Uccle in the suburbs of Brussels. Dogniaux has prepared a new review of the Belgian data (18) which has been harnessed to this study. As Appendix 2 explains, Dogniaux distinguishes between the Linke Turbidity Factor for mean cloudless conditions in the month, and the minimum Linke Turbidity Factor for exceptionally clear days in the month. The checks reported here were carried out using the monthly mean representative values. The data available for each hour of each month used in this check were

i) Measured monthly mean Linke Turbidity Factor $T_L(\gamma)$ for cloudless days

ii) Measured cloudless day hourly global irradiation on a horizontal surface

iii) Measured cloudless day hourly diffuse irradiation on a horizontal surface.

A program was written to use the monthly mean values of the declination to calculate the mid hour solar altitude for each month, then inserting the observed monthly mean hourly value of $T_L(\gamma)$, the predicted

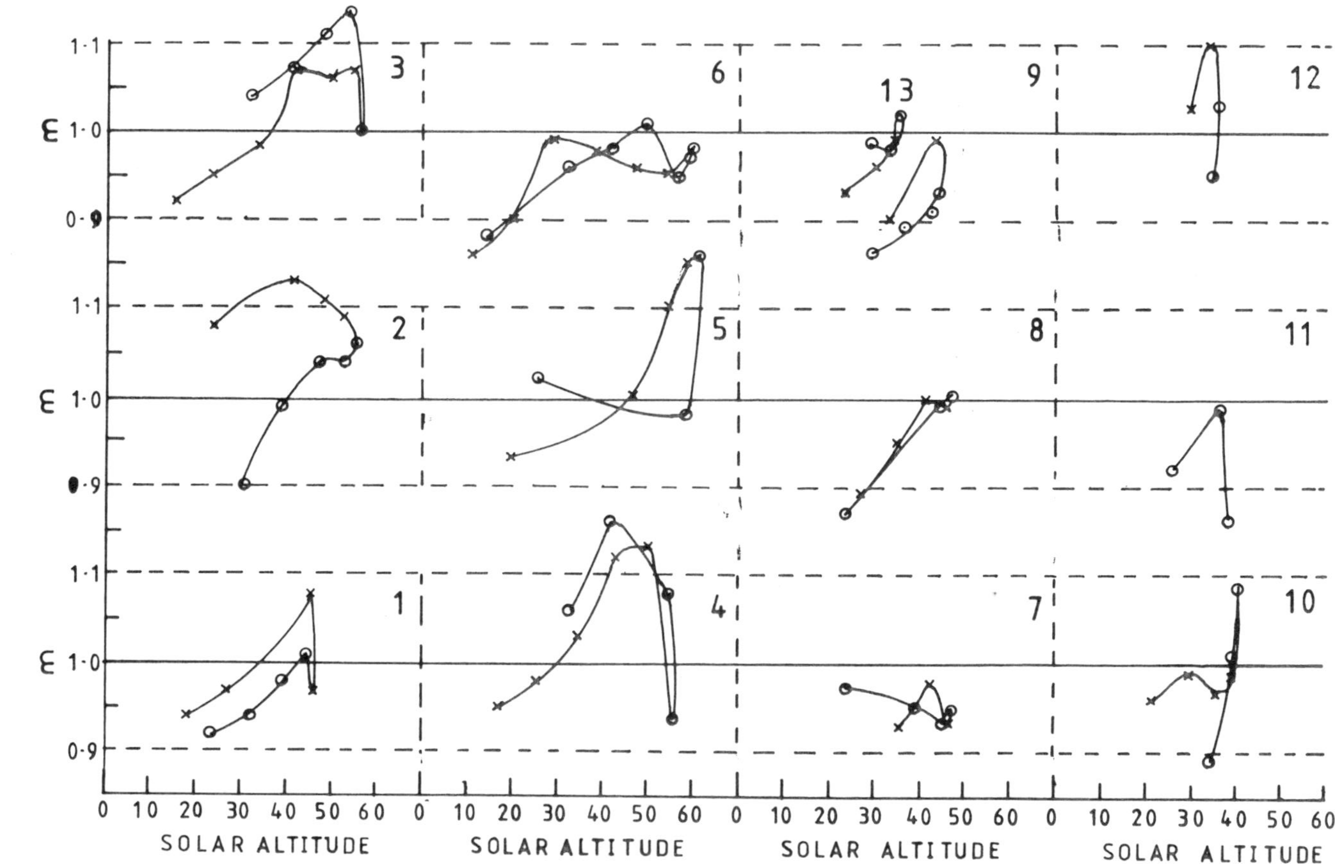

Figure 2.21 Diffuse error ratios hour by hour for 13 cloudless days at Cabauw plotted against solar altitude. Morning values marked with circles, afternoon values with crosses.

horizontal diffuse irradiance was estimated from the EC model, applying the mid month corrections to mean solar distance. The hourly error ratio (observed diffuse irradiance/predicted diffuse irradiance) was then calculated, making the assumption that the cloudless sky irradiance at the centre of each hour was defined by the hourly irradiation/hour. The results were classified by solar altitude bands as in the case of Bracknell and Cabauw, and means extracted. Table 2.11 gives these results.

It will be noted again that the mean values of $T_L(\gamma)$ are strongly dependent on solar altitude, but the range of values is much less than for individual days due to climatological averaging over a number of days. The EC model underpredicts the Uccle mean diffuse observations by 3% - 4%. The error ratio is fairly constant except in the band 10-20, where the error ratio is less than 1. The EC cloudless day model was drawn up using diffuse data from a range of European stations, some of which was shade ring corrected for non-isotropic skies, and some of which was not so corrected. As this additional correction is about 10% , one would expect a model drawn from such data sources to predict slightly low against a site where diffuse solar radiation observations are made with a pyranometer screened by an occulting disc. Nevertheless the errors are acceptably small.

28.2 Hamburg

Kasten, Golchert and Stolley (19), under CEC Contract ESF-004-80D as part of the CEC program also independently produced a cloudless day global solar radiation model as a function of Linke Turbidity Factor and solar elevation based on observations at Hamburg. 3799 cloudless hours were selected from the records for the 12 year period 1964-1975 using the following criteria

i) sunshine duration in hours, $S_h = 1$

ii) total cloud amount $<$ 1 okta at beginning and end of respective hour.

The model took the form

$$G_c = 0.83\ I_o\ .\ \sin\gamma\ \exp(-0.026\ T_L(\gamma)/\sin\gamma)$$

As $I_c(0,0)$ can be predicted from $T_L(\gamma)$, the diffuse component can be estimated by difference. Figure 2.22 shows the results of this computation compared with the CEC model at mean solar distance. It is clear that the Hamburg model breaks down for the prediction of diffuse sky irradiance at low turbidities and high solar elevations. However, the practical data sets do not normally contain such values, so the practical consequences are not serious. The agreement in general is very satisfactory, if the areas mentioned are excluded.

29. Comments on diffuse irradiance modelling tests for cloudless skies

In general the results give good confidence in the reliability of the CEC cloudless sky model for a wide range of Linke Turbidity Factors. At the individual hourly level the turbidity often changes rapidly. The synoptic air mass type is especially critical. Table 2.12 provides

Table 2.11 Accuracy of prediction of EC model of hourly diffuse solar radiation against observed monthly mean hourly diffuse solar radiation on a horizontal surface, Uccle, Belgium, for cloudless days, as a function of solar altitude.

Solar altitude band degrees	Number of means	Mean solar alt degrees	Mean $T_L(\gamma)$	Range of $T_L(\gamma)$	Horizontal diffuse Obs/Pred Mean
0-10	28	5.19	3.50	3.1 - 4.1	1.035
10-20	34	14.84	4.12	3.5 - 4.9	0.986
20-30	26	24.03	4.77	4.0 - 5.5	1.034
30-40	24	34.60	5.41	4.6 - 5.9	1.048
40-50	18	45.74	6.01	5.6 - 6.5	1.043
50-60	12	55.97	6.41	6.1 - 6.9	1.035
60-70	2	61.70	6.35	6.2 - 6.5	1.017

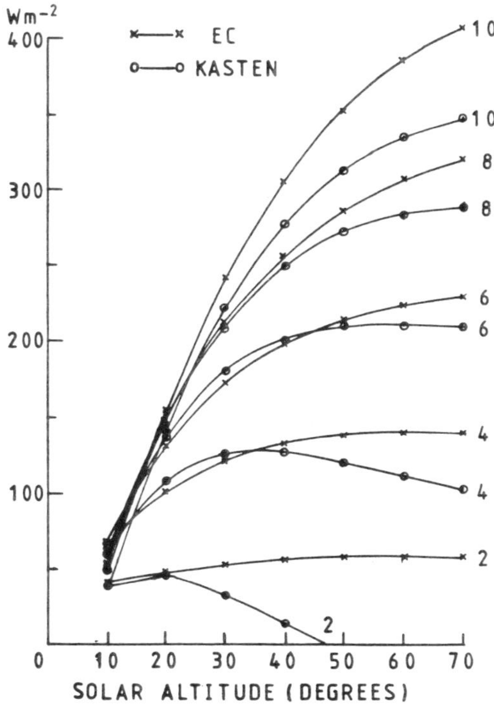

Figure 2.22 Comparison of predictions of CEC Sheffield model and the Kasten model for Hamburg as a function of solar altitude for $T_L(\gamma)$ between 2 and 10. In general the agreement is good, but the Hamburg model is not suitable for very low turbidities at higher solar elevations.

Table 2.12 Effect of synoptic air mass on Linke Turbidity Factor

Part I Linke Turbidity Factors in various synoptic air masses (average values), Potsdam, Germany. Source: Foitzik and Hinzpeter (20).

Synoptic air mass type	$T_L(\gamma)$
Continental arctic air	2.20
Maritime arctic air	2.10
Continental moderate air	2.89
Maritime moderate air	2.85
Continental tropical air	3.76
Maritime tropical air	3.66

Part II Characteristic Linke Turbidity Factors T_L estimated from Monteith and Unsworth (21) showing influence of synoptic air mass type and location in relation to pollution sources, estimated for 30° solar altitude, precipitable water vapour 15mm.

Site type	Air mass type	T_L
Northerly island site, minimum pollution from land sources	Polar	2.4
	Average	4.0
	Continental	5.6
Rural or coastal site exposed to natural aerosol pollution and small amounts of smoke	Polar	2.8
	Average	4.5
	Continental	6.0
Urban site within or close to a large town (say population exceeding 100,000)	Polar	4.5
	Average	6.0
	Continental	7.5

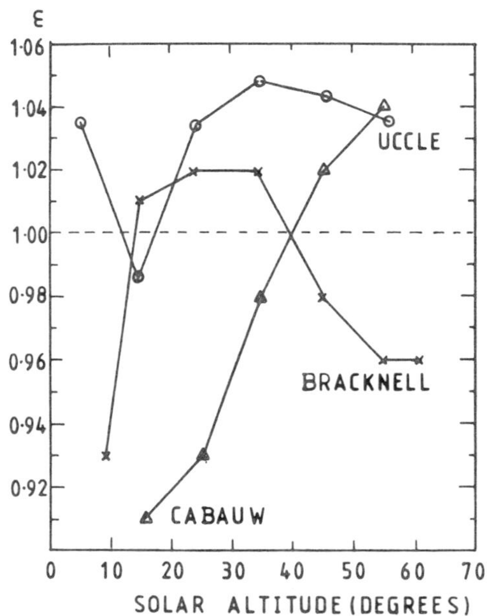

Figure 2.23 Diffuse error ratio ε, observed/predicted, as a function of solar altitude for Uccle, Bracknell and Cabauw using the CEC diffuse irradiance model for cloudless days.

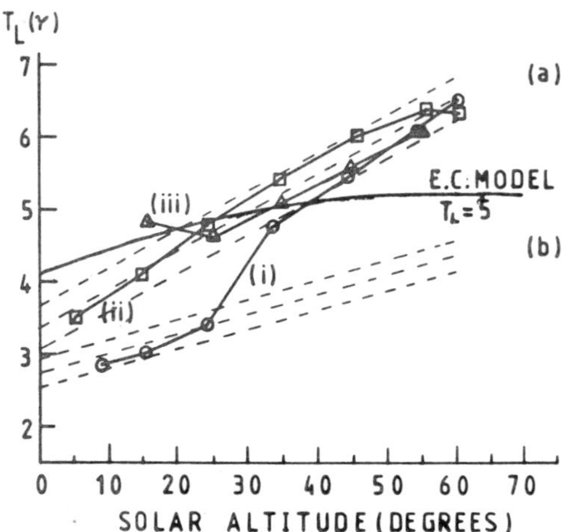

Figure 2.24 The relationship between the mean Linke Turbidity Factor and solar altitude at Bracknell, curve marked (i), Uccle, curve marked (ii), and Cabauw, curve marked (iii). The straight lines give the range of values predicted by the Dogniaux-Sneyers formula (a) mean cloudless conditions and (b) conditions of exceptional clarity.

observed data for Potsdam, Germany from Foitzik and Hinzpeter (20) and for the UK from Monteith and Unsworth (21). The values of Monteith and Unsworth which are higher than those reported for Potsdam seem to accord well with the three sets of detailed observations studied in these checks.

Figure 2.23 plots the mean diffuse errors for cloudless skies against solar altitude for the three sites. The differences are probably accounted for by the instrumental errors associated with the three approaches. The most accurate method of measuring diffuse radiation is likely to be the use of the shading disc following the sun, as at Uccle. The Bracknell data were obtained using a shading band and the results corrected for non-isotropic effects. The errors obtained will obviously depend on the accuracy with which the additional corrections can be made. This topic has been studied by another group in the CEC program, and the results published (22). The least accurate method of determining the diffuse radiation at lower angles of solar altitude is probably finding the diffuse by difference between the pyranometer reading and the resolved component of the pyrheliometer reading as at Cabauw due to cosine errors of pyranometers. These errors have been investigated under Action 1 of the CEC Solar Energy - solar radiation data programme (23). The studies of McGregor are particularly important in this respect (24).

30. Observed variation of Linke Turbidity Factor with solar altitude

Figure 2.24 plots the variation of the mean Linke Turbidity Factor with solar altitude for the three data sets studied. The formula of Dogniaux & Sneyers from Appendix 2 for mean cloudless conditions, and for exceptionally clear conditions are also plotted on the diagram. The Bracknell data set shows lower values at low solar altitudes compared with Cabauw, because Cabauw data were for the summer period only when the mean Linke Turbidity Factor in Europe increases. The three data sets are seen from Figure 2.24 to be reasonably consistent in relation to mean values of $T_L(\gamma)$. The results confirm Dogniaux's findings discussed in Appendix 2 that the solar altitude effect is greater than the seasonal effect when considering mean clear day Linke Turbidity Factors. The correction for solar altitude effects on $T_L(\gamma)$ adopted in the EC methodology is also superimposed on Figure 2.24.

31. Conclusion on diffuse solar radiation modelling on horizontal surfaces for cloudless days

A satisfactory inter-relationship has been established between the field measurements of diffuse radiation and the predictive models for cloudless sky conditions, so, provided the Linke Turbidity Factor can be properly established as a function of solar altitude, accurate irradiance predictions of global, beam and diffuse horizontal irradiance can be made for cloudless skies. The various European studies on diffuse irradiance carried out under the CEC Solar Energy R&D programme, Solar Radiation Data interlock well, and are scientifically consistent, one with another, though more attention needs to be given to the study of the best methods of accurately measuring diffuse radiation and recording the results in the best form for climatological studies of solar radiation.

CHAPTER 2 - PART IV

CONSTRUCTION OF A DAILY IRRADIATION MODEL FOR HORIZONTAL SURFACES FOR CLEAR DAYS, AND ITS APPLICATION TO THE PRODUCTION OF CLEAR DAY TABLES FOR THE EUROPEAN REGION USING AN OBJECTIVE METHOD FOR ESTIMATING LINKE TURBIDITY FACTORS FOR CLEAR DAYS

32. Use of the irradiance model to develop a daily global irradiation model

It is, of course, a simple matter to convert a direct beam and diffuse sky irradiance model to a daily global and diffuse irradiation model for horizontal surfaces by using numerical integration. The only further decision needed is the time interval to adopt to achieve the numerical integration. An integration interval of 1 hour was used for the systematic studies for the European region reported here. The astronomical daylength was first calculated, and then, working in local apparent (solar) time, the computational hours were set on the half hour, starting with the first half hour after sunrise. The last value was set at the half hour just before astronomical sunset. For example, if sunrise were to be at 0815 and sunset at 1545, the values used would be 0830, 0930, 1030, 1130, 1230, 1330, 1430, 1530. Appropriate values of the declination may be found either from Appendix 1 or using Chapter 6, Table 6.2. Thus, following the detailed methodology given in Chapter 6, the solar altitudes can be established at each half hour. Then, inputting the air mass 2 Linke Turbidity Factor, the value $T_L(\gamma)$ can be established at each half hour, and so a set of estimates of the horizontal beam and diffuse irradiance for the midpoint of each hour can be established. Assuming this irradiance to be representative of the mean irradiation in that hour, a daily numerical integration can then be carried out to find the daily beam diffuse and global irradiation on horizontal surfaces for cloudless days. As the direct beam irradiance and the diffuse irradiance are expressed in Wm^{-2}, it follows the units of irradiation will be Whm^{-2}. In the preparation of the CEC European Solar Radiation Atlas, Vol. II (25), the daily global and diffuse irradiation were expressed in $kWhm^{-2}$ ($Whm^{-2}/1000$) for convenience in engineering calculations.

A key problem, however, remains. Appropriate decisions have to be made for any site concerning the Linke Turbidity Factor to use in any month, but unfortunately there are relatively few systematic records of the Linke Turbidity Factor in Europe. Yet, to produce the inclined surface atlas, such datawere needed for 102 widely scattered sites from Norway to Greece. Considerable research had to go into finding a satisfactory solution to this problem.

33. Determining the Linke Turbidity Factor

33.1 Indirect approaches involving subjective judgement about site clarity

There are very few sites in Europe where the systematic pyrheliometric observations necessary to determine the Linke Turbidity Factor are made, so it was essential to find a suitable way of estimating the Linke Turbidity Factor. A number of different approaches are open.

33.2 Method 1 - Standard turbidity reference tables

One approach is to make judgements about the probable aerosol climate of any specific site, and then to use suitable tables, based on measurements at a range of similar sites, to select appropriate representative T_L values for that month for that site. This was the approach adopted by Krochmann and Aydinli of the University of Berlin in their contribution to the CEC Programme. They adopted the monthly mean values of the Linke Turbidity Factor suggested by Steinhauser (26) and Schulze (27), Table 2.13. The Berlin approach was also used by the University of Sheffield group at one stage in the development of the CEC Inclined Surface Atlas, but it is a subjective method involving judgement about sites, for the majority of which the Sheffield research group had no knowledge of atmospheric conditions nor site contact. As soon as one suggested certain sites might have a solar radiation climate typical of an industrial atmosphere, there was immediate protest from the meteorologists in that region that the subjective judgement was too harsh, and that the site should be moved one notch up the atmospheric clarity table. This subjective approach was obviously going to lead to serious distortions. Fortunately, it proved eventually possible to find an objective method of assessing site clarity using the actual observed data, as discussed later in this Chapter, which enabled this important difficulty to be totally circumvented.

33.3 Method 2 - Theoretical computational methods

Appendix 2 reviews in detail some of the methodologies that have been developed for estimating the Linke Turbidity Factor. Some of these methods are site specific, and hence are not suitable for generalisation. Others required inputs of precipitable water vapour and Angstrom Turbidity Coefficient, again requiring subjective judgement of atmospheric clarity at specific sites.

The most practicable of these methods for people with good knowledge of specific sites appears to be the method proposed by the World Meteorological Organisation (28). It has the added advantage, in relation to the clear day model eventually adopted for these CEC studies, that the correction factors to the air mass 2 T_L for solar altitude are identical. It is recommended, in the absence of any observed radiation data, that the WMO method be used as the most suitable subjective way for non-specialists to achieve reasonable estimates of the Linke Turbidity Factor in any region, provided they are familiar with the climate. As the method does demand subjective estimates of the colour of the sky at specific sites, it does not provide a general methodology. The experience required to estimate the typical colour of the sky across the whole year across the whole of Europe certainly did not reside in the research group in Sheffield, so this approach had to be abandoned too.

Table 2.13 Monthly mean values of the Linke Turbidity Factor for temperate climates. Source: Steinhauser (26) also Schulze (27).

Month	J	F	M	A	M	J	J	A	S	O	N	D	Yearly mean
Very clear atmosphere	1.5	1.6	1.8	1.9	2.0	2.3	2.3	2.3	2.1	1.8	1.6	1.5	1.9
Clear atmosphere	2.1	2.2	2.5	2.9	3.2	3.4	3.5	3.3	2.9	2.6	2.3	2.2	2.75
Urban atmosphere	3.1	3.2	3.5	4.0	4.2	4.3	4.4	4.3	4.0	3.6	3.3	3.1	3.75
Industrial atmosphere	4.1	4.3	4.7	5.3	5.5	5.7	5.8	5.7	5.3	4.9	4.5	4.2	5.00

Table 2.14 Dates used to select declination for studies of the relationship between T_L and G_{MAX}.

29th January	29th May	4th September
26th February	21st June	4th October
29th March	4th July	4th November
28th April	4th August	4th December

33.4 Method 3 - The development of an objective method for estimating the Linke Turbidity Factor for clear days in the European region

If observed values of clear day global and monthly mean global radiation are available, together with observed daily sunshine data, it becomes possible to develop more objective ways of determining the air mass 2 Linke Turbidity Factor. The CEC European Solar Radiation Atlas, Vol. I, Horizontal Surfaces (29). contains tables for 100 sites giving values of G_{max}, the monthly mean maximum global solar radiation, and S_{max}, the monthly mean maximum daily bright sunshine.* The Atlas also contains tabulated values of the coefficients a and b in the Angstrom regression equation

$$G_d/G_{od} = a + b\ S_d/S_o \qquad (2.34)$$

where G_d/G_{od} is the ratio of the daily global radiation received on a specific day to the corresponding extraterrestrial value

S_d/S_o is the ratio of the daily bright sunshine to the astronomical daylength.

Studies by Dogniaux and Lemoine (30) amongst others had demonstrated a relationship between T_L in any month and the monthly mean values of the sum (a + b). Refer Appendix 2, method 3. It was decided to attempt to seek objective relationships between T_L and (a + b) for days of monthly mean maximum horizontal surface radiation, G_{max} days using the detailed data in the Atlas.

34. Studies based on the assumption of totally cloudless conditions

An objective relationship between T_L and (a + b) for different months of the year in Europe was next developed as follows:

i) Using values of G_{max} in the horizontal tables in the CEC European Solar Radiation Atlas, Vol. I (29), estimates were prepared month by month, by successive approximation, of the input value of the AM2 Linke Turbidity Factor needed to match the cloudless day model outputs to the actual daily observed values of G_{max} in the European Solar Radiation Atlas, Volume I, Horizontal Surfaces, using the clear sky model described above. One basic problem is that the underlying extraterrestrial radiation is changing very fast in certain months. The dates selected for study had to be related to this fact. Based on the experience of previous clear day UK studies in the University of Sheffield, appropriate computational dates were identified, given in Table 2.14, as being statistically correct for examining the underlying solar geometry associated with the monthly mean maximum values of horizontal solar radiation. Data for 53 stations listed in Table 2.15 were processed. The Italian data were not processed because of doubts about the accuracy of the Robitsch pyranometers. No tabulated data estimated from sunshine alone were processed for any country.

ii) Monthly values of the sum of the Angstrom regression coefficients (a + b) for 45 sites were used. Statistical regressions were then sought between the values of T_L derived from the model and the input parameters, including latitude, S_d/S_o as well as a + b.

* Only observed data were used. The Atlas also contains estimated values derived from sunshine data.

Table 2.15 45 sites used for clear day in G_{max} radiation study.

Hojbakkegard
Lerwick
Eskdalemuir
Aldergrove
Cawood
Aberporth
London
Kew
Bracknell
Birr
Kilkenny
Valentia
Groningen
Den Helder
De Bilt
Vlissingen
Maastricht
Oostende
Melle
Uccle
Saint Hubert
Nordeney
Hamburg
Braunschweig
Braunlage
Wurzburg
Trier
Weihenstepan
Hohenpeissenberg
Trappes
Nancy
Macon
Limoges
Millau
Carpentras
Nice
Zurich
Davos
Locarno Monti
Wien
Salzburg
Innsbruck
Sonnblick
Klagenfurt
Athens

Table 2.16 Coefficients for estimating clear day Linke Turbidity Factor from the monthly values of the sum of the Angstrom coefficients (a + b) derived by linear regression.

	Jan	Feb	Mar	Apr	May	Jun	Jul	Aug	Sep	Oct	Nov	Dec
a_o	17.38	17.31	20.91	20.26	20.80	18.26	17.88	20.28	20.54	18.57	16.47	12.52
r	0.65	0.81	0.86	0.85	0.79	0.81	0.67	0.80	0.83	0.72	0.71	0.64
S.E.	0.92	0.66	0.59	0.62	0.73	0.60	0.88	0.75	0.66	0.79	0.75	0.82

iii) The simplest regression is the monthly regression of the derived values of T_L against the monthly mean values of a + b. It was noted that when $G = G_o$, $T_L = 0$, therefore (a + b) = 1 is a constrained point on the graph, i.e. there is no attenuation outside the atmosphere. The equation therefore takes the form:

$$T_L = a_o - a_o(a + b) \tag{2.35}$$

The values of a_o found in this way are given in Table 2.16 together with the correlation coefficient and the standard error of the estimate. The standard errors ranged from 0.92 in January to 0.59 in March.

iv) The use of multiple linear regression of monthly values of T_L with S_d/S_o (where S_d/S_o is the astronomically defined relative duration of sunshine) and (a + b) as the dependent variables was also explored and gave improved accuracy of estimation. The equation in this case is

$$T_L = a_o + a_1(S_d/S_o) + a_2(a + b) \tag{2.36}$$

The values of a_o, a_1 and a_2 obtained month by month proceeding in this way are given in Table 2.17a, together with the correlation coefficients and the standard error of the estimate. Table 2.17a shows a range of standard errors of 0.70 in December to 0.44 in March. Plotting out all the predictions showed the method was less accurate for stations above 500 m. The data for two stations, Hojbakkegard and Athens, also did not fit well. The multiple regressions, using (a + b) and S/S_o as dependent variables, were then recalculated eliminating Hojbakkegard, Athens, and the higher level stations, Saint Hubert, Hohenpeissenberg, Davos, Innsbruck, Sonnblick. The new multiple regression coefficients are given in Table 2.17b. The range of standard errors is reduced from 0.70 to 0.49 in December and from 0.44 to 0.38 in March. The correlation coefficients are also improved.

Separate multiple regressions were performed for all stations north of latitude 47°00'N. Some slight statistical improvement was achieved, but the difference was not great.

v) Finally, multiple linear regression of T_L with S/S, a + b, and latitude ϕ as the dependent variables was used. The accuracy was further increased, but not by a large amount.
The values of a_o, a_1, a_2 and a_3 obtained month by month proceeding in this way are given in Table 2.18 together with the correlation coefficients and the standard error of the estimate.
Table 2.18 shows the standard errors range from 0.70 in December to 0.43 in March. The extra complexity of introducing latitude appeared to offer little statistical improvement.

vi) At this stage of the programme development, it seemed that the formula from Table 2.17b should be used as the basis for the estimation of clear day radiation for the CEC inclined surface

Table 2.17 Coefficients for estimating clear day Linke Turbidity Factors from monthly relative sunshine duration S/S_o and monthly values of the sum of the Angstrom coefficients $(a + b)$ derived by multiple linear regression $T_L = a_o + a_1(S/S_o) + a_2(a + b)$

Table 2.17a Using all 45 stations

	Jan	Feb	Mar	Apr	May	Jun	Jul	Aug	Sep	Oct	Nov	Dec
a_o	14.46	17.97	19.05	21.55	21.93	19.85	18.42	22.06	20.26	15.40	14.14	11.58
a_1	-7.19	-4.66	-5.17	-3.99	-5.29	-3.75	-4.42	-4.61	-3.78	-4.44	-5.08	-3.67
a_2	-11.47	-16.37	-16.26	-19.86	-19.38	-18.14	-15.93	-19.79	-18.03	-12.14	-11.50	-10.11
r	0.89	0.84	0.89	0.83	0.81	0.80	0.76	0.84	0.83	0.80	0.81	0.70
S.E.	0.49	0.55	0.44	0.56	0.61	0.52	0.67	0.56	0.55	0.56	0.52	0.70

Table 2.17b Rejecting stations above 500m, Athens and Hojbakkegard

	Jan	Feb	Mar	Apr	May	Jun	Jul	Aug	Sep	Oct	Nov	Dec
a_o	13.23	17.01	18.71	22.02	22.43	21.03	19.28	23.43	22.18	14.82	14.13	11.04
a_1	-6.83	-5.82	-6.03	-5.05	-7.81	-5.33	-5.49	-5.96	-5.17	-4.79	-6.39	-4.09
a_2	-9.93	-14.68	-15.42	-19.87	-18.66	-18.73	-16.37	-20.75	-19.77	-11.18	-11.04	-9.16
r	0.88	0.86	0.91	0.90	0.86	0.89	0.89	0.92	0.89	0.81	0.85	0.79
S.E.	0.43	0.45	0.38	0.41	0.50	0.39	0.47	0.40	0.45	0.51	0.43	0.49

Table 2.18 Coefficients for estimating clear day Linke Turbidity Factors from monthly relative sunshine duration S/S_o, monthly values of the sum of the Angstrom coefficients $(a + b)$ and latitude ϕ derived by multiple linear regression
$T_L = a_o + a_1(S/S_o) + a_2(a + b) + a_3.$

	Jan	Feb	Mar	Apr	May	Jun	Jul	Aug	Sep	Oct	Nov	Dec
a_o	18.86	20.50	22.36	24.18	25.56	25.13	21.11	27.21	27.64	25.21	17.68	13.56
a_1	-9.56	-6.33	-7.42	-5.50	-7.03	-5.69	-5.45	-6.48	-7.41	-9.03	-6.81	-4.56
a_2	-11.11	-16.11	-16.06	-19.97	-20.31	-19.61	-16.13	-21.18	-19.16	-12.60	-11.62	-9.91
a_3	-0.083	-0.045	-0.054	-0.039	-0.043	-0.065	-0.041	-0.065	-0.099	-0.156	-0.060	-0.039
r	0.91	0.84	0.90	0.84	0.81	0.84	0.77	0.85	0.84	0.86	0.83	0.71
S.E.	0.45	0.55	0.43	0.56	0.61	0.48	0.67	0.55	0.54	0.48	0.50	0.70

radiation atlas. The data could then be normalised to the horizontal observations of G_{max} reported in the atlas and the normalisation errors recorded. For stations for which no observed values of G_{max} exist, the tabulated values would be as estimated. For stations where no values of (a + b) were available, the standard Dogniaux and Lemoine classification (30) could be used to obtain an estimate of (a + b) which could then be used to generate the clear day values.

35. Defects of cloudless day model for prediction of daily irradiation

The derived values of T_L were then examined and it was found that these derived values sometimes appeared rather high, especially at the sites where it was known there was typically a lot of cloud.

It was firstly observed that the data set was not statistically consistent between months, because between the winter and the summer solstice the extraterrestrial radiation at the end of the month is substantially greater than that at the beginning of the month. Thus, in this period, the monthly maximum of daily radiation tends to occur towards the end of the month. In the second half of the year the opposite is true; the maximum values tend to occur towards the beginning of the month. This phenomenon is a result of the change in solar declination across each month. In months where the solar declination is changing rapidly, the maximum global radiation is more likely to occur during the period of the month when the solar declination is greatest, while the months June and December are the exceptions. There is thus a bias in sampling which leads to apparently greater clarity in June and December.

This bias in sampling can be reduced by using a shorter sampling period than one month and hence reducing the range of solar declination within the sampling period. This effect is illustrated in Table 2.19. The table shows clearly that lower estimated values of turbidity result around June and December when the whole month is sampled. When the sampling period is reduced, this effect is reduced. However, it is still present, but to a lesser extent. In order to avoid this bias altogether it would be necessary to use sampling periods based on equal intervals of solar declination.

There is a second important statistical consideration. The more clear days in a month at any site, the more likely one is to sample an exceptionally clear cloudless day. Furthermore, in the more maritime northerly climates especially, it is hard to sample a day without any cloud. The consequence is that the clear day Linke Turbidity Factor derived from G_{max} is both a function of S/S_0 and (a + b).

There is also a difference in the average turbidity factor for all days and the turbidity factor for G_{max} days which also needs to be established. The problem of cloudiness on G_{max} days remained and also had to be resolved.

Table 2.19 Daily mean Linke Turbidity Factor at Kew calculated from daily global radiation exceeded for 5% of stated period in month.

	T_L 10 days sampled	T_L whole month sampled	Difference
Jan	4.90	5.74	0.84
Feb	5.04	6.06	1.02
Mar	4.04	5.36	1.32
Apr	5.26	6.28	1.02
May	5.44	5.61	0.17
Jun	4.73	4.70	0.03
Jul	5.05	5.58	0.53
Aug	6.23	5.81	-0.42
Sep	5.08	5.49	0.41
Oct	4.60	5.16	0.56
Nov	3.90	4.72	0.82
Dec	3.55	3.99	0.44

Table 2.20 Stations and periods of monthly maximum daily global radiation and sunshine data used in the final development of the EC inclined surface solar radiation prediction model - arranged in order of descending latitude.

	Latitude	Longitude	Elevation	Period
Lerwick, Scotland	60°08'N	01°11'W	82m	1966-75
Eskdalemuir, Scotland	55°19'N	03°12'W	242m	1966-75
Aldergrove, N. Ireland	54°39'N	06°13'W	68m	1969-75
Hamburg, Bundesrepublik Deutschland	53°38'N	10°00'E	14m	1964-73
Cambridge, England	52°13'N	00°06'E	23m	1966-71
Aberporth, Wales	52°08'N	04°34'W	133m	1966-75
Valentia, Ireland	51°56'N	10°15'W	20m	1964-74
London Weather Centre, England	51°31'N	00°07'W	77m	1967-75
Kew, England	51°28'N	00°19'W	5m	1966-75
Bracknell, England	51°23'N	00°47'N	73m	1966-75
Uccle, Belgique	50°48'N	04°21'E	105m	1966-75
Jersey, England	49°11'N	02°11'W	85m	1966-75
Paris/Trappes, France	48°46'N	02°01'E	168m	1967-75
Locarno-Monti, Switzerland	46°10'N	08°47'E	380m	1958-71
Carpentras, France	44°05'N	05°03'E	105m	1968-75

36. Improved clear day model taking account of actual observed values of monthly mean maximum sunshine at different sites

While the previously described model gave satisfactory results in the prediction of global irradiation, the Linke Turbidity Factors, derived by reiteration to make the model fit the observations were rather higher than seemed reasonable, especially at stations located in rather cloudy climates. This observation led on to a systematic examination of the relative duration of mean maximum monthly daily sunshine from the values of S_{max} reported in the Horizontal Atlas tables computed on the basis of the 4 degree sunrise/sunset daylength. It was found that the values of the relative duration of maximum bright sunshine σ_{max} for S_{max} days were often quite low in winter, especially in cloudy climates.

A new hypothesis was next introduced that the direct beam would be reduced in proportion to the actual observed values of S_{max}/S_{04max} where S_{04max} is the 4 degree sunrise/sunset daylength.

Values of S_{max} were available for the majority of the sites in the CEC Solar Radiation Atlas, Vol. I, Horizontal Surfaces, as well as values of G_{max}.

The final clear day method adopted for the CEC European Solar Radiation Atlas, Vol. II, Inclined Surfaces (25), uses three monthly inputs from the horizontal surface tables for each station whose data are tabulated in Volume I of the Atlas (20) to produce solar radiation estimates:

i) the observed values of the monthly mean maximum daily global radiation on a horizontal surface, G_{max}.

ii) The observed values of the monthly mean maximum daily duration of bright sunshine, S_{max}.

iii) The monthly mean values of the Angstrom (a + b). The horizontal Atlas mean values of (a + b) were derived from the daily global irradiation and the daily duration of bright sunshine on a month by month basis.

The final method adopted also operates using representative clear day dates for each month given in Table 2.14. The associated representative declinations chosen are given in Chapter 6, Table 6.2. Basically there is potentially most horizontal energy on clear days in any month when the sun is highest in the sky in that month. As already explained in paragraph 35, earlier studies had shown that the days of maximum global radiation on horizontal surfaces tended to occur statistically towards the end of calendar months when the declination was increasing during the course of that month, and towards the beginning of the month when the declination was decreasing during the course of that month, while, in June, the maximum values occurred around the summer solstice on 21st June.

It was then assumed that the days of maximum duration of bright sunshine were statistically associated with the days of high radiation. Thus the days with greatest hourly sunshine were assumed to follow the same temporal pattern as G_{max}. The problem was then to find an objective way of relating T_L to a + b taking account of actual values of S_{max}. The detailed method used was as follows:

1. The clear day relative sunshine duration was calculated for each month for each station, using a 4 degree altitude angle for defining sunrise and sunset. This altitude angle was used to calculate the possible daylength, S_{04max}. The declinations given in Chapter 6, Table 6.2 were adopted.

   Data was processed for 100 stations. Typical values of the ratio σ_{04max} were about 0.90. In mid winter when site obstructions have the greatest effect, several stations gave lower values, i.e. around 0.75. One station, Bolzano, gave exceptionally low values throughout the year. It appears from the evidence to be a relatively obstructed site. Appendix 3 gives further details of studies relating σ_{4max} to S/S_o.

2. 15 stations covering the Western European area were then selected for detailed study of their clear day properties. The site details are given in Table 2.20. Where possible a preference was placed on sites of known observatory standard, where the calibration standards are usually highest. The 15 stations used are arranged in order of decreasing atmospheric clarity in Table 2.21 using the measured mean annual value of the Angstrom $(\overline{a + b})$ to define clarity. The relative durations of monthly mean maximum sunshine calculated on a 4 degree sunrise basis for these 15 stations are given on a month by month basis in Table 2.22. The values of the ratio σ_{4max} for Carpentras are consistently higher than the rest. The type of sunshine recorder used there may be different. The maritime climates, Valentia and Lerwick, have the lowest values of σ_{4max}. A seasonal pattern, probably due to site obstruction, is present at some sites. For example, considerable site obstruction is known to exist at Locarno Monti.

3. The first estimates were prepared by using the monthly values of (a + b) for the above stations from the CEC European Solar Radiation Atlas, Vol. I, Horizontal Surfaces, 2nd edition, in a linear regression formula linking the monthly mean Linke Turbidity Factor at air mass 2 to monthly mean values of (a + b). This formula was developed for the CEC monthly mean day model and is discussed in detail in Chapter 5.* Then the daily horizontal radiation for a perfectly cloudless day was generated from this T_L value, using hour by hour calculations, and then summing on a daily basis. The results of the comparison are given in Table 2.23. In general the cloudless day model based on the monthly mean overestimates by a few percent, except in mid-winter.

4. This daily cloudless beam irradiation value was then reduced to allow for the clear day relative sunshine duration S_{max}/S_{04max} using the following relationship:

$$G_{max} = I_c(0,0) \times S_{max}/S_{04max} + D_c \qquad (2.37)$$

   This method was based on the observation of Kasten and Czeplak (31) that, under nearly cloudless conditions, the diffuse horizontal radiation is roughly independent of cloud cover, but the mean direct beam intensity is reduced due to interception by occasional patches of cloud. The predictions using the monthly mean turbidities were then compared with observations, using the monthly mean turbidities in Table 2.24. The results were better, but further improvements were sought.

* Chapter 5 deals with the model developed to produce monthly mean irradiation data.

Table 2.21 European stations selected for clear day study and associated annual mean values of Angstrom (a + b)

VERY CLEAR ATMOSPHERE $\overline{(a + b)} > .80$

	(a+b)
Valentia	.85
Lerwick	.82
Zurich	.81
Locarno-Monti	.80

CLEAR ATMOSPHERE $.76 < \overline{(a + b)} < .79$

Eskdalemuir	.79
Aberporth	.76

URBAN ATMOSPHERE $.73 < \overline{(a + b)} < .75$

Carpentras	.75
De Bilt	.75
Trappes	.75
Bracknell	.74
Hamburg	.74

LARGE CITY/INDUSTRIAL ATMOSPHERE $\overline{a + b} < .72$

Uccle	.72
Vienna	.71
Kew	.71
Athens	.68
London	.68

Table 2.22 Values of relative duration of monthly mean maximum sunshine S_{max}/S_{04max} for 15 European stations, computed assuming a 4 degree altitude of sunrise, using declinations in Table 6.2.

STATION \ MONTH	J	F	M	A	M	J	J	A	S	O	N	D
Valentia	.76	.81	.85	.91	.91	.93	.88	.92	.86	.75	.81	.77
Lerwick	.70	.86	.78	.80	.85	.90	.82	.84	.77	.77	.62	.79
Zurich	.77	.88	.85	.90	.89	.96	.96	.93	.89	.84	.77	.73
Locarno Monti	.92	.96	.94	.96	.93	.93	.93	.94	.96	.94	.93	.92
Eskdalemuir	.79	.89	.83	.88	.87	.92	.85	.89	.87	.83	.94	.99
Aberporth	.81	.89	.88	.91	.92	.97	.94	.90	.88	.83	.85	.95
Carpentras	1.03	1.00	.94	.98	1.00	1.00	1.01	1.00	.98	1.00	1.03	1.06
De Bilt	.80	.84	.86	.93	.91	.95	.92	.93	.86	.86	.81	.95
Trappes	.84	.90	.92	.94	.93	.94	.95	.96	.89	.90	.90	.92
Bracknell	.84	.90	.83	.90	.89	.94	.91	.91	.86	.85	.86	.87
Hamburg	.87	.92	.90	.91	.95	.99	.95	.94	.92	.84	.88	.96
Uccle	.89	.90	.88	.95	.93	.93	.97	.95	.93	.90	.94	1.03
Vienna	.84	.88	.89	.92	.94	.96	.96	.93	.91	.92	.92	.95
Kew	.84	.89	.84	.90	.91	.95	.95	.91	.84	.89	.87	.92
Athens	.98	.97	.92	.97	.97	.96	.96	.98	.91	.95	1.01	1.00
London	.84	.86	.84	.91	.92	.94	.95	.93	.85	.90	.88	.90

Table 2.23 Percentage errors in predicted monthly mean maximum daily global radiation in comparison with observed monthly mean maximum daily global radiation on a horizontal surface predicted using mean day model Linke Turbidity Factors and assessing cloudless conditions.

(Observed/Predicted - 1) x 100%

STATION	JAN	FEB	MAR	APR	MAY	JUN	JUL	AUG	SEP	OCT	NOV	DEC
Valentia	1.5	-0.1	-5.0	-3.2	-4.9	-0.7	-2.9	-1.9	-3.4	-5.5	1.8	3.8
Lerwick	6.5	1.0	-4.1	-5.6	-1.4	-1.7	-3.8	-4.1	-3.0	-3.2	-0.4	12.1
Zurich	6.4	2.2	-1.9	-3.9	-2.7	4.2	-2.5	0.2	-2.4	1.1	6.4	2.1
Locarno Monti	-2.8	-4.5	-1.5	-3.9	-4.3	-4.0	-8.0	-4.1	-0.4	-1.6	-2.5	-4.8
Eskdalemuir	10.4	1.8	-1.8	-2.0	0.3	-1.1	-3.6	-0.1	1.2	3.2	8.0	8.6
Aberporth	7.4	-3.4	-3.3	-6.3	-6.3	-2.7	-3.2	-4.2	-4.7	-3.3	6.8	13.4
Carpentras	-0.2	-3.7	-5.2	-2.8	-5.1	-3.7	-6.2	-3.8	-2.6	-1.6	-0.1	3.0
De Bilt	8.5	-0.5	-1.6	-2.3	-4.6	-1.8	-1.1	-0.1	2.3	0.1	3.3	7.0
Trappes	-0.7	-4.8	-3.0	-3.2	-3.9	-3.4	-6.0	-0.7	-3.0	2.5	4.8	2.3
Bracknell	5.7	2.7	-5.8	-2.6	-4.7	-4.2	-7.9	-0.6	0.9	-0.8	-0.1	6.6
Hamburg	7.8	0.5	-1.7	-2.7	-5.4	-3.3	-3.0	-3.3	1.4	1.6	9.5	6.4
Uccle	8.4	2.1	-2.1	-3.3	-0.9	-0.4	-1.4	2.0	2.4	2.7	10.5	9.7
Vienna	8.9	3.0	1.3	-2.5	-1.9	-0.5	-4.4	-0.4	1.9	-1.1	-1.8	7.1
Kew	9.0	1.8	-4.7	-0.6	-5.9	-2.4	-3.2	-2.1	-2.9	2.6	1.4	9.4
Athens	7.1	6.0	-1.8	0.7	-2.4	-2.6	-3.9	-0.1	-0.7	1.7	5.6	7.5
London	10.1	5.1	-4.3	1.7	-3.5	-0.3	-2.9	1.3	0.0	2.7	1.7	3.6

Table 2.24 Unimproved clear day prediction model (clear model Linke Turbidity Factors assumed to be same as average day values). Percentage errors in the prediction of monthly mean maximum daily global radiation on a horizontal surface (G_{max}) from values of monthly mean maximum daily duration of bright sunshine (S_{max}) for 15 stations selected from CEC Solar Radiation Atlas, Vol. I (2nd edition)

(Observed/Predicted - 1) x 100%

Station	Jan	Feb	Mar	Apr	May	Jun	Jul	Aug	Sep	Oct	Nov	Dec
Valentia	-1	0	5	3	5	1	3	2	4	6	-2	-4
Lerwick	-6	-1	4	6	1	2	4	4	3	3	0	-11
Zurich	-6	-2	2	4	3	-4	3	0	3	-1	-6	- 2
Locarno Monti	3	5	2	4	4	4	9	4	0	2	3	5
Eskdalemuir	-9	-2	2	2	0	1	4	0	-1	-3	-7	-8
Aberporth	-7	3	3	7	7	3	3	4	5	3	-6	-12
Carpentras	0	4	6	3	5	4	7	4	3	2	0	- 3
De Bilt	-8	0	2	2	5	2	1	0	-2	0	-3	- 7
Paris/Trappes	1	5	3	3	4	3	6	1	3	-2	-5	- 2
Bracknell	-5	-3	6	3	5	4	9	-1	-1	1	0	- 6
Hamburg	-7	0	2	3	6	3	3	3	-1	-2	-9	- 6
Vienna	-8	-3	-1	3	2	1	5	0	-2	1	2	- 7
Uccle	-8	-2	2	3	1	0	1	-2	-2	-3	-10	- 9
Athens	-7	-6	2	-1	2	3	4	0	1	-2	- 5	- 7
London	-9	-5	5	-2	4	0	3	-1	0	-3	- 2	- 3

5. The new model was then operated iteratively on a month by month basis by systematically changing the input values of the air mass 2 Linke Turbidity Factor T_L until the predicted values of G_{max} matched the observed values of G_{max}, using an automated program to do this.

   The consequent matching input value of T_L was recorded for each month for each of the 15 stations. The values of the clear day atmospheric turbidity derived in this way are given in Table 2.25 together with the associated values of the Angstrom (a + b) taken from station tables for horizontal surfaces in the European Solar Radiation Atlas, Vol. I.

6. Linear regressions were then sought on a month by month basis between the monthly derived values of T_L for clear days and the monthly mean values of Angstrom (a + b). Good monthly correlations were found using the G_{max} data for the 15 stations used,as Table 2.26 shows.

7. The regression equations relating the air mass 2 Linke Turbidity Factor T_L to the Angstrom (a + b) values in Table 2.26 were then substituted into the clear day model to produce an improved clear day model for predicting daily values of G_{max} from S_{max} and (a + b).

Table 2.27 gives the errors using the improved clear day turbidity model based on Table 2.26. The improved accuracy will be noted. The biggest errors for the selected sites occurred on sites obstructed in winter, for example, Locarno Monti.

37. Computation of standard clear day tables for the European region

A computer program was then written to estimate and store in computer file hourly values of the direct beam and diffuse irradiance on a horizontal surface for clear days, using as initial inputs values for the CEC European Solar Radiation Atlas, Vol. I, Horizontal Surfaces of:

i) site latitude and altitude above sea level

ii) the observed values of the monthly mean maximum daily duration of bright sunshine S_{max},

iii) the monthly mean values of Angstrom (a + b) from the Horizontal Surface Atlas tables.

In addition the declination for G_{max} days was input for each month. These hourly global irradiance values were added up to obtain preliminary estimates of the daily global irradiation. Then, to produce standard radiation tables for clear days, the predicted daily global radiation for each month was then compared with the observed daily value of monthly mean maximum global radiation, G_{max}. The hourly data for G_c and D_c were then normalised by multiplying each value by the ratio (observed maximum global radiation/predicted maximum daily global radiation). These corrected hourly values of G_c and D_c on a horizontal surface were then retained on computer file for the subsequent slope prediction calculations described in detail in Chapter 3. Basically, in the next stage, the slope values are also estimated hour by hour. The direct beam irradiance on inclined planes is computed trigonometrically. The diffuse sky irradiance is estimated

Table 2.25 Clear day air mass 2 Linke Turbidity Factors derived using observed monthly mean maximum values of global radiation G_{max} together with monthly mean values of S_{max} with associated values of monthly Angstrom (a + b).

STATION	JANUARY a+b	JANUARY AM2 T_L	FEBRUARY a+b	FEBRUARY AM2 T_L	MARCH a+b	MARCH AM2 T_L	APRIL a+b	APRIL AM2 T_L	MAY a+b	MAY AM2 T_L	JUNE a+b	JUNE AM2 T_L
Valentia	.83	2.1	.85	2.0	.85	1.6	.84	2.3	.86	1.9	.86	2.3
Lerwick	.79	3.2	.80	2.8	.84	1.8	.82	2.1	.87	2.2	.84	2.5
Zurich	.78	3.3	.79	3.1	.81	2.6	.81	2.6	.85	2.4	.85	3.2
Locarno Monti	.78	2.0	.79	2.2	.83	2.4	.82	2.5	.85	2.2	.84	2.1
Eskdalemuir	.78	3.6	.80	2.9	.81	2.7	.80	3.1	.81	3.6	.79	3.4
Aberporth	.73	3.8	.75	2.9	.80	2.6	.79	2.5	.79	2.7	.80	3.0
Carpentras	.70	3.2	.74	3.0	.78	2.6	.80	3.0	.79	3.0	.79	3.0
De Bilt	.77	3.5	.77	3.0	.76	3.4	.77	3.6	.77	3.4	.76	3.7
Trappes	.78	2.3	.75	2.7	.76	3.2	.77	3.4	.77	3.5	.76	3.4
Bracknell	.76	3.2	.75	3.7	.78	2.4	.74	3.9	.74	3.8	.74	3.4
Hamburg	.71	3.9	.77	3.1	.77	3.3	.77	3.5	.77	3.3	.74	3.8
Uccle	.70	4.1	.73	3.9	.73	3.8	.75	3.7	.76	4.3	.75	4.2
Vienna	.70	4.3	.69	4.6	.76	4.0	.74	4.0	.75	4.2	.75	4.1
Kew	.70	4.3	.71	4.1	.73	3.3	.73	4.5	.72	3.8	.72	4.3
Athens	.64	4.8	.67	5.5	.73	3.8	.72	5.0	.71	4.8	.71	4.3
London	.66	4.8	.67	5.2	.68	4.0	.71	5.4	.70	4.7	.72	4.7

STATION	JULY a+b	JULY AM2 T_L	AUGUST a+b	AUGUST AM2 T_L	SEPTEMBER a+b	SEPTEMBER AM2 T_L	OCTOBER a+b	OCTOBER AM2 T_L	NOVEMBER a+b	NOVEMBER AM2 T_L	DECEMBER a+b	DECEMBER AM2 T_L	ANNUAL (a+b)
Valentia	.87	2.0	.84	2.5	.84	1.9	.86	1.0	.84	2.1	.82	2.3	.85
Lerwick	.84	2.3	.81	2.5	.82	2.2	.80	2.1	.78	2.4	.80	3.5	.82
Zurich	.81	3.1	.80	3.4	.80	2.6	.80	2.7	.79	3.4	.75	2.7	.81
Locarno Monti	.81	2.0	.81	2.6	.82	2.6	.79	2.5	.78	2.2	.73	2.0	.80
Eskdalemuir	.79	3.1	.77	3.9	.78	4.5	.79	3.2	.77	3.3	.75	3.0	.79
Aberporth	.81	2.9	.76	3.3	.76	2.8	.76	2.6	.73	3.9	.72	3.8	.76
Carpentras	.77	3.0	.76	3.4	.77	3.1	.74	3.1	.70	3.4	.69	3.2	.75
De Bilt	.75	4.3	.73	4.6	.75	4.2	.76	3.1	.75	3.3	.77	2.8	.75
Trappes	.76	3.2	.73	4.4	.74	3.4	.73	3.9	.74	3.5	.77	2.4	.75
Bracknell	.73	3.1	.73	4.7	.73	4.2	.72	3.4	.72	3.2	.72	3.3	.74
Hamburg	.72	4.5	.72	4.1	.74	4.2	.75	3.5	.76	3.7	.72	3.2	.74
Uccle	.74	4.5	.71	5.4	.72	4.7	.70	4.3	.71	4.5	.66	4.0	.72
Vienna	.75	3.7	.73	4.5	.72	4.6	.68	4.0	.69	3.3	.65	4.0	.71
Kew	.71	4.6	.69	4.8	.71	3.8	.71	4.2	.68	3.8	.71	3.6	.71
Athens	.69	4.6	.68	5.4	.69	4.5	.66	4.7	.63	5.0	.62	4.5	.68
London	.70	4.8	.68	5.8	.70	4.5	.67	4.7	.64	4.3	.65	3.7	.68

Table 2.26 Monthly mean values of the regression constant f_m in the formula developed for estimating the appropriate monthly air mass 2 Linke Turbidity Factor to use to compute for any European direct beam and diffuse irradiances on horizontal surfaces on days of monthly mean maximum daily global radiation using the tabulated values of the Angstrom (a + b) from the European Solar Radiation Atlas, Vol. I, 2nd edition.

$$T_L(AM2) = f_c(1 - (a + b))$$

Derived from observed data on monthly mean maximum daily global radiation and monthly mean maximum duration of bright sunshine for 15 European stations.

Month	f_c	Correlation Coefficient
Jan	13.47	-0.853
Feb	13.91	-0.916
Mar	13.38	-0.901
Apr	15.38	-0.956
May	15.39	-0.917
Jun	15.41	-0.895
Jul	14.87	-0.890
Aug	16.27	-0.939
Sep	14.61	-0.904
Oct	13.15	-0.944
Nov	12.83	-0.806
Dec	11.57	-0.752

Table 2.27 Improved clear day prediction model. Percentage errors in the prediction of monthly mean maximum daily global radiation on a horizontal surface (G_{max}) from values of monthly mean maximum daily duration of bright sunshine (S_{max}) for 15 stations selected from the CEC European Solar Radiation Atlas, Vol. I (2nd Edition).

(Observed/Predicted -1) x 100%

Station	Jan	Feb	Mar	Apr	May	Jun	Jul	Aug	Sep	Oct	Nov	Dec
Valentia	1	1	2	1	2	-1	0	1	4	7	0	-1
Lerwick	-3	1	1	4	-1	0	1	4	3	5	4	-8
Zurich	-2	-1	-1	2	0	-6	-1	-1	2	0	-4	1
Locarno Monti	9	6	-2	2	1	3	5	4	0	3	6	11
Eskdalemuir	-5	0	-1	0	-4	-1	0	-1	-1	-2	-3	-1
Aberporth	-1	5	0	4	3	1	0	4	5	4	-3	-5
Carpentras	8	5	2	0	1	2	2	3	2	3	4	4
De Bilt	-3	2	-2	0	1	0	-3	-1	-2	1	0	-2
Paris/Trappes	6	6	-1	1	0	2	2	0	3	-1	-1	2
Bracknell	0	-1	3	0	1	2	4	-1	-1	2	4	0
Hamburg	0	1	-2	0	2	1	-1	3	-1	0	-5	1
Vienna	-2	-2	-5	0	-2	-1	0	0	-2	2	6	1
Uccle	-1	-1	-2	1	-3	-2	-3	-3	-2	-1	-5	-1
Athens	0	-5	-1	-3	-1	1	0	-1	1	-1	-1	-1
London	-2	-3	1	-4	0	-1	-1	-2	0	-1	3	4

using the clear sky slope algorithms which allow for the non-isotropic irradiance of the clear sky. The ground contribution is estimated from the global irradiance, assuming an isotropic ground with an albedo of 0.2. The daily global radiation on inclined planes is obtained by adding the three components hour by hour over the length of the day.

The method thus developed was totally objective and needed no subjective judgement whatsoever about specific site conditions. It is the actual monthly mean maximum daily observations, G_{max} and S_{max}, that determine the specific values of air mass 2 T_L associated with G_{max} observed for any site. This objectivity made the final production of the clear day tables for the CEC European Solar Radiation Atlas, Vol. II, Inclined Surfaces, a straightforward process, as there was neither a need for the Sheffield research group to make their own subjective judgements about the site clarity of sites they did not know, nor was there any need to enter into extensive correspondence or discussions about the validity or otherwise of the clarity decisions for different sites, for the clarity characteristics of the atmosphere are implicit in the model and emerge mathematically from the observed inputs.* No longer was there a need to discuss whether any city had an industrial type of atmosphere or a clearer type of atmosphere. The data supplied from the observing stations answered the issues objectively and specifically for each site.

38. Developments consequent on the completion of the CEC Inclined Surface Atlas

One of the fundamental difficulties throughout this project has been the lack of availability of observed European data on the diffuse irradiation associated with global radiation. This has made it impossible to check the accuracy with which the two components of the global radiation can be estimated for G_{max} conditions. Ideally one would like to have observed values of D specifically associated with G_{max} i.e. the daily diffuse irradiation on G_{max} days. Values of D_{max}, the monthly mean maximum diffuse radiation, are, of course, not the figures required. This associated diffuse information was simply not available in the European data set, so the methodology necessary to produce the European inclined surface studies had to be based on global radiation measurements considered on their own. If such data is available improved approaches can be adopted.

39. New UK methodology for clear day predictions

The UK Department of Energy placed a contract in the University of Sheffield, Department of Building Science, to assist in the preparation of a 'Designers' Handbook of UK Data for Solar Energy Applications' (32). As part of the process of producing this book, new studies were made using a data set of monthly values of G(95%) with associated simultaneous values of D, where G(95%) is the mean of all monthly daily global irradiation values for all days where the daily global irradiation exceeded 90% of the total values in the month. This data was available for four UK sites, Kew, Aberporth, Eskdalemuir and Lerwick. The cloudless sky model discussed above was further modified to allow for some back scattering of the lost radiation from the beam due to the small amount of cloud present. By comparison with the diffuse observations associated with G(95%), it was found reasonable to put back about 46% of the radiation lost from the direct beam due to occasional clouds in the form of diffuse radiation,

* There were about 4 sites for which estimates of (a + b) had to be made. This was done by comparison with adjacent sites.

assuming a Moon and Spencer radiance distribution, as described in Chapter 4. It was also found more accurate to apply a general site clarity correction based on the annual mean value of $\overline{a+b}$ at the site. This annual site correction factor f_8 took the form

$$f_8 = (7.12889 - 14.9747(\overline{a + b}) + 9.10674(\overline{a + b})^2 \qquad (2.37)$$

where $(\overline{a + b})$ is the annual mean value of the sum of the Angstrom regression coefficients. This correction was the same as that used for the monthly mean model whose detailed development is discussed in detail in Chapter 5. Thus the model, developed to produce clear day tables for the UK, was based on the following modification to the model used for the CEC tables:

$$G_c = I_c(0,0) \cdot \sin\gamma \cdot \sigma_{4max} + (0.46(1-\sigma_{4max}) \, I_c(0,0) \sin\gamma + D_c)/f_8 \qquad (2.38)$$

The first term, as before, represents the direct beam irradiance on a horizontal surface. The second term represents the diffuse irradiance on a horizontal surface for G_{max} days corrected for slight cloudiness.

Conclusions

A validated European methodology for estimating the clear sky horizontal irradiance has been developed. An objective method for estimating the appropriate monthly values of the air mass 2 Linke Turbidity Factor T_L from observed data in the CEC European Solar Radiation Atlas, Vol. I, Second edition, has been produced to use with this clear day model. Thus, using this data, basic solar radiation files for horizontal surfaces could be objectively prepared for the hourly beam and diffuse irradiance on horizontal surfaces for days of mean maximum monthly radiation (G_{max} days) for the whole geographical area covered by the Atlas. These basic horizontal values of direct beam and diffuse irradiance were then used to prepare the clear day tables for inclined surfaces using the methodologies described in the next Chapter. The method can be applied to any site where appropriate input data are available.

The full method requires two basic inputs, the value of S_{max} and the value of the sum of the Angstrom regression coefficients. The hourly irradiances predicted are then normalised against the observed values of G_{max}. Where such observed data do not exist, hourly estimates of global and diffuse irradiance can still be prepared, provided there are values of the monthly mean bright sunshine. Appendix 3 shows there is a relationship between S_{max} and S_m. The monthly values of the Angstrom (a + b) can be estimated by comparison with nearby sites with similar atmospheric characteristics, using the CEC European Solar Radiation Atlases as the key data source for European values of (a + b). An alternative is to use the methodology of Dogniaux and Lemoine discussed in Appendix 2. The resulting values, of course, cannot be normalised to G_{max} values in such cases, as the observed values will not exist.

CHAPTER 2 - References

1. Frohlich, C, (1981), Solar radiation at the earth's surface in Proc. Solar World Forum, Brighton, (eds. Hall, D.O. and Morton, J.), Pergamon Press, pp.2313-2321.
   Refer also Frohlich, C. and Shaw, G., (1980), Applied Optics, 19, pp.1773-1775.

2. Young, A.T., (1980), Revised depolarisation correction for atmospheric extinction, Applied Optics, 19, p.3427.
   Refer also Young, A.T., (1981), On the Rayleigh scattering optical depth of the atmosphere, J. of Applied Meteorology, 20, p.3278.

3. Robinson, N., (1966), Solar Radiation, Elsevier, p.114.

4. Page, J.K. and Rodgers, G.G., Mathematical basis of programs for computing average hourly irradiance and daily irradiation on sloping surfaces, Volume 1 of final report for Contract 292-77-UK in CEC Solar Energy Programme, Project F, Action 3.2, Internal Report of Department of Building Science, University of Sheffield.

5. Unsworth, M.H. and Monteith, J.L., (1972), Aerosol and solar radiation in Britain, Q. J. R. Met. Soc., 98, 778-797.

6. Krochmann, J., (1979), Calculation method and its comparison with measuring results, Final report for contract 290-77-ESD in CEC Solar Energy Programme, Project F, Action 3.2, Internal Report of Institut für Lichttechnik, Berlin.

7. Kasten, F., (1965), A new table and approximation formula for the relative optical air mass, Arch. Met. Geoph. Bibl., B, 14, pp.206-223.

8. Kasten F., (1980), A simple parameterisation of the pyrheliometric formula for determining the Linke Turbidity Factor, Meteorologische Rundshau, 33, pp.124-127.

9. Dogniaux, R. and Doyen, P., (1968), Analyse Statistique du Trouble Atmosphérique à Uccle; [à partir d'observations radiométriques. Period de Référence 1951-65], Institut Royal Météorologique à Belgique Publications Série A, No. 65.

10. Stagg, J.M., (1950), Solar radiation at Kew Observatory, Met. Off. Geophysical Memoir No. 86, HMSO, London.

11. Armstrong, P., UK Meteorological Office, (1981), Personal communication.

12. WMO, (1981), Meteorological aspects of the utilisation of solar radiation as an energy source, Technical Note No. 172, WMO, No. 557, WMO Geneva.

13. Ibid, reference 4.

14. Ibid, reference 6.

15. Aydinli, S., (1981), Uber die Berechnung der zur Verfugung stehenden Solarenergie und des Tageslichtes, Dissertation TU Berlin, Fortschrittsberichte der VDI Zeitschriften-Reihe 6, No. 79, p.24-27.

16. Painter, H.E., (1981), The shade ring correction for diffuse irradiance methods, Solar Energy, 26, 361-363.

17. Slob, W.H., (1983), Private communication. Refer also Climatological values of solar irradiation on the horizontal and several inclined surfaces at De Bilt, in Solar Energy R&D in the European Community, Series F, Vol. 2, Solar Radiation Data, (Ed. W. Palz), D. Reidel Publishing Company, p.120-125.

18. Dogniaux, R., (1984), Algorithmes pour l'estimation du facteur total de trouble atmosphérique et leur application aux données radiométriques de l'IRM à Uccle, Privately communicated.

19. Kasten, F., Golchert, H.J. and Stolley, M., (1983), Parametisation of radiation fluxes as a function of solar elevation cloudiness and turbidity, in Solar Energy R&D in the European Community, Series F, Vol. 2, Solar Radiation Data (Ed. W. Palz), D. Reidel Publishing Company, p.108-112.

20. Foitzik, L. and Hinzpeter, H., (1958), Sonnenstrahlung und Lufttrübung, (Probleme der kosmischen Physik, Band 31), Akad. Verlagsges, Leipzig.

21. Unsworth, M.H. and Monteith, J.L., (1975), Aerosol and solar radiation in Britain, Q.J. Royal Met. Soc., 101, 778-797, Corrected table, (1976), 102, 598.

22. Kasten, F., Dehne, K. and Brettschneider, W., (1983), Improvement of measurement of diffuse solar radiation, in Solar Energy R&D in the European Community, Series F, Vol. 2, Solar Radiation Data (Ed. W. Palz), D. Reidel Publishing Company, p.221-225.

23. Plazy, J.L., (1983), Action leaders progress report on "Calibration and characteristics of radiometers" in Solar Energy R&D in the European Community, Series F, Vol. 2, Solar Radiation Data (Ed. W. Palz), D. Reidel Publishing Company, p.2-8.

24. McGregor, J., (1983), The implications for calibration and field use of non ideal cosine behaviour in Kipp and Zonen CM-5 pyranometers, in Solar Energy R&D in the European Community, Series F, Vol. 2, Solar Radiation Data (Ed. W. Palz), D. Reidel Publishing Company, p.9-15.

25. Commission of the European Communities, (1984), European Solar Radiation Atlas, Volume II, Inclined Surfaces, TUV, Verlag.

26. Steinhauser, F., (1934), Die mittlere Trübung der Luft an verschiedenen Orten, Gerlands Beiträge zur Geophysik, 42, pp.110-121.

27. Schulze, R.W., (1970), Strahlenklima der Erde Steinkopff Verlag, Darmstadt.

28. Ibid, reference 12.

29. Commission of the European Communities, (1984), European Solar Radiation Atlas, Volume 1, 2nd Edition, Global radiation on horizontal surfaces, TUV, Verlag.

30. Dogniaux, R. and Lemoine, M., (1983), Classification of radiation sites in terms of different indices of atmospheric transparency, in Solar Energy R&D in the European Community, Series F, Volume 2, Solar Radiation Data, (Ed. W. Palz), D. Reidel Publishing Company, p.94-107.

31. Kasten, F. and Czeplak, G., (1980), Solar and terrestrial radiation dependent on the amount and type of cloud, Solar Energy, 24, 177-189.

32. Lebens, R. and Page, J.K., (1984), Designer's Handbook of UK Data for Solar Energy Applications, Draft ready for publication submitted to Department of Energy, Energy Technology Support Unit, August 1984.

CHAPTER 3

THE RADIANCE OF CLOUDLESS SKIES AND THE GROUND IN EUROPE IN RELATION TO THE PREDICTION OF THE DIFFUSE RADIATION FROM THE CLOUDLESS SKY AND GROUND ON SLOPES AND THE CONSTRUCTION OF A CLOUDLESS DAY SLOPE RADIATION MODEL

Abstract

This Chapter is divided into four parts. The first part describes the complex field measurements of the actual radiance of cloudless skies and the ground under such skies in association with measurement of the global and diffuse irradiance on slopes carried out by Dr. P. Valko. The measurements were mainly made in Switzerland, but some were made in France. The prediction of slope irradiances from observed radiance patterns is discussed.

The second part of this Chapter describes the theoretical model developed in the University of Sheffield to predict the radiance distribution of cloudless skies, and its use to develop practical methods for estimating the diffuse sky radiation from clear skies falling on inclined planes. The inter-relationship of the theoretical approach and P. Valko's observations is demonstrated, and the results of testing the model against other field observations at Bracknell and Cabauw are presented. The considerable improvement in accuracy over the isotropic model is demonstrated. The excellent results of these checks have confirmed the value of adopting a fundamental scientific approach to the prediction of solar radiation on slopes based on the modelling of the radiance properties of the cloudless sky as a function of the Linke Turbidity Factor and position in the sky.

In the third section, the problem of modelling the radiance of the ground is reviewed. The accuracy of the current methods used to estimate the clear day ground reflected radiation is considered in relation to the ground radiance measurements of P. Valko in Switzerland. Preliminary results of the study of effective ground albedo are presented. This review suggests that this area requires more detailed scientific attention in future programmes of work.

The final section describes how the EC cloudless day model was used for predicting the irradiance of slopes for the inclined surface solar radiation tables for clear days published in the EC, European Solar Radiation Atlas, Vol. II, Inclined Surfaces.

CHAPTER 3 - PART I

THE OBSERVED RADIANCE DISTRIBUTION OF CLOUDLESS SKIES AND THE PREDICTION OF THE ASSOCIATED COMPONENTS OF THE DIFFUSE IRRADIANCE ON INCLINED PLANES

1. The Goals

Systematic measurements of the short wave solar radiance over the whole 4π steradian sphere, supplemented by pyranometer and pyrheliometer measurements, are essential to develop tested analytical models to predict the diffuse irradiance on tilted planes for different times of day and year, for different states of the sky at different sites. Although such measurements are of great interest for basic atmospheric science and for practical engineering applications, adequate data have for long been sparse and are still incomplete.

Considering the intricate angular dependence of the scattered and reflected radiation fluxes on atmospheric and local ground surface parameters, a measurement programme was developed in Switzerland with three specific aims:

- to measure the instantaneous angular components of both radiance and irradiance over both the upper (sky) and the lower (ground) hemisphere, together with relevant atmospheric parameters,
- to relate empirically observed radiance distribution patterns and slope irradiances to atmospheric and site specific quantities, as a function of the angular geometry,
- to integrate the multivariate instantaneous relationships thus established over hours and days to compute the irradiation of any specified exposed surfaces for any conditions of sky and site.

This chapter is specifically concerned with cloudless conditions. Data for four main sites are discussed. EMPA is a semi-suburban roof top site in Zurich, 445 m above mean sea level. Locarno Monti is a roof top site on the south side of the Alps in Switzerland at 380 m above mean sea level. Carpentras is in the South East of France at 99 m above mean sea level. The Weissfluhjoch is a high level mountain site at 2670 m above mean sea level. At Carpentras, the measurements were made on the roof of the Observatory of the Météorologie Nationale and at Weissfluhjoch on the roof of the Federal Institute for Snow and Avalanche Research. Data from a fifth site have also been used and included only in Figure 3.16. These measurements were made in April 1984 under heavy snow-covered field conditions in Le Locle, a mountain plateau of 1240 m a.s.l. in the western part of Switzerland. The wide range of station elevations covered by the measurement programme is thus evident.

2. Instrumentation and measurements

Starting in the early seventies, a number of field measuring devices have successively been developed and integrated into a mobile solar radiation research system (Bener, Frohlich and Valko (1), Valko and Heimo (2), WMO (3), Heimo (4)). The whole equipment is lodged in a trailer (6.3 x 2.3 x 2.6 m, 4 tons) allowing transportation and operation at selected sites in any season. A DEC PDP 8A computer on board controls and co-ordinates the measuring programme of the separate instruments and provides for initial processing of data. The mobile system is a state-of-the-art platform for data acquisition during short term measurement campaigns and is not designed to be located at one site for compiling long period records of solar radiation climate. The detailed selection of measuring components of the system and the programme of their operation corresponds with the research goals described above.

The following measurements are made:

i) Observations of the direct solar beam and global irradiance normal to the beam

a) Direct solar irradiance at normal incidence is recorded by a Davos type absolute radiometer (Brusa and Frohlich (5)) over the total solar wavelength range as well as in the broad spectral bands of the Schott filters OG 1, RG 2 and RG 8.

b) Spectral solar irradiance at normal incidence at 16 discrete wavelengths in the visible and near infra red is also recorded using an instrument developed by Frohlich and Bener (6). Data obtained from these two devices serve among other purposes for getting detailed information on atmospheric turbidity (Schubiger (7)).

These instruments are mounted on a sun-tracker together with a pyranometer. This pyranometer records the global irradiance on a plane normal to the sun. The diffuse irradiance on a surface normal to the beam may thus be found by subtracting the direct beam irradiance from the global reading. Figure 3.1 shows the instrumental arrangement on the suntracker.

ii) Global irradiance on inclined planes

Global irradiance on inclined planes is recorded by a system of four pyranometers driven by step motors, providing measurements on altogether 77 different tilts and aspects within about six minutes, Figure 3.2. A recent modification of the control unit permits a choice between different angular configuration schemes, but, until this change was introduced, measurements were taken in the horizontal position as well as at the tilt angles 18°, 39°, 60°, 75° and 90° (vertical surface) in a number of azimuth steps. The horizontal surface global irradiance is separately recorded by a sixth pyranometer throughout each series of measurements.

Figure 3.1 The instruments following the sun and mounted on the automatic sun-tracker are (i) pyranometer, top left, (ii) silicon diode radiance detector, centre top, (iii) absolute radiometer with broad band Schott-filters OG1, RG2 and RG8, top right, and (iv) spectroradiometer measuring at 16 different wavelengths, bottom.

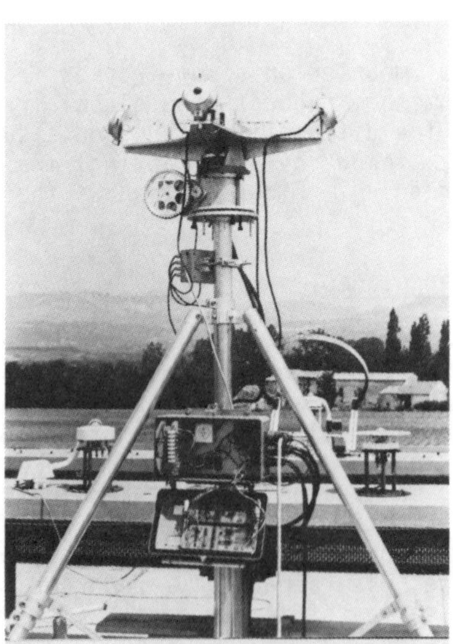

Figure 3.2 The four pyranometers for measuring global irradiance on slopes are driven by step motors. 77 measurements on different tilts and directions are made in about 6 minutes.

Figure 3.3 The four silicon diode sensors with 5° apertures scan the radiance of the sky at 121 points in about 2 minutes. Their spectral response is different from the pyranometer.

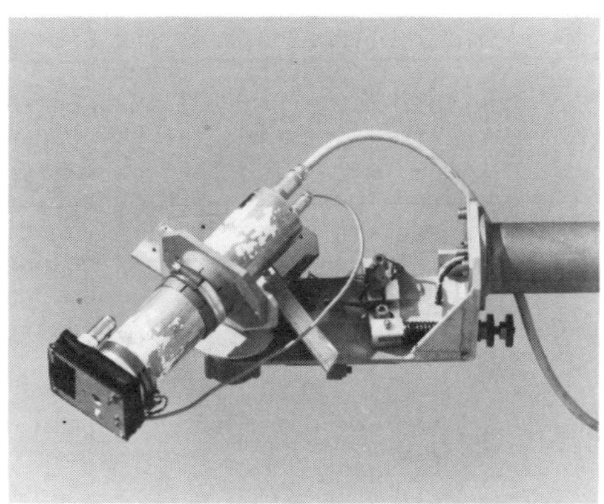

Figure 3.5 The radiance detector branch. The sensor is a silicon diode. It measures radiance in the total spectrum out to 1100 nm as well as the spectral radiances through interference filters at 368, 500 and 778 nanometres.

Figure 3.4 For measurements in the lower hemisphere one pyranometer and a 5° full view angle radiance detector are mounted on either end of a cross-beam at the top of a 10 m telescopic mast. Rotation of the beam on the mast and tilting of the pyranometer and radiance detector are accomplished with step motors, measuring, in this way, radiance and irradiance in 77 different instrument positions in about 26 minutes.

iii) Sky radiance

In the original programme the sky radiance was measured at 121 points altogether: at the elevation angles 0° (horizontal), 18°, 36°, 54°, 72° and 90° (zenith), using azimuth steps of 15 degrees. A complete set of 121 measurements takes about two minutes. The unit used for scanning the sky consists of four silicon diode radiometers of 5° full view angle, driven by step motors to face the different required positions (Figure 3.3). In February 1983 the programme was extended to scan the sky radiance in more detail across the lowest belt of the sky, traversing once at the elevation angles 3, 5, 8 and 12 degrees and, in a second run, at 2, 4, 6, 9 and 12 degrees, again using azimuth steps of 15 degrees, making 60 or 75 additional points respectively.

iv) Radiance of the ground, and the global irradiance of overhanging surfaces

Radiance reflected from the ground as well as global irradiance of overhanging surfaces - i.e. surfaces exposed between vertical and downward facing horizontal - are scanned from a telescopic mast at 10 m height above ground. The instruments, one silicon diode radiometer and one pyranometer, are mounted on each end of a cross beam, Figure 3.4. By rotation around the mast at 10 m height, the irradiance is recorded by moving the pyranometer to different tilts giving pyranometer records at 77 positions within about 26 minutes. Simultaneously the reflected radiance is scanned at 65 points at the elevation angles of 0° (horizon), -18°, -36°, -54°, -72° and -90° (nadir), Figure 3.5.

The diffuse irradiance for all pyranometer positions is automatically computed by subtracting the respective resolved component of the recorded direct beam from the global pyranometer readings. By integrating the measured sky radiances over the azimuth and elevation angles with respect to any pyranometer position, the total diffuse irradiance can be split into its sky- and ground-reflected components for all (2 x 77) surfaces giving 154 values. In this way the complete system provides both irradiance and radiance data over the whole 4π steradian space.

Digital information on the type, amount and angular distribution of clouds over the sky dome is gained by computer processing of all sky photographs taken at regular time-intervals throughout the observations by a Nikon fish-eye camera. The coding of data pays particular attention to sky conditions in the immediate vicinity of the sun's disk. Data compilation is completed by measuring temperature, pressure and humidity of the air.

3. Routine processing of data

Before proceeding to detailed analysis, some listings, tables and graphs - such as plots of radiance and irradiance against time of day or against angular parameters - are prepared first to get a survey of the data. As an example, Figure 3.6 demonstrates the angular distribution of measured and derived quantities by a set of polar diagrams for a cloudless sky of medium turbidity at EMPA. EMPA is a semi-suburban roof-site near Zurich, with both buildings and green areas in its vicinity. The key site and radiation characteristics belonging to this selected case are given within the heavy frame in Table 3.1, i.e. Example No. III.

Table 3.1 compares cloudless day observed results for 5 different conditions of atmospheric turbidity. Both the Schuepp B_s and the Linke Turbidity Factor are given. $G_c(\beta, \alpha_s)_{max}$ refers to the maximum global irradiance on a slope normal to the beam, while G_c refers to the horizontal surface. $D_c(\beta, \alpha_s)_{max}$ refers to the maximum slope value of the diffuse irradiance from sky and ground. This maximum does not occur on a surface normal to the beam, but on a more steeply tilted surface, or even vertical surface, partly due to the influence of the ground reflections. D_c is the observed diffuse irradiance on a horizontal surface. Its magnitude increases sharply with increase in atmospheric turbidity. The ratio D_c/G_c. follows the same trend. D_c (90, α_s) is the diffuse irradiance from the sky alone on a vertical surface, which may be expressed as a ratio to the total incident diffuse irradiance from the sky and ground together. The greater the turbidity, the greater is the proportional sky contribution. R_c is the reflected component reaching a downwards facing horizontal surface. The ratio R_c/G_c is the cloudless day albedo. It will be noted, due to the nature of the ground cover, which includes the roof at Zurich, and Locarno Monti, the values are relatively low compared with the value of 0.20 used as standard in preparing the CEC Inclined Surface tables. There are no observations of R_c for the Weissfluhjoch example, but the ground was snow covered in most cases. The clear day zenith radiance L_{cz} increases sharply with increase in turbidity. L_{cn} is the clear day nadir radiance, i.e. the radiance of the ground looking directly downwards. The ratio L_{cn}/L_{cz}. decreases with increase of turbidity, because the energy to be reflected falls, while at the same time the zenith radiance increases.

Although, due to lack of space, it is not possible to reproduce here graphs for all the conditions, Table 3.1 allows comparisons to be made, taking account of the influence of turbidity on radiance and slope irradiance. Table 3.2 summarises the range of measurements taken with the system.

Typical graphical outputs for the measurements

Figure 3.6 gives eight graphical outputs resulting from one set of observations for cloudless skies at EMPA, Zurich, for 0923 on 29/7/81.

Figure 3.6a - Polar diagram - global irradiance on slopes

For clear skies, the global irradiance on inclined planes shows always regular, almost circular patterns, such as in Figure 3.6a, with maximum for the surface perpendicular to the sun ($\beta = 90° - \gamma$, $\alpha_s = 0$). Row 11 of Table 3.1 gives the sun-tracking pyranometer irradiance value. This is not entered on the diagram. As Table 3.1 indicates, the values depend on solar height and atmospheric turbidity. They, of course, also depend on the reflective properties of the ground and surroundings. Figure 3.6a shows the sun-facing vertical surface received 10.8 times more radiation than a surface facing opposite the sun. For the other cases described in Table 3.1, the corresponding diagrams would show a decrease of this ratio both for smaller and for stronger turbidities, 5.2 for the mountain site

Weissfluhjoch and 7.5 for the EMPA suburban example V with the highest turbidity. Table 3.1, example 1, for Weissfluhjoch is for snow covered ground and landscape. But, even without snow, for a solar height of 58°, again in very clean air, this ratio still amounts to 3.9. Figure 3.7a shows the diffuse component only. It should, however, be noted that the ground rises to the north of the Weissfluhjoch which influences this ratio considerably.

Figure 3.6b - Polar diagram - diffuse irradiance on slopes

In contrast to the global distribution, the values in Figure 3.6b indicate a steady decrease in the diffuse irradiance from sunward vertical through horizontal, towards the opposite vertical exposure. Figure 3.6b is of a transitional character between the graphs in Figure 3.7a and Figure 3.7b. which give the corresponding diffuse irradiance results for exceptionally clear, and relatively turbid cloudless skies. The difference of the patterns in Figure 3.7 is remarkable, especially the opposite directions of the gradients. Even for high turbidities, the surface receiving most diffuse radiation, seems to be steeper than normal to the sun. In row 13 of Table 3.1, the term in brackets (β°) gives the inclination angle of the surface of maximum diffuse irradiance. One may also immediately get D_c $(90° - \gamma, \alpha_s = 0) = G_c(\beta,\alpha_s)_{max} - I_c$, from Table 3.1.

The ratio of the sum of the sky-diffuse and ground reflected components on a vertical surface facing away from sun (Figure 3.6b) decreases to less than 3 compared with 10.8 for the global irradiance.

It is of interest, that, by using 1700 clear sky values out of a five years continuous records at Locarno-Monti, maximum diffuse (sky + ground) irradiance of sun-facing vertical surfaces D_c (90, 0) was found for a solar height of 40°, if the Schuepp turbidity coefficient B_s was > 0.100. For very clean air, B_s < 0.050 the maximum was reached at 60° (8). On average, considering every single hourly value over a 9 year period, thus including all kinds of cloudy and overcast skies too, $D_c(90,0)$ always increased with solar height. At γ = 60°, the all weather average and the clear sky diffuse values for B_s = 0.150 were the same at 215 Wm^{-2}. Such features result from interactions of decreased ground reflection with increased turbidity (and increased cloudiness), increased sky radiance with increased turbidity, and decreased relative contribution of augmented aureole radiance with increasing solar altitude which also implies a decreasing cosine of the incidence angle on a vertical surface.

Figure 3.6c - Polar diagrams for normalised sky radiance

In Figure 3.6c, radiances $L_c(\Theta,\alpha_s)$ measured at the 121 points in the sky have been normalised to the simultaneous radiance L_{cz} in the zenith. The absolute values of L_{cz} are given in row 19 of Table 3.1. Typical for cloudless sky distributions of this kind is the rapid increase of radiance near the sun and a minimum sky radiance region in the opposite sky at about 90° angular distance from the sun. A comparison with similar graphs for the other examples in Table 3.1, not shown here, indicates the strongest radiance gradients for very clean air. The radiance gradients decrease with increasing turbidity, on both the solar side of the sky and on the opposite side of the sky between zenith and horizon.

Table 3.1 Characteristic values for five selected distributions of radiance and irradiance over the full 4π steradian space for cloudless skies of different atmospheric turbidity.

Row No	Example No.:	I	II	Fig. 3.6a -3.6g	IV	V
1	Site	Weissfluhjoch	Locarno-Monti	EMPA	EMPA	EMPA
2	Geogr. Latitude	46°49'	46°10'	47°18'	47°18'	47°18'
3	Geogr. Longitude	9°50'	8°47'	8°38'	8°38'	8°38'
4	Altitude m a.s.l.	2670	380	445	445	445
5	Date	3/9/1980	13/10/1981	29/7/1981	5/8/1981	19/8/1981
6	True Solar Time	10^{33}	12^{12}	09^{23}	10^{35}	10^{04}
7	Solar Height $\gamma°$, Azimuth $\alpha°$	47 -31	36 +5	48 -61	55 -36	48 -44
8	Turbidity Coefficient B_S (Schuepp)	0.020	0.040	0.130	0.210	0.380
9	Turbidity Factor T_L (Linke)	2.5	2.7	4.0	4.9	6.9
10	I_c Direct normal $[Wm^{-2}]$	1066	948	824	789	619
11	$G_c(\beta,\alpha_s)_{MAX}$ $[Wm^{-2}]$	1133	1058	987	987	943
12	G_c $[Wm^{-2}]$	835	637	741	821	728
13	$D_c(\beta,\alpha_s)_{MAX}$ $[Wm^{-2}]$ $(\beta°)$	224 (90)	133 (90)	184 (90)	207 (67)	347 (60)
14	D_c $[Wm^{-2}]$	55	82	124	172	257
15	$[D_{cs}(\text{Vert.})/D_c(\text{Vert.})](\alpha_s=0° / \alpha_s=180°)$. 100%	30/30	56/35	61/43	73/50	81/53
16	D_c/G_c ("Turbidity")	0.07	0.13	0.17	0.21	0.35
17	R_c $[Wm^{-2}]$	-	103	87	100	96
18	R_c/G_c "Albedo"	-	0.16	0.12	0.12	0.13
19	L_{cz} $[Wm^{-2}$ steradian$^{-1}]$	10.5	13.1	23.7	42.1	57.3
20	L_{cn} $[Wm^{-2}$ steradian$^{-1}]$	-	20.8	21.1	20.6	19.8
21	L_{cn}/L_{cz}	-	1.6	0.89	0.50	0.34
22	$\pi L_{cz}/D_c$	0.60	0.50	0.60	0.77	0.70

Table 3.2 Summary of the mobile system measurements[1]

Year	Number of Days with Measurements	Total Number of Data Sets[2]	Number of Data Sets for ⩽ 1/8 cloudiness[3]	Number of Data Sets including mast-measurements
1976-1978[4]	72	903	354	
1979[5]	23	451	200	
1980[6]	42	535	200	56
1981	50	655[7]	226	518 EMPA: 359 Loc.M: 159
Total[8]	187	2,544	980	574

1. Detailed survey also with statements for single sites etc. is given in Reference 15.

2. One set (series) includes as a rule complete scans of all instruments listed in Chapter 3, Part I, Section 2.

3. On the basis of fish-eye sky photographs, only 551 series out of the 1979-1981 total of 626 have been qualified as representative of clear sky conditions.

4. Sky radiance scanner and mast-system was not yet operating.

5. Mast-system not yet operating: 399 out of the 451 series were measured at Carpentras.

6. Mast-system operating from 21st October, 1980; 519 out of the 535 series were measured at Weissfluhjoch.

7. 415 series out of the 655 were measured at EMPA, 240 at Locarno Monti.

8. At EMPA 69 series have been compiled for 1982 which are already processed. Further, more than 400 series, hitherto compiled (about 200 from these at the mountain site Le Locle in 1984) are not yet processed. The new scans of sky radiance within the 12° belt close to the horizon are not included in the figures in this Table.

Figure 3.6d - Polar diagrams for the irradiance from the sky alone

Figure 3.6d demonstrates the angular distribution of the sky component $D_{cs}(\beta,\alpha_s)$ of the total diffuse surface irradiances. The values of $D_{cs}(\beta,\alpha_s)$ are found by integration of the measured sky radiances. The integration process is discussed in Appendix 4. For Figure 3.6d, and also for all other Table 1 examples I to V, the irradiance patterns have roughly circular shape around the sun's position. Skimming through a greater number of graphs, it seems that for $\gamma > 20°$, $D_{cs}(\beta,\alpha_s)_{max}$ occurs on the surface normal to the sun ($\beta = 90° - \gamma$), but for $\gamma < 20°$ at $\beta < (90° - \gamma)$. This indicates that aureole radiance around a sinking sun can no longer compensate the loss of irradiance from the shrinking share of the sky seen by an ever steeper receiving surface. For low values of γ, of course, the aureole is, in part, also disappearing beneath the local horizon. The magnitude of these influences depends on the turbidity. The dependence of this threshold solar elevation on turbidity and cloudiness is under current study.

Figure 3.6h - Polar diagrams showing proportion of total diffuse received from sky

Using the data in Figure 3.6b and Figure 3.6d, it becomes possible to estimate the proportion of the diffuse irradiance associated with the sky alone. For the surface facing perpendicular to the sun, the sky component amounts to up to 90% and more of the total diffuse radiation on the surface for all turbidities; refer Figure 3.6h for the selected case for EMPA. As may also be seen from Figure 3.6h, a vertical surface facing the sun with the sun at 48° elevation receives 61% of its diffuse irradiance from the sky, and the vertical surface facing away from the sun towards the opposite horizon receives 43% of the total diffuse irradiance on the slope from the sky. Corresponding values of this ratio for the other examples are listed in row 15 of Table 3.1. Since, for the horizontal surface, $D_{cs} = D_c$, the figures indicate a decreasing gradient in $D_{cs}(\beta,\alpha_s)$ with increasing turbidity. For example I, valid for the snow covered landscape, all vertical surfaces seem to receive about the same amount of sky diffuse radiation, irrespective of the azimuth. However, it should be emphasised that the sky component D_{cs} of the total diffuse irradiance D_c of a horizontal surface is related to the whole 0 - 90° elevation interval, irrespective of possible mountains or other obstacles which may protrude beyond the ideal horizon into the sky. For example, at the Weissfluhjoch, there are such obstacles, especially to the north of the measuring site.

Figures 3.6e and 3.6f - Polar diagrams of irradiances on downward facing surfaces

Figure 3.6e and Figure 3.6f of the irradiances on downward facing surfaces have been drawn using the pyranometer measurements from the instruments mounted on the mast at 10 m height above ground. In Figure 3.6f the direct beam contribution has been removed using the pyrheliometer observations. Comparing the entries for the vertical surfaces in Figure 3.6e (around the horizontal circle) with the corresponding vertical surface values in Figure 3.6a, it appears that observed global irradiances along the same line of sight are not equal. Measurements at about 2 m above ground, i.e. height of the system of four pyranometers, exceed those at 10 m height for small surface solar azimuths, i.e. surfaces facing the sun, but

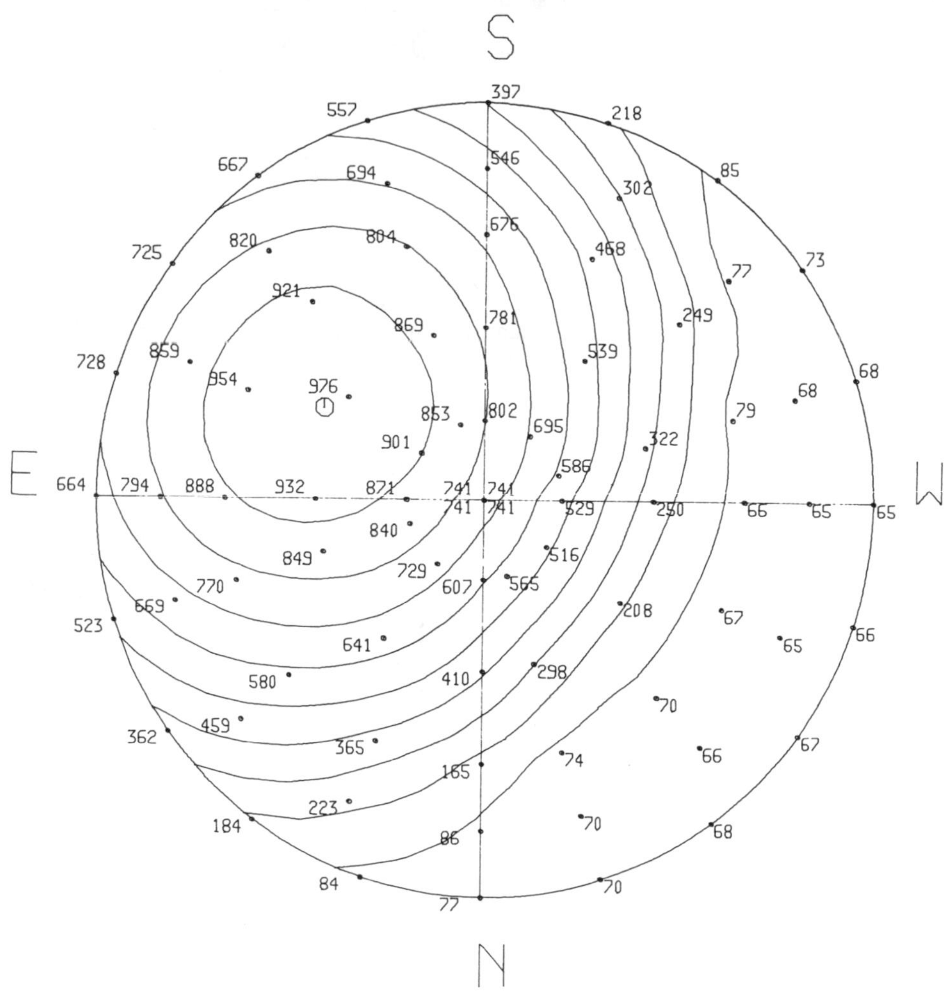

Figure 3.6a Angular distribution at EMPA, Zurich, of global irradiance on surfaces inclined between upward facing horizontal and vertical positions, $G_c(\beta, \alpha_s)$; $B_s = 0.130$. Units: Wm^{-2}, isolines 100 Wm^{-2} steps. Sun's position marked with ⊙ .

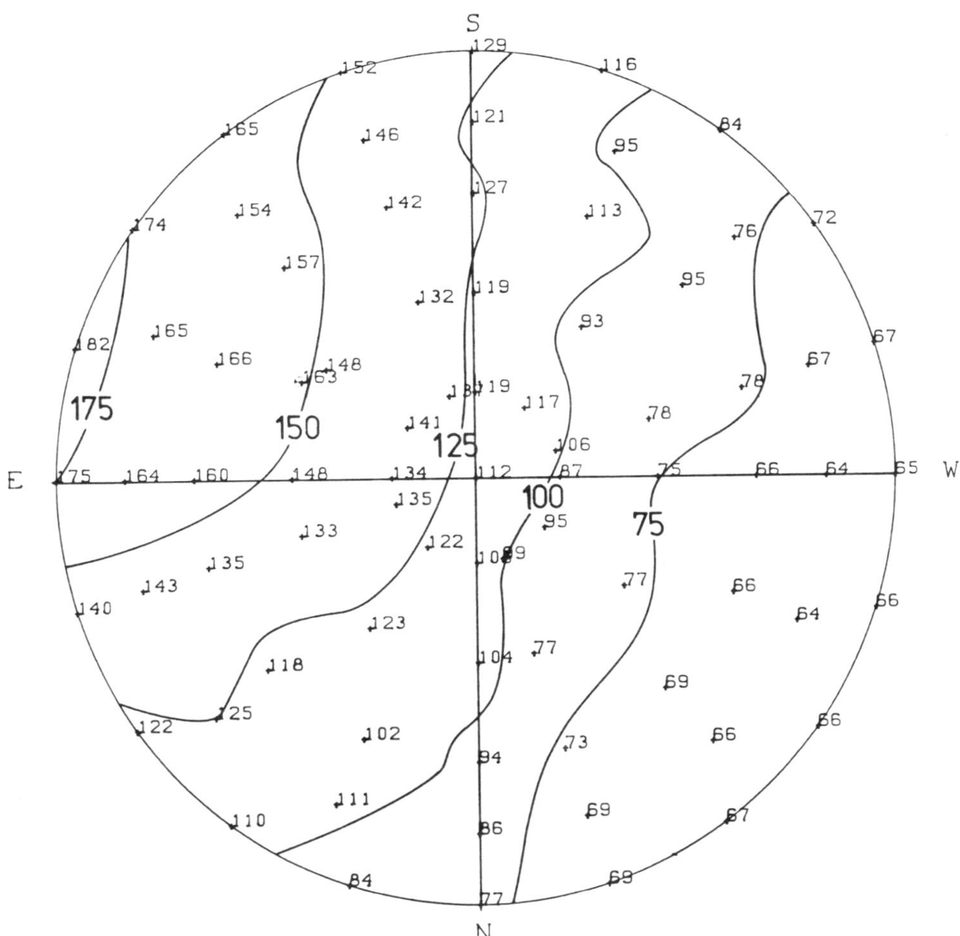

Figure 3.6b Angular distribution at EMPA, Zurich, of the diffuse component, including both sky-diffuse and ground-reflected components on upward facing planes between the horizontal and vertical positions, $D_c(\beta, \alpha_s)$, B_s = 0.130. Units: Wm^{-2}, isolines at 25 Wm^{-2} intervals.

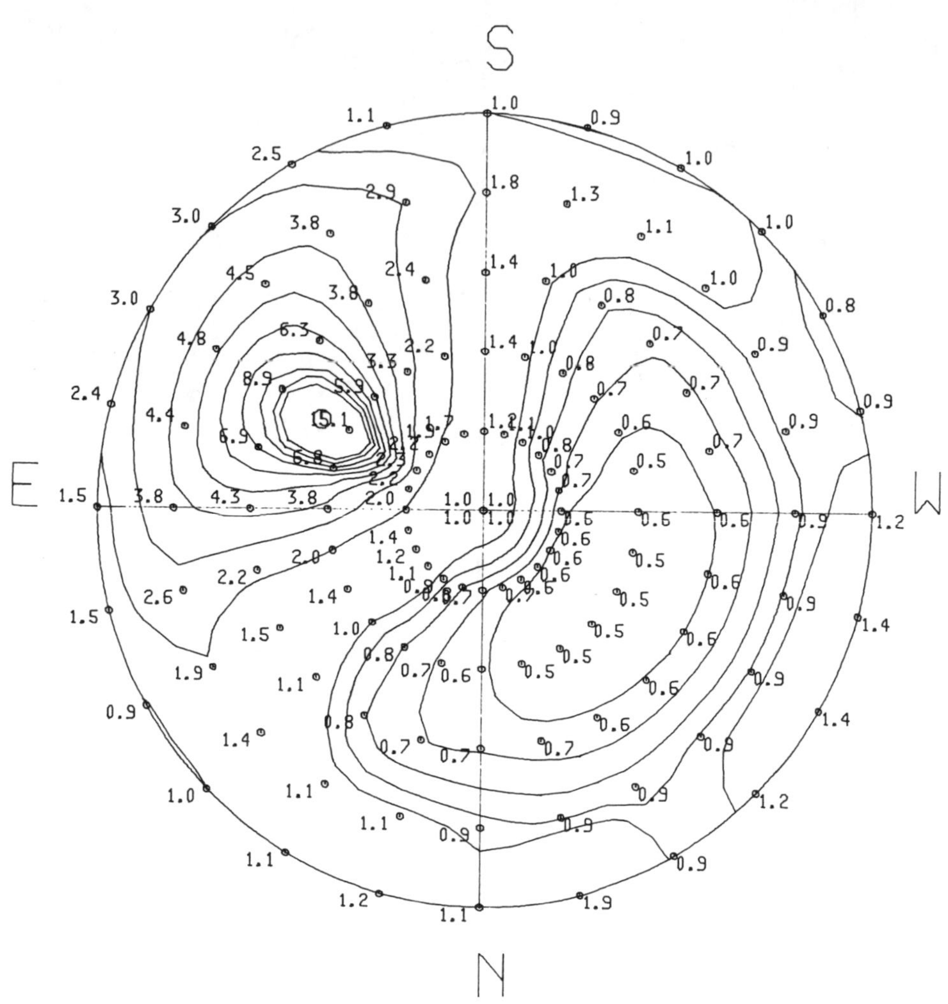

Figure 3.6c Angular distribution at EMPA, Zurich, of sky radiance normalised to the radiance of the zenith, $L_c(\Theta, \alpha_s)/L_{cz}$; B_s = 0.130. Units: dimensionless, isolines at 0.1 intervals for values below 1.0 and at 1.0 for values above 1.0. Sun's position marked with ⊙ .

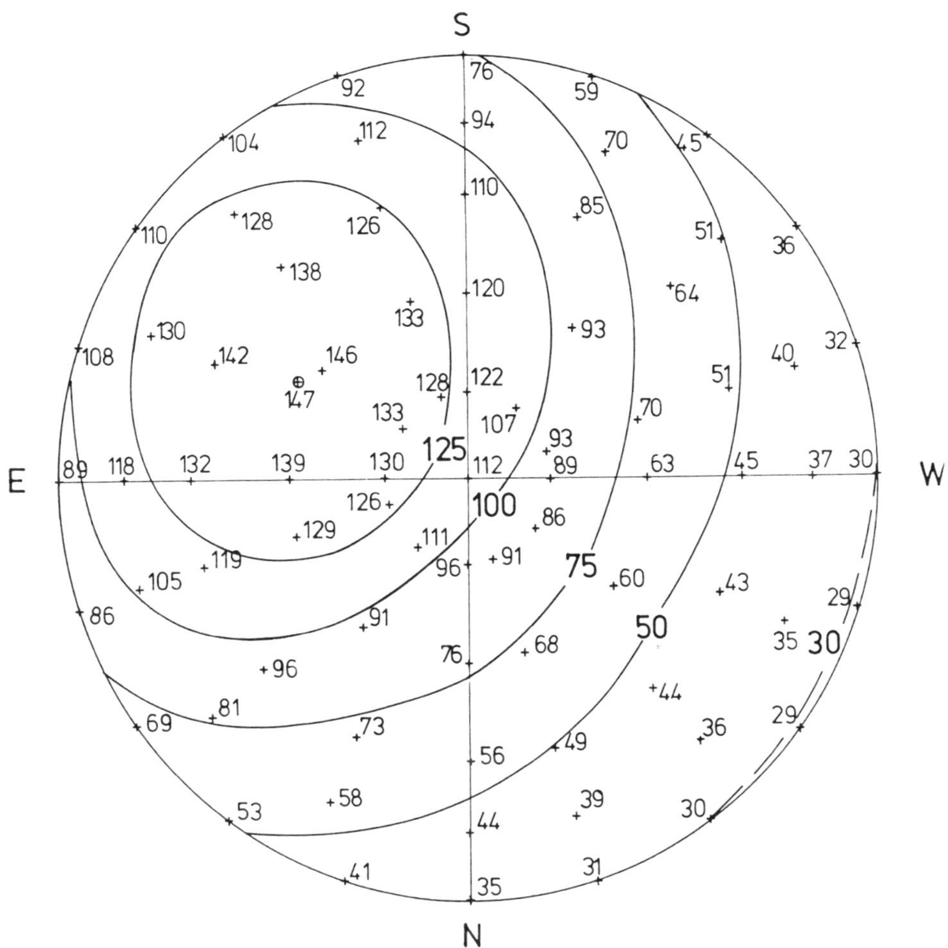

Figure 3.6d Angular distribution at EMPA, Zurich, of diffuse irradiance from the sky alone, $D_{cs}(\beta,\alpha_s)$, i.e. sky component of Figure 3.6(b) obtained from Figure 3.6(c) by integration over the sky hemisphere with respect to each of the 77 inclined pyranometer positions. Units: Wm^{-2}, isolines at 25 Wm^{-2} intervals.

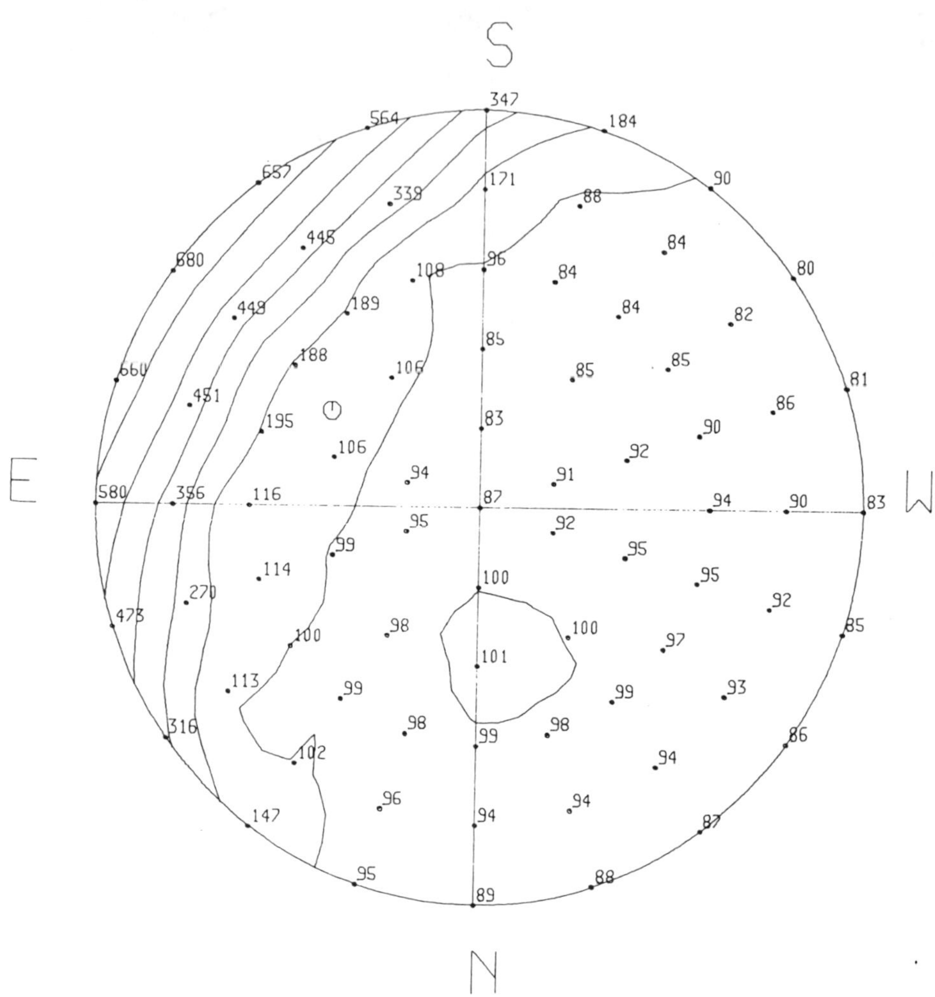

Figure 3.6e Angular distribution at EMPA, Zurich, of the global irradiance on planes facing downwards between the vertical and horizontal downward facing (overhangs), $G_c(-\beta, \alpha_s)$. $B_s = 0.130$. Units: Wm^{-2}, isolines at 100 Wm^{-2} intervals. Reflected image of sun's position marked ⊙.

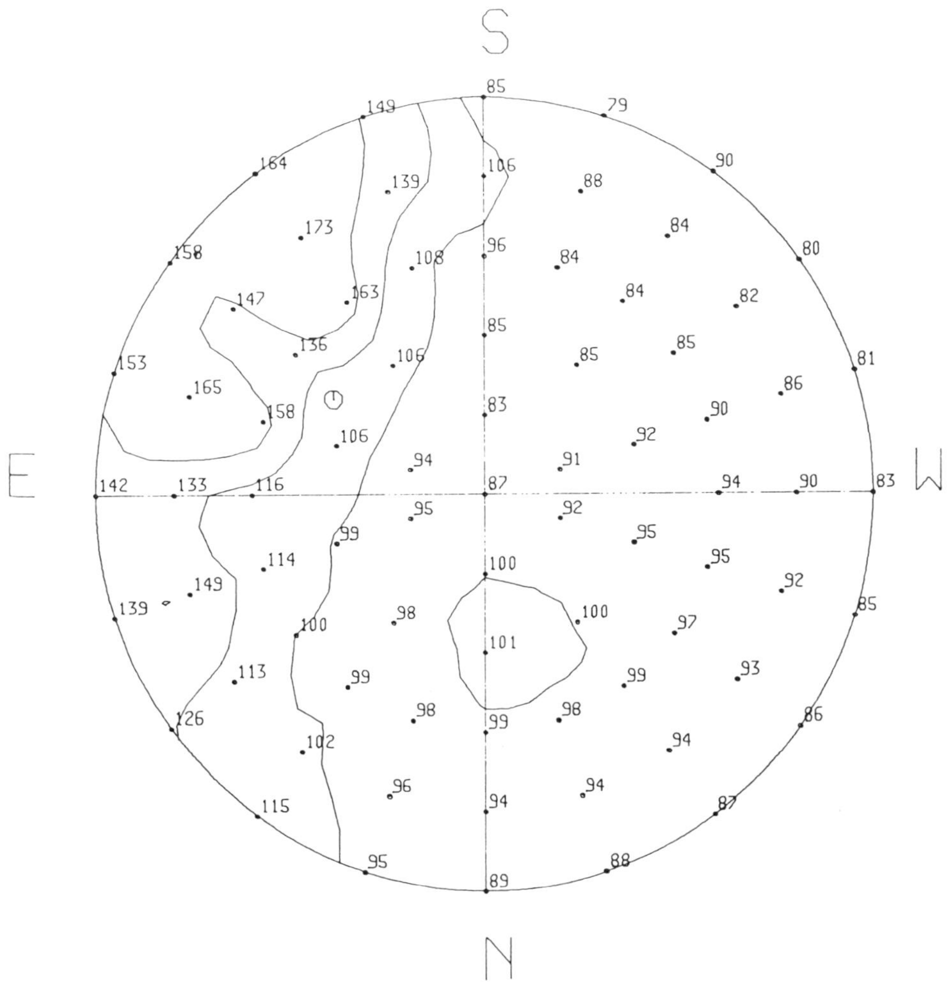

Figure 3.6f Angular distribution at EMPA, Zurich, of the diffuse component (sky-diffuse and ground reflected component) on planes facing between 90° and horizontal downwards, $D_c(-\beta, \alpha_s)$. Units: Wm^{-2}, isolines at 25 Wm^{-2} intervals. Reflected position of sun's image marked ⊙.

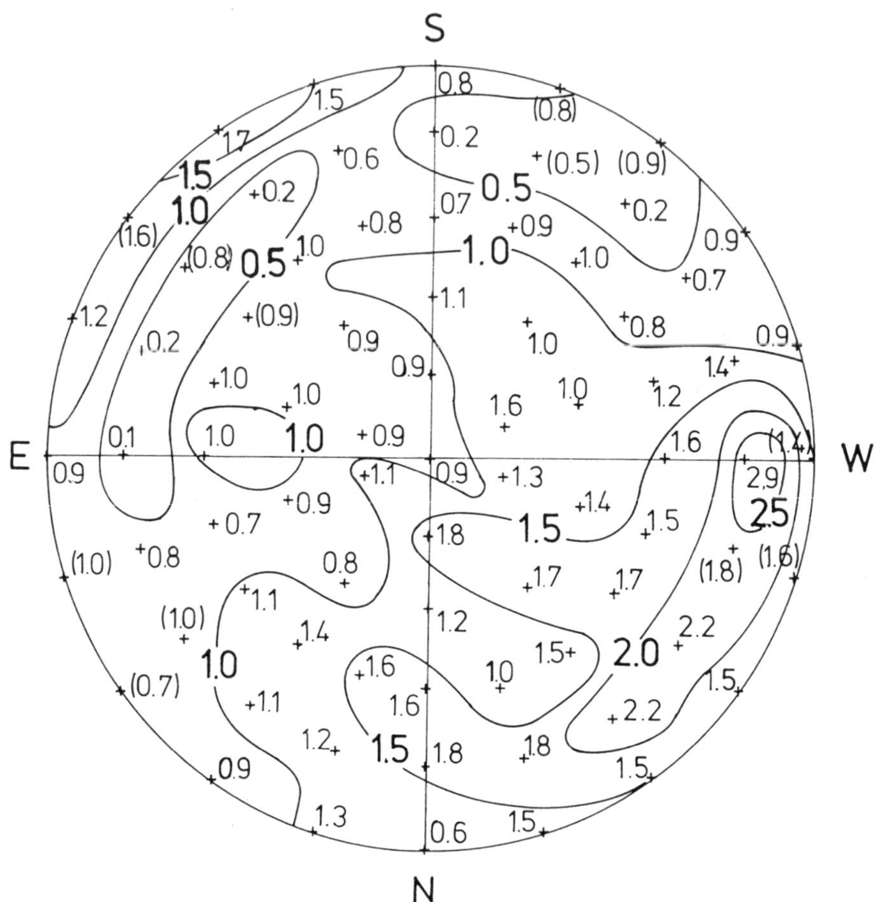

Figure 3.6g Angular distribution at EMPA, Zurich, of reflected radiance normalised to the radiance of the zenith showing the ground radiance field is far from isotropic. The graph implies the downwards facing albedo can be very different from the vertical surface albedo: for example, compare the values in the NW. Dimensionless; isolines at 0.5 steps based on original entries to 2 decimal places.

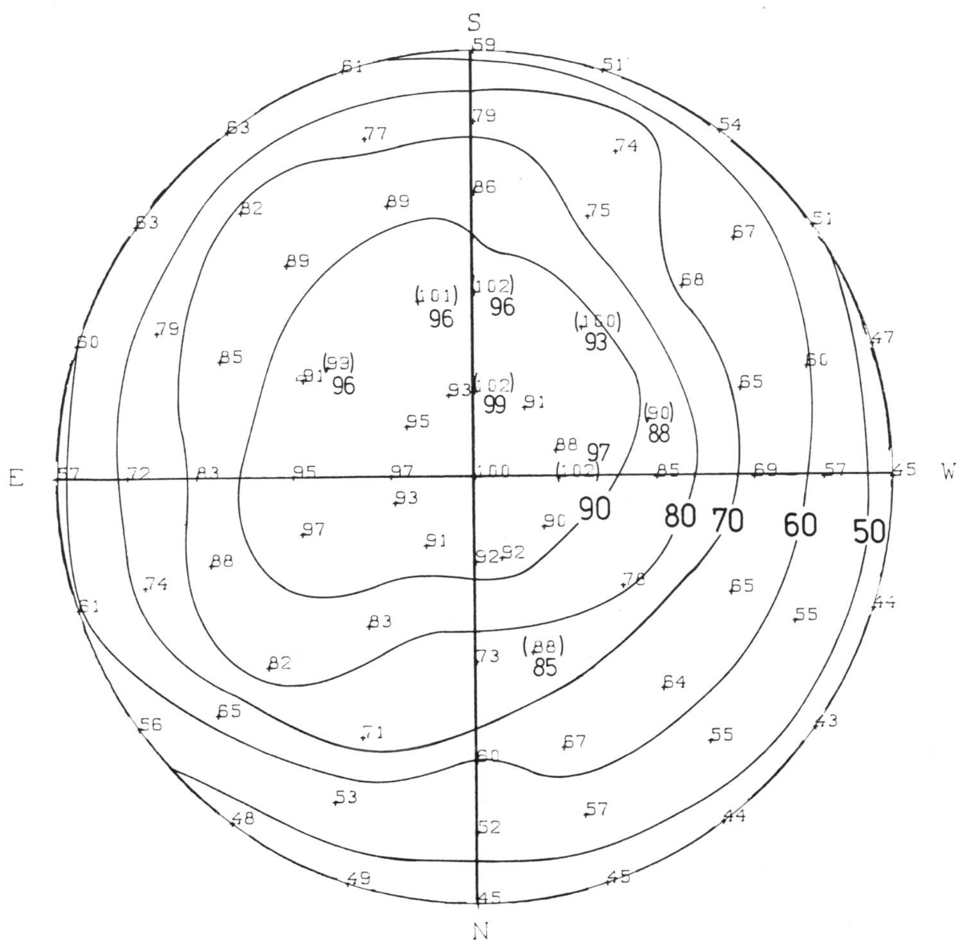

Figure 3.6h Angular distribution of the sky component $D_{cs}(\beta,\alpha_s)$ at EMPA, Zurich, expressed as a percentage of total (sky + ground) diffuse irradiance. Note the distortion of the diagram in the direction of the sun. Units: %, isolines at 10% intervals.

are smaller than those at 10 m height for $\alpha_s > 90°$, i.e. facing away from the sun. Since ground reflectivities around the two devices are different (fish-eye photographs taken downwards from 10 m are given in Figures 3.22 to 3.25 (9)), further interpretation of such differences is needed.

Figure 3.6g - Polar diagrams of reflected radiances

Reflected radiances are normalised to zenith radiance rather than to the nadir radiances for the sake of achieving comparability between Figures 3.6g and 3.6c. Because of the differences at the installation site at two roof levels, the horizontal radiances plotted in Figure 3.6c at 2 m and in Figure 3.6g at 10 m, scanned in pairs at equal azimuths around the horizon (full view angle of the scanners is 5°), are not really comparable with each other. In spite of the irregularities, radiance patterns in Figure 3.6g show a steady increase of reflected radiance along the roof from a minimum in south-east towards a maximum in north-west. Values in parenthesis (only 65 points could be scanned instead of 77) have been interpolated, both by using the horizon-measurements of the four-sensors system and also by consulting the fish-eye photographs taken from the roof.

The nadir/zenith radiance ratio L_{cn}/L_{cz} in row 21 of Table 3.1 indicates a definite decrease of the ratio with increase of turbidity. This dependence on turbidity is observable for the whole roof surface - which is the same in the cases III, IV and V; using the north-west maximum-strip value, for instance, the ratio sinks from 2.2 - 2.9 (example III) to 1.5 (example IV) and finally to 1.1 (example V). For case I the mast was unfortunately out of operation, but measurements above snow cover for clean air at $\gamma = 25°$ yield L_{cn}/L_{cz} of about 10. Snow, for obvious reasons, has a very strong influence on this ratio.

4. Work reported in detail elsewhere

Further details of various other factors perceivable from the graphs and from Table 3.1 are discussed more fully elsewhere (9, 10, 11, 12, 13). These publications contain graphs which, in addition to representing different cloudless sky conditions, also include diagrams for partially clouded skies with different cloud types of differing amount with a discussion of their properties. Other useful tools for routine analysis of results, like azimuthal cross sections at different elevation angles, or along vertical circles at constant azimuths, plots of irradiance against cosine of the incidence angle, etc. are also included. For clear skies the ratio $D_c(\beta, \alpha_s)/D_c$ i.e. total diffuse irradiance of inclined surfaces to that of the horizontal surface as function of solar altitude γ, surface inclination β, surface-solar azimuth α_s and atmospheric turbidity B_s have been analysed graphically and have been reported in reference (12). The corresponding analysis for the sky component alone is currently under study. Meanwhile, the clear sky radiance $L_c(\Theta, \alpha_s)$ has been related empirically to solar height, elevation angle Θ and azimuth with respect to the sun, α_s, of the respective point in the sky, as well as to atmospheric turbidity. The ratio D_c/G_c was used as a measure of turbidity.

5. Relationship between Schuepp Turbidity coefficient and the Linke Turbidity Factor

The analysis that follows is based on the use of the Schuepp Turbidity Coefficient B_s. The CEC Inclined Surface prediction methodology is based on the use of the Linke Turbidity Factor. As the Schuepp Turbidity Coefficient specifically takes account of the absorption effects of water vapour as a separate factor, while the Linke Turbidity Factor includes the effects of water vapour absorption within the factor, the relationship between the two factors is dependent on the precipitable water vapour content of the air (14). Figure 3.8 shows the relationship between the Linke Turbidity Factor T_L and the Schuepp Turbidity Coefficient B_s as a function of w/m, where w is the precipitable water vapour in centimetres, and m is the optical air mass. Taking a representative solar altitude of 30°, i.e. a sea level air mass of 2, the value of w/m under winter conditions with w = 0.5 cm would be 0.25, and under summer conditions with w = 2.0 cm, w/m = 1. The analytic relationship between T_L and B_s is

$$T_L = (B_s + 0.54)(1.75 \log_{10}(w/m + 0.1) + 14.5) - 5.4 \qquad (3.1)$$

In (14) there is also a more general graph on the regression between B_s and T_L, extended up to $B_s = 1.0$ and $T_L = 16$.

6. Regression analysis of clear sky radiances

Following the regression analysis carried out between slope irradiance and horizontal irradiance to find relationships of the form:

$$D_c(\beta,\alpha_s)/D_c = f(B_s, \gamma,\alpha_s, \beta) \qquad (3.2)$$

a similar graphical analysis, using cloudless day data described below, was applied to relate sky radiance normalised to the zenith value to atmospheric turbidity and the two angular parameters Θ, and α_s:

$$(L_c(\beta,\alpha_s)/L_{cz}) = f(D_c/G_c, \gamma,\alpha_s, \Theta) \qquad (3.3)$$

It will be noted that the Schuepp Turbidity Coefficient B_s in formula 3.2 has been replaced by the horizontal diffuse/global irradiance ratio for pragmatic reasons. The access to this ratio is far easier and immediate for any place with observations of D_c and G_c than deriving the value of B_s. Best fit curves in empirical scattergrams relating the ratio $L_c(\beta,\alpha_s)/ L_{cz}$ to (B_s, γ) or to $(D_c/G_c,\gamma)$ can be found without significant differences in statistical scattering between the two formulations (12).

The functional relationship between D_c/G_c and the Schuepp Turbidity Coefficient B_s and solar altitude has been explored in detail for three sites in Switzerland, Locarno-Monti, Davos, and the Weissfluhjoch (15). Because of the relatively small turbidity ranges of the alpine stations, Davos and Weissfluhjoch, this comparison could not be extended for $B_s >$ 0.120 for them; for Locarno-Monti, however, the $D_c/G_c = f(\beta, \gamma)$ regression could be extended up to $B_s = 0.600$. Figure 3.9 compares the results for the 3 sites. In general the agreement between the three sites is good, except at low solar altitudes. The majority of the observations at very low turbidities made at the Weissfluhjoch at 2670 m asl were made with snow covered ground so the effects of inter-reflection between snow and sky can

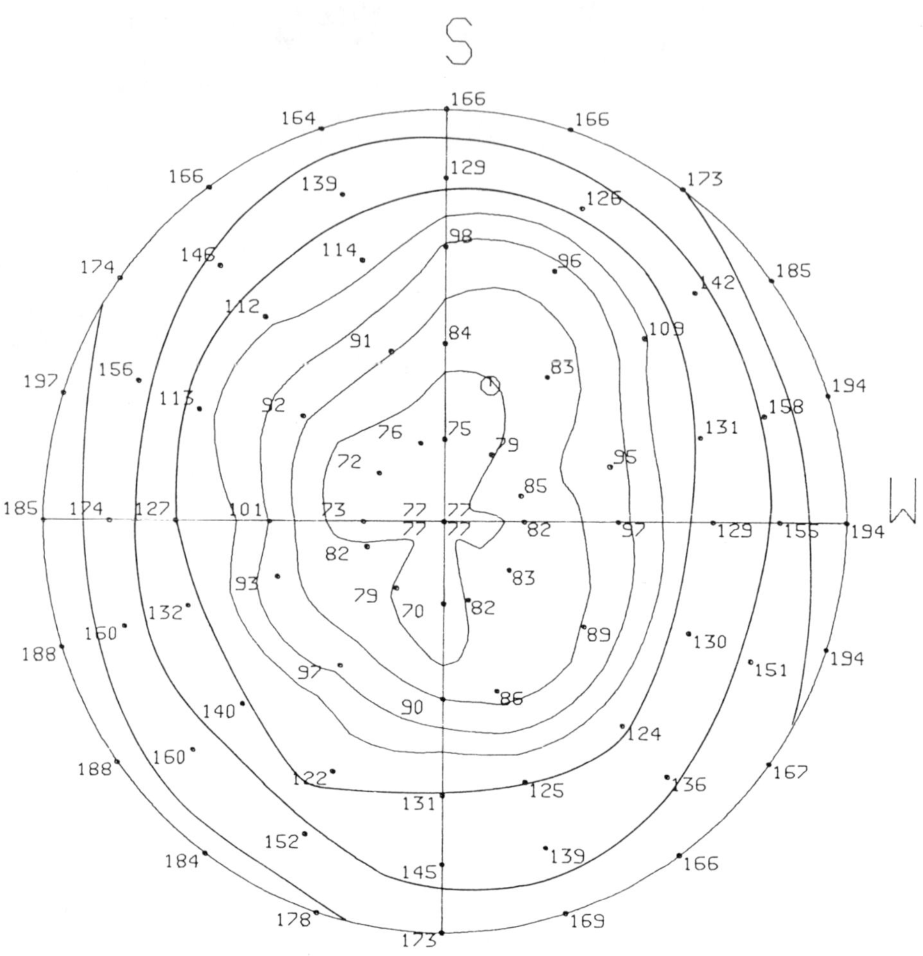

Figure 3.7a Angular distribution at the Weissfluhjoch mountain top of diffuse component on inclined planes, $D_c(\beta, \alpha_s)$ for very low values of Schuepp atmospheric turbidity coefficient, B_s = 0.030.
Units: Wm^{-2}. Isolines drawn at intervals of 10 Wm^{-2} for $D_c(\beta, \alpha_s) < 100$ and at intervals of 25 Wm^{-2} for $D_c(\beta, \alpha_s) > 100 Wm^{-2}$

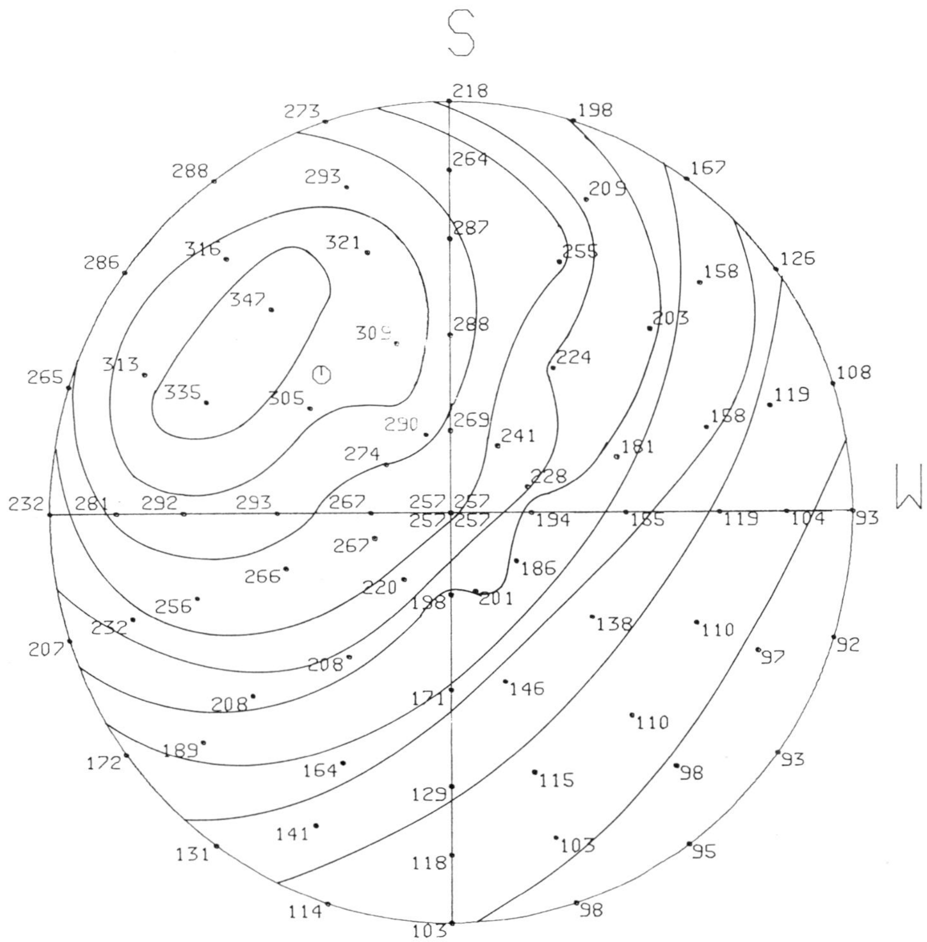

Figure 3.7b Angular distribution at EMPA, Zurich of diffuse component on inclined planes $D_c(\beta,\alpha_s)$ for very high values of atmospheric turbidity on the 19th August, 1981. Schuepp turbidity coefficient $B_s = 0.380$. This graph is for example V in Table 3.1.
Uni s: Wm^{-2}, isolines at 10 Wm^{-2} intervals for $D_c(\beta,\alpha_s) < 100$ Wm^{-} and at intervals of 25 Wm^{-2} for $D_c(\beta,\alpha_s) > 100$ Wm^{-2}. Position of sun marked with ☉.

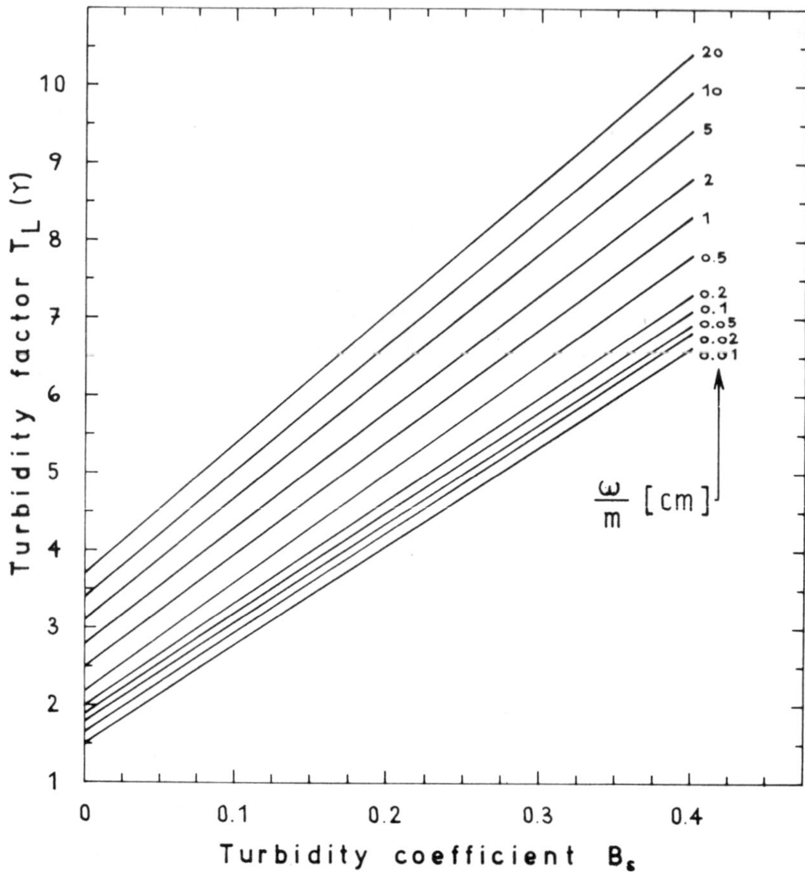

Figure 3.8 Relationship between Linke Turbidity Factor $T_L(\gamma)$ and the Schuepp Turbidity Coefficient B_s as a function of the ratio w/m where w is the precipitable water vapour in cm and m is the optical air mass.

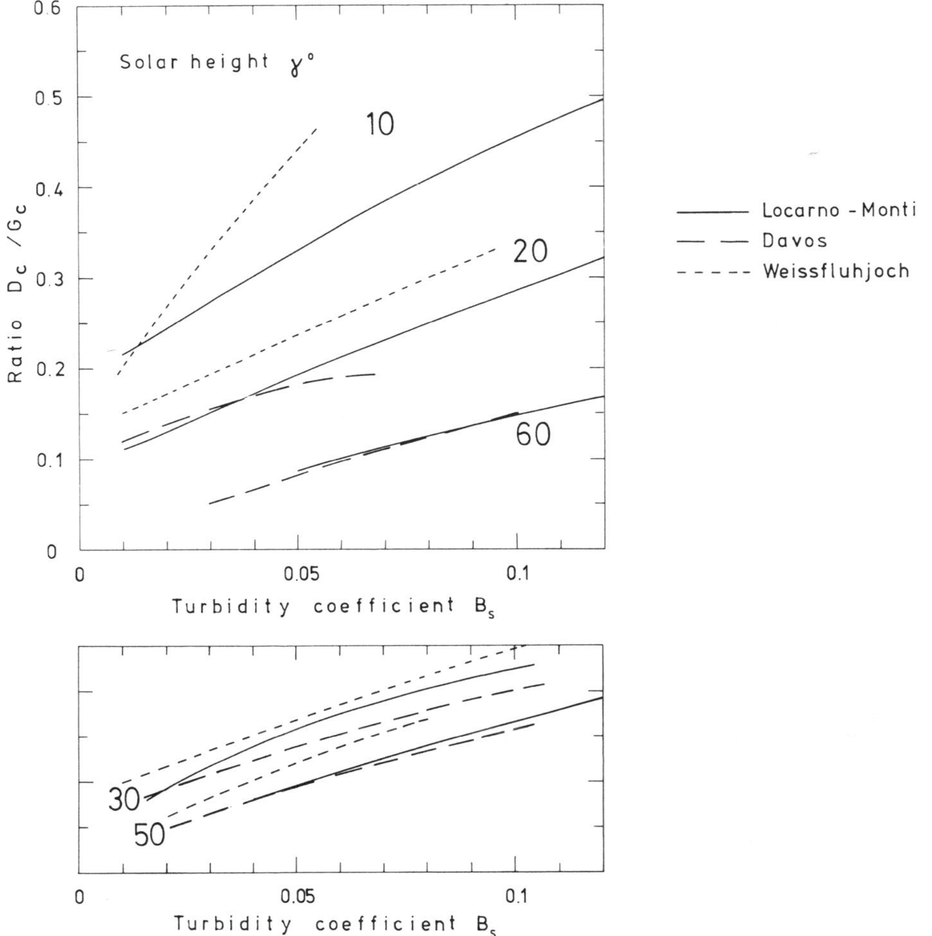

Figure 3.9 The relationship between D_c/G_c and the Schuepp Turbidity Coefficient B_s as a function of solar altitude γ. Snow was present during some of the Weissfluhjoch measurements. The increase in D_c due to snow is evident.

be identified, especially at lower solar elevations, typical of winter. Some of the Weissfluhjoch measurements are thus not representative of snow-free ground. The key issue, however, is that satisfactory functional relationships can be established between D_c/G_c and B_s, so enabling D_c/G_c to be substituted in the subsequent analysis.

Using the collection of fish-eye sky-photographs and also visual observations (noted according to the synoptic code) made with each data set of 121 radiance measurements, some sets of observations with only traces of clouds, that are small and far enough from the sun, have also been included. 551 cases were selected representing clear sky conditions covering 37 separate days of measurement. The measurements have been compiled for the three sites involved in Table 3.1 as well as for Carpentras in Southern France (latitude 44°02', longitude 5°01', altitude 99 m a.s.l.). For the immediate analysis described below, only 182 selected data sets out of the 551 cloudless day data sets were used. In this further selection process, judgement was used to ensure that each of the 37 days were represented by data sets which suitably covered the available observed ranges of both turbidity and solar altitude. Polar diagrams like Figure 3.6c are attached to reference 12 for all these 182 selected cloudless day distributions.

The dates of the measuring campaigns and the number of cloudless day data sets computed for each site were as follows:

Carpentras: 05/6/79 - 29/6/79, 51 distributions

EMPA: 15/6/81 - 28/8/81, 58 distributions

Locarno-Monti: 30/9/81 - 29/10/81, 23 distributions

Weissfluhjoch: 06/8/80 - 10/12/80, 50 distributions

Using this material, graphs illustrating either the partial regression on solar altitude

$$(L_c(\Theta, \alpha_s)/L_{cz}) = f_1(\gamma) \qquad (3.4)$$

or a fuller regression on γ, and D_c/G_c

$$(L_c(\Theta, \alpha_s)/L_{cz}) = f_2(\gamma, D_c/G_c) \qquad (3.5)$$

or sometimes both, have been prepared for the constant elevation angles Θ = 18°, 36°, 54° and 72° at the constant azimuths α_s = 0° (facing sun), 15°, 30°, 60°, 90°, 135° and 180°. (The arithmetic mean was used for each pair of values at $+\alpha_s$ and at $-\alpha_s$.) Figures 3.10, 3.11 and 3.12 which are for surfaces facing within 30° of the sun's direction belong to the first regression type, described by equation 3.4, while Figures 3.13 and 3.14 for the remaining directions belong to the second type described by equation 3.5. First type scattergrams could be prepared for sky zones where no definite dependence of relative radiance on atmospheric turbidity could be detected. This was the case for surfaces facing the sun, i.e. $\alpha_s \leqslant 30°$. For $\alpha_s \geqslant 60°$, normalised radiance shows increasing dependence on turbidity and hence on D_c/G_c with increasing α_s. Figure 3.15, prepared by combining the essential features of the four graphs in Figure 3.13 into a common presentation, belongs to the second type of regression graphs.

As may be seen from Figures 3.10, 3.11, 3.12, the influence of solar altitude on normalised sky radiance within the azimuth range $\alpha_s \leqslant 30°$ is strong at small elevation angles in the sky and vanishes as one moves towards the zenith, where the limiting value is of course 1. For points facing the sun, i.e. $\alpha_s = 0$, sharp maxima appear at $\Theta = \gamma$ i.e. the centre of the aureole is being sampled. The phenomenon is also present, but less sharply, when $\alpha_s = 15°$. Even when $\alpha_s = 30°$, the phenomenon is still present to a very small extent and L_c/L_{cz} shows still a slight hump around $\Theta = \gamma$.

For the range $\alpha_s \geqslant 60°$, the normalised radiance decreases with turbidity, and, of course, with solar altitude. The values depend to a smaller extent on α_s. The graphs also indicate the augmented radiance of sky zones near the horizon, i.e. the horizon brightening effect for cloudless conditions.

7. New radiance observations near the horizon

The augmented radiance of the sky zones near the horizon is illustrated in more detail by Figure 3.16a - 3.16f. Data for the preparation of these graphs were gained from the additional scanning program which concentrated on measurements in the lowest belt of the sky, as mentioned in section 2.iii of this Chapter.

At the EMPA site measurements were made in February and April 1983 for low but mostly medium turbidities, and at the snow covered mountain site Le Locle (1240 m a.s.l.) in April 1984, mostly for small turbidities. The sky zone near the horizon is, as Figure 3.16 demonstrates, brighter than the zenith up to a solar elevation of about 40°-50°. About 15 series of July measurements taken at the EMPA site (15), not included in the graphs, would show for $\alpha_s > 90°$ and $\gamma > 35°$ relative radiances < 0.5 as a consequence of the effect of stronger turbidities (see also Figure 3.13). Though the number of measurements is relatively small, some further features may also be recognised:

- The Le Locle graph for $\Theta = 18°$ (Figure 3.16f) shows good consistency with Figures 3.10, 3.11, 3.12, 3.13 and 3.15, thus confirming the site independent general validity of the results.

- Comparing the single graphs of Figure 3.16 with each other, it may be noted that L_c/L_{cz} at EMPA is greater for $\Theta = 9°$ than for $\Theta = 3°$ for all azimuths and almost all solar elevations. At Le Locle the ratio is greater for $\Theta = 8°$ than for $\Theta = 5°$ for small values of γ, but it is smaller for $\Theta = 8°$ than for $\Theta = 5°$ for larger values of γ (probably as an effect of the snow cover). Furthermore, the ratio is greater for $\Theta = 8°$ than for $\Theta = 12°$ for almost all solar heights. Finally, L_c/L_{cz} for $\Theta = 18°$ is both smaller than for $\Theta = 12°$ at Le Locle, as well as for the belt $\Theta = 9°$ to 15° at EMPA, again for almost all values of Θ, in both cases. It thus seems that maximum relative radiance of the cloudless sky near the horizon appears at about 8° to 10° above the horizon.

- Comparing Figures 3.16a and 3.16b with Figures 3.16c and 3.16d, the difference of the shape of the curves for the two places is apparent. For small solar elevations relative radiances at EMPA are greater than

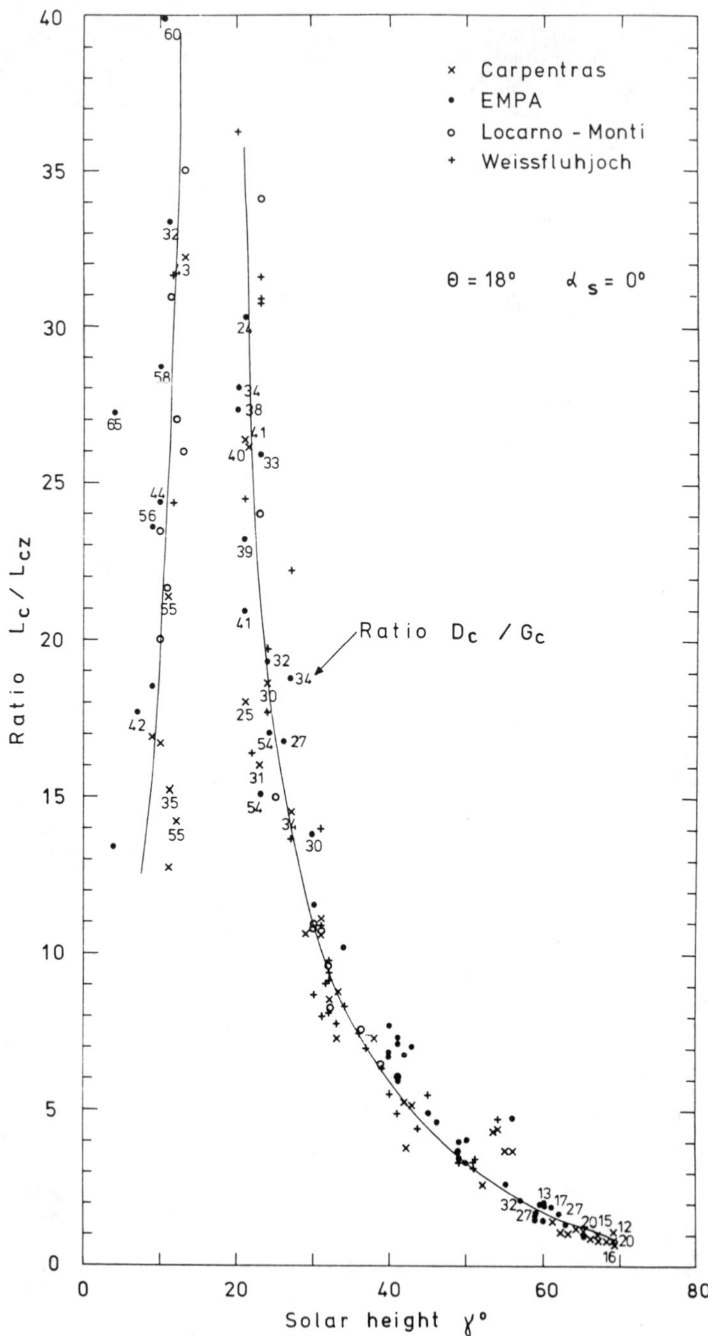

Figure 3.10a Scatter diagram relating the ratio L_c/L_{cz} to the solar altitude γ for points on the sun's vertical meridian, i.e. $\alpha_s = 0°$, for elevation angle $\Theta = 18°$. Symbols for the four measurement sites are different, refer legend.

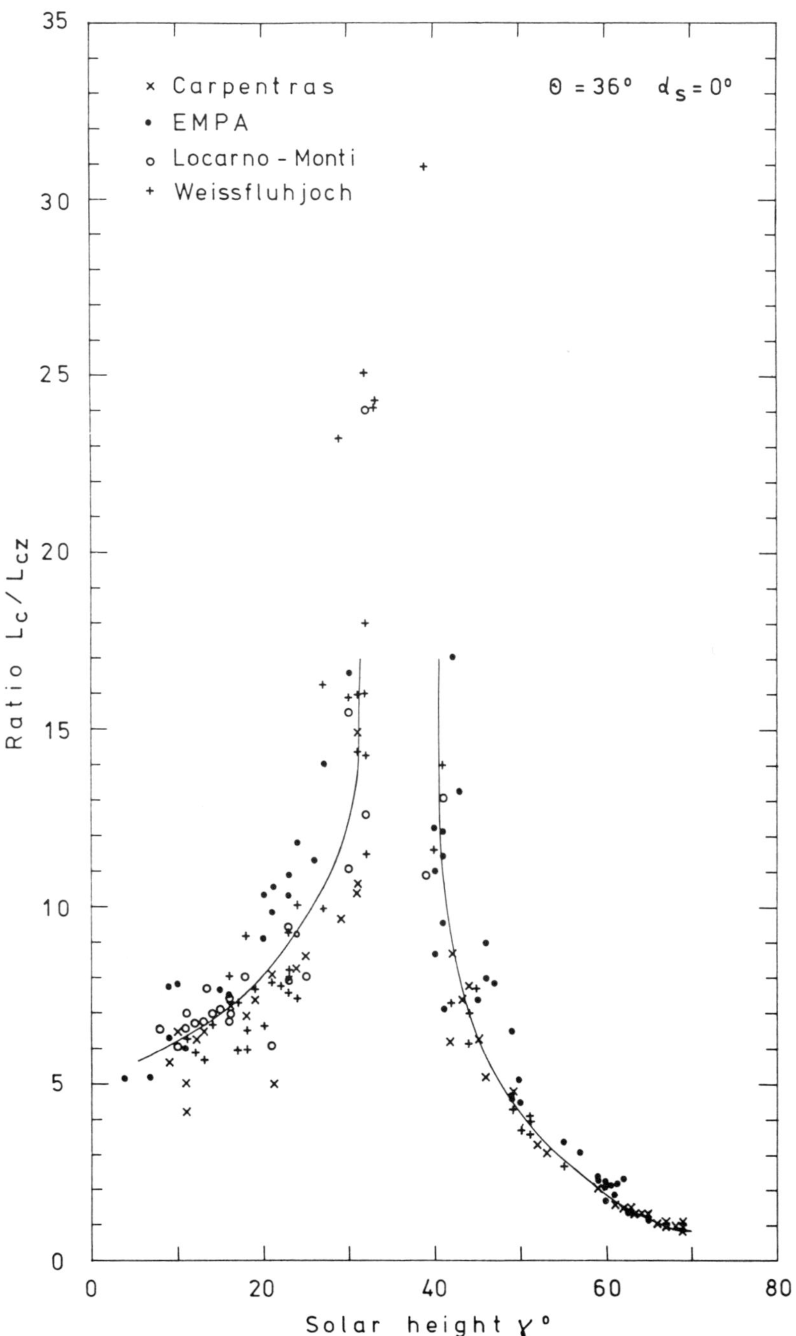

Figure 3.10b Scatter diagram relating the ratio L_c/L_{cz} to the solar altitude γ for points on the sun's vertical meridian, i.e. $\alpha_s = 0°$, for elevation angle $\Theta = 36°$. Symbols for the four measurement sites are different, refer legend.

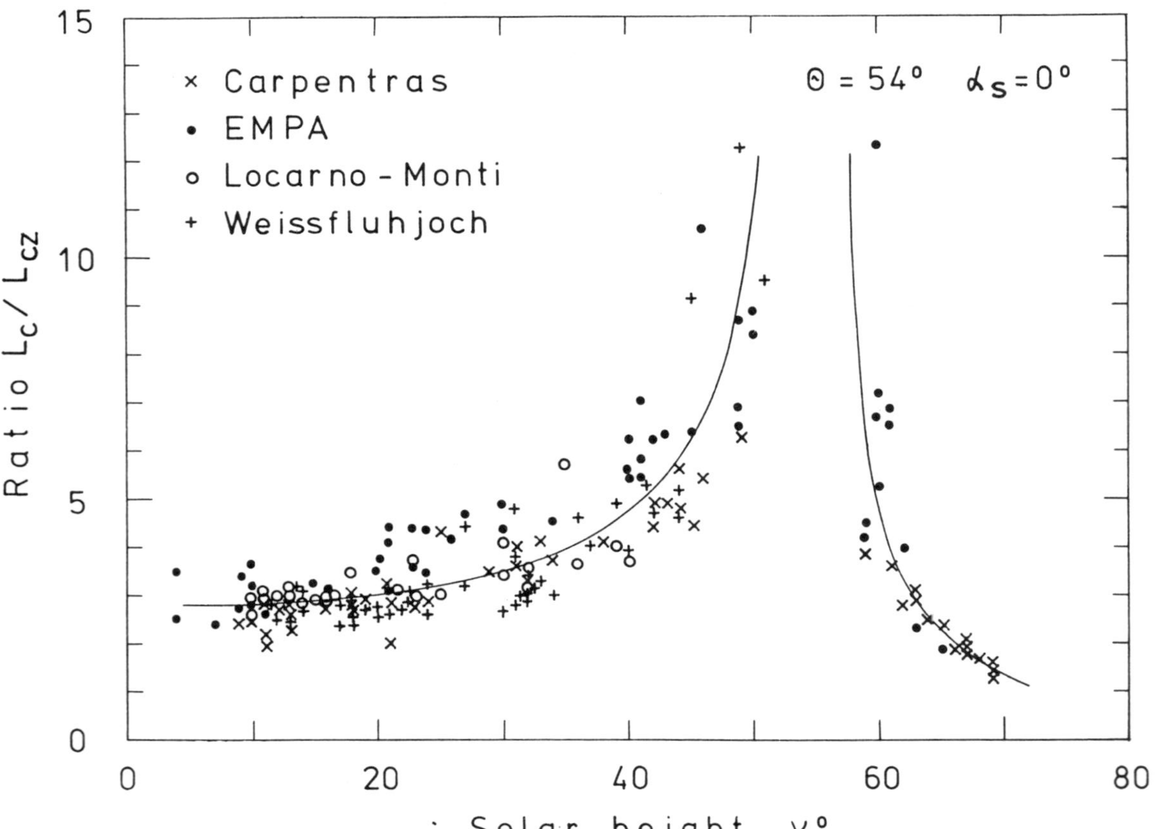

Figure 3.10c Scatter diagram relating the ratio L_c/L_{cz} to the solar altitude γ for points on the sun's vertical meridian, i.e. $\alpha_s = 0°$, for elevation angle $\Theta = 54°$. Symbols for the four measurement sites are different, refer legend.

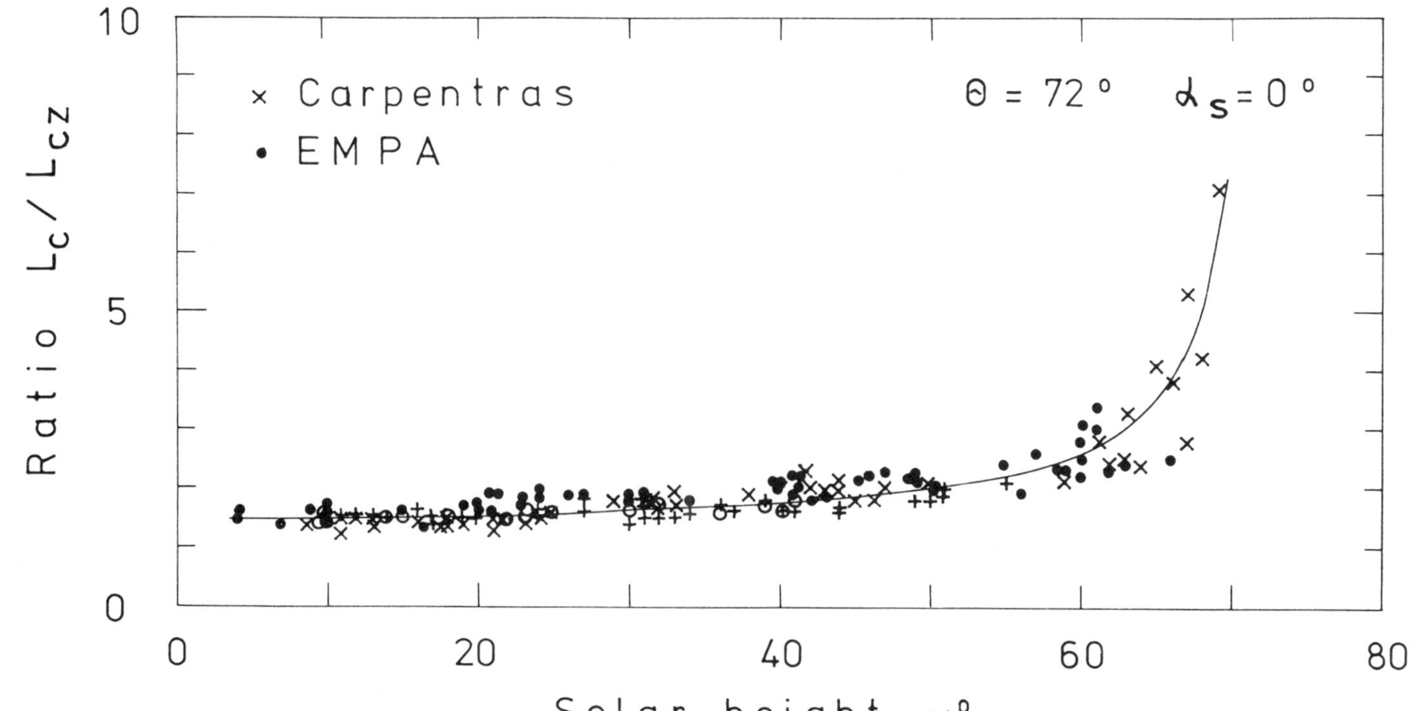

Figure 3.10d Scatter diagram relating the ratio L_c/L_{cz} to the solar altitude γ for points on the sun's vertical meridian, i.e. $\alpha_s = 0°$, for elevation angle $\Theta = 72°$. Symbols for the four measurement sites are different, refer legend.

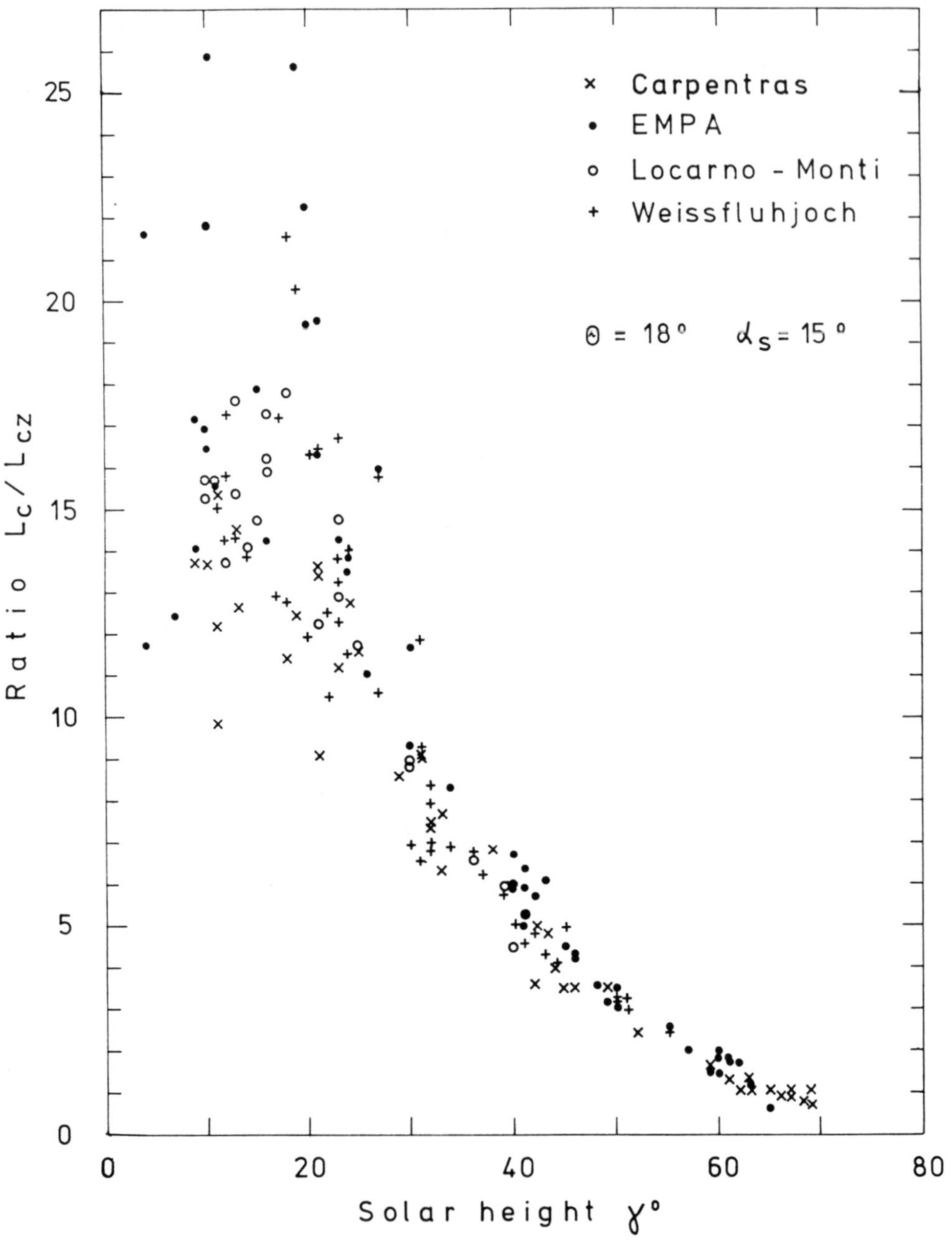

Figure 3.11a Scatter diagram relating the ratio L_c/L_{cz} to the solar altitude γ for α_s = 15°, and elevation angle = 18°. Symbols for the four measurement sites are different, refer legend.

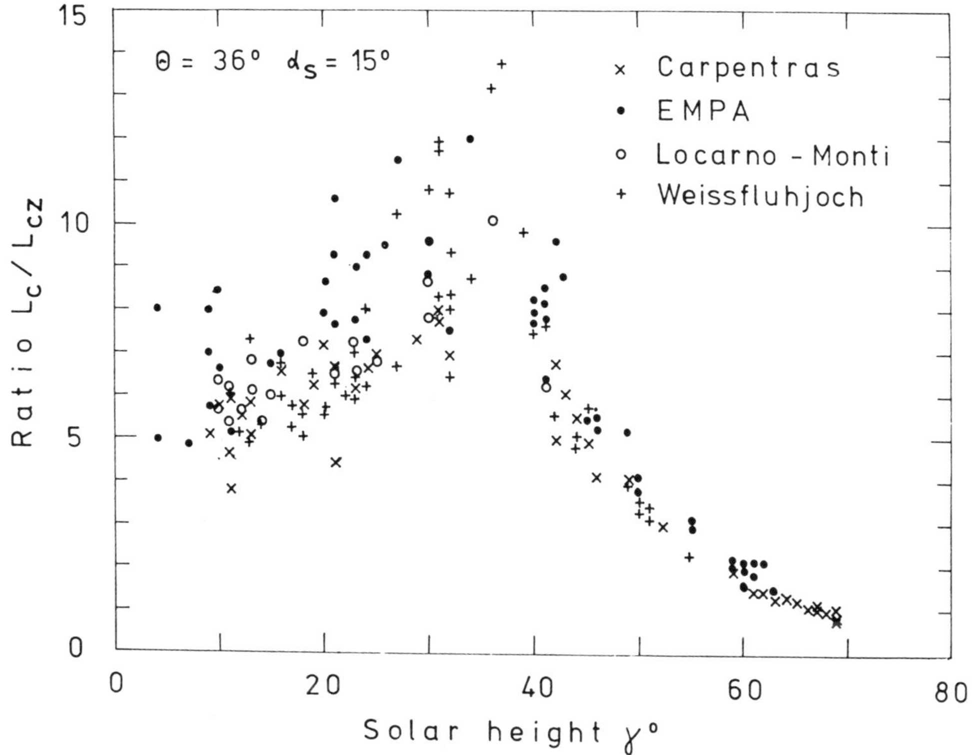

Figure 3.11b Scatter diagram relating the ratio L_c/L_{cz} to the solar altitude γ for α_s = 15°, and elevation angle = 36°. Symbols for the four measurement sites are different, refer legend.

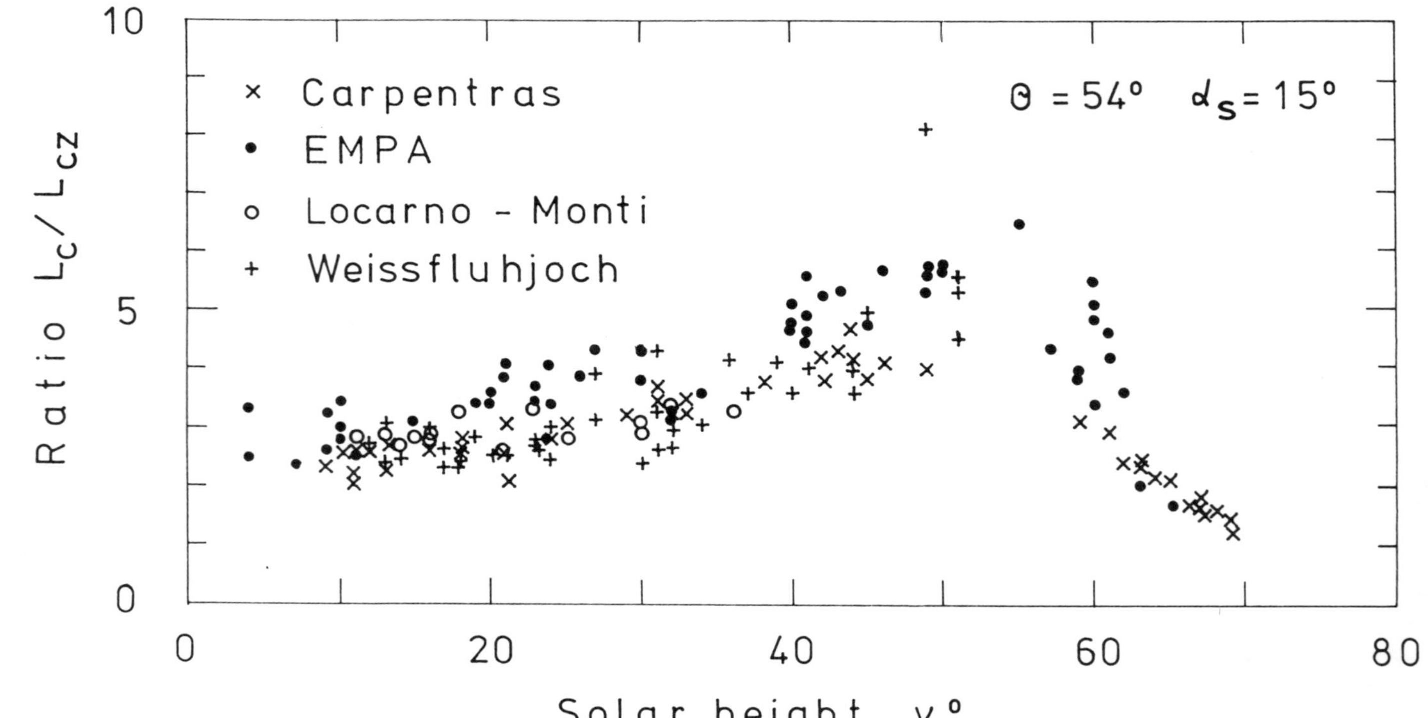

Figure 3.11c Scatter diagram relating the ratio L_c/L_{cz} to the solar altitude γ for $\alpha_s = 15°$, and elevation angle $\Theta = 54°$. Symbols for the four measurement sites are different, refer legend.

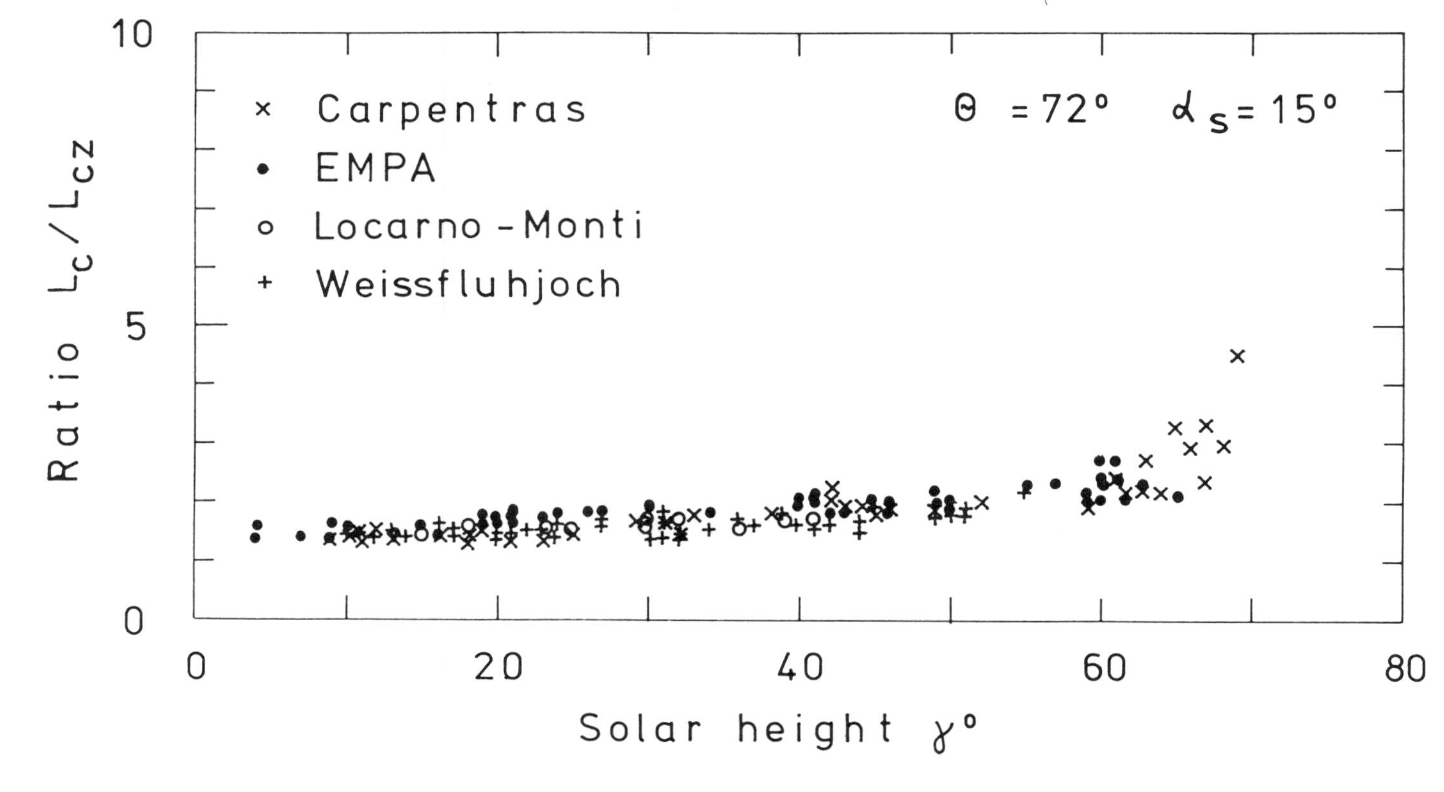

Figure 3.11d Scatter diagram relating the ratio L_c/L_{cz} to the solar altitude γ for $\alpha_s = 15°$, and elevation angle $\Theta = 72°$. Symbols for the four measurement sites are different, refer legend.

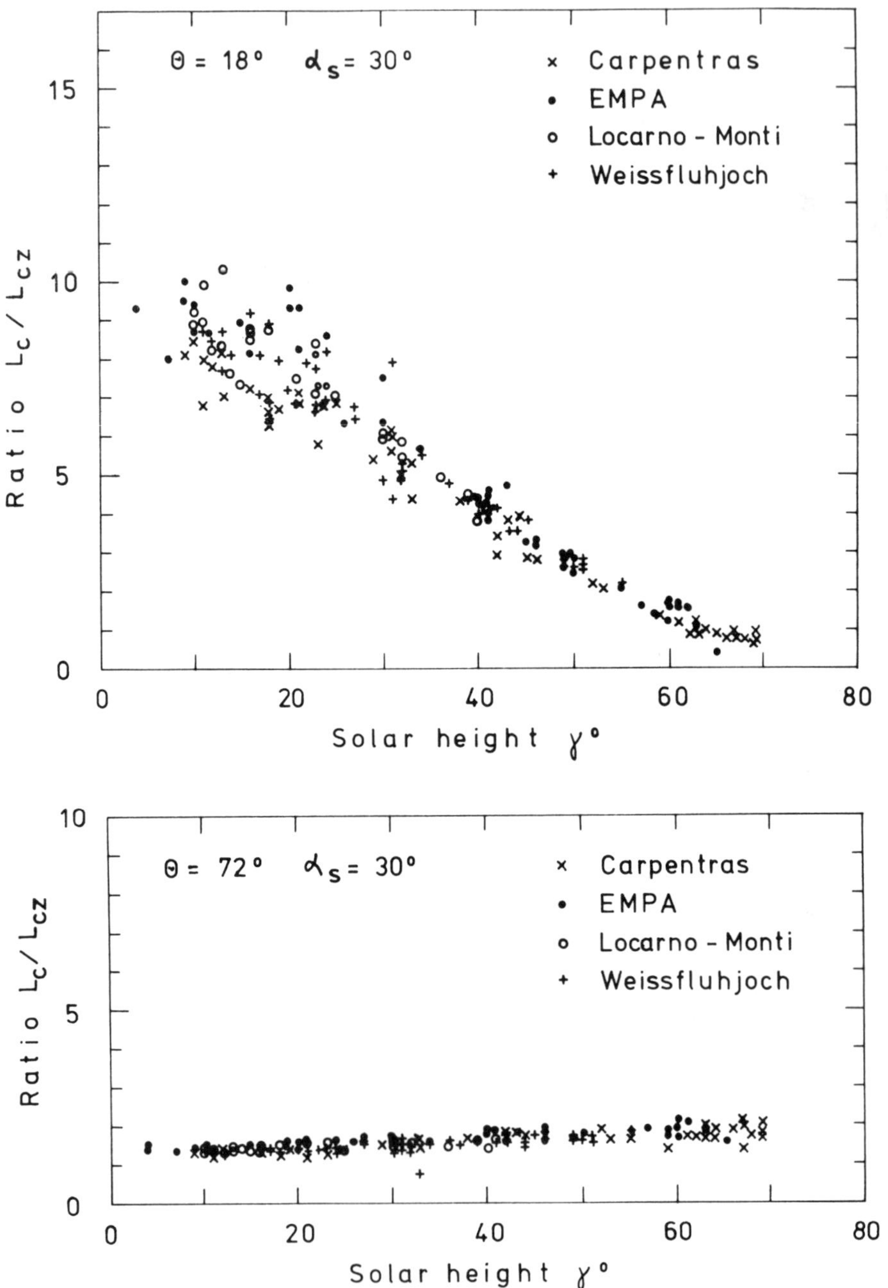

Θ = 18° α_s = 30°
× Carpentras
• EMPA
o Locarno - Monti
+ Weissfluhjoch
Ratio L_c / L_cz
Solar height γ°
15
10
5
0
0
20
40
60
80
Θ = 72° α_s = 30°
× Carpentras
• EMPA
o Locarno - Monti
+ Weissfluhjoch
Ratio L_c / L_cz
Solar height γ°
10
5
0
0
20
40
60
80

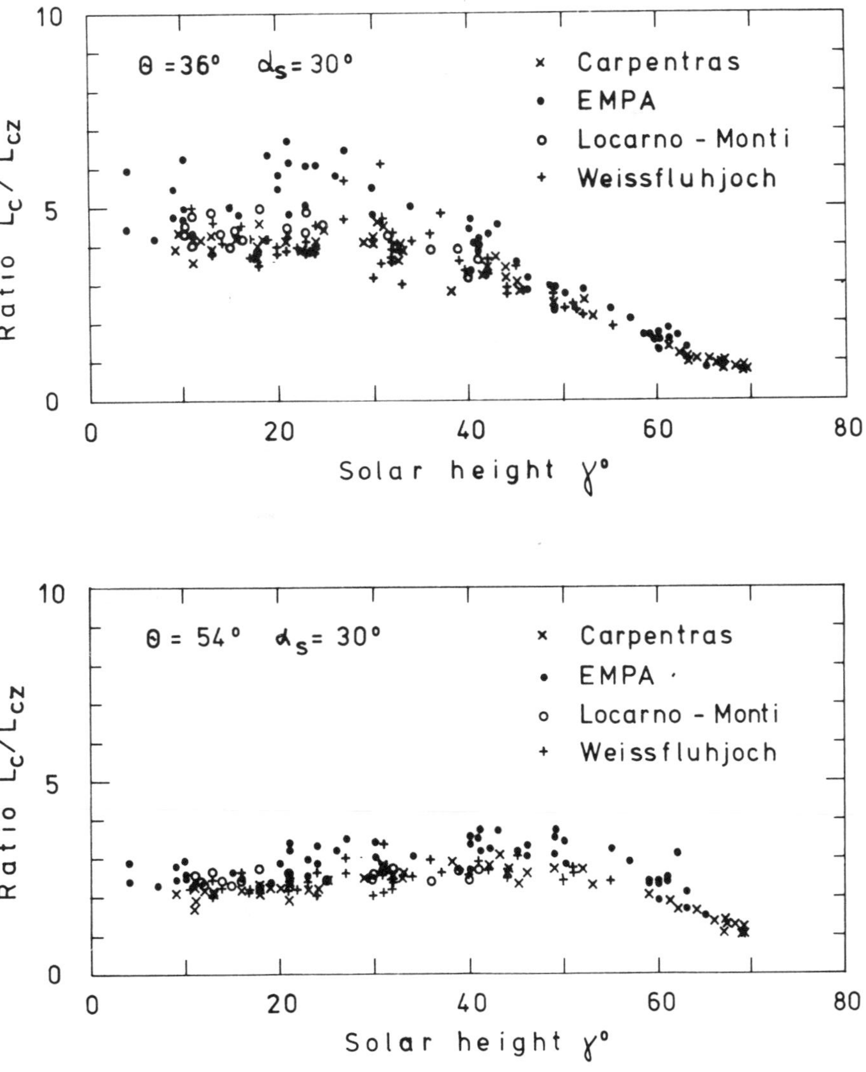

Figure 3.12 Family of scatter diagrams relating the ratio L_c/L_{cz} to solar altitude γ for $\alpha_s = 30°$, for elevation angles $\Theta = 18°$ (top left), 36° (top right), 54° (bottom right) and 72° (bottom left). Symbols for the four measurement sites are different (see legend).

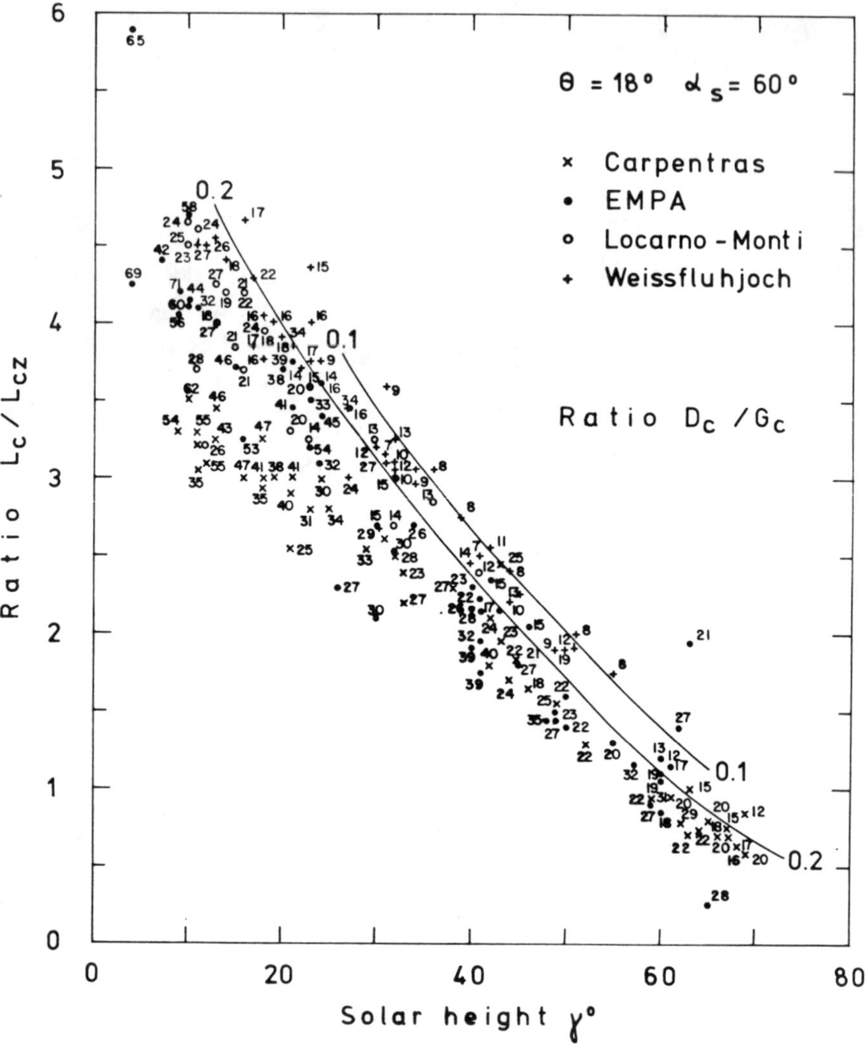

Figure 3.13a Scatter diagram relating the ratio L_c/L_{cz} to the solar altitude γ for elevation angle Θ = 18° for points in the sky at α_s = 60°. The small numbers give the D_c/G_c ratio for each entry. Isolines of this ratio have been drawn by eye in steps of 0.1.

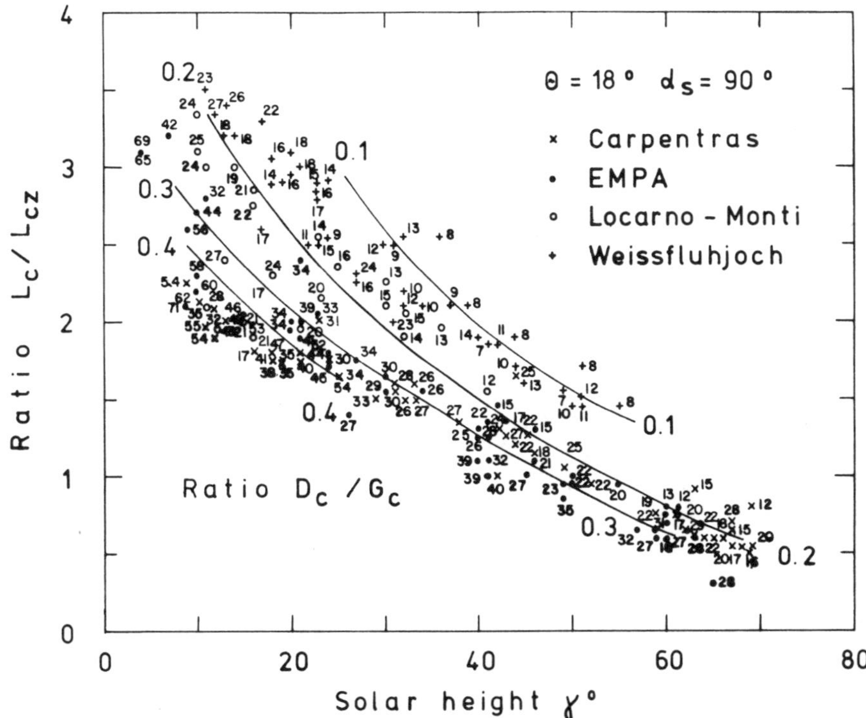

Figure 3.13b Scatter diagram relating the ratio L_c/L_{cz} to the solar altitude γ for elevation angle Θ = 18° for points in the sky at α_s = 90°. The small numbers give the D_c/G_c ratio for each entry. Isolines of this ratio have been drawn by eye in steps of 0.1.

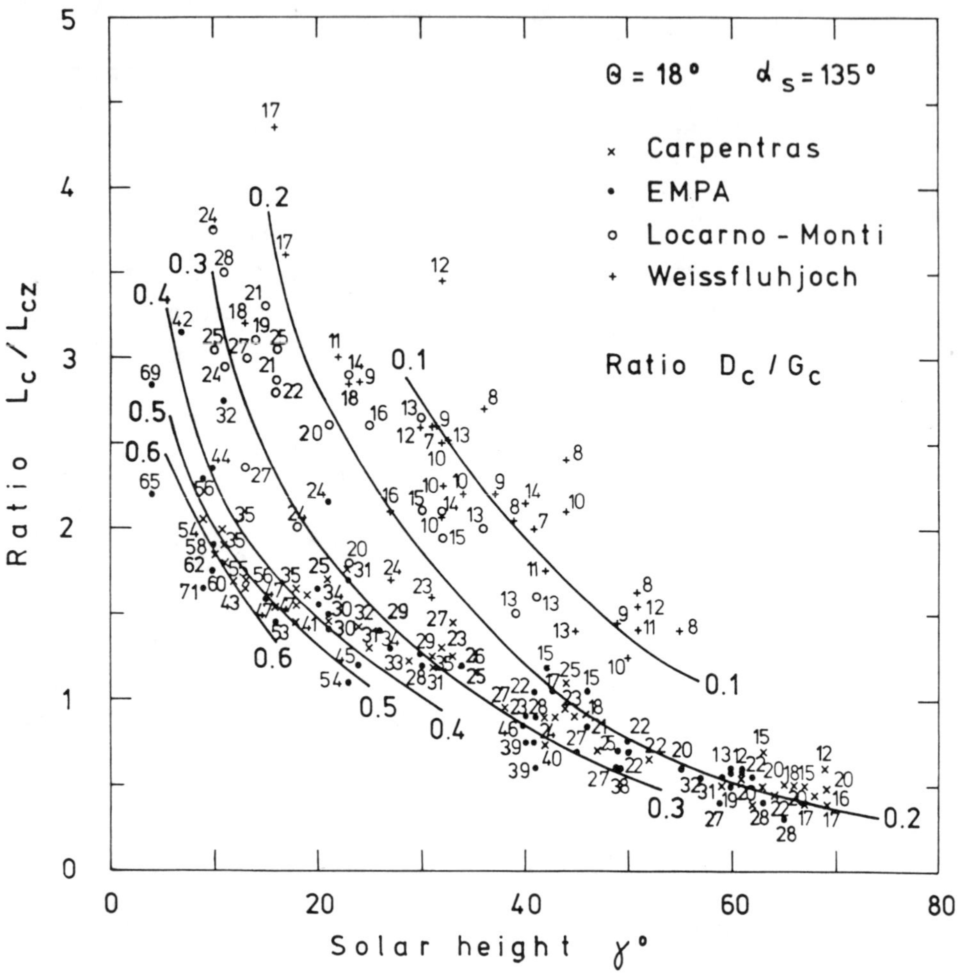

Figure 3.13c Scatter diagram relating the ratio L_c/L_{cz} to the solar altitude γ for elevation angle Θ = 18° for points in the sky at α_s = 135°. The small numbers give the D_c/G_c ratio for each entry. Isolines of this ratio have been drawn by eye in steps of 0.1.

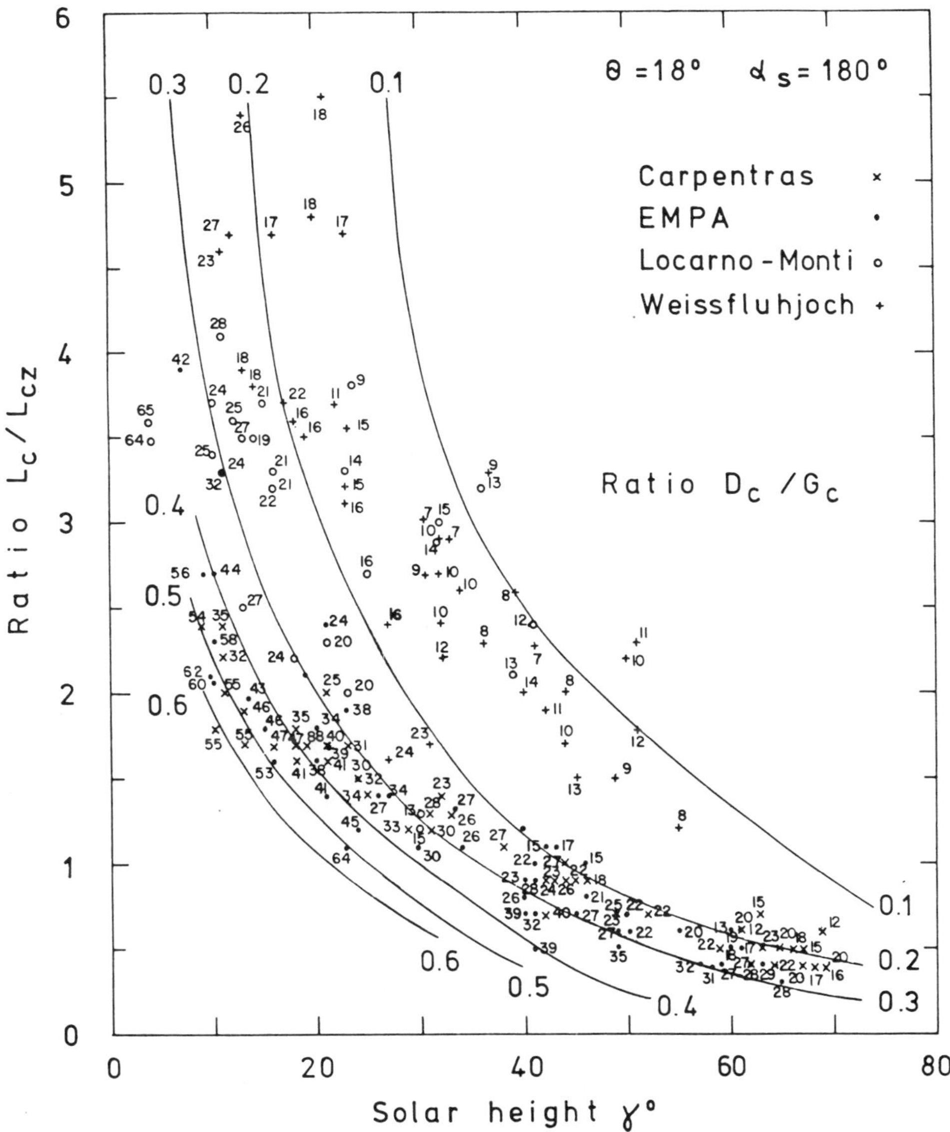

Figure 3.13d Scatter diagram relating the ratio L_c/L_{cz} to the solar altitude γ for elevation angle Θ = 18° for points in the sky at α_s = 180°. The small numbers give the D_c/G_c ratio for each entry. Isolines of this ratio have been drawn by eye in steps of 0.1.

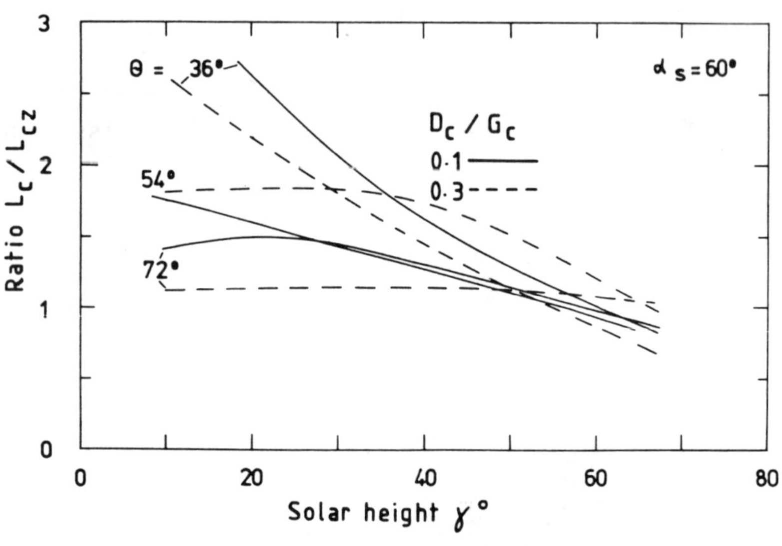
θ = 36°
54°
72°
$\alpha_s = 60°$
D_c / G_c
0·1
0·3
Ratio L_c / L_{cz}
Solar height γ°
0
1
2
3
0
20
40
60
80

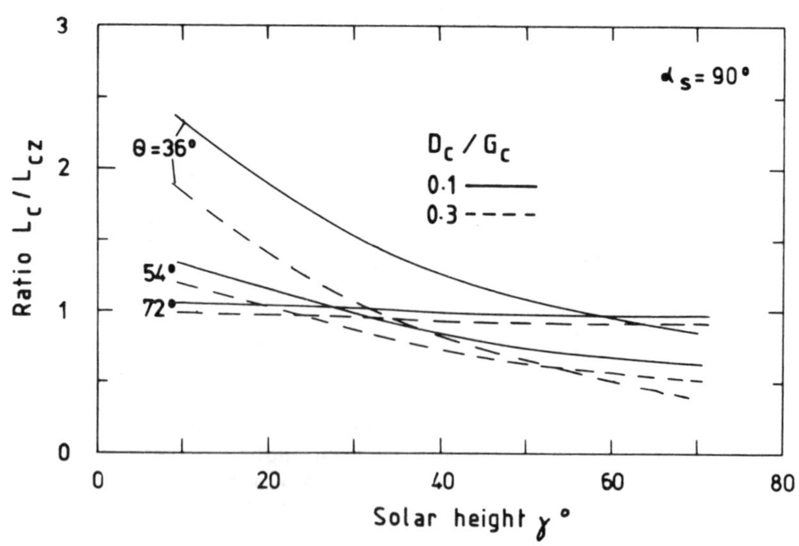
θ = 36°
54°
72°
$\alpha_s = 90°$
D_c / G_c
0·1
0·3
Ratio L_c / L_{cz}
Solar height γ°
0
1
2
3
0
20
40
60
80

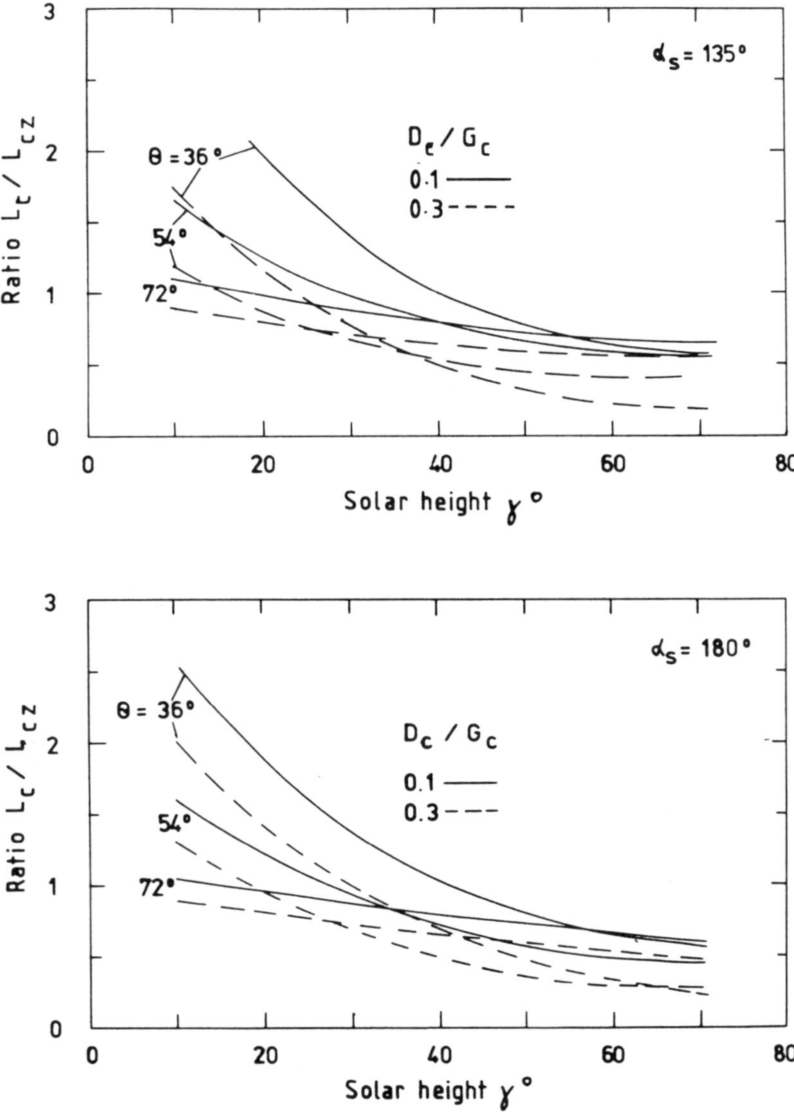

Figure 3.14 Family of scatter diagrams relating the ratio L_c/L_{cz} to solar altitude γ as in Figure 3.13, but for elevation angles Θ = 36°, 54° and 72° without showing the individual entries. Only the pair of isolines D_c/G_c = 0.1 and 0.3 have been drawn to indicate the bandwidth of the scattering.

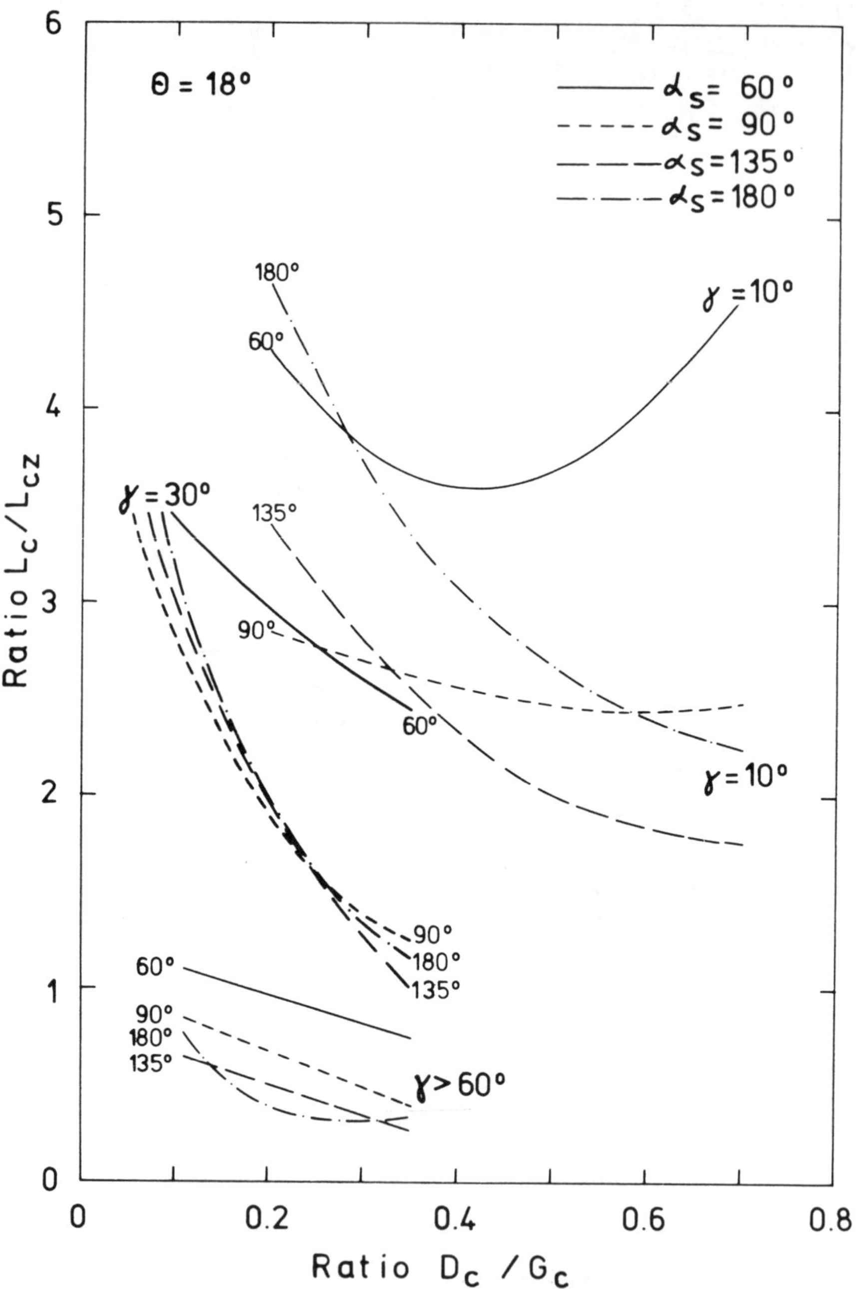

Figure 3.15 Same as Figure 3.13, for a constant sky patch elevation Θ = 18°, but replotted with the solar height γ and the irradiance ratio D_c/G_c interchanged. Individual entries are omitted. Instead of presenting four single graphs as in Figure 3.13, isolines were drawn for the four α_s values 60°, 90°, 135° and 180° for solar elevations γ = 10°, 30° and 60°.

at Le Locle. For medium values of the solar altitude γ the Le Locle radiances are greater than those of EMPA and for about $\gamma > 50°$ no significant differences can be discovered. For small γ this may be attributed to the generally higher sky radiance for lowland air compared with clean mountain air; on the other hand, for the medium solar altitude range of clear mountain air, the decrease in the relative sky brightness of the horizon belt may no longer be offset for skies above a highly reflective snow cover in contrast with the EMPA cases. For $\gamma > 50°$ it finally seems, that rapidly increasing zenith radiance appears to mostly wipe out the influence of differences in the reflective properties of the underlaying ground on the radiance of the sky close to the horizon.

8. Relationships between zenith radiance and diffuse irradiance on a horizontal surface

In order to relate the radiance observations to the diffuse irradiance observations on a horizontal plane, some adjustments had to be made to the radiance observations to allow for the fact that the waveband observed in the radiance measurements was different from the waveband observed by the pyranometer. These spectral adjustments are described in Appendix 4. 267 clear sky cases were examined in detail. After the zenith radiance values had been adjusted spectrally, the ratio L_{cz}/D_c was extracted. The examination of the ratio L_{cz}/D_c helps indicate to what extent the sky irradiances resulting from real clear skies differ from those resulting from an isotropic sky. The observed ratios are entered in Row 22 of Table 3.1. The departures from the isotropic value of 1 are quite significant.

Figure 3.17 which shows the relationship between L_{cz} and D_c, was developed using the 267 clear sky cases mentioned. Some dependence of the ratio on site and turbidity certainly exists, but is not systematically recognisable. It is, in any case, much smaller than the dependence on solar altitude. Considering the ratio as a function of the solar altitude alone and expressing D_c in Wm^{-2}, the zenith radiance L_{cz} thus can be estimated from the mean trend curve in Figure 3.17 as

$$L_{cz} = (1/\pi)\, D_c \cdot f(\gamma) \qquad Wm^{-2}\ \text{steradian}^{-1} \qquad (3.6)$$

Figure 3.17, enables $f(\gamma)$ to be estimated for any solar elevation.

9. Irradiance on inclined surfaces

Appendix 4 explains how the irradiance on inclined planes can be estimated by numerical integration of the radiance function for sloping surfaces. One can thus proceed systematically from value of D_c to estimate L_{cz} from Figure 3.17, and then using appropriate integration methods to the determination of $D_{cs}(\beta,\alpha_s)$. An estimate of the ground reflected component has then to be made to obtain $R_c(\beta,\alpha_s)$.

By adding the two components, one can achieve estimates of $D_c(\beta,\alpha_s)$ for any slope for any given value of the solar elevation γ, and Schuepp Turbidity Coefficient B_s, or Linke Turbidity Factor $T_L(\gamma)$.

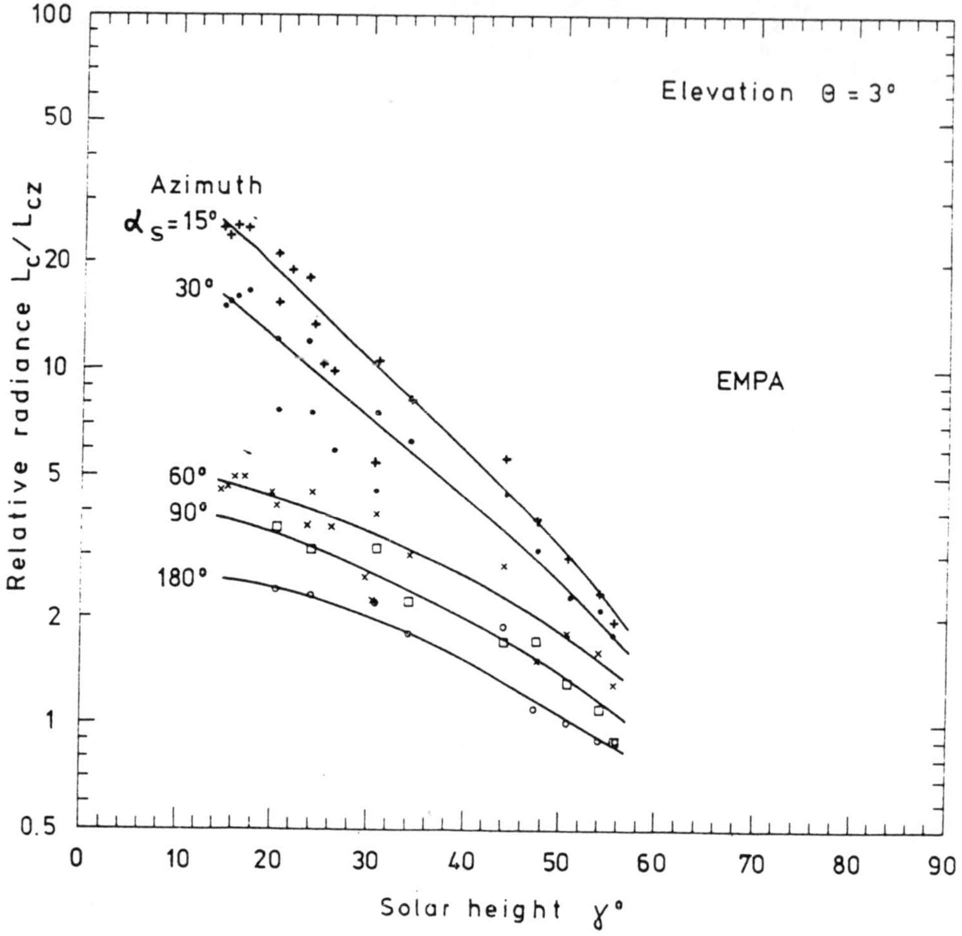

Figure 3.16a Normalized clear sky radiance L_c/L_{cz} at a low elevation in the sky as a function of solar height for different values of the azimuth α_s measured at an elevation angle of 3° at the lowland station EMPA.

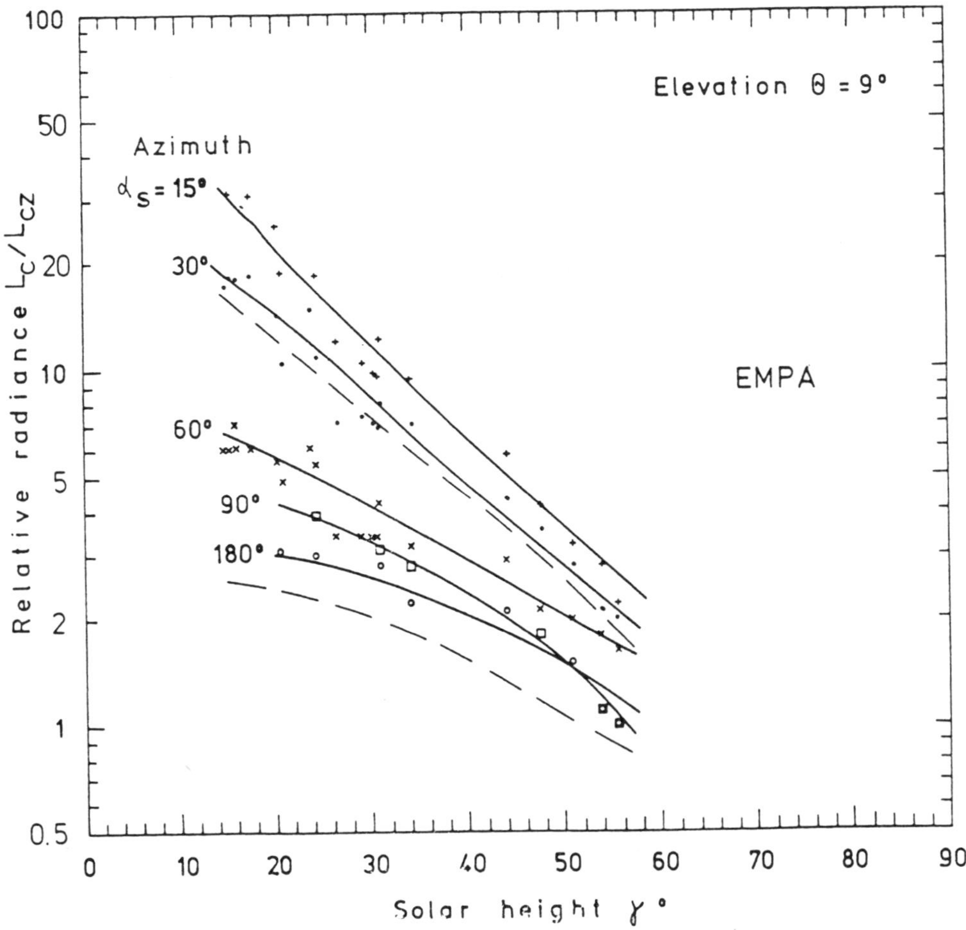

Figure 3.16b Normalized clear sky radiance L_c/L_{cz} at a low elevation in the sky as a function of solar height for different values of the azimuth α_s measured at an elevation angle of 9° at the lowland station EMPA. The dashed curves on this graph are the values for α_s = 30 and α_s = 180 taken from Figure 3.16a.

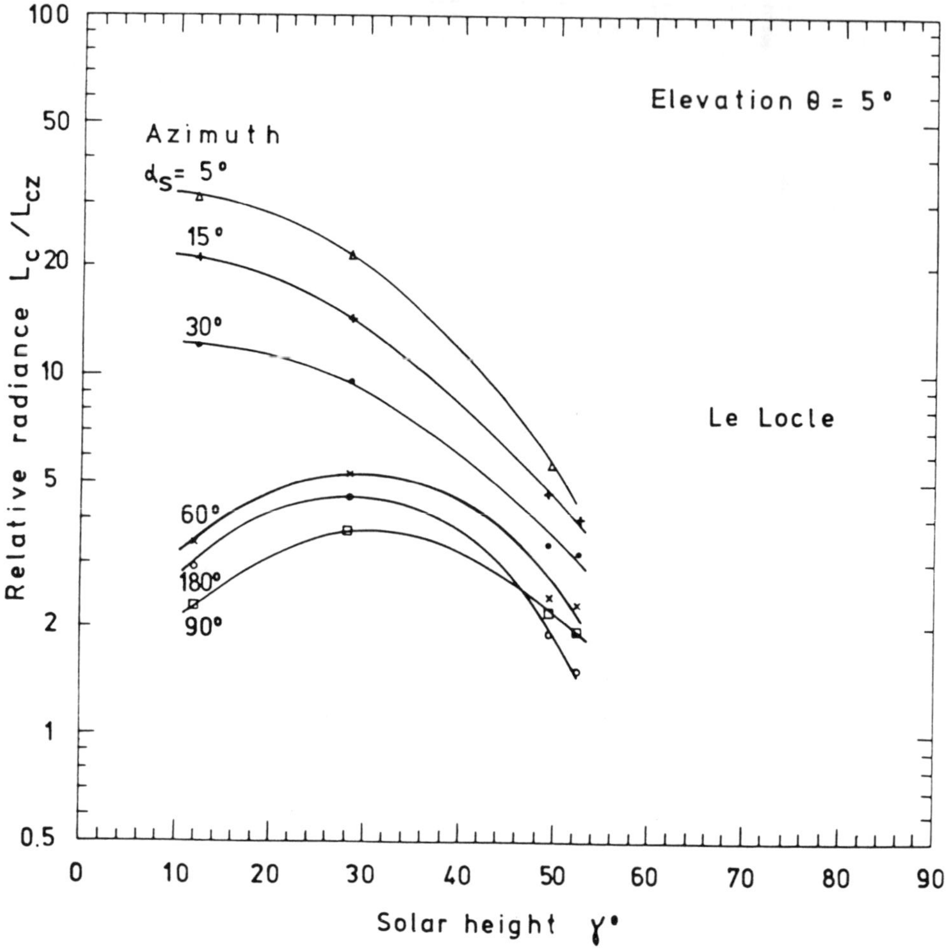

Figure 3.16c Normalized clear sky radiance L_c/L_{cz} at a low elevation in the sky as a function of solar height for different values of the azimuth α_s measured at an elevation angle of 5° at the snow covered mountain site Le Locle (1240 m a.s.l.).

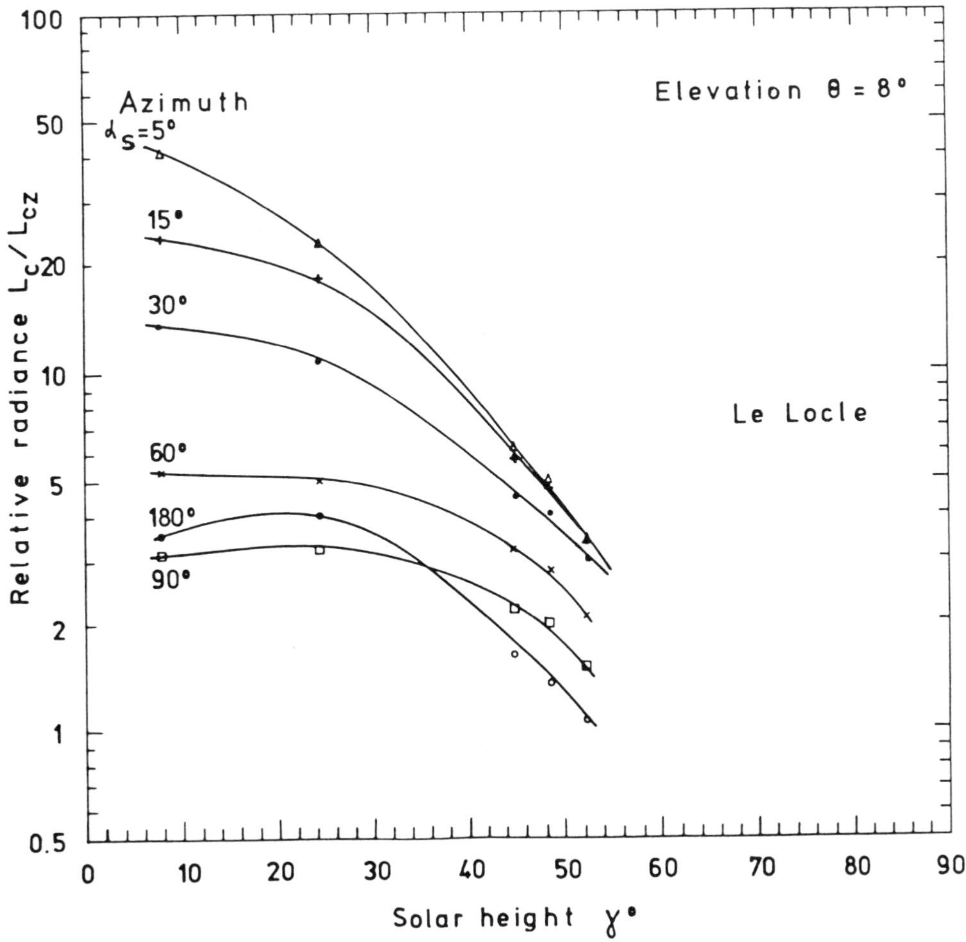

Figure 3.16d Normalized clear sky radiance L_c/L_{cz} at a low elevation in the sky as a function of solar height for different values of the azimuth α_s measured at an elevation angle of 8° at the snow covered mountain site Le Locle (1240 m a.s.l.).

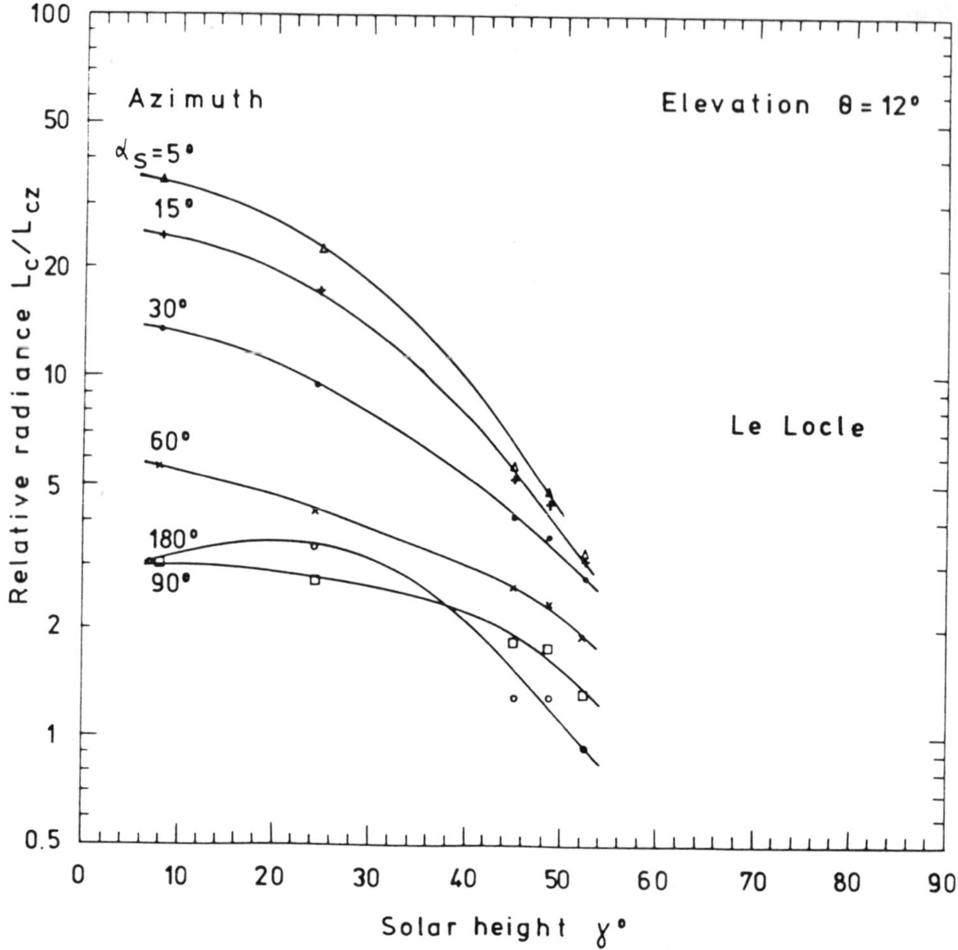

Figure 3.16e Normalized clear sky radiance L_c/L_{cz} at a low elevation in the sky as a function of solar height for different values of the azimuth α_s measured at an elevation angle of 12° at the snow covered mountain site Le Locle (1240 m a.s.l.).

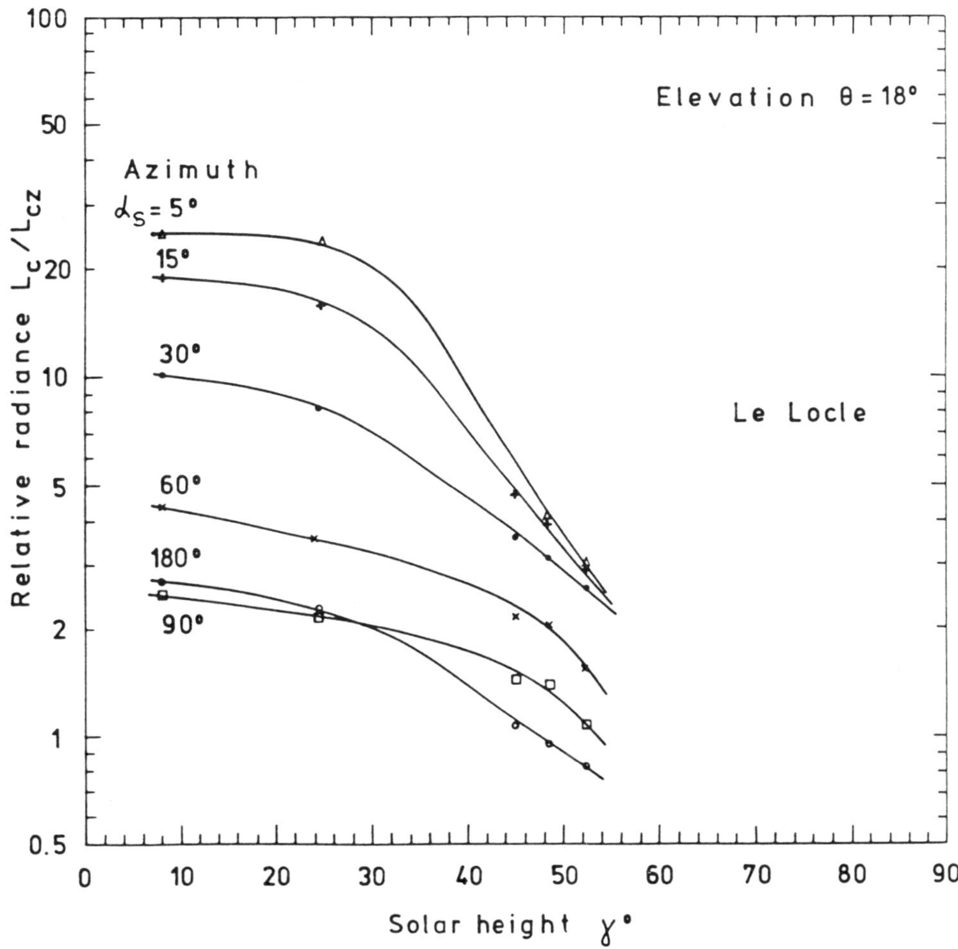

Figure 3.16f Normalized clear sky radiance L_c/L_{cz} at a low elevation in the sky as a function of solar height for different values of the azimuth α_s measured at an elevation angle of 18° at the snow covered mountain site Le Locle (1240 m. a.s.l.).

10. Current work on sky radiance distributions

Actually the problem stands close to a rather complete solution (15). Using the system of four pyranometers, more than 300 sets of measurements made under clear sky conditions at the four sites mentioned, could be included in the development of the multi-variate relationship of equation 3.2. Likewise the sky component alone should be capable of description using a relationship of the power:

$$\frac{D'_{cs}(\beta,\alpha_s)}{D'_c} = f(D_c/G_c,\ \gamma,\ \alpha_s,\ \beta) \tag{3.7}$$

where D'_{cs} and D'_c are defined by analogy as in Appendix 4, Equation A4.1.

It was found using the same more than 300 sets of clear sky cases that such relationships could be detected by studying the radiance data scanned by the silicon diode system.

Equations 3.2 and 3.7 contain only ratios, viz. normalised diffuse irradiances and these are, to a good approximation, free of any calibration effect of different instruments used (see Appendix 4). Therefore, since

$$\frac{D_{cs}(\beta,\alpha_s)}{D_c} = \frac{D'_{cs}(\beta,\alpha_s)}{D'_c}.$$

the relative contribution of the reflected component $R_c(\beta\ ,\ \alpha_s)$ may immediately be computed as

$$\frac{R_c(\beta,\alpha_s)}{D_c} = \frac{D_c(\beta,\alpha_s)}{D_c} - \frac{D'_{cs}(\beta,\alpha_s)}{D'_c} \tag{3.8}$$

The values of $R_c(\beta,\alpha_s)$ may thus be obtained for the spectral range of the pyranometers as

$$R_c(\beta,\alpha_s) = D_c\left[\frac{D_c(\beta,\alpha_s)}{D_c} - \frac{D'_{cs}(\beta,\alpha_s)}{D'_c}\right]\ \mathrm{Wm^{-2}} \tag{3.9}$$

and for the spectral range of the silicon diodes as

$$R'_c(\beta,\alpha_s) = D'_c\left[\frac{D_c(\beta,\alpha_s)}{D_c} - \frac{D'_{cs}(\beta,\alpha_s)}{D'_c}\right]\ \mathrm{Wm^{-2}} \tag{3.10}$$

Empirical solutions of equation 3.2 for the four sites differ from each other due to different local conditions of ground reflectivity. Results of these studies are given in Section III of this chapter.

In addition to deriving a bidirectional albedo index $\rho g(\beta,\alpha_s)$, a bidirectional index $N(\beta\ ,\ \alpha_s)$, characterising the angular distribution of clouds over the sky dome has been computed with respect to each of the 77 pyranometer surfaces. Based on the fish-eye photographs, this latter was

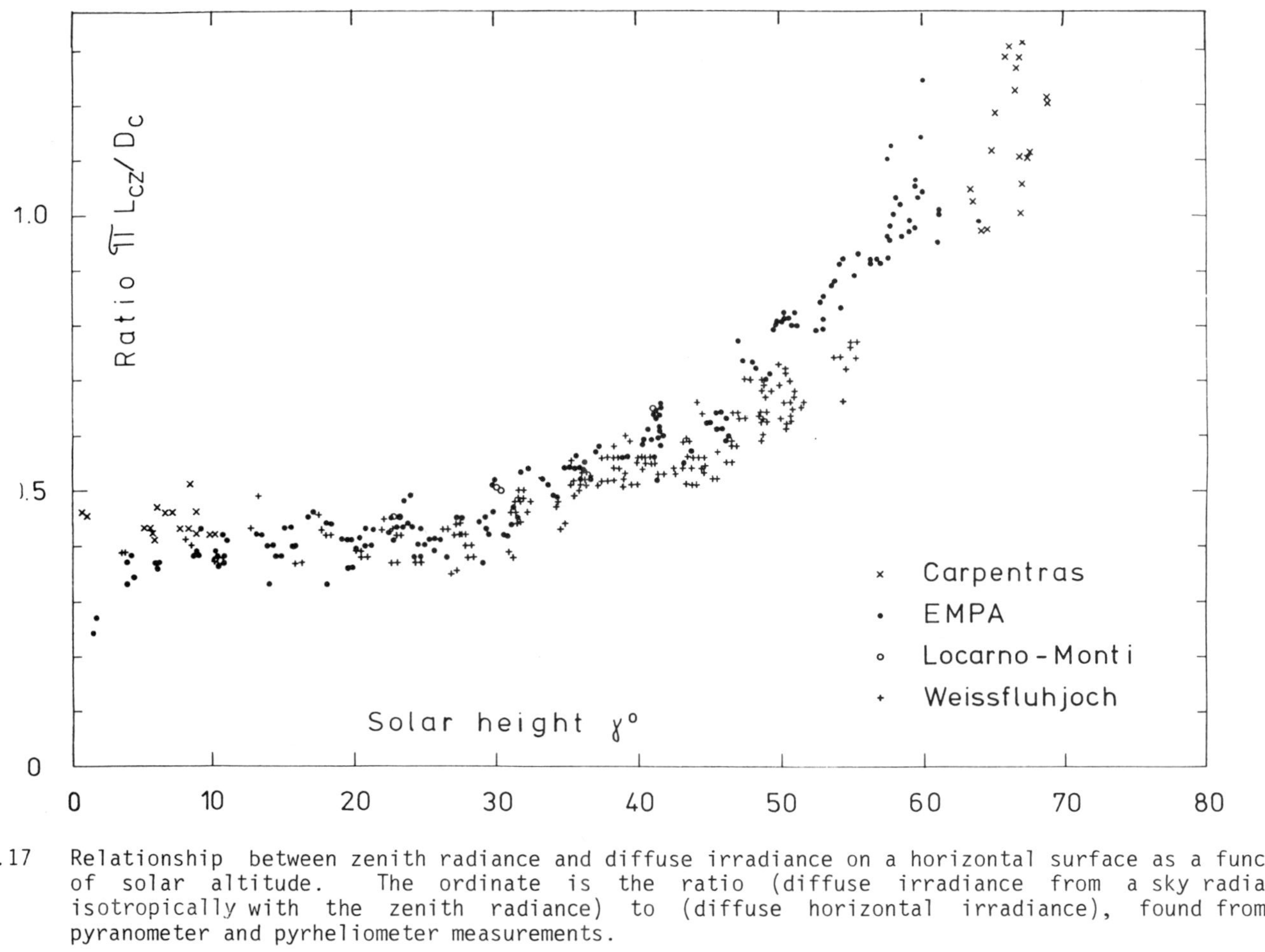

Figure 3.17 Relationship between zenith radiance and diffuse irradiance on a horizontal surface as a function of solar altitude. The ordinate is the ratio (diffuse irradiance from a sky radiating isotropically with the zenith radiance) to (diffuse horizontal irradiance), found from the pyranometer and pyrheliometer measurements.

computed in the same geometric way as the double integral for the sky radiances, but the $L_S(\Theta,\alpha_s)$ values have been replaced by cloud amount data for 24 angular segments of the sky dome. Fractions of the 24 segments, due to different intersections of the 77 inclined planes with the sky dome, have also been considered when deriving $N(\beta,\alpha_s)$.

The extension of equation 3.2 both by $N(\beta,\alpha_S)$ and also by the normalized index $N(\beta,\alpha_s)/N(\beta=0)$ is under way in order to develop irradiance distribution functions for cloudy skies, working separately for different types of clouds. Overcast sky irradiance distribution functions need only the cloud type as a further parameter.

11. Current work - effects of obstruction of horizon by terrain

The programme will be completed by computing an index which will take into account the impact of the natural horizon. This index must consider the different angular weight of the obscured parts of the sky with respect to the 77 pyranometer planes.

12. Conclusions from Swiss clear sky observational program

If there is access to the cloudless sky values of the global and diffuse irradiance on a horizontal surface, one can determine, for that place, for cloudless skies

- the angular distribution of sky radiance both in absolute units, as well as in relative units normalised to the radiance of the zenith;

- the sky component of diffuse irradiance on any surface of selected inclination to the horizontal and of selected orientation, by integrating numerically the estimated radiance over the sky hemisphere related to that surface.

CHAPTER 3 - PART II

THE CLEAR SKY RADIANCE MODEL USED AS THE BASIS FOR THE CEC PREDICTION PROCESS FOR ESTIMATING DIFFUSE IRRADIANCE ON INCLINED SURFACES ON CLEAR DAYS

13. General principles

The accurate prediction of the diffuse irradiance on slopes from the clear blue sky is a difficult problem scientifically because the radiance distribution of the clear sky is so complex as Part I of this Chapter has made very clear. As Figure 3.6c shows, the radiance of the brightest part of the sky can be over 20 times as high as the radiance of the dullest part of the sky. It is not surprising that the isotropic approximation,which assumes a uniform sky radiance, produces somewhat inaccurate estimates of diffuse slope irradiance from the clear sky.

Figure 3.6c shows there is a zone of relatively high radiance immediately around the solar disc, the radiance of which falls off approximately exponentially as one moves away from the edge of the solar disc The lowest radiance occurs at a point at 90° to the sun along a line facing directly away from the sun, i.e. on a meridian of 180° to the sun's meridian. As the atmospheric turbidity increases, the radiance of the sky increases, but the zone of high brightness around the solar disc, in particular, increases proportionally more. At lower turbidities, scattering close to the horizon produces a band of relatively high radiance around the horizon which is observed as a horizon whitening relative to the deeper blue above. This relative horizon brightening effect decreases as the turbidity increases, and, at high turbidities, actually reverses, i.e. the horizon becomes a zone of low radiance relative to sections of the sky above at greater angles of elevation.

Any scientifically based computational method for estimating clear day diffuse irradiance requires that some suitable relative sky radiance distribution model is available for assessing the integrated impacts of the radiance patterns encountered in practice in order to make it possible to relate diffuse slope irradiance values accurately to the observed or predicted horizontal irradiance values.

14. The original Berlin model - use of the CIE clear sky model

The original Berlin model for estimating clear day diffuse sky irradiance on slopes was based on international clear sky daylighting practice and used the Kittler model (16) which was originally evolved for relatively clear conditions of low turbidity. A model covering more turbid conditions was developed by Gusev (17). The problem of both these models is that they are not turbidity sensitive models.

The Kittler formulation is:

$$\frac{L_c(\Theta, \alpha_s)}{L_{cz}} = \frac{(0.91 \quad + 10e^{-3\eta}+0.45 \cos^2 \eta)(1 - e^{-0.32/\sin \Theta})}{(0.91 + 10e^{-3\xi}+ 0.45 \cos^2 \xi)(1 - e^{-0.32})} \qquad (3.11)$$

stated to be for Linke Turbidity Factors below 5.

The Gusev formulation is:

$$\frac{L_c(\Theta, \alpha_s)}{L_{cz}} = \frac{(0.856 + 16e^{-3\eta} + \ 0.3 \cos^2 \eta)(1 - e^{-0.32/\sin \Theta})}{(0.856 + 16e^{-3\xi} \ + \ 0.3 \ \cos^2 \xi)(1 - e^{-0.32})} \qquad (3.12)$$

stated to be for Linke Turbidity Factors around 8, i.e. rather polluted environments

where $L_c(\Theta, \alpha_s)$ = Luminance at specific point p in the sky at angle Θ from the horizontal plane at azimuth angle of α_s from the sun's direction

L_{cz} = Clear sky zenith luminance

η = Angle in radians between the element of the sky at point p and the centre of the solar disc

ξ = Zenith distance in radians of sun (angle between the centre of the solar disc and the zenith)

Θ = The angle between the element of the sky at point p and the horizontal plane.

α_s = Angle between the vertical plane containing the normal to the surface, and the plane of the sun's meridian.

The exponential term in each line is associated with the forward scattering within the solar aureole. The factor weighting this term increases in the Gusev formula which is associated with a high turbidity. The exponential term in the second bracket allows for multiple scattering within the atmosphere which becomes more important towards the horizon.

15. Liebelt turbidity sensitive sky luminance model for cloudless skies

Liebelt (18) has produced an alternative luminance formulation in terms of Linke Turbidity Factor T_L which appears more rational in terms of physical concepts. It is also more suitable in terms of mathematical modelling.

Liebelt's formulation is:

$$\begin{aligned}\text{Log}(L_c(\Theta,\alpha_s)/L_{cz}) = {} & \log(1 - e^{-0.088m.T_L(\gamma)}) + \log(1 + P(20)(e^{-3\eta} - 0.009) \\ & + P(21)\cos^2\eta) - \log(1 - e^{-0.088\ .T_L(\gamma)}) \\ & - \log(1 + P(20)(e^{-3\xi} - 0.009) + P(21) \cos^2 \xi) \qquad (3.13)\end{aligned}$$

where the main symbols are the same as for equations above and where P(20) and P(21) are parameters dependent on turbidity, m is the optical air mass for element (Θ, α_s) of the sky and $T_L(\gamma)$ the Linke Turbidity Factor. One should note in making comparisons with equations 3.11 and 3.12 that m = 1/sin Θ. The optimal parameters reported by Liebelt were:

Turbidity Interval	Optimal Parameter P(20)	P(21)
$3.50 < T_L(\gamma) < 3.95$	8.117	0.276
$4.50 < T_L(\gamma) < 4.70$	9.673	0.491
$5.40 < T_L(\gamma) < 5.70$	11.409	0.338
$6.40 < T_L(\gamma) < 6.70$	14.265	0.457

By fitting continuous functions to Liebelt's values for P(20) one may obtain a suitable function for P(20) to aid computation. Thus:

$$P(20) = 0.6155 + 1.9687\ T_L(\gamma)$$

Allowing some extrapolation one might put on limits as $2 < T_L(\gamma) < 8$. The problem with Liebelt's original formulation is the random variation of the parameter P(21) which makes it unsuitable for mathematical modelling purposes. It should also be stressed that these models are all based on luminance observations and not on radiance observations. They provided, however, the ideas for the development of the model used to produce clear day tables in the CEC Inclined Surface Atlas for Europe.

16. Development of a sky radiance model for cloudless skies

A thorough scientific review of the literature on relative radiance distribution was first carried out and a new clear sky relative radiance model evolved which used the principles of the Liebelt formulation, but drew heavily on the work of Steven and Unsworth (19) on the observed relative radiance distribution of the clear sky at Sutton Bonington, near Nottingham in the UK to assess the values of the constants. The proposed relationship took the form

$$\frac{L_c(\Theta,\alpha_s)}{L_{cz}} = \frac{(1 - e^{-0.088\ T_L(\gamma)/\sin\Theta})(a_1 + a_2(\exp(-3\eta\ .\pi/180) + a_3 \cos^2 \eta)}{(1 - e^{-0.088\ T_L(\gamma)})(a_1 + a_2(\exp(-3\xi\ .\ \pi/180) + a_3 \cos^2\xi)} \tag{3.14}$$

where $L_c(\Theta,\alpha_s)$ = radiance at specific point of sky at angle Θ from horizontal plane at an azimuth angle α_s from sun's meridian

L_{cz} = zenith radiance

η = angle in degrees between sun's position and selected point of sky

$T_L(\gamma)$ = Linke Turbidity

ξ = Zenith distance in degrees (angle between centre of solar disc and zenith)

α_s = Azimuth angle of surface from the sun's meridian

Θ = Angle between the element of the sky at point p and the horizontal plane

and where $a_1 = 0.8995 - 0.0053\ \gamma$, $a_2 = 0.6155 + 1.9687\ T_L(\gamma)$ and $a_3 = 0.409 - 0.0096\ \gamma$.

17. Use of radiance model to estimate ratios of the diffuse radiation on slopes for the diffuse radiation on a horizontal plane

Chapter 2 set out details of the model adopted to assess values of the clear sky diffuse irradiance on the horizontal plane. Formula 3.14 may be used to calculate the ratio of the clear sky diffuse sky radiation on an inclined surface of slope β, and azimuth angle α_S in relation to the sun, to the horizontal value, $D_{cs}(\beta,\alpha_s)/D_c$.

The principles of the basic integration process are set out in Appendix 4, but the integration pattern used was different. In the numerical integration process adopted to calculate the ratio $D_{cs}(\beta,\alpha_s)/D_c$. the sky was divided into 5° zones of azimuth angle α_s and 5° zones of sky patch altitude , i.e 36 x 2 x 18 i.e. 1296 patches: as there is a symmetry about the solar meridian, the integration had only to be carried out for 648 patches of sky. Nevertheless the process of numerical integration is a long one and unsuitable for rapid computation.

With this difficulty in mind, a far more rapid but nevertheless still fairly complex method for estimating the ratio $D_{cs}(\beta,\alpha_s)/D_c$ from the values of $T_L(\gamma)$ was developed. This is described in Appendix 5, which gives full details of the algorithms developed to produce the CEC inclined surface solar tables for Europe.

As $T_L(\gamma)$ is a defined function of T_L and γ, it is possible to develop the relationship $D_{cs}(\beta, \alpha_s)/D_c$ as a function of T_L, the air mass 2 Linke Turbidity Factor.

Chapter 6 contains tables of the ratio $D_{cs}(\beta,\alpha_s)/D_c$. computed using air mass 2 Linke Turbidity Factors of 1.5, 2, 3, 4, 5, 6, 7, 8, 9 and 10, which represents the upper practical limit for clear skies. These values of $D_{cs}(\beta, \alpha_s)/D_c$ are given in Chapter 6, Tables 6.14(a) to 6.14(j), which are designed for practical computation. The basic solar geometry has first to be established to use the tables, refer Appendix 1. The azimuth angle of surface α has then to be converted to an azimuth angle related to the direction of the sun's meridian α_s as follows:

$$\alpha_s = \alpha - \psi \qquad (3.15)$$

where ψ is the solar azimuth angle as defined in Appendix 1,

α is the surface azimuth angle measured from due south, east negative, west positive.

The ratio $D_{cs}(\beta, \alpha)/D_c$ has been designated as the correction function f_2, where

$$D_{cs}(\beta,\alpha) = f_2\ D_c \qquad (3.16)$$

18. Comparison with the isotropic model

One may compare the results in Tables 6.14(a) - 6.14(j) with the predictions of an isotropic model which assumes the sky of uniform radiance. The differences are greatest for vertical surfaces. The isotropic value of the ratio f_2 for vertical surfaces will be 0.5 for all orientations. Figure 3.18 compares the difference for a typical air mass 2 Linke Turbidity Factor of 4. At low solar altitudes, the value of f_2 is about 3 times the isotropic value for vertical surfaces facing the sun falling to less than half the isotropic value on surfaces facing away from the sun. Even at 60° solar elevation the value on surfaces facing the sun is .73 i.e. a ratio increase of (.73/.50) while on surfaces facing away from the sun the value is .31, i.e. a ratio decrease of (.31/.50). It clearly makes a considerable difference to take proper account of the sky radiance distribution. The next section reports tests of the model results against actual field observations. However, to make such checks, one has first to assemble a global radiation model for clear days from the components discussed to this stage.

19. Checking the slope model against observations for cloudless days

No observations of diffuse radiation from the sky alone were available for Europe, so the diffuse model could not be checked directly. The checks had, therefore, to be carried out using global irradiance/irradiation measurements. Combining the results of Chapter 2 and the preceding discussion in this Chapter, and adding a component for ground reflected radiation, the global irradiance at an inclined plane at any instant can be estimated for cloudless skies as:

$$G_c(\beta,\alpha) = I_c \cos \nu + f_2 D_c + 0.5 \rho_g G_c(1 - \cos \beta) \qquad (3.17)$$

where I_c is the clear sky beam irradiance normal to the beam, Wm^{-2}

ν is the angle of incidence (refer Appendix 1)

f_2 is the ratio $D_{cs}(\beta,\alpha)/D_c$.

D_c is the clear sky diffuse irradiance on the horizontal plane, Wm^{-2}

ρ_g is the ground albedo

G_c is the clear sky global irradiance falling on the ground, Wm^{-2}

β is the slope tilt from the horizontal plane

α is the azimuth angle of the slope measured from due south.

The issue of the ground contribution is discussed in greater detail in Part III of this Chapter. The computation at this stage assumes a ground of uniform brightness.

Once a global model is available, one can check the value $G_c(\beta,\alpha)$ predicted with the value $G_c(\beta,\alpha)$ observed. On surfaces facing away from the sun the first term of equation 3.17 disappears, so the modelling checks the accuracy of the assumptions about the sky and the ground in combination. This introduces an ambiguity in the checking process, as Part III of this chapter makes clear.

Three sets of measurements were used for checking the clear sky slope model.

1. A set of 13 selected clear day hourly data for Cabauw, Netherlands for the summer of 1979.

2. A set of 41 selected clear day hourly data for Bracknell, UK, for days in January, April, July, October. The vertical Bracknell data were measured with pyranometers screened from ground reflections.

3. Various results from Valko, both those reported in part one of this chapter, and elsewhere.

20. Clear day slope checks of the CEC model for Cabauw

As part of the CEC program on solar radiation data, an extensive measurement network was set up at Cabauw (20) to observe solar radiation on inclined planes and on horizontal surfaces. Hourly data for 13 selected clear days for Cabauw were supplied for testing purposes by Slob (21). The selected data from the data sets used in these checks were the horizontal and diffuse irradiation measurements made with pyranometers, the direct beam irradiation measurements made with a pyrheliometer with the associated derived Linke Turbidity Factors, and the observed data of hourly global irradiation on vertical south, east, west and north facing surfaces and 45° inclined surfaces facing east, south and west. The position of the sun expressed as solar altitude and azimuth angle were also included in each data sub-set. The data set also contained information about the proportion of each period during which the sunshine calculated from the pyrheliometer reading exceeded an intensity of 140 Wm^{-2} and also 200 Wm^{-2} during that hour as well as actual sunshine observations from a nearby site. As a first stage in the data selection process, only hours with a relative sunshine fraction of 1.0 computed for 140 Wm^{-2} and at least 0.9 computed for 200 Wm^{-2} threshold were accepted. This selection process produced 136 hourly sets of observations, each set consisting of 13 hourly pieces of information,

i) G_c, I_c, D_c, $T_L(\gamma)$,

ii) γ and ψ, $G_c(90,0)$, $G_c(90,90)$, $G_c(90,180)$, $G_c(90,270)$,

iii) $G_c(45,90)$, $G_c(45,180)$ and $G_c(45,270)$,

There were observed data for other slopes but they were not used as such checks are very time consuming.

The first step was to test the horizontal surface cloudless day model against the derived values of D_c. The observed diffuse irradiance at the start of the data series was estimated in one way only, by subtracting the resolved component of the direct beam irradiation measured by the pyrheliometer from the global horizontal surface irradiation measured by the pyranometer. For the second part of the data set, the diffuse irradiation was also observed by a pyranometer shielded by a shadow band, so, for about half the data set, there were two independent estimates of the diffuse horizontal irradiation available. All the selected hourly data were typed into a microcomputer file, and a program produced to compare

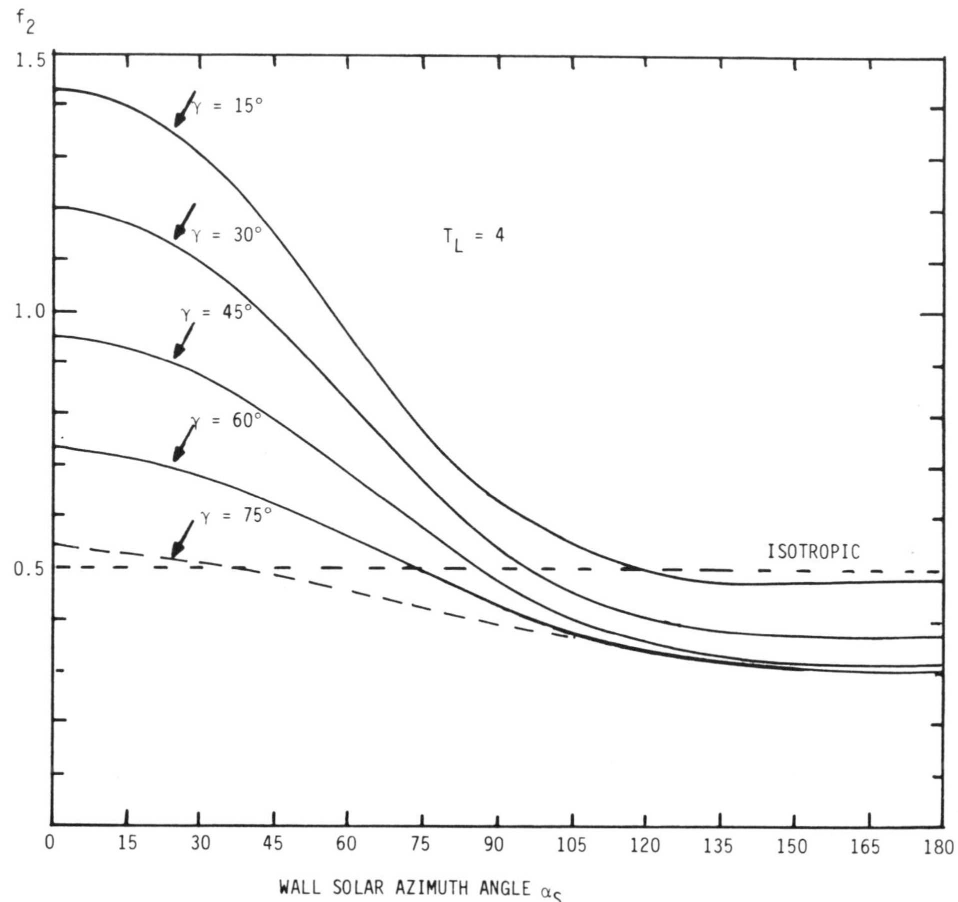

Figure 3.18 The relationship between $D_c(90,\alpha_s)$ and D_c predicted, using the CEC clear sky model as a function of wall solar azimuth angle for solar elevations of 15°, 30°, 45°, 60° and 75° for T_L = 4. The 75° degree values are tentative. The difference from the isotropic approximation is clear.

observation with prediction using the standard CEC model adopted to produce the final slope irradiation tables for the Atlas. The first checks were made by using the observed value of the Linke Turbidity Factor $T_L(\gamma)$ to estimate D_c. These values were then compared with the observed values of D_c. The ratio (D_c predicted/D_c observed) was computed. The general results of this part of the study have already been discussed in Chapter 2. It appeared, however, that some of the values of D_c derived were unreasonably low. For part of the series, however, independent shading band pyranometer measurements of D_c were also available. It was found that these apparently erroneous readings showed up very clearly in the ratio of D_c (derived from global - pyrheliometer resolved beam)/D_c (shadow band pyranometer), i.e. the shadow band correction factor. This ratio should have a value of 1.10 to 1.20 or so. However, sometimes it was less than 1.0, which was clearly wrong. All hours were then rejected where the shadow band ratio was less than 1.06. Further inspection of the data showed some hours in the daily diffuse data series were exceptionally high compared with their adjacent values, and it appeared some cloud must have been present for short times during these hours. These hours of excessively high diffuse radiation were also eliminated. This left a data set for 98 hours out of 136 selected cloudless hours originally entered into the computer. Details about the selected days are given in Table 3.3.

The next stage was to use the observed value of $T_L(\gamma)$ together with information about the solar geometry, direct beam irradiance and diffuse sky horizontal irradiance to calculate the expected global radiation on any of the inclined planes for which observations had been put into the computer. The predicted global estimate was thus the sum of three components, i) the measured beam resolved onto the plane in question, ii) the computed sky diffuse computed from the observed horizontal diffuse radiation and the slope correction factor for the appropriate slope, value of $T_L(\gamma)$ and solar altitude, and iii) the ground reflected component. Slob had reported albedos of around 0.23 at this site, using downward facing pyranometers. It was decided at first to use 0.20, but, further on in the vertical surface checks, an albedo value of 0.25 was adopted as realistic for the grass covered site in question.

Each predicted value of $G(\beta,\alpha)$ was compared with the observed value and their ratios systematically stored in a computer file together with the other basic input data. Another program was then written to sort the data by 10 degree bands of solar altitude, and 45° zones of wall solar azimuth angle, 0-45°, 45°-90°, 90°-135° and 135° to 180°. The means of the errors and the standard deviations for each level of altitude and azimuth angle were then extracted, so the precise pattern of errors could be examined.

The results of these studies are presented in Table 3.4 and Table 3.5 and are illustrated in Figure 3.19 and Figure 3.20. In order to compare the results with the isotropic predictions, a similar set of files for vertical surfaces were built up using the isotropic assumption for estimating the sky diffuse, i.e. $D_c(\beta,\alpha) = \frac{1}{2}D_c$.

The analysis shows the EC cloudless day model used to produce the inclined surface tables is relatively accurate. There is a tendency to slightly underestimate the diffuse radiation in sectors facing away from the sun compared with measurements with these particular pyranometers. The errors are greatest at lowest solar altitudes as might be expected.

Table 3.3 Basic information about Cabauw clear day data set for summer of 1979 used for clear day slope checks.

Day ref. number	Date	Max. solar alt. degrees	Min* $T_L(\gamma)$	Max* $T_L(\gamma)$	Number of observations selected	Number meeting refined criteria
1	14/4/79	46.9	4.1	7.7	10	8
2	10/5/79	55.2	4.2	7.0	11	9
3	15/5/79	56.5	5.8	9.0	11	11
4	19/5/79	57.4	3.7	6.2	13	9
5	16/6/79	61.1	3.9	7.4	12	8
6	12/7/79	59.7	3.9	5.2	14	13
7	29/8/79	47.5	5.8	6.3	11	7
8	30/8/79	47.1	4.8	5.4	11	7
9	6/9/79	44.6	4.1	4.8	10	6
10	16/9/79	40.9	3.4	4.3	10	7
11	21/9/79	38.9	4.0	5.0	7	3
12	29/9/79	35.8	5.2	8.3	7	4
13	30/9/79	35.5	3.8	5.0	9	6
					136	98

* For hours selected on basis of refined criteria.

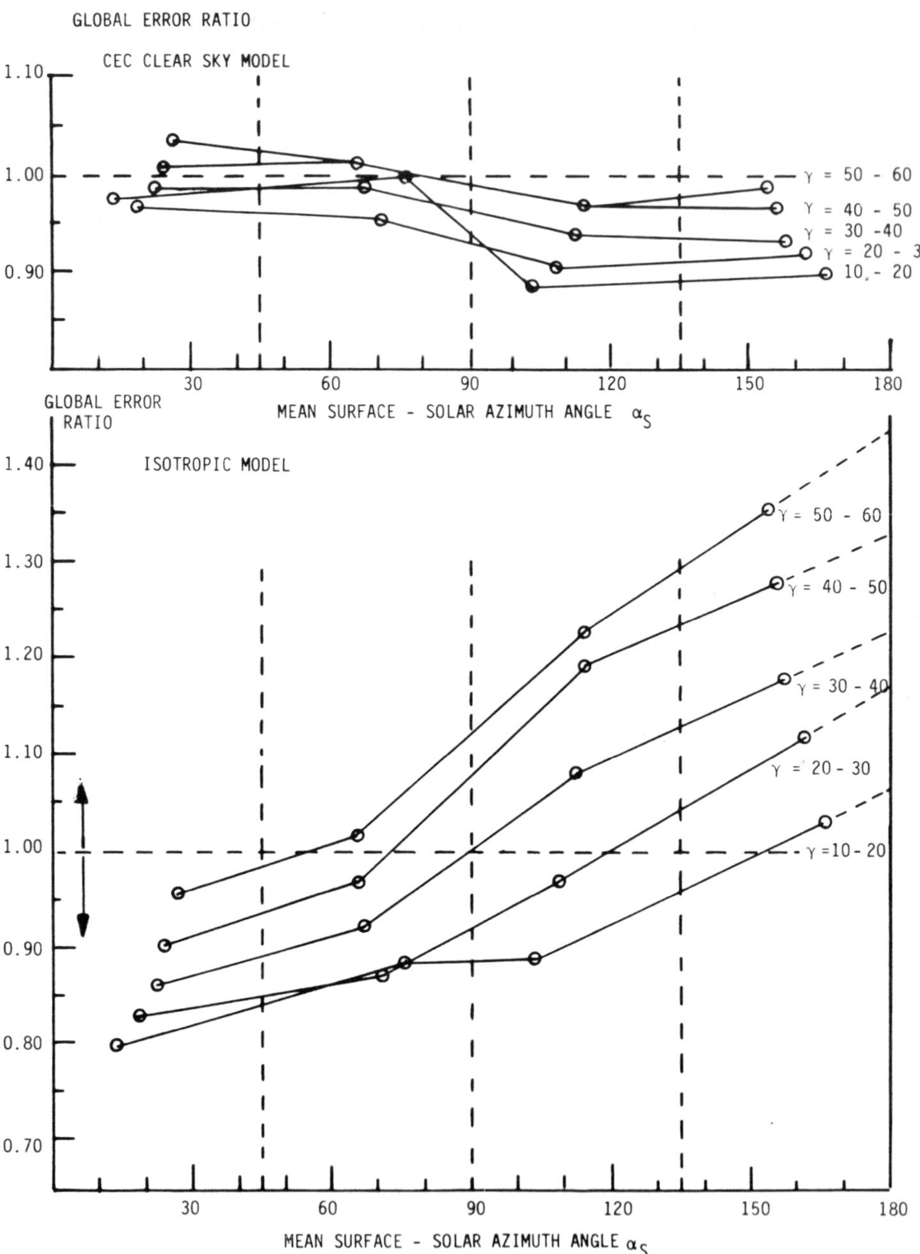

Figure 3.19 Prediction errors in global radiation on vertical surfaces for selected cloudless hours at Cabauw as a function of wall solar azimuth angle α_s for different solar altitudes. Top CEC model. Bottom isotropic model.

Table 3.4 Error analysis of predicted hourly global solar radiation on vertical surfaces for cloudless days at Cabauw. ε = mean error for altitude and azimuth band (predicted hourly global radiation/observed hourly global radiation) on slope
a) 90° vertical, all cloudless hours, albedo = 0.25
b) 90° vertical, selected cloudless hours, albedo = 0.25
c) 90° vertical, isotropic, albedo = 0.25.

Solar altitude band degrees	Mean solar altitude degrees		Azimuth 0-45°			Azimuth 45-90°			Azimuth 90-135°			Azimuth 135-180°		
			Mean solar azimuth degrees	ε	S.D.	Mean solar azimuth degrees	ε	S.D.	Mean solar azimuth degrees	ε	S.D.	Mean solar azimuth degrees	ε	S.D.
10-20	15.7	a	15.0	.949	.039	75.0	.950	.081	105.0	.842	.101	165.0	.856	.089
	16.0	b	13.8	.975	.044	76.2	.997	.062	103.8	.883	.029	166.2	.899	.063
	16.0	c	13.8	.800	.035	76.2	.884	.062	103.8	.887	.053	166.2	1.034	.142
20-30	25.2	a	19.8	.958	.038	70.2	.939	.079	109.8	.873	.082	160.2	.884	.078
	25.7	b	18.8	.969	.028	71.3	.952	.065	108.8	.903	.045	161.3	.920	.050
	25.7	c	18.8	.828	.031	71.3	.870	.045	108.8	.971	.096	161.3	1.115	.133
30-40	34.9	a	23.8	.983	.028	66.2	.986	.055	113.8	.920	.072	156.2	.918	.064
	35.1	b	22.3	.987	.022	67.7	.989	.053	112.3	.937	.063	157.7	.931	.057
	35.1	c	22.3	.864	.043	67.7	.921	.054	112.3	1.081	.119	157.7	1.178	.104
40-50	45.0	a	25.1	1.013	.025	64.9	1.015	.050	115.1	.967	.057	154.9	.961	.062
	45.0	b	24.1	1.009	.021	65.9	1.012	.048	114.1	.971	.057	155.9	.965	.062
	45.0	c	24.1	.904	.030	65.9	.966	.063	114.1	1.192	.120	155.9	1.275	.092
50-60	55.2	a	25.4	1.038	.018	66.2	1.009	.062	113.8	.963	.097	154.6	.984	.092
	55.3	b	26.1	1.037	.019	65.7	1.011	.060	114.3	.966	.100	153.9	.989	.092
	55.3	c	26.1	.955	.037	65.7	1.016	.066	114.3	1.229	.106	153.9	1.353	.104
Means for all solar altitudes	34.9	a	21.6	.987	.043	68.6	.980	.071	111.4	.915	.092	158.4	.921	.087
	38.8	b	27.0	.999	.033	68.3	.994	.058	111.7	.944	.070	158.0	.949	.069
	38.8	c	27.0	.882	.058	68.3	.941	.075	111.7	1.105	.147	158.0	1.217	.141

Table 3.5 Error analysis of predicted hourly global solar radiation on 45° planes for cloudless days at Cabauw. ε = predicted hourly global radiation/observed hourly radiation. Albedo .20.

Solar altitude band degrees	Sector	Mean solar altitude degrees	Mean solar azimuth degrees	Error G	Standard deviation
10-20	0- 45	16.0	13.8	.967	.033
	45- 90	18.1	81.0	1.059	- 1 obs only 75.6 .9501
	90-135	15.6	104.8	.911	.061
	135-180	16.0	166.2	1.057	.045
20-30	0- 45	25.2	19.8	.954	.036
	45- 90	26.1	69.7	.990	.040
	90-135	25.1	107.5	.927	.126
	135-180	25.4	162.4	1.004	.062
30-40	0- 45	35.1	22.3	.965	.016
	45- 90	35.1	67.7	.990	.035
	90-135	35.8	114.3	.973	.080
	135-180	34.1	160.1	.909	.053
40-50	0- 45	45.0	24.1	.985	.023
	45- 90	45.0	65.9	1.001	.027
	90-135	45.2	110.5	1.027	.045
	135-180	44.6	150.5	.909	.117
50-60	0- 45	55.3	26.1	.993	.009
	45- 90	55.3	65.7	1.007	.018
	90-135	55.7	113.7	1.017	.026
	135-180	52.7	135.5	1.024	- 1 obs only 138.3 .992
Mean of all hours	0- 45	38.8	22.0	.976	.022
	45- 90	40.5	67.7	.998	.029
	90-135	39.5	111.0	.970	.107
	135-180	32.3	158.3	.961	.091

However, one must remember this is the situation which produces the biggest pyranometer errors. Not only are the mean errors much less than with an isotropic model but the range of standard deviations is much less as well. The accuracy achievable with the EC model if the direct beam intensity is known, as it was here, is much greater for surfaces facing the sun than on surfaces facing away from the sun. Thus, on surfaces where there is most energy, the prediction errors are least. The adoption of a radiance based model for cloudless sky slope radiation predictions is thus fully supported by the excellent results emerging from these detailed comparisons of prediction and observations.

The isotropic model, on the other hand, gives large errors in the global radiation on slopes under cloudless conditions. The errors on surfaces facing the sun are greatest with the low sun. The errors on surfaces facing away from the sun are greatest with the high sun. The relatively large standard deviations encountered with the isotropic model have already been commented upon.

21. Bracknell vertical surface observation checks - cloudless days

Section 2.7.2 discussed the selection of the Bracknell cloudless day data used to check the horizontal surface diffuse irradiance model. This data set also included corresponding observed values of the hourly global irradiation on vertical surfaces facing north, east, south and west. The vertical irradiation was measured with Kipp and Zonen CM2 pyranometers which do not have a particularly good cosine response. These pyranometers are screened from ground reflections with a blackened shading mask which was assumed to have an albedo of 0.02. The data set used is outlined in Table 3.6. The horizontal diffuse values obtained using shading bands were corrected using the Painter methodology discussed in Section 2.7.2.

The value of $T_L(\gamma)$ was evaluated using $(G_c - D_c)/\sin\gamma$ where γ is the solar altitude at the middle of the hour in question. Then, using the EC methodology, the ratio $D_c(\beta, \alpha)/D_c$ was calculated in order to find $D_c(\beta,\alpha)$ from the shadow band corrected value of D_c. Finally, the direct beam component $(G_c - D_c) \cos \nu /\sin \gamma$, and the reflected component $.01/G_c$ were added to obtain the estimated global radiation on the vertical plane for each of the four vertical surfaces. This value was then compared with the corresponding observed value. The data set of error ratios was then divided into 10° bands of solar altitude, 10° - 20°, 20° - 30°, 30° - 40°, 40° - 50°, 50° - 60° and 60°+, and 4 bands of wall solar azimuth angle 0° - 45°, 45° - 90°, 90° - 135°, 135° - 180° and the mean errors (predicted/observed) extracted together with the mean values of the solar altitude and wall solar azimuth for each band.

Table 3.7 gives the numerical results of these vertical surface checks. The model appears to over-estimate slightly the global irradiation on surfaces facing the sun. This may be the result of applying the Painter additional shadow band correction which is less than the EC recommended correction for cloudless conditions. The results are plotted in Figure 3.4. It will be seen that the range of errors increases for surfaces facing away from the sun. Here the effects of solar altitude are systematic. The albedo of the shading screens was set at 0.02. Back-reflections, however, could be of some importance for vertical surfaces facing away from the sun, especially with a high sun. If the

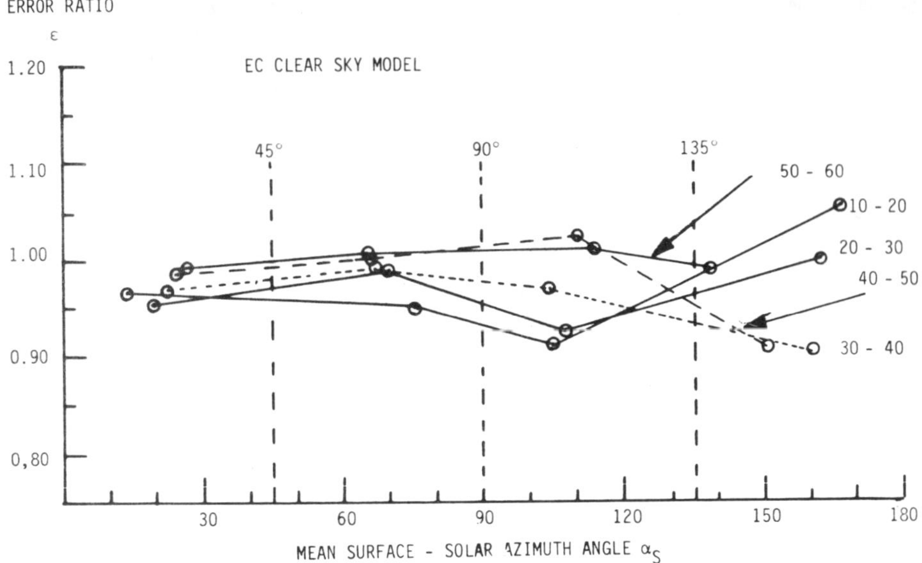

Figure 3.20 Prediction errors of global solar radiation on 45° planes for selected cloudless hours at Cabauw as a function of wall solar azimuth angle α_s for different solar altitudes.

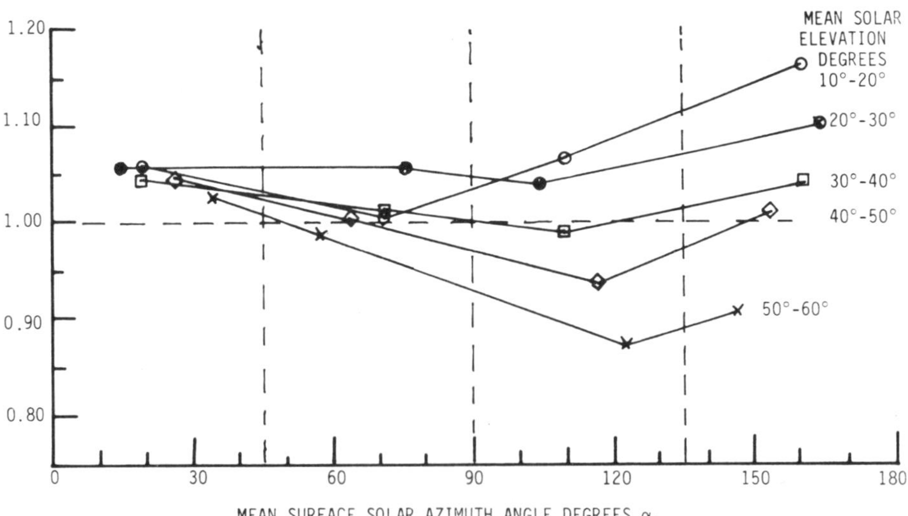

Figure 3.21 Prediction errors in global solar radiation on vertical surfaces for selected cloudless hours at Bracknell as a function of wall solar azimuth angle α_s for different solar altitudes.

Table 3.6 Data set used for checking vertical surface predictions with observations at Bracknell for cloudless conditions, using observed vertical hourly irradiation on vertical north, vertical east, vertical west and vertical south surfaces.
Selected criteria: Sunshine in hours = 1.0 hours
Cloud amount = 0 or 1 okta

January			
1/1/1967	4 hours	22/1/1973	4 hours
3/1/1967	5 hours	24/1/1971	4 hours
6/1/1970	6 hours	29/1/1974	5 hours
21/1/1972	6 hours		
		Total	34 hours
April			
6/4/1969	10 hours	20/4/1976	12 hours
14/4/1968	3 hours	24/4/1975	10 hours
16/4/1974	4 hours	25/4/1972	8 hours
17/4/1971	3 hours	27/4/1973	5 hours
		Total	55 hours
July			
2/7/1976	14 hours	9/7/1971	10 hours
3/7/1967	3 hours	10/7/1967	7 hours
4/7/1969	6 hours	30/7/1967	5 hours
9/7/1967	4 hours	30/7/1975	11 hours
		Total	60 hours
October			
1/10/1970	3 hours	22/10/1967	3 hours
14/10/1972	4 hours	27/10/1971	8 hours
21/10/1972	8 hours		
		Total	26 hours
Overall total:			175 hours

screens back-reflect 4% of the incident global radiation at a solar elevation of 60° instead of the 2% assumed, the underprediction at a global horizontal irradiance of 800 Wm^{-2} elevation would be 8 Wm^{-2} compared with a typical observed vertical value of 78 Wm^{-2} on a surface facing away from the sun, i.e. an underprediction value of 0.90. The results, therefore, seemed satisfactory in the light of the relative early model of pyranometer used, and the limitations imposed by the use of blackened shading masks. Once again the importance of shadow band corrections in diffuse radiation climatology is underlined.

22. Checks of EC model against Valko's results presented in Table 3.1

The final checks on the validity of the clear sky radiance model used to produce the EC inclined surface tables were carried out on the data presented by Valko in Table 3.1 of this Chapter.

The EC model was used to produce estimated values which corresponded to the various observed values summarised in Table 3.1 by Valko. A standard ground albedo of 0.2, as in the EEC method was adopted, except for the mountain site of Weissfluhjoch which was reported to have been partially snow covered. An estimate was made for the ground albedo on the Weissfluhjoch as this data was missing from the Valko data set. A value of 0.38 was used for reasons explained later.

The pyrheliometer observation was first used to estimate the Linke Turbidity Factor $T_L(\gamma)$ using the EC methodology. These values proved slightly different from Valko's which were indirectly derived from the Schuepp B_s. The EC clear sky model was then used, adopting this value of $T_L(\gamma)$, to produce all the corresponding quantities to those in Table 3.1. Examination of Table 3.8 shows the results of this comparison between observation and prediction.

The main differences appear in the estimates of the horizontal diffuse irradiation D_c at higher turbidities, and in the assessed radiance of the ground. The low estimate of D_c is possibly the consequence of the fact that the CEC clear sky model described in Chapter 2 was developed empirically from a set of data obtained with pyranometers located in various parts of Europe shaded with shading bands, and for the most part corrected using only the iostropic correction, a process which would produce a slight underestimate of clear sky diffuse radiation.

The third section of this Chapter discusses the issues of the radiance of the ground in some detail. While Valko provides values of R_C/D_C, the downward facing albedo, these results were achieved over roofs of buildings. The albedo at the nadir can be estimated from the stated radiance and the global horizontal irradiance G_C. The nadir albedo values found for Locarno Monti are 0.10 and for EMPA, 0.089, 0.079, 0.085 for cases III, IV and V respectively. These nadir albedos are very typical values for a dark roofs. It will be noted that the albedo at the nadir is less than the overall downward facing albedo in Row 18 of Table 3.1. Examination of the downlooking fish-eye pictures of the two sites shows the horizon view field is quite different in nature from the downward view field, Figures 3.22 - 3.25. One can, in fact, estimate a vertical surface albedo for surfaces facing the sun from Valko's data, by first calculating the ground component on vertical surfaces from rows 13 and row 15 of Table 3.1, cases II and III, and then, for a vertical surface facing sun, setting

$$\rho_g(90, \alpha_s = 0) = 2\ R_g(90, \alpha_s = 0)/G_c \qquad (3.18)$$

Table 3.7 Error analysis of predicted hourly global solar radiation on vertical surfaces for cloudless days at Bracknell. ε = mean error for altitude and azimuth band (predicted hourly global radiation/observed hourly global radiation/observed hourly global radiation on slope).

90° vertical selected cloudless hours, albedo = 0.02.

Solar altitude band degrees	Mean solar altitude degrees	No. of hours	Azimuth 0-45°			Azimuth 45-90°			Azimuth 90-135°			Azimuth 135-180°		
			Mean solar azimuth degrees	ε	S.D.	Mean solar azimuth degrees	ε	S.D.	Mean solar azimuth degrees	ε	S.D.	Mean solar azimuth degrees	ε	S.D.
10-20	15.2	56	19.2	1.056	.053	70.8	1.005	.106	109.2	1.066	.173	160.8	1.165	.111
20-30	24.2	33	14.1	1.056	.049	75.9	1.063	.084	104.1	1.042	.113	165.9	1.102	.075
30-40	33.5	25	19.9	1.044	.059	70.1	1.010	.082	109.9	.994	.130	160.1	1.043	.105
40-50	44.6	31	26.2	1.050	.051	63.8	1.005	.082	116.2	.939	.110	153.8	1.016	.129
50-60	54.6	10	33.4	1.056	.024	56.6	.988	.051	123.4	.874	.118	146.6	.909	.127
60+	60.7	7	14.5	1.056	.018	75.5	.915	.054	104.5	.786	.073	165.5	.861	.120

Proceeding in this way, a value of $\rho_g(90,\alpha_s=0)$ of .18 was found for Locarno Monti, which compares with an overall downward facing albedo of .16. The corresponding value of $\rho_g(90, \alpha_s = 0)$ for EMPA was .17, compared with the downward facing albedo of 0.12. Both these values are very close to 0.2, the standard value used in the CEC Tables. The vertical surface albedo at the Weissfluhjoch was estimated as 0.38, reflecting the influence of a landscape partially snow covered, but with bare rock protruding through the snow. This value was used for the Weissfluhjoch calculations. The issues concerning the radiance of the ground are further explored in Part III of this Chapter.

Table 3.9 shows the results of adjusting the predicted horizontal diffuse irradiance to match the actual observed diffuse irradiance, D_c, and setting the various albedos to match the observed values given above. Very close agreement with observation is then achieved. Obviously, for cloudless sky conditions, the combination of a pyrheliometer observation with a global pyranometer observation enables a greater prediction accuracy to be achieved than prediction using the pyrheliometer observation alone.

An important advantage of working with a radiance model as the fundamental basis for solar radiation modelling is that information about the radiance distribution and the slope irradiance can be simultaneously achieved within one computational structure. This makes the EC model a very powerful tool for applied solar energy studies, because it enables the performance of semi-concentrating and concentrating devices to be scientifically studied within the same modelling structure as is used for inclined planes.

The radiance model in its present form is obviously very satisfactory, though with Valko's new structured results described in Part I of this Chapter, it is now possible to further improve the model as the next section explains.

23. Reappraisal of radiance distribution formula to match systematised results of Valko

Subsequent to the finalisation of the main body of work in the CEC European Solar Radiation Atlas, Vol. II, Inclined Surfaces, by the Department of Building Science, University of Sheffield, Valko presented the systematised results of his observations. This Swiss review was prepared within the framework of this programme.* A study of this data shows some changes are desirable in equation 3.14 to align it even better with the field radiance observations of Valko. It must be stressed that all clear sky diffuse predictions for the CEC European Solar Radiation Atlas, Vol. II, Inclined Surfaces, were however based on the use of Equation 3.14 with the constants given immediately beneath Equation 3.14.

The detailed results of 551 series of measurements carried out at four stations, Carpentras in France, and EMPA, Locarno Monti and the Weissfluhjoch, a mountain site at 2675m, in Switzerland are now available (Valko, P., 1982, Empirical study of the angular distribution of sky radiance and ground reflected radiation fluxes, Final Report of work accomplished till August 1982, CEC Contract No. ESF-020-80-D), as well as the new work presented in Chapter 3, Part I. This new published observed

Footnote: The final report will be available shortly, refer reference 15.

Table 3.8 Comparison between EC clear sky model predictions and Valko's observations presented in Table 3.1.

			I Values	I Ratio CEC/Valko	II Values	II Ratio CEC/Valko	III Values	III Ratio CEC/Valko	IV Values	IV Ratio CEC/Valko	V Values	V Ratio CEC/Valko
	Site:		Weissfluhjoch		Locarno-Monti		EMPA		EMPA		EMPA	
9	Turbidity Factor T_L	Valko	2.5		2.7		4.0		4.9		6.9	
		CEC	2.38	0.95	2.47	0.92	3.91	0.98	4.67	0.95	6.32	.916
10	I_c Direct normal Wm^{-2}	Valko	1066		948		824		789		619	
		CEC	1066	1.00	948	1.00	824	1.00	789	1.00	619	1.00
11	$G_c(\beta,\alpha)_{MAX}$ Wm^{-2}	Valko	1133		1058		987		987		943	
		CEC	1165	1.028	1068	1.006	993	1.006	982	.995	892	.946
12	G_c Wm^{-2}	Valko	835		637		741		821		728	
		CEC	825	.988	627.1	.984	734.7	.991	800	.974	669.8	.920
13	$D_c(\beta,\alpha)_{MAX}$ Wm^{-2} (B)	Valko	224		133		184		207		347	
		CEC	201	.897	140.9	1.059	185	1.005	207.5	1.002	277.6	0.80
14	D_c Wm^{-2}	Valko	55		82		124		172		257	
		CEC	45.5	.827	69.6	.849	119.6	.965	153.8	0.894	210	0.817
15	D_{cs}(Vert.)/D_c(Vert.) $\alpha_s = 0$	Valko	30/30		56/35		61/43		73/50		81/53	
	$\alpha_s = 180$	CEC*	22/12		55/34		60/44		60/35		73/42	
16	D_c/G_c	Valko	0.07		0.13		0.17		0.21		0.35	
		CEC	0.06		0.11		0.17		0.19		0.31	
17	R_c	Valko	-		103		87		100		96	
		CEC	-		125	1.21	147	1.69	160	1.60	131.6	1.37
18	R_c/G_c	Valko	-		0.16		0.12		0.12		0.13	
		CEC	0.38		0.20		0.20		0.20		0.20	
19	L_{cz}	Valko	10.5		13.1		23.7		42.1		57.3	
		CEC	9.0	.857	11.62	.887	26.3	1.11	41.2	0.978	47.86	0.835
20	L_{cn}	Valko	-		20.8		21.1		20.6		19.8	
		CEC	-		39.8	1.91	46.7	2.21	50.9	2.47	41.8	2.11
21	L_{cn}/L_{cz}	Valko	-		1.6		0.89		0.50		0.34	
		CEC	-		3.4		1.77		1.235		0.87	
22	$\pi L_{cz}/D_c$	Valko	0.60		0.50		0.60		0.77		0.70	
		CEC	0.62	1.03	0.524	1.65	0.69	1.15	0.84	1.09	0.716	1.022

* A rock stands to north which influences values.

data has been used to re-examine, in a preliminary way, the basic sky radiance formula developed earlier from the much more limited work of Steven in England, which enabled the initial methodology to be developed. Substituting b_o for 0.088 in the first term on the top and bottom lines of Equation 3.14, the basic problem is to choose the most appropriate values of the constants, b_o, a_1, a_2, a_3 to give a good match to the Valko data.

a_2 determines the magnitude of the contribution of the aureole. Previously, a value of $(.6155 + 1.9687\ T_L(\gamma))$ had been used. Valko's observations show a higher aureole effect and a better match was achieved by multiplying the above expression by 1.35, i.e. $a_2 = (0.8309 + 2.658\ T_L(\gamma))$.

a_1 and a_3 determine the contribution for non-isotropic scattering of the sky which is of primary importance away from the aureole. For the pure Rayleigh sky $a_1 = 1$ and $a_3 = 1$. The additional aerosol scattering alters the balance between a_1 and a_3. On the Rayleigh sky assumption when $T_L(\gamma) = 1$, then $a_3 = 1$. Using the previously used data of Liebelt, a_3 was re-evaluated as a polynomial function of Linke Turbidity Factor passing through $T_L(\gamma) = 1.2$ and $a_3 = 1$, instead of as a linear function of T_L. 1.2 was adopted to make some allowance for water vapour absorption effects. The following polynomial was found appropriate for a_3:

$$a_3 = 2.3843 - 1.7087\ T_L(\gamma) + 0.54709(T_L(\gamma))^2 - 0.07481(T_L(\gamma))^3 + 0.003631\ (T_L(\gamma))^4 \quad (3.19)$$

It was found this expression was suitable to fit the Valko data without alteration.

The value of a_1 was estimated from the Steven data as a function of turbidity using Steven's values of a_1 and making the assumption on Rayleigh scattering principles when $T_L = 1$, $a_1 = 1$. The following expression was found:

$$a_1 = 1.0744 - 0.0744\ T_L(\gamma) \quad (3.20)$$

(Unfortunately, the relationship had previously been expressed in terms of solar altitude which happened accidentally in the case of the Steven data set used, to closely correlate with the Linke Turbidity Factor. This was a scientific misjudgement).

Liebelt had suggested a value of 0.088 for b_o. He was working on luminance distribution and this value is roughly the value of the Rayleigh optical depth of the atmosphere at the wavelength at the centre of the visible spectrum. It was found pragmatically that the value of b_o is critical in determining the horizon brightening effect as a function of turbidity, but it exerts very little influence on the relative values of the radiance in the higher sky. At the time of preparing these specific revisions, Valko had not yet prepared his radiance measurements for low elevations in a systematic form, so it was not possible to comment on the appropriate value of b_o which best fits the observed data.* The actual value of b_o has very little influence on the ratio of the zenith radiance to the horizontal diffuse irradiance, but it does exert an important influence on the vertical surface diffuse sky irradiance because of the

* Appendix 9 presents a review of modelling of the low sky elevation data.

Table 3.9 Comparison between EC clear sky model predictions for high turbidity conditions and Valko's observations at EMPA, with diffuse horizontal irradiance normalized to observed D_c. Assumptions: Albedo for surfaces with tilt > 90° 0.17. Nadir albedo 0.085. Downward facing albedo 0.13.

			Values	Ratio CEC/ Valko
	Site:		EMPA	
9	Turbidity Factor T_L	Valko	6.9	
		EC	6.32	.916
10	I_c Direct normal Wm^{-2}	Valko	619	
		EC	619	1.00
11	$G_c(\beta,\alpha)_{MAX}$ Wm^{-2}	Valko	943	
		EC	949	1.006
12	G_c Wm^{-2}	Valko	728	
		EC	717	.984
13	$D_c(\beta,\alpha)_{MAX}$ Wm^{-2} (B)	Valko	347	
		EC	330.5	0.952
14	D_c Wm^{-2}	Valko	257	
		EC	257	1.00
15	$D_{cs}(\text{Vert.})/D_c(\text{Vert.})$ $\alpha_s = 0$ $\alpha_s = 180$	Valko	81/53	
		EC	78/49	
16	D_c/G_c	Valko	0.35	
		EC	0.36	
17	R_c	Valko	96	
		EC	93	0.969
18	R_c/G_c	Valko	0.13	
		EC	0.13	1.00
19	L_{cz}	Valko	57.3	
		EC	58.5	0.835
20	L_{cn}	Valko	19.8	
		EC	19.4	.979
21	L_{cn}/L_{cz}	Valko	0.34	
		EC	0.33	
22	$\pi L_{cz}/D_c$	Valko	0.70	
		EC	0.715	1.02

Figure 3.22 Fish-eye view of lower hemisphere at EMPA 10m above old roof from radiometer mast. Note complex variations of surfaces and materials, which make the assessment of the ground reflected solar irradiance difficult. Mast measurements were made at this site until the end of 1981.

Figure 3.23 Fish-eye photograph of the lower hemisphere from 10m height above the new EMPA roof, where the mast measurements have been made since January 1982.

Figure 3.24 Fish-eye photograph of the lower hemisphere from 10m height above the roof of the Osservatorio Ticinese, Locarno-Monti, where measurements were made during the period 30 September - 29 October, 1981. Note Lake Maggiore to south.

Figure 3.25 Fish-eye view of the lower hemisphere at the Weissfluhjoch. Note the bare rock to the north east and south east.

large relative contribution of the zone close to the region of the horizon to the irradiation of vertical surfaces. Provisionally, pending further study of Valko's radiance work at low sky elevations, it was decided to retain Liebelt's value of 0.088 for b_o. Further revision must lie in the field of future work.

The suggested revised values of constants to fit the Valko data using equation 3.14 are thus:

$$a_1 = 1.0744 - 0.0744\ T_L(\gamma)$$

$$a_2 = 0.8309 + 2.658\ T_L(\gamma)$$

$$a_3 = 2.3843 - 1.7087\ T_L(\gamma) + 0.54709(T_L(\gamma))^2 - 0.07481(T_L(\gamma))^3. + 0.003631(T_L(\gamma))^4 \qquad (3.21)$$

where $T_L(\gamma)$ is the Linke Turbidity Factor at solar altitude γ.

It is stressed once again that this was not the relationship adopted for the CEC Inclined Surface Atlas Tables, as the preparation of these Tables preceded the availability of Valko's recent detailed findings.

Figure 3.26 shows the predictions of the new algorithm for Linke Turbidity Factors of 2, 4 and 6 superimposed on the corresponding Valko graphs for the section facing the sun. The main features are well represented but the aureole is not fully accounted for by formula 3.14 using the constants given in Equation 3.21.

Figure 3.27 gives the corresponding values for a sector set at 30° to the sun. In this region the agreement is very good.

Figure 3.28 shows the predictions for the region outside the aureole for solar altitude 10°, superimposed on the Valko curves. The complex inter-relationships are reasonably well represented.

Figure 3.29 shows the predicted relationship between solar altitude and the ratio $L_c(\Theta,\alpha_S)/L_{cz}$ for a sky patch elevation of 18° for different values of solar altitude for a wall solar azimuth of 135°. Isolines of D_c/G_c are plotted to compare with Valko. These values of D_c/G_c were generated using Linke Turbidity Factors of 2, 4 and 6. Similar diagrams were drawn for the data series, all giving broadly satisfactory results.

By carrying out the appropriate integration of equation 3.14 numerically, using the new constants in 3.21, the predicted values of $\pi L_z/D_{c1}$ can be obtained. These are superimposed on Valko's data in Figure 3.30. Again the agreement is good.

These studies were carried out assuming the site was at sea level. While the radiance distribution for a given Linke Turbidity may not vary very much with altitude, the ratio D_c/G_c does change considerably. It would be better if the very high mountain data were excluded from data sets for the lower level sites, as this may obscure studies correlating theory and practice.

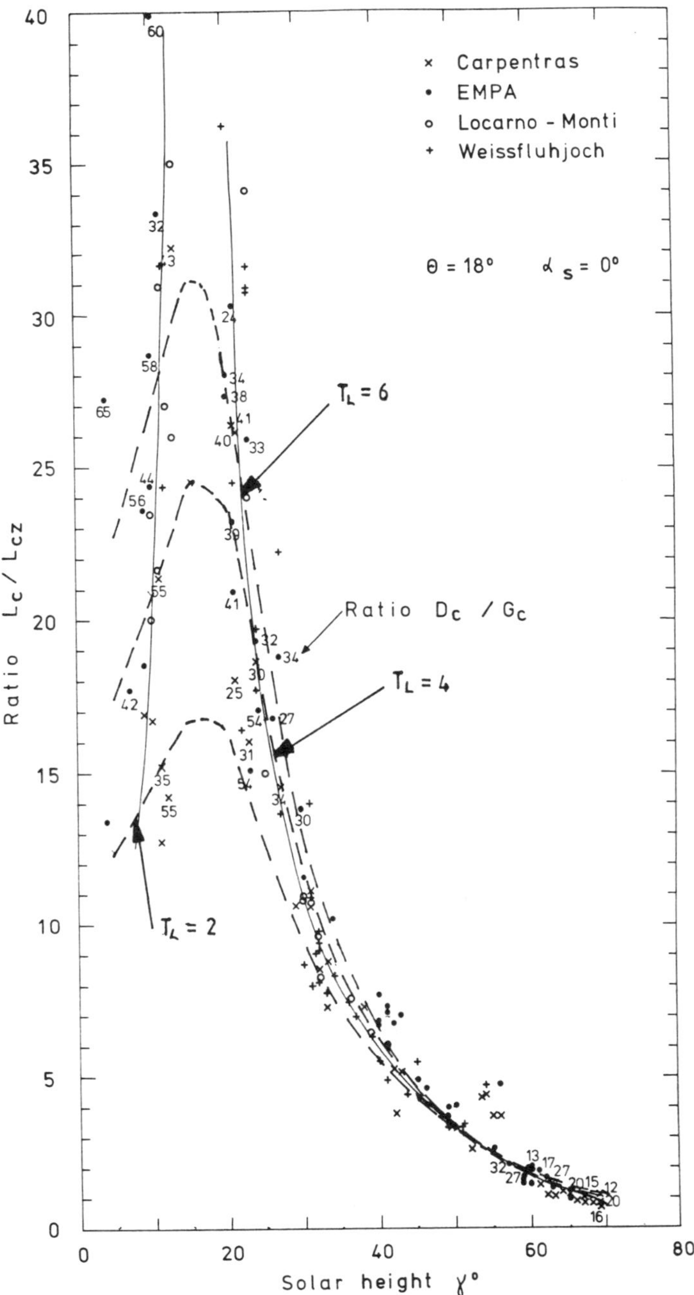

Figure 3.26a The dotted curves give the values of $L_c(\Theta,\alpha_s)/L_{cz}$ computed from the revised radiance distribution algorithm superimposed on Valko's diagrams for $\alpha_s = 0°$ for a sky patch elevation of 18° for Linke Turbidity Factors of 2, 4 and 6. The formula does not fully reproduce the aureole effect.

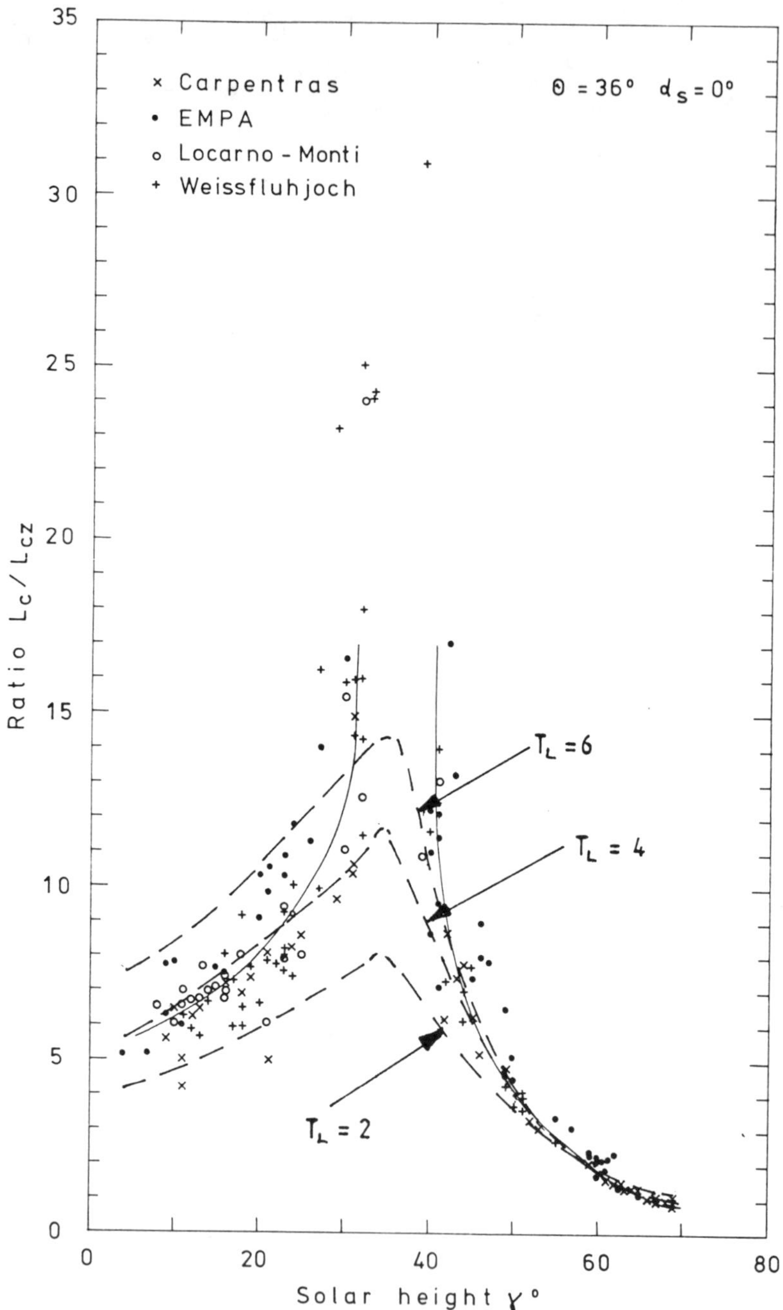

Figure 3.26b The dotted curves give the values of $L_c(\Theta, \alpha_s)/L_{cz}$ computed from the revised radiance distribution algorithm superimposed on Valko's diagrams for $\alpha_S = 0°$ for a sky patch elevation of 36° for Linke Turbidity Factors of 2, 4 and 6. The formula does not fully reproduce the aureole effect.

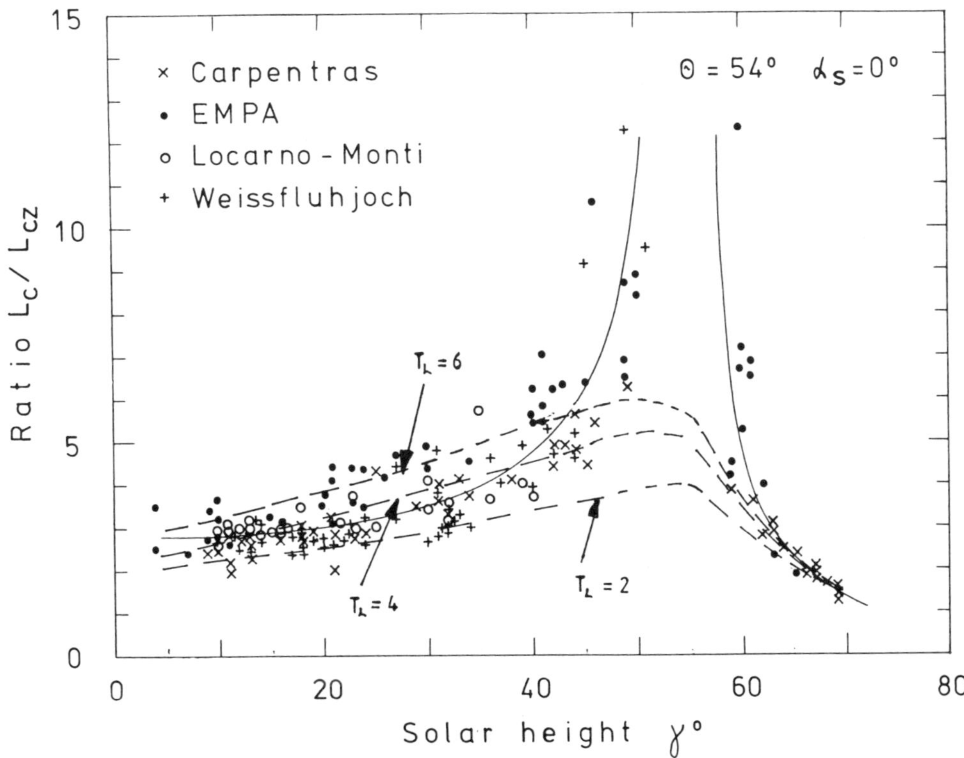

Figure 3.26c The dotted curves give the values of $L_c(\Theta,\alpha_s)/L_{cz}$ computed from the revised radiance distribution algorithm superimposed on Valko's diagrams for α_s = 0° for a sky patch elevation of 54° for Linke Turbidity Factors of 2, 4 and 6. The formula does not fully reproduce the aureole effect.

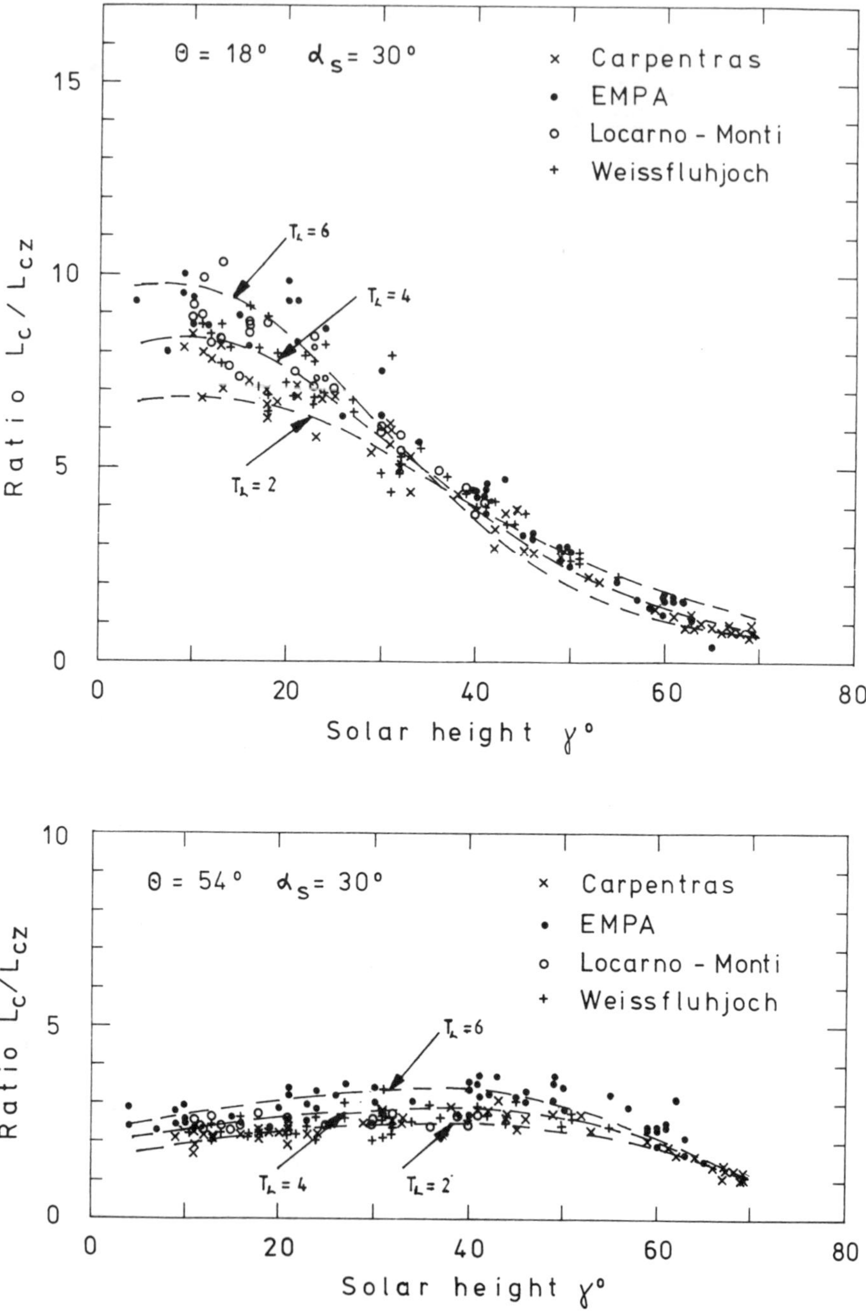

Figure 3.27 The values of $L_c(\Theta\ ,\ \alpha_s)/L_{cz}$ computed from the revised algorithm superimposed on Valko's diagrams for α_s = 30°, for sky patch elevations of 18° and 54° for Linke Turbidity Factors of 2, 4 and 6.

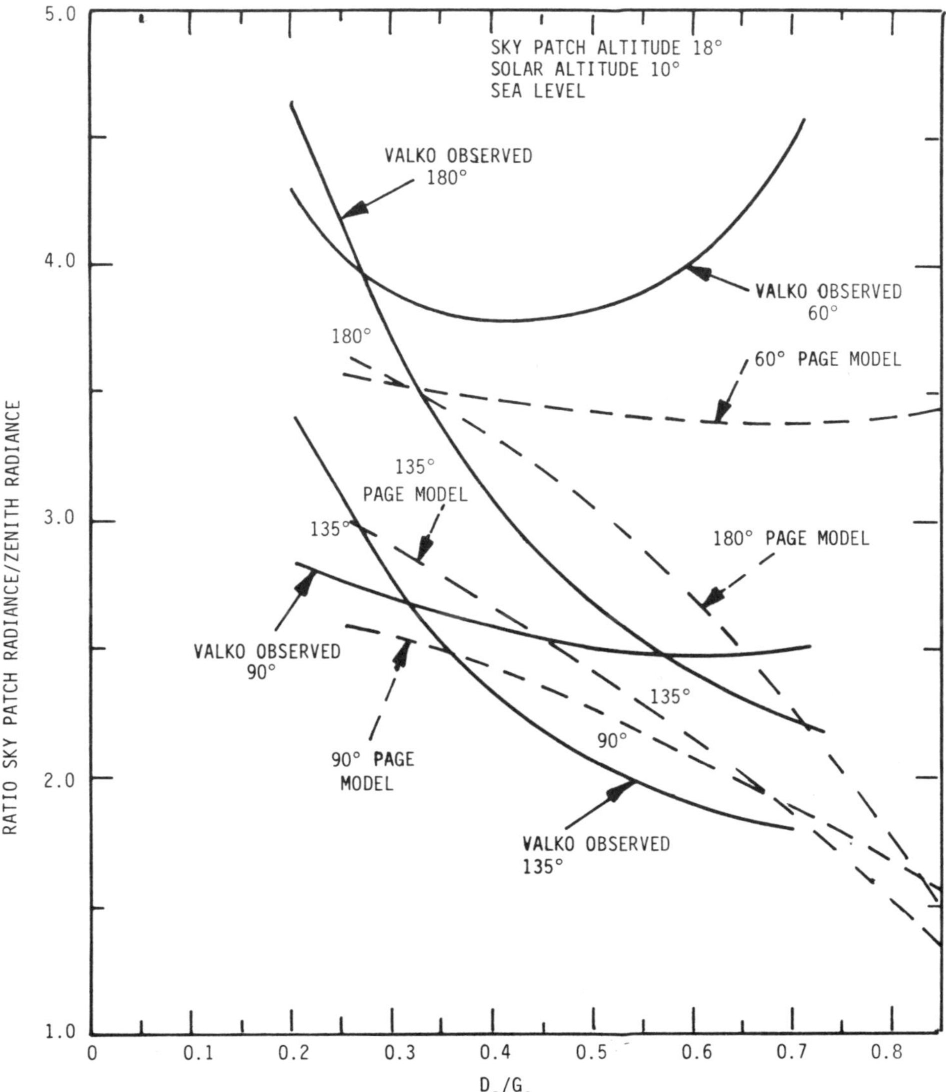

Figure 3.28 Comparison of EC radiance ratio predictions with Valko's observations as a function of D_c/G_c for a sky patch altitude of 10°. The complex relationships are reasonably represented by the EC model. Solar altitude 10°.

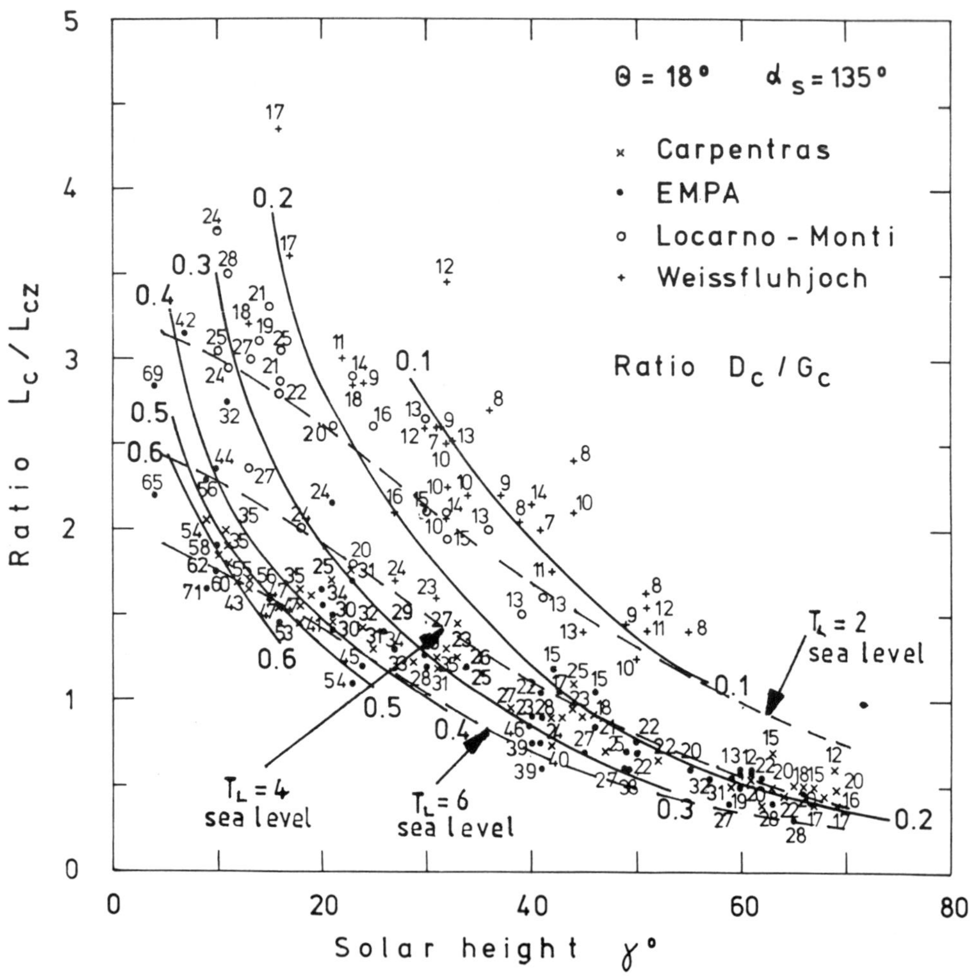

Figure 3.29 The values of $L_C(\Theta,\alpha_S)/L_{CZ}$ predicted using Linke Turbidity Factors of 2, 4 and 6 superimposed on Valko's diagram for a sky patch albedo of 18°, with a value of α_S = 135.

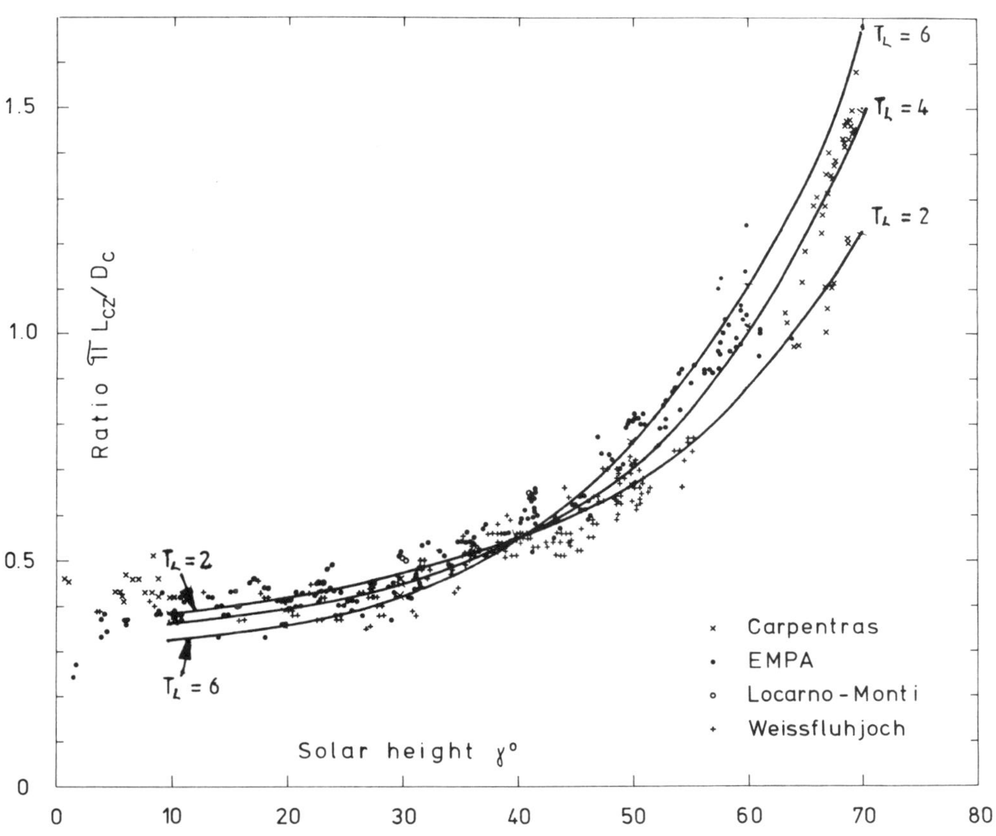

Figure 3.30 The computed ratio L_{cz}/D_c for Linke Turbidity Factors of 2, 4 and 6, superimposed on Valko's observed values. The predicted ratio appears too low compared with observation for small solar elevations.

Using the analytic techniques already developed, it will not be a difficult task in any future research programme to recompute the various clear day diffuse algorithms including the tables of values of f_2 given in Chapter 6, Tables 6.14a to 6.14g and the associated algorithms given in Appendix 5, though this task could not be achieved within the framework of time in which this book was prepared.

Even though improvements are possible, the results achieved with the preliminary EC model remain very good when checked against field observation.

25. Prediction of zenith radiance

Appendix 6 considers in detail the relationship between the zenith radiance and the horizontal surface diffuse irradiance for cloudless skies predicted by the various models, and provides appropriate mathematical methods for predicting the relationship. The linking function is important for the rapid estimation of radiance from diffuse irradiance values, using equation 3.14 with the constants given in 3.21. If observations of G_c and D_c are available together with knowledge of the solar elevation and azimuth, the Linke Turbidity Factor can be estimated from the estimated direct beam normal irradiance $(G_c - D_c)\sin\gamma$.

CHAPTER 3 - PART III

ESTIMATION OF REFLECTED RADIATION FROM THE GROUND

26. Introduction

The work of P. Valko described in Chapter 3. Part I, did not deal with the radiance of the ground which varies considerably. Several factors affect the radiance of the ground. For a start, there are often substantial variations in the actual irradiance falling on the ground due to obstructions, for example the shadow of a tall building can substantially reduce the irradiance of the ground on the side of a building opposite to the sun. Furthermore the ground may not be level. A second important factor is that albedo of the surface can vary very widely across the field of ground surface view of any inclined plane. For example snow may cover part of the field of view, and bare rock may protrude in other parts of the field of view. The albedo is sometimes very sensitive to the angle of incidence of the sun on the surface. Water provides a well known example, but similar factors apply in the case of surfaces like grass when the sun is low. Table 3.10 provides typical values of the ground albedo, including the directional effects over water.

The basic problem for practical applications of solar energy is that the ground reflected component is so site specific. The only accurate methodology for dealing with such site specific factors is the numerical integration of the radiance function i.e. applying the techniques of Appendix 4 specifically to the lower hemisphere rather than the upper hemisphere, and then using the results of this integration to assess the effective albedo in different directions. The illustrations of the fish-eye camera view of the different sites at which ground radiance measurements have been made with the Swiss equipment illustrate how complex the field of view may be (Figures 3.22, 3.23, 3.24 and 3.25). The arrow marks the north direction. However, before considering these points, the conventional approach to ground reflected solar energy will be reviewed. This was the approach that had to be adopted in the production of the inclined surface tables for the CEC Solar Radiation Atlas, Vol. II, Inclined Surfaces, because no detailed information was available about local site conditions for the various sites for which Tables were being prepared.

27. Methodology for estimating ground reflected radiation adopted in the CEC European Solar Radiation Atlas, Vol. II, Inclined Surfaces

In addition to direct and diffuse sky components, a significant amount of solar radiation reaches inclined surfaces after reflection from the ground, which had to be considered in the estimated tables. The method chosen for the prediction of ground reflected irradiance which follows is common to all sky conditions. It may be applied either to irradiance predictions or to daily irradiation totals.

Table 3.10 Albedo of typical ground surfaces

Surface	ρ_s	Source
*Grass (July, August)	0.25	Monteith (1961)
Lawns	0.18-0.23	Dogniaux (1973)
Dry grass	0.28-0.32	Dogniaux (1973)
Uncultivated fields	0.26	Dogniaux (1973)
Bare soil	0.17	Monteith (1961)
**Macadam	0.18	Dogniaux (1973)
**Asphalt	0.15	Holder & Greenland (1951)
**Concrete clean	0.30	Teneluis (1960)
**Concrete weathered (Liverpool)	0.20	Page (1960)
Fresh snow	0.80-0.90	Dogniaux (1973)
Old snow	0.45-0.70	Dogniaux (1973)
Water surfaces solar altitude > 45°	0.05	Deacon (1969)
" " > 30°	0.08	
" " > 20°	0.12	
" " > 10°	0.22	

* Monteith found a variation in the albedo of short grass from about 0.22 at a solar altitude of 60° to 0.28 at a solar altitude of 20°. At 10° Roach found albedo values as high as 0.3 to 0.4.

** Climatic factors, pollution, etc., produce considerable variations. Local assessments of the albedo are desirable to take account of the actual surface conditions in that region.

It is assumed that the radiance of the ground is uniform in all directions, i.e. obeys Lambert's law. Two variables affecting the amount of ground reflected irradiance reaching any sloping surface are considered, the ground albedo which determines the proportion of incident short wave radiation which is reflected by the ground and secondly the inclination of the receiving surface which determines the proportion of this reflected radiation that will actually fall on the surface per unit area. The ground reflected diffuse component was calculated using the following formula:

$$R_g(\beta, \alpha) = 0.5(1 - \cos \beta)\ \rho_g\ G \quad Wm^{-2} \tag{3.22}$$

where $R_g(\beta, \alpha)$ is the ground reflected irradiance received on an inclined surface

β is the angle of the inclined surface

ρ_g is the ground albedo

G is the global irradiance on a horizontal surface, Wm^{-2}.
(Add subscript c for clear, m for mean, and b for overcast conditions).

For clear sky conditions the value of G_c as defined in Chapter 2 was used and an allowance made for the value of S_{max}/S_{04max}. In the computation of the CEC Inclined Surface Solar Radiation Tables, a ground albedo value of 0.2 was adopted as the standard value. This value is representative of typical surfaces seen by most buildings. The effects of other surfaces can be explored by changing the actual values to match the proposed surface conditions. Typical values of the albedo are given in Table 3.10. In practice the most important variations in the ground reflected component are due to snow which may, when clean, have an albedo exceeding 0.80. Sometimes climatologically weighted monthly mean values of the albedo are used, for example monthly mean snow cover 10 days out of 30 days, mean albedo = 2/3 x 0.2 + 1/3 x 0.8 = 0.4. It will be noted that it is always necessary to estimate the global radiation on a horizontal surface in order to estimate the reflected radiation on an inclined surface. However, the basic methodology for estimating the ground reflected component is the same for clear, overcast and monthly average conditions. It will also be noted it is an implicit assumption in the method that the ground surface is horizontal.

28. Non-isotropic ground radiance

If the ground radiance is not isotropic, one has to recognise that the effective albedo in equation 3.22 is a function of slope tilt, orientation and time of day. Thus we may write, for clear sky conditions:

$$R_c(\beta, \alpha) = 0.5\ \rho_c(\beta,\alpha,t)(1 - \cos \beta)\ G_c \tag{3.23}$$

where $\rho_c(\beta,\alpha,t)$ refers to effective albedo for a specific surface of given tilt and orientation for cloudless sky conditions at a given time t.

Time has to be introduced, because the position of the sun and shadow patterns vary with time. It is not possible to find a general solution for all sites, but it is possible to identify in general terms those zones of the ground which potentially will contribute most to the ground reflected irradiance and this can help identify the appropriate albedo to choose for simplified computation in any specific situation.

Figure 3.31 sets out the general geometry of the lower hemisphere, and identifies geometrically in angular terms an element lying between $-\Theta + \frac{1}{2}\delta\Theta$ and $-\Theta - \frac{1}{2}\delta\Theta$ from the horizon in a direction from due south lying between $\alpha - \frac{1}{2}\delta\Theta$ and $\alpha + \frac{1}{2}\delta\Theta$. If the element is assumed to be horizontal, the global irradiance falling on the element is G_C, and the albedo in the direction of P of the element is $\rho_C(-\Theta, \alpha)$, the radiance of the element will be:

$$\frac{1}{\pi} \times \rho_C(-\Theta, \alpha) \times G_C$$

The area of the element is $(\cos - \Theta) \times \delta\Theta \times \delta\alpha$. The corresponding contribution to the irradiance of a downward facing horizontal element at P will be:

$$\frac{1}{\pi} \times \rho_C(-\Theta, \alpha) \times G_C \times \sin(-\Theta) \cos\Theta \,\delta\Theta \cdot \delta\alpha \qquad (3.24)$$

while the irradiance contribution on a vertical element at P with a wall-element solar azimuth angle of α_S is:

$$\frac{1}{\pi} \times \rho_C(-\Theta, \alpha) \times G_C \times \cos^2(-\Theta) \cos\alpha_S \,\delta\Theta \cdot \delta\alpha, \qquad (3.25)$$

provided $\alpha_S > 0$

The function $\sin(-\Theta)\cos\Theta$ has its maximum value of 1 at $\Theta = -45°$ while the function $\cos^2(-\Theta)$ has its maximum value at 0°. Therefore, for a downward looking surface, the radiance of the ground around -45° exerts the greatest impact, while, for vertical surfaces, it is the radiance properties of the horizon zone which exert the greatest impact. Figure 3.32 shows, for a ground of uniform radiance, the proportion of the energy from a quarter sphere contributed to a vertical surface which comes from below any stated angle $-\Theta$. The significant factor is the slope of the curves. With a vertical surface, the slope of the curves is greatest closest to the horizontal plane and about 62% of the contribution comes from the ground surface with values of Θ of less than -30°. In contrast, for the horizontal downward looking surface, the slope is greatest at 45°, and it is the middle zone between $\Theta = -27°$ and $\Theta = -63°$ which contributes 62% of the reflected energy from a ground surface of uniform radiance.

If suitable judgement is exercised concerning the choice of albedo in the case of non-uniform ground conditions, then more accurate estimates of reflected radiation can be made without resorting to full numerical integration. Obviously, for upward tilting surfaces, the radiance properties of the horizon zone become even more important for accurately assessing the impacting reflective irradiance than is the case with vertical surfaces but, as the value of $(1 - \cos\beta)$ decreases with decrease of β, the overall diffuse ground contribution rapidly becomes less important than the diffuse sky contribution, as Figure 3.6h from Part I of

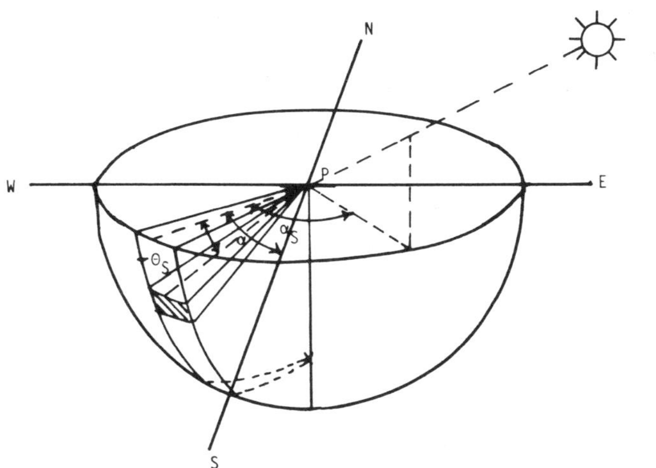

Figure 3.31 Geometry used to describe the position of down facing elements of ground. The azimuth angle is measured from due south. The surface sun azimuth angle α_S is defined as indicated.

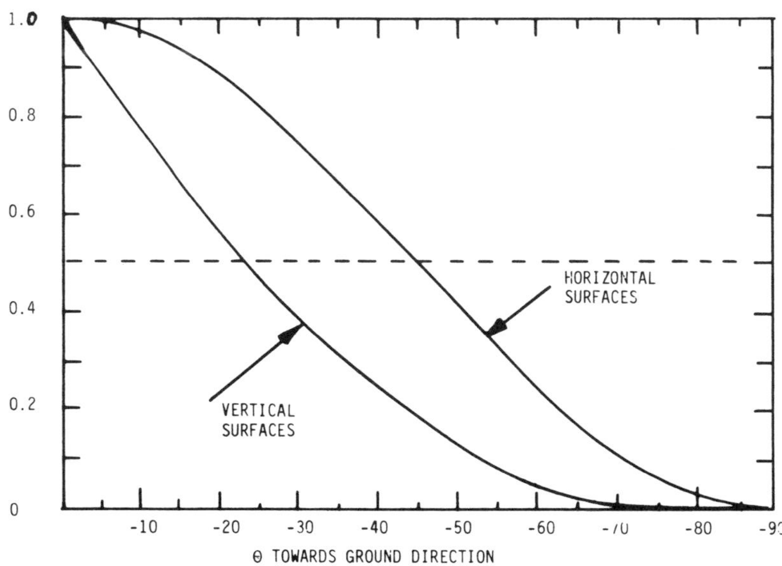

Figure 3.32 Proportion of energy contributed from a downward facing quarter sphere of uniform radiance falling on a vertical surface and on a downward facing horizontal surface. Most of the reflected radiant energy on a vertical surface comes from the sector within 40° of the horizon.

this chapter clearly illustrates. One important consequence of the Swiss observational work is to underline the need to think more carefully about the ground contribution in estimating inclined surface irradiation. This factor is especially important in snowy climates. The directional nature of the reflecting properties of ground surfaces also deserve more attention, as the field observations discussed in the following sections demonstrate.

As the effective ground albedo is not constant, the use of the downward facing albedo can lead to significant errors in the estimation of the reflected solar radiation on inclined planes, especially those tilted upwards. In general, as most solar collectors face upwards, it is the albedo close to the horizon that it is the most important to assess accurately. As this is often vegetation, the choice of 0.2 for computing the CEC inclined surface tables seems entirely appropriate to describe typical conditions. In the case of snow covered ground, the ground reflected component will be far greater. No allowance for snow was made in the computation of the CEC European Solar Radiation Atlas, Vol. II, Tables, but the Atlas explains how to make appropriate adjustments where necessary to allow for changes in ground albedo.

29. Survey of measurements of reflected radiation from the ground

The data from the mast of reflected radiance and of irradiance on overhangs have not yet been processed to the stage which would allow their general presentation in a systematised form, as was done for the sky radiance in Part I of this Chapter. Nevertheless, typical site specific differences of reflected radiance patterns - presented as quasi-instantaneous distributions of the consequent albedo for inclined planes $\rho_c(\beta,\alpha)$ - can be demonstrated using selected examples (Figures 3.33a to 3.36).

A greater amount of both normalized total diffuse irradiance ratio data, $D_c(\beta,\alpha_S)/D_c$, as well as normalized sky diffuse irradiance ratio data, $D_{CS}(\beta, \alpha_S)/D_C$, is already structured in such a way, that the ground-reflected share $R_C(\beta,\alpha_S)/D_C$, computed according to equation 3.8, can be displayed as a function of solar altitude γ, surface-solar azimuth α_S, surface inclination β and turbidity index D_C/G_C (Figures 3.37 to 3.41). $D_{CS}(\beta, \alpha_S)$ was, as mentioned in Part I, Section 10, evaluated by solving the double integral numerically as described in Appendix 4.

Figure 3.37 plots, for a plane of slope 60° facing the sun ($\alpha_S = 0$), the ratios $D_{CS}(60, 0)/D_C$ and $R_C(60, 0)/D_C$ for a range of solar elevations at EMPA. The single number plotted against each point in the top half is the turbidity ratio D_C/G_C. These numbers are carried into the bottom half of the Figure and additionally the value of the downward facing ground albedo is entered immediately below. Where no second figure is entered, the value was not available. The value of $\rho_g(-0)$ varies between 7% and 22%.

Figure 3.38 plots, for a plane of slope 60° at an angle of 120° to the sun, the ratios $R_C(60, 120)/D_C$ using the same conventions as described in the previous paragraph. A fall off in the ratio with decrease of solar altitude will be noted, compared with the sharp increase in the previous Figure 3.37.

Figure 3.39 generalises the relationships as a function of wall solar azimuth angle and solar altitude for a plane inclined at 60° to the horizontal for EMPA. For sun facing surfaces the ratio $D_{CS}(60,\alpha_S)/D_C$ is at a maximum when the solar elevation is around 20°. This observed fall off at low solar elevations is not predicted by the EC theoretical model, Tables 6.14a-j. However, the full Swiss data were not available when the model was developed.

Figure 3.40 compares the values of the ratio $R_C(39, \alpha_S)/D_C$ and $R_C(90, \alpha_S)/D_C$ for a solar altitude of 20° for 3 sites, EMPA, Locarno Monti and the Weissfluhjoch. The Weissfluhjoch data are divided into two groups, ground snow covered, and ground non-snow covered.

30. Computation of ground albedo from the Mobile System Measurements

In this section the horizontal surface global irradiance was chosen as the basis of normalization, keeping consistency in this way with the commonly adopted albedo concept, as the ratio of reflected energy to incident energy, defined here for a horizontal surface. In contrast, in Figure 3.6g, to achieve comparability with Figure 3.6c, the reflected radiances were normalized to the sky radiance at the zenith.

There are three ways of using the observed data for computing albedo values over the whole ground field of view for the 77 pyranometer positions. All computations have to consider the spectral adjustment of data measured by the different instruments described by equations A4.2 and A4.8 of Appendix 4.

Rejecting isotropic reflection, the simplest way is to compute an albedo for clear sky conditions for each point $(-\Theta,\alpha)$ of the lower hemisphere using the following equation

$$\rho_C(-\Theta,\alpha) = \frac{\pi\, L_C(-\Theta,\alpha)}{G_C} \tag{3.26}$$

in analogy to the computations for Figure 3.17. In this case the 77 reflected observed radiances $L_C'(-\Theta,\alpha)$ (65 measured and 123 interpolated values, as mentioned with Figure 3.6g) had first to be adjusted spectrally, using the relationship:

$$C_R = \frac{R_C(-0)}{\iint_{-0} L_C'(-\Theta,\alpha)\, d\Theta \,.\, d\alpha} \tag{3.27}$$

$R_C(-0)$ denotes the pyranometer measurement of reflected irradiance on the inverted horizontal surface and $\iint_{-0} L_C'(-\Theta,\alpha)d\Theta . d\alpha$ is the double integral of reflected radiance before spectral correction with respect to the inverted horizontal surface. The full expression of the latter is analogous to equation A4.1 in Appendix 4 and has been computed in analogy to equation A4.5. The process matches the radiance observations to the spectral response of the pyranometers, before the ground albedo, in a specific direction $\rho_C(-\Theta,\alpha)$, is calculated using the following formula:

$$\rho_C(-\Theta,\alpha) = \pi\, C_R\, L_C'(-\Theta,\alpha)/G_C \tag{3.28}$$

An implicit assumption in deriving an effective albedo is that the ground is flat and horizontal. $\rho_C(-\Theta,\alpha)$ thus means the albedo of that surface in a specific direction on the assumption that the surface irradiance is the horizontal surface irradiance. Except Figures 3.33b and 3.33c, all polar diagrams within the series, Figures 3.33 to 3.36 demonstrate distributions of $\rho_C(-\Theta,\alpha)$ defined in terms of the horizontal surface irradiance.

As a more sophisticated approach, routine computations of total diffuse irradiance on inverted surfaces $D_C(-\beta,\alpha)$ - see Figure 3.6(f) for example - were combined with the sky radiances integrated with respect to the 77 inverted (mast) pyranometer surfaces $\iint_{-\beta\alpha} L_C'(\Theta,\alpha)\,d\Theta.d\alpha$ to compute as the first step, $R_C^*(-\beta,\alpha)$ the ground reflected part of the total diffuse irradiance $D_C(-\beta,\alpha)$ falling on the 77 inverted pyranometer surfaces of orientation $(-\beta,\alpha)$. These values were obtained as normalized variables as follows:

$$\frac{R_C^*(-\beta,\alpha)}{R_C(-0)} = \frac{D_C(-\beta,\alpha)}{R_C(-0)} - \frac{\iint_{-\beta\alpha} L_C'(\Theta,\alpha)\,d\Theta\,.\,d\alpha}{\iint_{-0} L_C'(-\Theta,\alpha)\,d\Theta\,.\,d\alpha} \qquad (3.29)$$

$R^*(-\beta,\alpha)$ indicates that this term was computed roundabout through equation 3.29 in contrast to $R_C(-\beta,\alpha)$, which is computed directly as the respective double integral of the ground reflected radiances $L_C'(-\Theta,\alpha)$.

Using the values of $R_C^*(-\beta,\alpha)$ the albedo was computed as

$$\rho^*(-\beta,\alpha) = \frac{R_C^*(-\beta,\alpha)}{G_C}\,.\,F_\beta \qquad (3.30)$$

The factor $F_\beta = \frac{\pi}{\pi-\beta}$ normalizes $R_C^*(-\beta,\alpha)$ to reflected irradiances from the full hemisphere "seen" by the surface of inclination $-\beta$ radians. F_β varies between the limits of 2 (vertical surfaces) and 1 (inverted horizontal surface). Figure 3.33b is an example for the distribution of $\rho^*(-\beta,\alpha)$.

Figure 3.33c is, for the present, the only example, where the mean effective albedo could immediately be computed from observed ground radiances as:

$$\rho(-\Theta,\alpha) = \frac{R_C(-0)}{G_C}\,\frac{\iint_{-\beta\alpha} L_C'(-\Theta,\alpha)\;d\Theta\,.\;d\alpha}{\iint_{-0} L_C'(-\Theta,\alpha)\;d\Theta\,.\;d\alpha}\,.\,F_\beta \qquad (3.31)$$

The double integral $\iint_{-\beta\alpha} L_C'(-\Theta,\alpha)\,d\Theta.d\alpha$ with respect to the 77 inverted pyranometer surfaces $(-\Theta,\alpha)$ is not yet computed on a routine basis.

Referring back to Part I, Section 10 of this Chapter, it is evident that both expressions $\rho^*(-\beta,\alpha)$ and $\rho(-\Theta,\alpha)$ have to take account of the spectral matching adjustment; this is unavoidable, since the albedo has to be normalized to the pyranometer value of G_C. In interpreting the albedo distributions, it should be remembered that the measurements made by the different instruments in their different positions follow a prescibed temporal sequence (see Part I, Section 2 of this Chapter). Though this is not of critical influence under clear sky conditions, except very low solar elevations, albedo values have necessarily to be computed using data which is not strictly simultaneous. It should be mentioned in particular that:

1. One scan of the 77 reflected radiances lasts 25 minutes 20 seconds; the nadir position is measured once during each 25 minutes 20 second scan, 7 minutes after start.

2. The mast-pyranometer scanning sequence coincides exactly with that of the reflected radiance scanner. The inverted horizontal value is observed once during each scan, 7 minutes after the start. This value of $R_c(-0)$ and a simultaneously (within 10 seconds) measured G_c are used in equations 3.28, 3.30 and 3.31.

3. The system of four silicon diodes (Figure 3.3) scan the sky in 121 points in about 2 minutes. In fact a double scan (there and back) takes place during the first 4 minutes 20 seconds of the 25 minutes 20 second mast-cycle. All further $L'(\Theta,\alpha)$ scans during the remaining 21 minutes are made by interference filters and can, therefore, not be used for computing total spectrum albedo values. Data used in equation 3.29 can, therefore, neither be simultaneous, nor temporally symmetric.

31. Presentation of albedo distributions for different sites

The fish-eye photographs of the lower hemisphere (Figures 3.22, 3.23, 3.24 and 3.25) give some impression of the unhomogeneous ground reflectivity of the measurement sites. In order to take the pictures, one of the instruments at the end of the branch on the top of the mast at 10 m height (see Figures 3.4 and 3.5) was removed and replaced by the camera. This was always done on the last day of each excursion. The fish-eye views, therefore, are not strictly representative of the precise times the measurements were made. This aspect has especially to be considered for the variations of snow cover at the mountain site, Weissfluhjoch. Since the mast was not operating before October 1980, the corresponding fish-eye picture and, of course, measurements of reflected solar radiation are not available for Carpentras (excursion in June 1979).

In each polar diagram of the albedo-distributions, presented in Figures 3.33 to 3.36, the reflected image of the sun's position has also been marked (ϴ). Values of the isolines and the numerical entries are expressed as percentages. Except in Figure 3.35b, all distributions are for clear skies. Figure 3.35b represents measurements under a sky heavily overcast by stratocumulus clouds.

Date, time and most important characteristics of the presented distributions are summarised in Table 3.11. As may be recognised, Figures 3.33a, b and c belong to example III of Table 3.1 and thus complete the set of polar diagrams 3.6a - 3.6h. Similarly, Figure 3.35a belongs to example II of Table 3.1. The values in the first ten rows of Table 3.11 for these two cases have already been included in Table 3.1 as well. Row 11 contains the clear sky nadir value of the albedo ρ_{cn} (ρ_{bn} for the overcast day for Figure 3.35b). In row 12 the maximum value of the respective 77 reflected radiances is entered. Row 13 gives the angular co-ordinates of points corresponding to these values. The maximum of reflected radiances may intuitively be compared with the average sun + sky radiance G_c/π, entered in row 14. The ratio of the two quantities in row 15 gives the maximum clear sky albedo $\rho_c(-\Theta,\alpha)(\max)$ $\{\rho_b(-\Theta,\alpha)$ (max) for Figure 3.35b$\}$. Except that the numerical values differ by a constant ratio $G_c/\pi \, . \, C_R \, . \, L'_{cz}$,

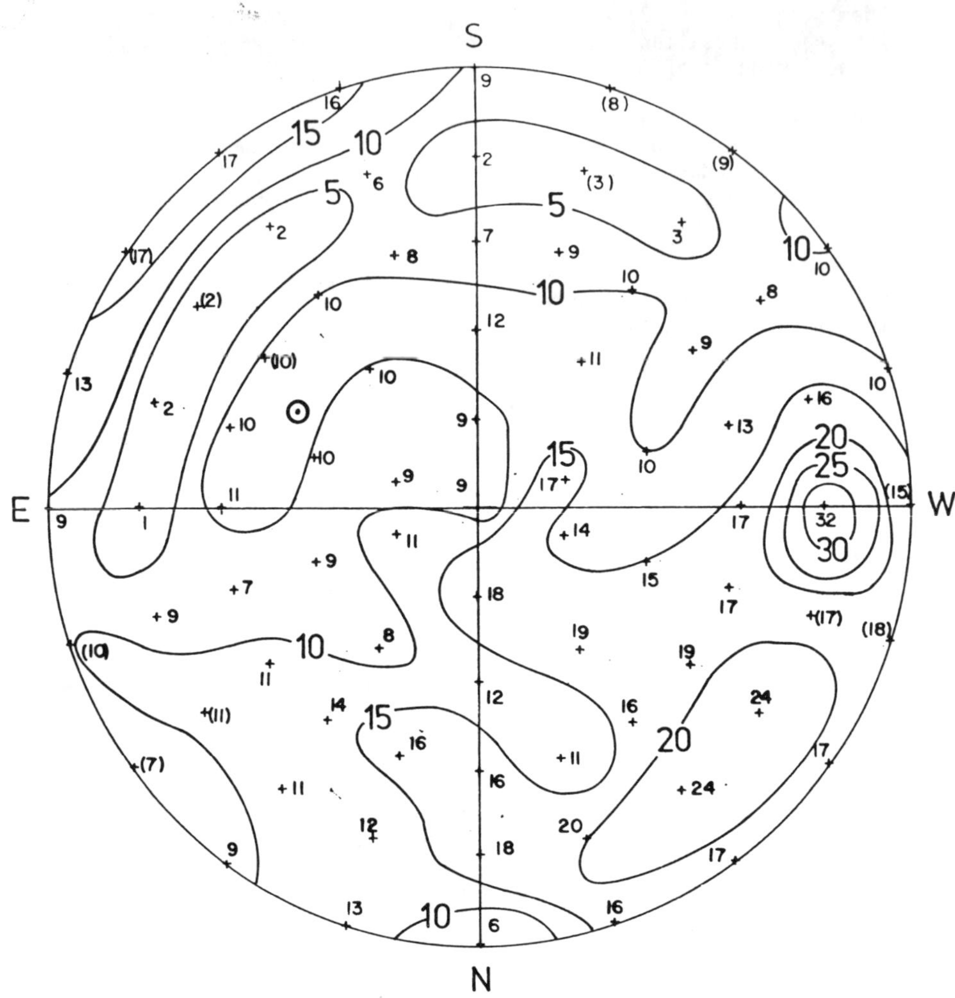

Figure 3.33a Angular distribution at EMPA, Zurich, old roof (Figure 3.22) of ground albedo $\rho_c(-\Theta,\alpha)$ computed by Equation 3.28 for the 77 instrument positions $(-\Theta,\alpha)$ where reflected radiances $L'(-\Theta,\alpha)$ have been measured from 10m height above the roof. Dimensionless; isolines at steps of 5%. Reflected image of sun's position marked ⊙.

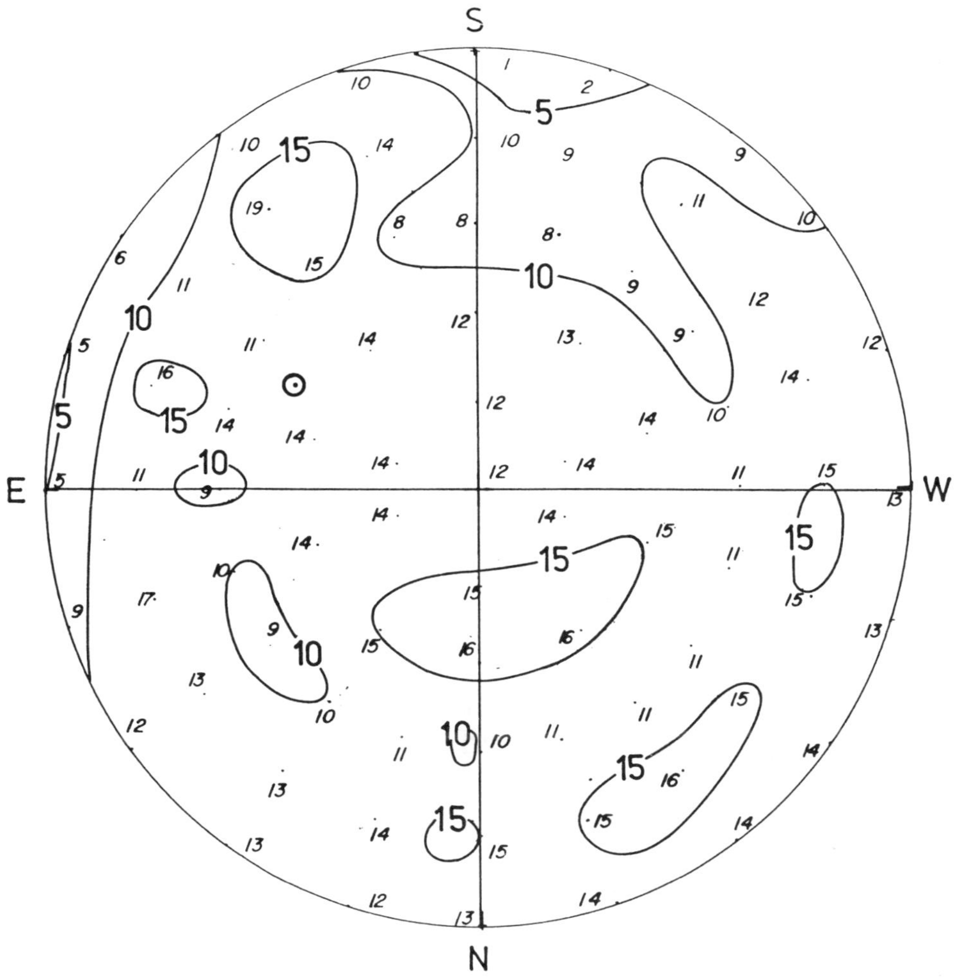

Figure 3.33b Same as Figure 3.33a, but mean directional values $\rho(\beta,\alpha)$ have been computed by Equation 3.30 using both data of reflected radiances and sky radiances.

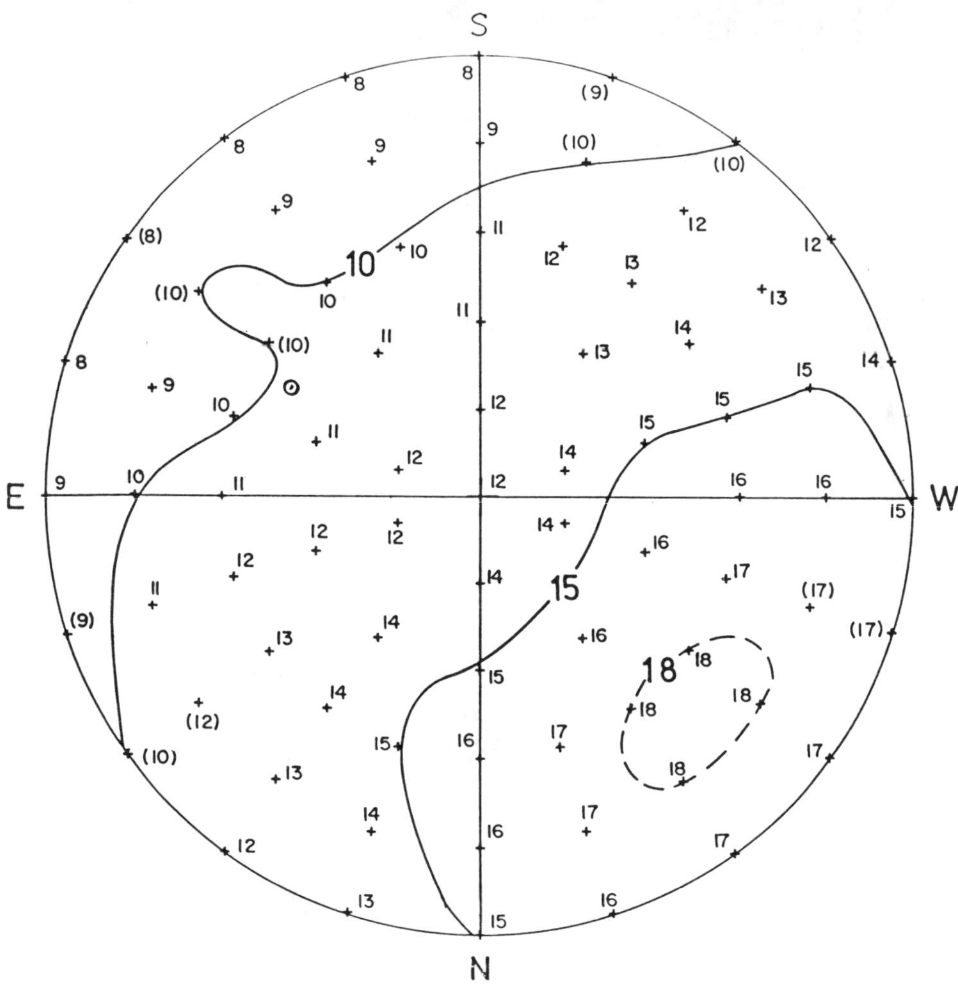

Figure 3.33c Same as Figure 3.33a, but mean directional albedo values $\rho(\beta, \alpha)$ have been computed according to Equation 3.30 by integrating the 77 measured reflected radiances themselves to obtain the irradiance with respect to each of the 77 inverted mast-pyranometer positions. Note maximum mean albedo values are 50% above nadir values, and minimum values 33.3% below.

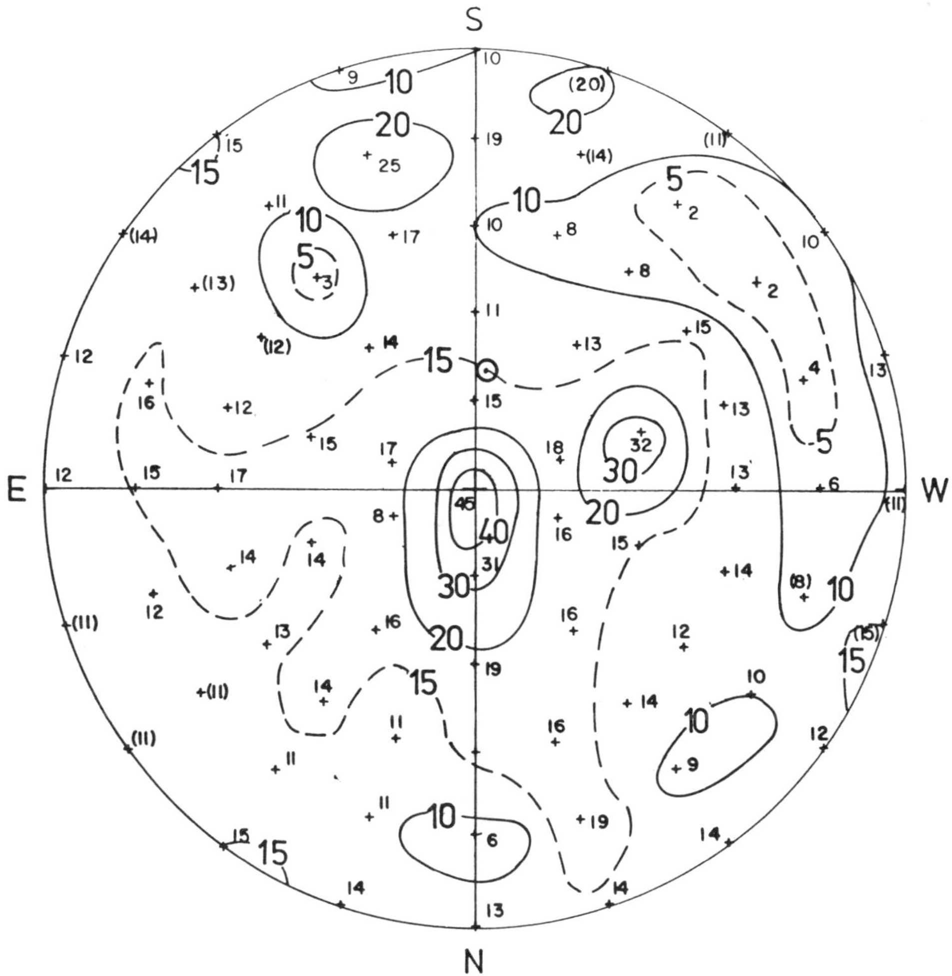

Figure 3.34 Angular distribution at EMPA, Zurich, new roof (Figure 3.23) of ground albedo $\rho_c(-\Theta,\alpha)$ computed as for Figure 3.33a. Dimensionless; isolines at steps of 10%, additionally in some regions at steps of 5%. Reflected image of sun's position marked ⊙.

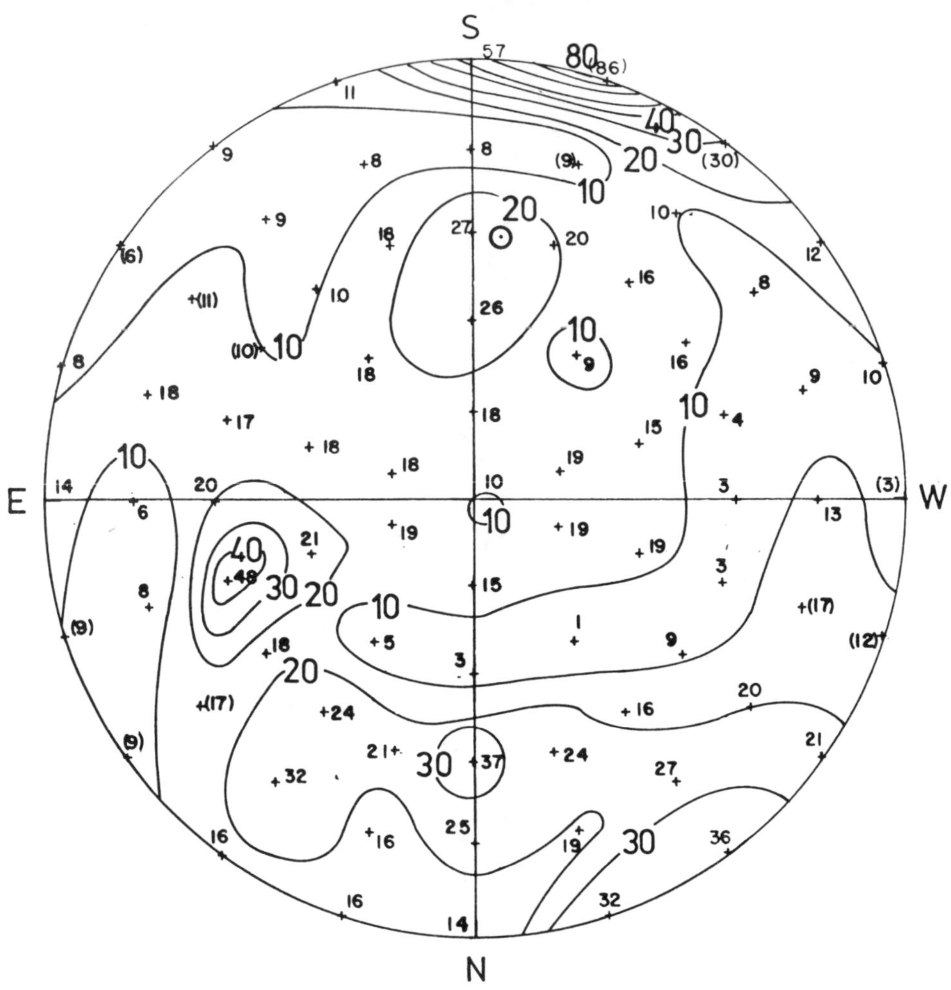

Figure 3.35a Angular distribution at Locarno-Monti, roof of the Osservatorio Ticinese (Figure 3.24) of ground albedo $\rho_c(-\Theta,\alpha)$ computed as for Figure 3.33a. Dimensionless; isolines at steps of 10%. Reflected image of sun's position marked ⊙.

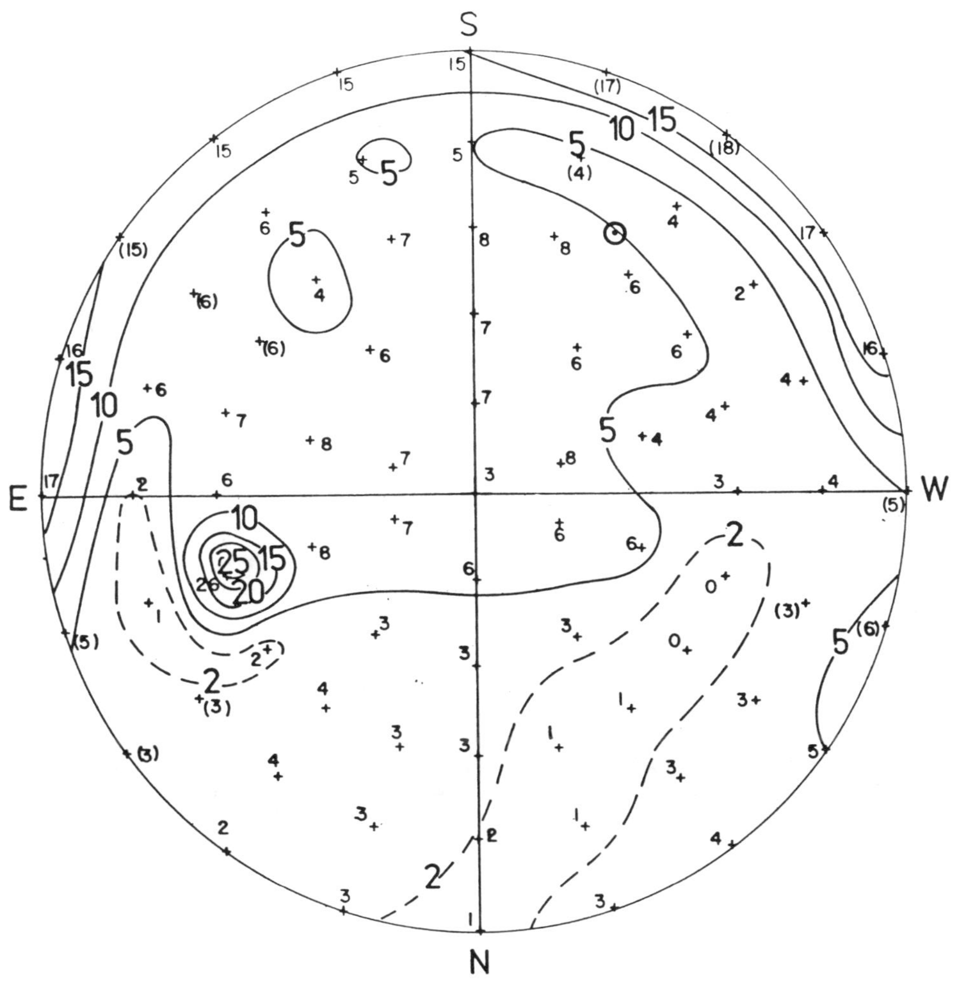

Figure 3.35b Same as Figure 3.35a, but the sky was overcast by stratocumulus clouds (10/10 Sc). The angular distribution of $\rho_b(-\Theta,\alpha)$ was computed as for Figure 3.33a. Dimensionless; isolines at steps of 5%, additionally (dashed) contours of $\rho_b(-\Theta,\alpha)$ = 2% for minimum regions. Reflected image of sun's position marked ⊙.

Figure 3.33a is evidently identical with Figure 3.6g. The patterns show correspondingly irregular and strong variations through the angular field with albedo increasing from south-east towards the maxima in west and north-west. The mean directional albedo field $\rho^*(\beta,\alpha)$ is much smoother as Figure 3.33b shows. This was computed using equation 3.29. Data used to compute the field of $\rho_c(\beta,\alpha)$ in Figure 3.33c are the best harmonised in time. This method yields, as expected, the most smooth and regular patterns. The general trend, detectable already from Figure 3.6g, appears here with remarkable clarity. It is advantageous to compute integrated albedo distributions instead of point values, and furthermore to use exclusively mast obtained data in contrast to the mixed procedure used to prepare Figure 3.33b, particularly as the term in brackets in equation 3.31 considers the measurements of ground reflection alone for the spectral adjustment while the existence of the second term within the brackets of equation 3.29 implies the simplifying assumption that both sky radiance and ground reflected radiance may spectrally be adjusted by a common factor.

As mentioned when discussing Figure 3.6g earlier, the instruments at EMPA have not been continuously operated from the same roof. This was not the case for the other sites. (The EMPA new mast installation site is about 50 m towards east-south-east from the old site and lies about 3 m higher than the old roof.) Figure 3.33b fits the effective distribution of mean surface albedo much better than the point-albedo field values of Figure 3.33a.

Corresponding polar diagrams of $\rho_c(-\Theta,\alpha)$ for the other EMPA examples IV and V from Table 3.1 have also been prepared, but due to lack of space they are not reproduced here. It, however, is of interest to know whether increasing turbidity modifies the albedo field by some assignable amount or not. After all, the site is the same and the position of the sun is similar in each case (see Table 3.1). When related to the values in Table 3.1 III, valid for Figure 3.33a, the turbidity changes for two cases by factors of 1.62 and 2.92 in terms of B_s and the turbidity factor D_c/G_c by 1.24 and 2.06. While neither R_c nor R_c/G_c (rows 17 and 18 in Table 3.1) seems to be sensitive enough to show a clear decrease, the ratio of the extreme albedos $\rho_c(-\Theta,\alpha_s)(\max)/\rho_c(-\Theta,\alpha_s)(\min)$ decreases from 32.3/0.79 = 40.9 (Figure 3.33a) to 29.0/1.25 = 23.2 and 32.7/1.79 = 18.3 respectively (the albedo diagrams contain only rounded percentages). A possible smoothing effect of turbidity on the albedo mosaic will be tested later using a great amount of data. It should be mentioned, that while $\rho_c(-\Theta,\alpha_s)(\min)$ appears at $\Theta = -18°$, $\alpha = -90°$ in all three cases, $\rho_c(-\Theta,\alpha_s)(\max)$ at $\Theta = -18°$, $\alpha = +90°$ of case III shifts to $\Theta = -36°$, $\alpha = 126°$ in case IV and to $\Theta = -18°$, $\alpha = 126°$ in case V. Figures 3.34 - 3.36 are, as already emphasised, point albedo distributions and are, therefore, comparable only with Figure 3.33a. The most apparent difference between Figure 3.33 and 3.34 is the value of the nadir albedos. The reason is immediately apparent if one compares the fish-eye photographs with the pattern of observations. The concrete platform beneath the mast on the new EMPA roof appears, under intensive sunlight, like bright sand. The white object with strong reflection in the west of the old roof is no longer visible from the new roof. The $\rho_c(-\Theta,\alpha_s)$ distribution for Locarno-Monti (Figure 3.35a) also shows good consistency with the corresponding fish-eye picture, Figure 3.24. The lightest part of the concrete roof to the east-north-east, as well as the narrow strip of the Lake Maggiore visible

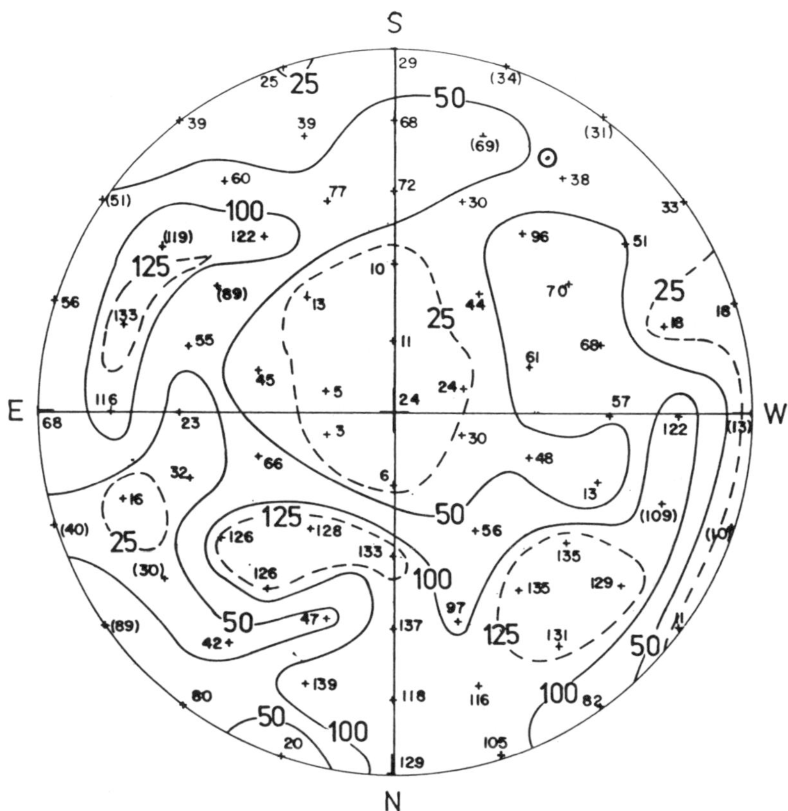

Figure 3.36 Angular distribution at Weissfluhjoch, roof of the Federal Institute for Snow and Avalanche Research (Fig. 3.25) of ground albedo $\rho_c(-\Theta, \alpha)$, computed as for Figure 3.33a. Dimensionless; isolines at steps of 25%. Reflected image of sun's position marked ⊙.

on the horizon in the sun's direction show the highest reflectivity (this latter certainly contains a strong specularly reflected component), likewise high values appear along the road and on the light walls of the building north of the Institute. On the other hand, the shadow of the Institute building on the grass causes the lowest effective albedo values along the $\Theta = -54°$ belt north and west from the mast's site.

Figure 3.35b for Locarno-Monti has been included in the set of distributions to demonstrate the remarkable difference between the albedo values for clear and overcast skies for the same site. The $\rho_b(-\Theta,\alpha)$ values close to the horizon in the sun's direction represent now the albedos of the relatively dark lake surface, when compared with the clear sky conditions, but the albedos are still higher than those for the other parts of the landscape. The bright spot in the east of the roof seems to keep its relative maximum value under the overcast sky.

The most striking features of the Weissfluhjoch distribution (Figure 3.36) are the albedo values exceeding 1.0 (100%) for coherent snow covered parts of the scenery (compare with the fish-eye view, Figure 3.25). It will be noted that these surfaces are down sun. It is evident that, in some directions, wide deviations from the downward facing horizontal albedo $\rho_c(-0)$, which may in general be considered as best approximation of the average landscape albedo, occur. Inclined surfaces facing the sun will have a higher irradiance than a horizontal surface when the sun is low as is the case for the Weissfluhjoch example. The albedos here, however, have been estimated from the horizontal irradiance. Consequently the computed surface albedos can exceed 100% and may reach 180% or more. Further, $R_c(-0)/G_c$ values > 90% have often been measured for horizontal surfaces for fresh snow. Since the Weissfluhjoch scenery is made up for a great part of sloping terrain, such high albedo values at certain azimuths computed from horizontal irradiances are not surprising, especially for low solar elevations. It is a visual experience that parts of a snow field or of a sea surface often appear far brighter than the most brilliant regions of the sky. The comparison of the figures in rows 12 and 14 of Table 3.11 help to illustrate this fact. The patterns of the Weissfluhjoch radiance distribution correspond fairly well with the dark and white areas of the fish-eye view. The relatively low albedo values for the nadir and points in its vicinity are in part caused by the presence of dark objects. These low values may also indicate that parts of the concrete roof are free of snow.

32. Preliminary results - ground reflected diffuse irradiance on inclined surfaces - clear skies

The ground reflected component of total diffuse irradiance on inclined surfaces is, as mentioned, not currently processed as an immediate integral of measured reflected radiances in the Swiss observational programme. This component may, however, be estimated from the difference between total and sky diffuse irradiances on positively inclined surfaces. In computing this difference, one must remember that the measuring programme of the two instrument systems (Figures 3.2 and 3.3) is not strictly simultaneous. The pyranometer measurements in the 77 positions takes 6 minutes, 40 seconds and the double scan of the 121 sky radiances lasts 4 minutes, 20 seconds. Associating, therefore, only the second half of the double scan with the pyranometer measurements, the time intervals covered by the two data sets are both centred around the third minute after the common start (in fact they are centred around the 194th and 200th second respectively).

Table 3.11 Characteristic values for the albedo distributions presented in Figures 3.33 to 3.36.

Row No.	Figure No:	3.33a	3.33b	3.33c	3.34	3.35a	3.35b	3.36
1	Site		EMPA		EMPA	Locarno-Monti		Weissfluhjoch
2	Date		29.7.1981		30.6.1982	13.10.1981	20.10.1981	25.11.1980
3	Time Solar Time		0923		1205	1212	1341	1402
4	Solar Height $\gamma°$, Azimuth $\alpha°$		48 -61		65 7	36 8	28 32	16 32
5	Turbidity Coefficient B_s (Schuepp)		0.130		0.125	0.040	(Sc10)	0.015
6	Turbidity Factor T_L (Linke)		4.0		4.0	2.7	(Sc10)	2.2
7	Turbidity Ratio D_c/G_c		0.17		0.16	0.13	1.00	0.18
8	$R_c(-0)$ or $R_b(-0)$ $[Wm^{-2}]$		87		132	103	4	212
9	G_c or G_b $[Wm^{-2}]$		741		923	637	77	305
10	Pyranometer Albedo $R_c(-0)/G_c$		0.12		0.14	0.16	0.06*	0.70
11	Nadir Albedo $\rho_{cn}(-0)$ or $\rho_{bn}(-0)$	0.09			0.45	0.10	0.03	0.24
12	$L_c(-\Theta,\alpha)_{max}$ $[Wm^{-2}\ sterad^{-1}]$		75		132	170	6.5	137
13	$(-\Theta,\alpha)$ for Row 12)		-18 90		-90-	0 18	-36 -108	-36 180
14	Average sun + sky radiance G_c/π $[Wm^{-2}\ sterad^{-1}]$		236		294	203	25	97
15	$\rho_c(-\Theta,\alpha)$ (max) = $L_c(-\Theta,\alpha)_{max}/(G_c/\pi)$		0.32		0.45	0.84	0.26	1.37

* $R_b(-0)/G_b$

Each set of the $R_c(\beta,\alpha)$ irradiances computed this way has first been processed on its own. Within such individual series, normalized values $R_c(\beta,\alpha_S)/D_c$ show a fair link with surface solar azimuth α_S for each of the measured inclinations β = 18°, 39°, 60°, 75° and 90°. It was, therefore, easy to interpolate the ratio $R_c(\beta,\alpha_S)/D_c$ for round values of α_S = 0°, 30°, 60°, 90°, 120°, 150° and 180°. Each of these 5 x 7 = 35 (β,α_S) sub-divisions is available for the 551 clear sky cases discussed in Chapter 3, Part I, Section 6 (Refer Table 3.2). This material was thus available for the statistical analysis of the regression

$$\frac{R_c(\beta,\alpha_S)}{D_c} = f(\frac{D_c}{G_c}, \frac{R_c}{G_c}, \gamma, \alpha_S, \beta) \tag{3.32}$$

Since, due to staff limitations, the mast system could often not be kept operating parallel to the other instruments (see Table 3.2), the downward facing mast albedo $\rho_g(-0) = R_c/G_c$ could not always be included as an additional parameter in the regression analysis.

The same amount of data of the same basic structure is, of course, also available for analysing the sky diffuse irradiance component according to the regression equation 3.7.

Similar scatter diagrams were prepared for solving equation 3.32 as used for solving equations 3.2, 3.3 and 3.7. As examples of the partial relationship between $R_c(\beta, \alpha_S)/D_c$ and the solar height γ, Figure 3.37, bottom half, shows the scatter diagram of the ratio for the constant surface geometry β = 60°, α_S = 0° and Figure 3.38 for β = 60°, α_S =120°. For both Figures 3.37 and 3.38, the same selection of EMPA site clear sky data sets has been chosen as used earlier for solving equation 3.3. For α_S = 0°, i.e. Figure 3.37, and 180°, only one value is observed but, for the azimuths in between, always a pair of values for + α_S and for $-\alpha_S$ were observed and plotted on the graph (marked x and . respectively). If the entries were near enough to each other, the values of the parameters D_c/G_c and, where available, - $\rho_c(-0)$ were entered only once.

As may be seen, all three groups of scattering points show some dependency on the solar height. No influence of the turbidity parameter D_c/G_c on the ratio $D_{cS}(\beta,\alpha_S)/D_c$ may be detected, in consistency with the results of Part I, Section 6. Any influence of turbidity on the ratio $R_c(\beta,\alpha_S)/D_c$ for α_S = 0 is also questionable; however, for α_S = 120° a trend is recognisable showing the expected decrease of the reflected share with increasing turbidity. No clear influence of albedo on the mean values seems to appear for $\gamma > 30°$ both for α_S = 0° and for α_S = 120°, but for $\gamma < 30°$, $R_c(\beta, \alpha_S)/D_c$ tends to increase with increasing downward facing albedo for both azimuth directions. Since D , D /G and ρ (0) also depend on the solar elevation and, as this dependence is different for the three quantities, other groupings of data may eventually prove more successful in isolating single regressions within the multi-parametric relationship (for instance refer Figure 3.41).

Figure 3.39 presents the mean trends of the functions $D_{cS}(\beta,\alpha_S)/D_c = f(\gamma)$ and $R_c(\beta, \alpha_S)/D_c = f(\gamma)$ from Figure 3.37 and 3.38 together with the corresponding regression curves as a function of α_S and solar altitude γ. The regression curves have been drawn by eye through the scattering points (all 5 x 7 single scatter diagrams, both for $D_{CS}(\beta,\alpha_S)/D_C$, as well as for $R_C(\beta, \alpha_S)/D_C$). By changing the proportion of the ordinates against that used for the former "processing" graphs, the function type has been made much clearer. Data for the other sites are available in reference 15.

Due to geometry alone, the sky-diffuse contribution to the total diffuse irradiance of positively inclined surfaces is in general greater than that of ground reflection. However, for clean air and high ground albedo, the effect of geometry may be overweighed by the high ground reflected contribution (the dependence of the critical ratio $R_C(\beta,\alpha)/D_{CS}(\beta,\alpha) = 1$ on the other parameters is currently under study). The form of all curves in Figure 3.39 reflects clearly the interaction of the individual dependencies of $D_{CS}(\beta,\alpha_S)$ and $R_C(\beta,\alpha_S)$ on γ and α_S, and D_C on γ in the different ranges of these variables. For sun facing slopes and small values of γ, $D_{CS}(\beta,\alpha)$ increases evidently faster with solar elevation than D_C does. In the range of about $20° \leqslant \gamma \leqslant 50°$, all sky diffuse curves show the opposite behaviour, while, for $\gamma > 50°$, slopes facing opposite the sun show the impact of the known effect of horizon brightening. The ratio $R_C(\beta,\alpha)/D_C$ behaves in more or less a complementary way to the $D_{CS}(\beta,\alpha)/D_C$ functions. The decrease of $D_{CS}(\beta,\alpha)/D_C$ with γ which is strong for $\alpha_S = 0$, blunts with increasing α_S and eventually turns into a slight increase for $\alpha_S > 90$. As a consequence, the curves for α_S = constant show an opposite hierarchy for large values of γ than for small values of γ, with $\gamma = 40°$ being a critical solar elevation which gives a value of $R_C(\beta,\alpha_S) = 0.17\ D_C$ for all azimuths. Similar turning points have also been found earlier for the total diffuse irradiance on vertical surfaces (for example refer Figure 8.28 in Reference 11). A more detailed interpretation of these functions will be published in due course (15).

Figure 3.40 gives an impression of the partial relationship $R_C(\beta,\alpha_S)/D_C = f(\alpha_S)$ for different sites. The left part of the graph is for a surface inclination, $\beta = 39°$, and the right part for $\beta = 90°$, both at the relatively low solar elevation $\gamma = 20°$. In preparing the graphs, the data sets selected have been restricted to the solar elevation limits $19° < \gamma < 21°$. Within this interval only 6 series could be found for EMPA, 5 for Locarno-Monti and also 5 for Weissfluhjoch - 3 with and 2 without snow cover, to test. Plotting the single values from within individual data sets, one set for clockwise and one set for anti-clockwise observations in the range $0° < \alpha_S < 180°$, yielded curves of looping and meandering shape dependent on the absolute azimuth of the sun's position, Figure 3.41. To preserve the survey data, Figure 3.40 presents the zones of this meandering (see legend). The following features, all valid for $\gamma = 20°$ only, may be observed (the different scales of the left and right ordinates should be noted):

1. For the same azimuth, $R_C(\beta,\alpha_S)/D_C$ is, of course, greater for all sites for vertical surfaces $\beta = 90°$ than for $\beta = 39°$, but for Weissfluhjoch by a far greater amount than for the other sites. A rough check of the arithmetic means of $R_C(\beta,\alpha_S)/D_C$ at $\alpha_S = 90°$ (rows 1 and 2) yields the following table:

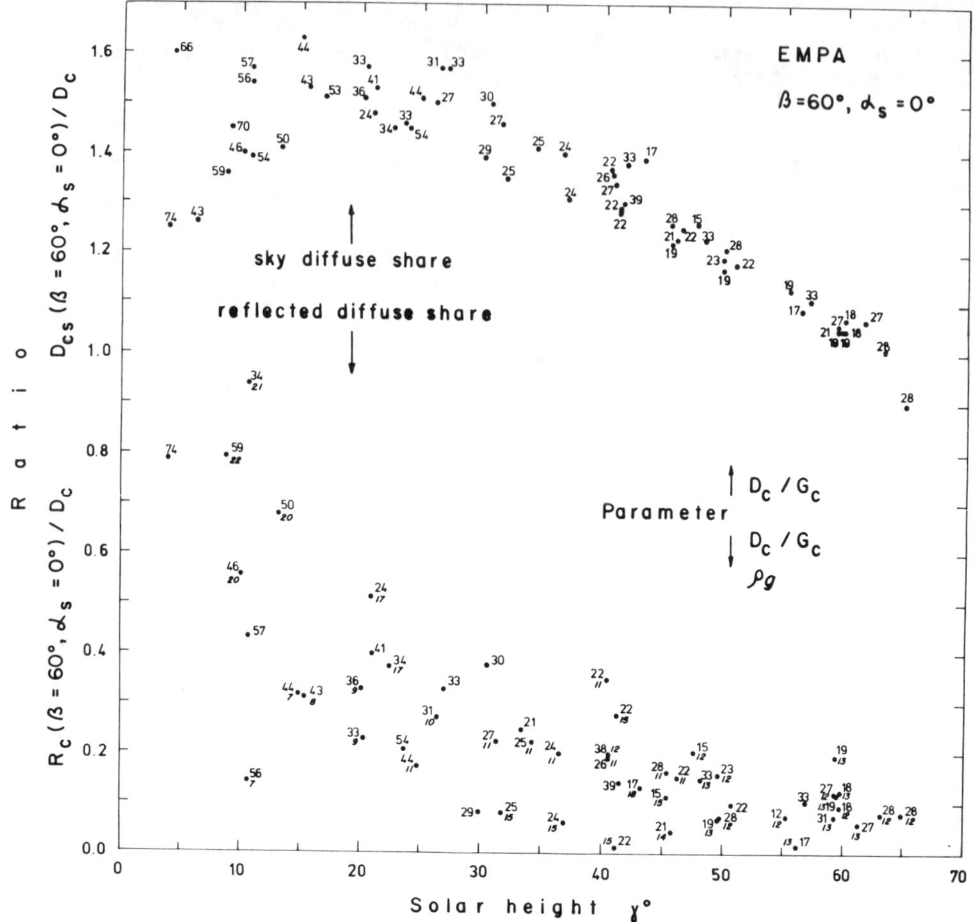

Figure 3.37 Scatter diagram relating the ratio $D_{cs}(\beta,\alpha_s)/D_c$ (upper part), as well as the ratio $R_c(\beta,\alpha_s)/D_c$ (bottom part) to solar altitude γ for a sun facing surface (α_s = 0) inclined at β = 60° to the horizontal at EMPA. The turbidity parameter D_c/G_c is ascribed to the entries in both parts of the graph (straight figures), the horizontal albedo ρ_g only to the $R_c(\beta,\alpha_s)/D_c$ entries (smaller figures beneath those for D_c/G_c).

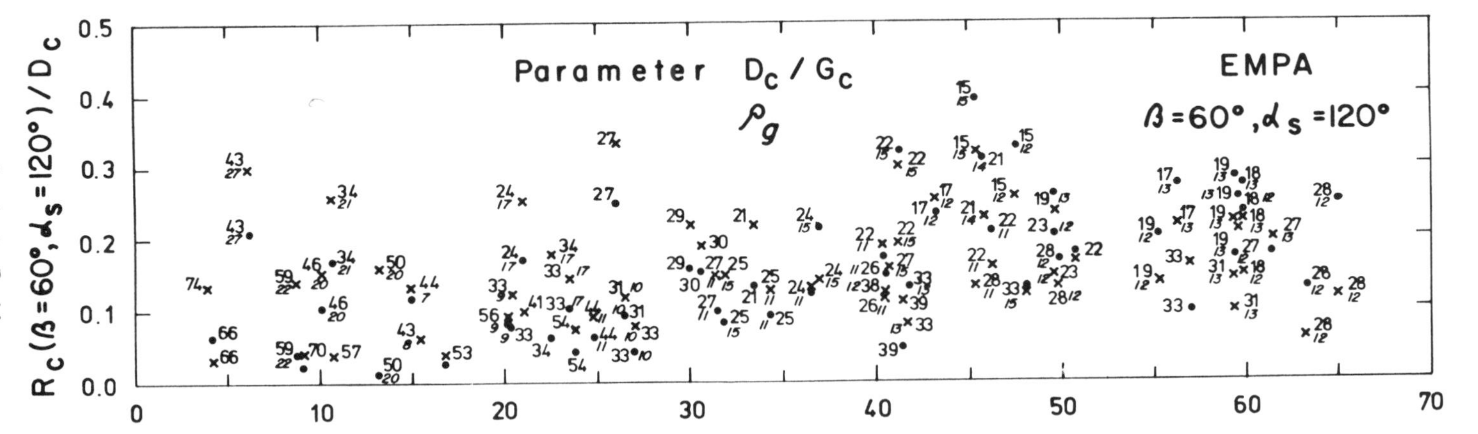

Figure 3.38 Same as Figure 3.37, but for the ratio $R_c(\beta, \alpha_s)/D_c$ alone with the surface of β = 60° inclination at sun surface azimuth angle α_s = 120° relative to the sun's position.

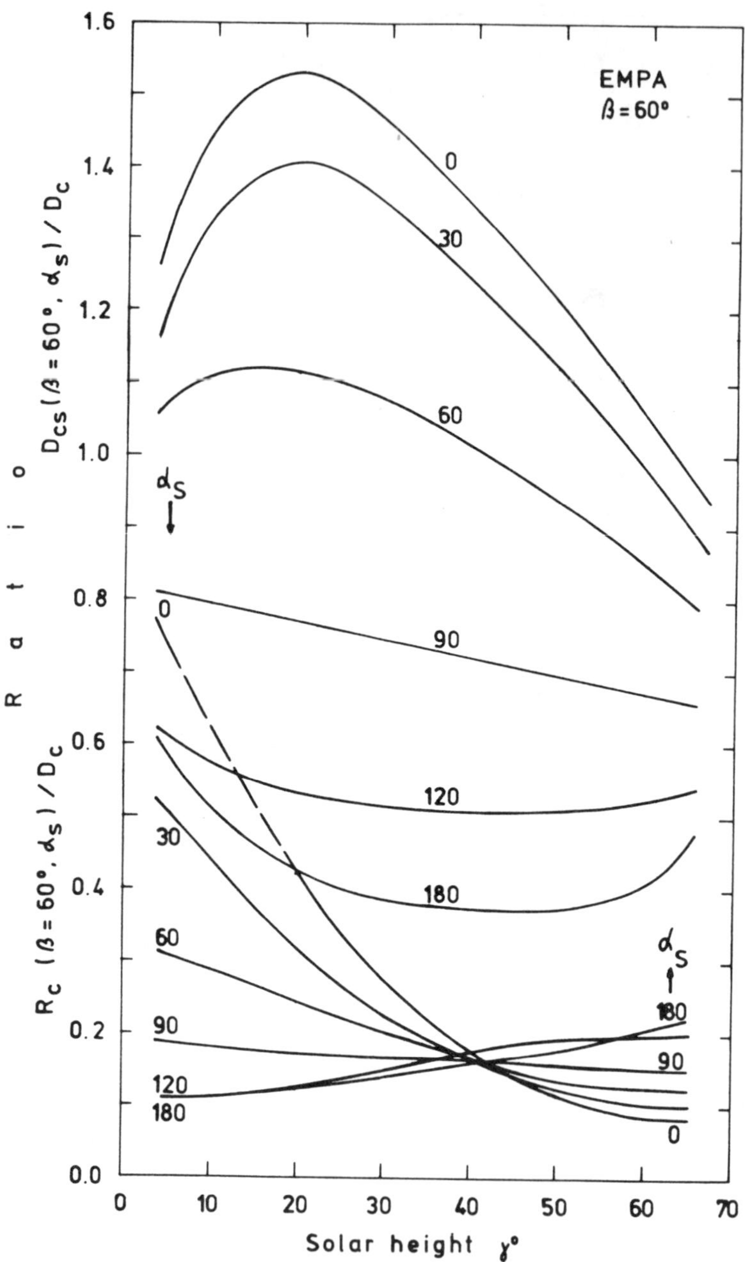

Figure 3.39 Mean relationship between the ratio $D_{cs}(\beta,\alpha_s)/D_c$ and the solar height γ for constant values of the surface-solar azimuth α_s (parametric curves labelled along their centre), as well as between the ratio $R_c(\beta,\alpha_s)/D_c$ and the solar height γ for the same values of α_s (parametric curves labelled left and right). Both families of curves valid for surface inclination β = 60°. Data of the EMPA-site were used.

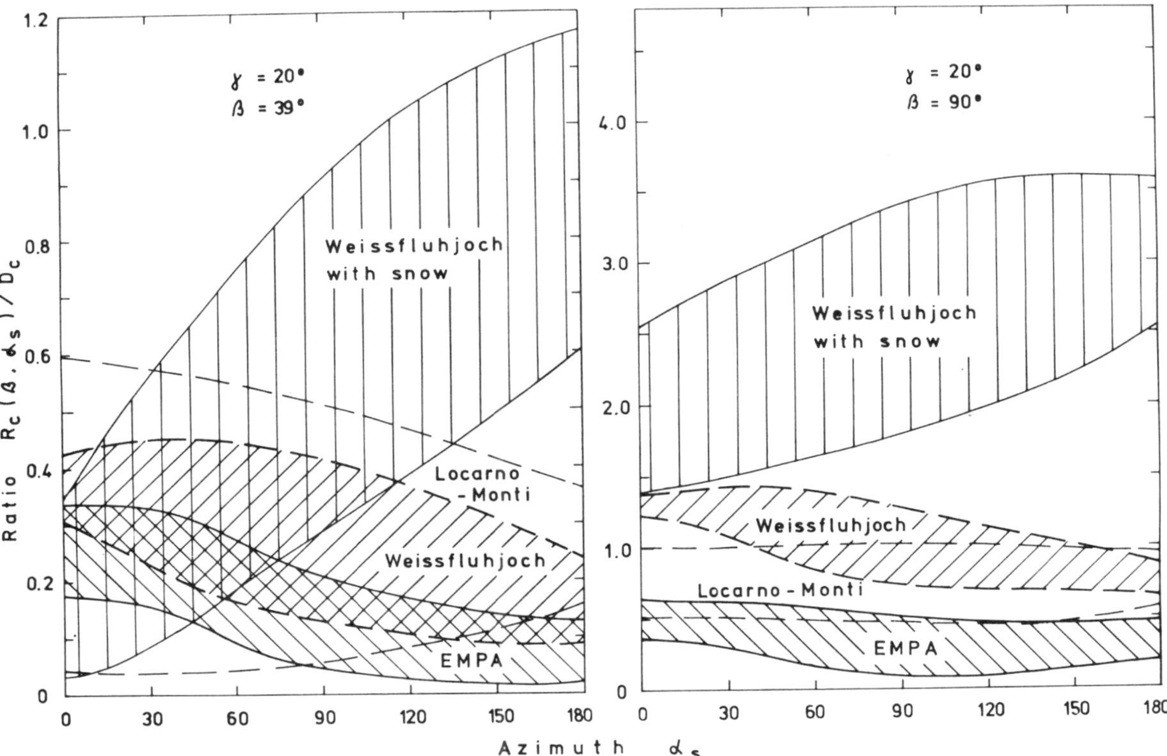

Figure 3.40 Zones indicating functional shape and width of scattering of the relationship between the ratio $R_c(\beta, \alpha_s)/D_c$ and surface-solar azimuth α_s, using measurements from different sites. Left graph is for surface inclination $\beta = 39°$, right graph for $\beta = 90°$; both for a solar height $\gamma = 20°$.

Legend: i) Weissfluhjoch with snow: continuous boundaries, vertical hatching;
ii) Weissfluhjoch: fat dashed boundaries, hatching from top right to bottom left;
iii) Locarno-Monti: thin dashed boundaries, no hatching.
iv) EMPA: continous boundaries, hatching from top left to bottom right.

Row No.	Tilt	$R_c(\beta,\alpha_S)/D_c$ Weissfluhjoch with snow (1)	no snow (2)	Ratio (1)/(2) (3)	$R_c(\beta,\alpha_S)/D_c$ Locarno Monti (4)	EMPA (5)
[1]	$\beta = 39°$	0.57	0.25	2.28	0.30	0.12
[2]	$\beta = 90°$	2.50	1.00	1.81	0.72	0.29
[3]	Ratio [2]/[1]	4.39	4.00		2.40	2.42

The isotropic ratio has a value of 4.49.

Comparison of the figures in row 3 show that reflected radiation has a much greater weight under dark blue mountain skies than under lowland conditions even in the absence of snow. Column 3 indicates on the other hand that, under the given surface-solar geometry, snow roughly doubles the ratio $R_c(\beta,\alpha_S)/D_c$.

2. Without snow cover $R_c(\beta,\alpha_S)/D_c$ either decreases or does not change with α_S the relative decrease is more pronounced for $\beta = 39°$ than for $\beta = 90°$. For $\beta = 90°$, $R_c(\beta,\alpha_S) = D_c$ for Weissfluhjoch on average, and for Locarno-Monti only as the upper limit. This indicates, that under alpine conditions, average reflected radiance over the lower hemisphere is about twice as much as the average radiance of the sky.

3. For Weissfluhjoch, with snow cover, $R_c(\beta,\alpha_S)/D_c$ increases with α_S, more strongly for $\beta = 39°$ than for $\beta = 90°$. The rocky slope to the north of the roof is part of the cause of this increase, since it affects fully the pyranometer value $D_c(\beta,\alpha_S)$ but is only partially caught in $D_{cs}(\beta,\alpha_S)$ (scanning grid is not dense enough), thus causing a too high difference for the observed $R_c(\beta,\alpha_S)$. However, this increase is partly caused by the angular distribution of radiance over the highly reflective terrain itself and the slope of the terrain. By closer inspection of the $\rho_c(-\Theta,\alpha_S)$ distributions (Figures 3.33a to 3.36) and from earlier studies of the daily change of reflected radiances (9), it may be concluded that the terrain appears brighter when viewing with the sun behind than when observing against the sun. The impact of terrain slope on surface irradiance is especially important for surfaces facing the sun.

Figure 3.41 presents the effect of turbidity expressed through the ratio D_c/G_c on the function $R_c(\beta,\alpha_S)/D_c = f(\alpha_S)$ at a solar height $\gamma = 60°$, for a surface inclination $\beta = 39°$ (left part of the graph) and for a vertical surface $\beta = 90°$ (right part of the graph). 14 clear sky EMPA site series of measurements were involved in preparing the continuous lines on the graph, all fulfilling the condition $58° < \gamma < 62°$. Attention should be paid to the different ordinate scales on the left and right parts of the graphs.

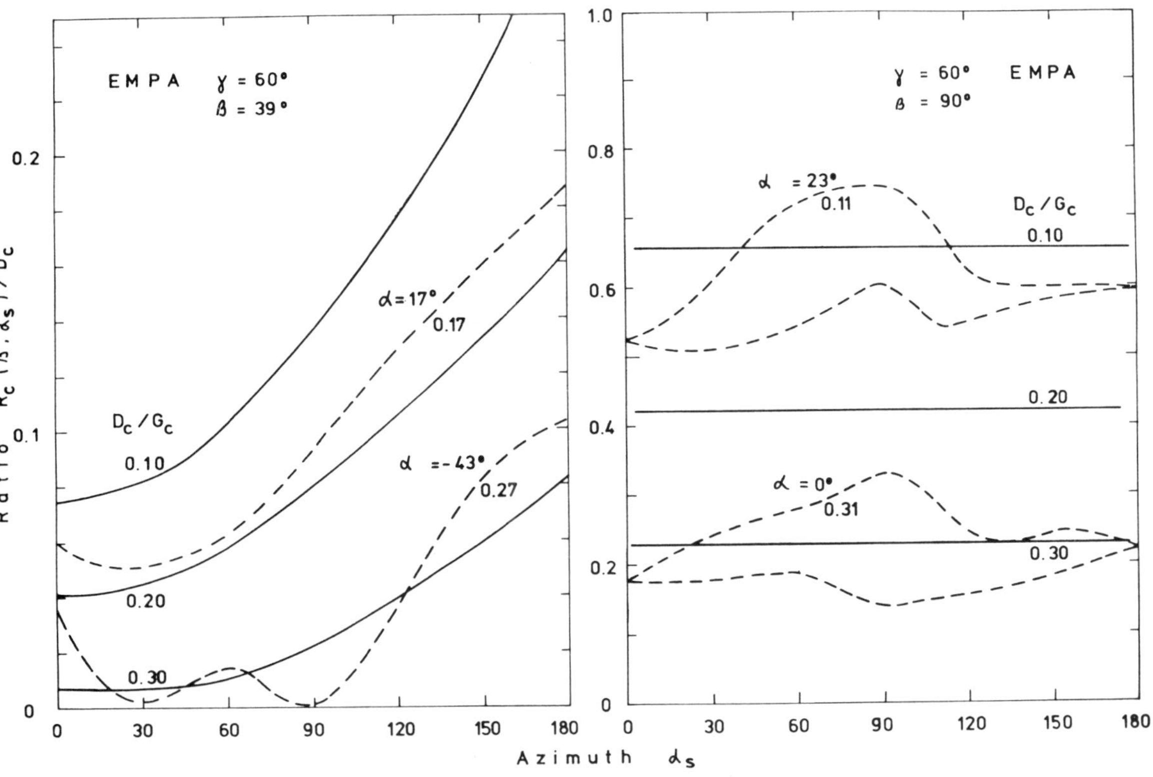

Figure 3.41 Mean relationship between the ratio $R_c(\beta,\alpha_s)/D_c$ and surface-solar azimuth α_s for different values of the turbidity index (parametric curves in the left graph, straight horizontal lines in the right graph, both for D_c/G_c = 0.10, 0.20 and 0.30). Left graph is for surface inclination β = 39°, right graph for β = 90°, both for a solar height γ = 60°. In both parts of the diagram also two individual examples are drawn (dashed curves, each labelled by respective values of the sun's azimuth α_s and of D_c/G_c). EMPA-site measurements are used.

Two examples of the 14 pairs of meandering curves for specific observations have also been drawn (dashed curves) on each part of the graph; in the left part, for β = 39°, only values for the $+\alpha_S$ range have been plotted while in the right part both the α_S+ and α_S- ranges have both been plotted. The azimuth of the sun with respect to South and the value of D_c/G_c have been ascribed to each set of curves. Some asymmetry is evident in the two cases plotted for β = 90°. The curves drawn by continuous lines mark the mean regression for constant values of the turbidity index D_c/G_c.

As may be seen, the ratio $R_c(\beta,\alpha_S)/D_c$ shows a definite increase with α_S for β = 39°, but no clear change with α_S could be detected for β = 90°. Increasing turbidity represented by increase in the ratio D_c/G_c causes a clear decrease of the ratio $R_c(\beta,\alpha_S)/D_c$ for both slopes. As the turbidity increases, D_c rises and G_c falls. As $R_c(\beta,\alpha_S)$ is strongly influenced by G_c, so the ratio $R_c(\beta,\alpha_S)/D_c$ falls. It is apparent from the shape of the distribution curves that the $R_c(\beta,\alpha_S)$ values for vertical surfaces are much higher compared with those for β = 39° for sun facing slopes than is the case for slopes facing away from the sun. For turbid air (D_c/G_c = 0,3) the vertical values even for α_S = 90° are about 10 times as large as the values on the 39° slope. (The isotropic ratio for these two slopes is 4.49). Intercomparisons of the two parts of the graph with each other, as well as with the EMPA data in Figure 3.40, indicate that the angular distribution of reflected radiance, especially its variation between nadir and horizon, significantly depends both on the sun's position and also on turbidity.

A systematic survey and interpretation of these and other features needs a more complete analysis of the full multi-variate relationships involved. This, however, implies the immediate processing of the mast data of reflected radiation data. Nevertheless, the preliminary results discussed in this section emphasise both the importance of ground reflection for the prediction of solar radiation on inclined surfaces, and the complexity of the issues involved.

CHAPTER 3 - PART IV

CLEAR DAY MODEL ADOPTED TO PRODUCE INCLINED SURFACE TABLES FOR CEC EUROPEAN SOLAR RADIATION ATLAS, VOL. II.

33. Clear sky model adopted for computation

A complete description has been given of the way in which the three components of the clear sky model have been developed, for horizontal surfaces in Chapter 2 and for inclined surfaces in the first three parts of this Chapter. This concluding section sets out the main details concerning the preparation of the clear day part of the slope tables published in the CEC European Solar Radiation Atlas, Vol. II, Inclined Surfaces. Chapter 6 provides more detailed specific computational instructions.

The horizontal surface model used was described in detail in Chapter 2. The first stage in producing clear day tables for slopes was to prepare monthly files of hourly clear day horizontal surface radiation for days of monthly mean maximum global radiation for each site. In order to compute the solar geometry, the first decision that had to be made was to select an appropriate design date in each month. Dates in the middle of the month were not chosen for the reasons already explained. It was a matter of scientific judgement to estimate the most appropriate day in each month to be representative of the days when G_{max} is most likely to be observed. The standard dates selected are given in Chapter 6, Table 6.2.

Once the appropriate monthly declinations had been established, by using the formulae in Appendix 1, the astronomical daylength could be established, and also the length of time the sun was above 4 degrees altitude on the selected day of each month. (Also refer Chapter 6, Table 6.16a and 6.16b and Table 6.17a and 6.17b).

The monthly air mass 2 Linke Turbidity Factor for clear days was then derived objectively from the monthly values of the Angstrom a + b for each site. These values of (a + b) are published in the CEC European Solar Radiation Atlas, Vol. I, Horizontal Surfaces. The method for obtaining the air mass 2 Linke Turbidity Factor from them has been described in detail in Chapter 2. The detailed relationship between the air mass 2 Linke Turbidity and the sum of the Angstrom regression coefficients used is given in Table 6.15.

Numerical values of the beam irradiance $I_c(0,0)$ and the diffuse irradiance on horizontal surfaces D_c were estimated on each half hour of the day between astronomical sunrise and sunset, working in solar time, i.e. if the sun rose at 0815, at 0830, 0930, ... etc.

The observed hourly values of $I_c(0,0)$ were then multiplied by the ratio S_{max}/S_{40max} where S_{max} is the observed monthly mean maximum daily sunshine, and S_{40max} is the 4 degree sunrise daylength. The cloudless day value of the diffuse on the horizontal surface was left unaltered.

The estimated daily global irradiation on horizontal surfaces for each month was obtained by adding up the hourly values over the day. This predicted value was compared with the observed value of G_{max} published in the CEC European Solar Radiation Atlas, Vol. I, to obtain a daily normalisation factor for each month at that site (predicted G_{max}/observed

G_{max}). All the predicted hourly values of $I_c(0,0) \times S_{max}/S_{40max}$ were then divided by the normalisation ratio as were all the values of D_c. The consequence of this process is that the sum of hourly values of the two components exactly matches the observed values of G_{max} on the horizontal plane given in the Horizontal Surface Atlas. These readjusted hourly values were then held on file for the subsequent slope calculations, together with the hourly solar geometry generated to compute them.

The monthly inclined surface tables were prepared from the hourly horizontal files. Using the stored geometrical data for the solar position, the cosine of the angle of incidence could be generated for any hour for any surface. The hourly direct beam slope irradiance was estimated as $I'_c(0,0) \cos \nu / \sin \gamma$ where $I'_c(0,0)$ is the normalised hourly beam irradiance of the horizontal surface, $\cos \nu$ is the cosine of the angle of incidence, and γ is the solar elevation. The clear sky slope correction function for diffuse radiation f_2 was then evaluated for each specific slope and orientation, and the consequent values of $D_c(\beta,\alpha)$ established as $f_2 D'_c$ where D'_c is the normalised diffuse irradiance on a horizontal surface. Summing over the day, the daily slope global and diffuse irradiation were estimated in kWh m^{-2}. A suitable printing routine to output the data from file completed the table production process. A sample table is included as Table 3.12. 102 tables for specific sites were produced, and are included in the CEC European Solar Radiation Atlas, Vol II, Inclined Surfaces.

Programs to carry out this operation were developed both on the University of Sheffield Prime mainframe computer, and on Apple microcomputers. The automation of the process is very desirable, as there is a lot of numerical calculation to be performed.

Chapter 6 provides a description of a manual approach which throws further light on the fine details of the process to be followed in the computations.

Where values of G_{max} were not available for sites in the Atlas, the data could not be normalised. For a few sites, data on S_{max} were also missing and estimates were prepared for these values, using the data about the relationship between S_{max} and S_m/S_o given in Appendix 3.

Table 3.12 Sample of lower half of European Solar Radiation Atlas Table with the associated heading. In the atlas G_c is used to describe $G_c(\beta,\alpha)$ and D_c to describe $D_c(\beta,\alpha)$.

IRELAND VALENTIA

Latitude : 51°56'N Longitude : 10°15'W Altitude : 20 m

Estimated clear day means of daily global radiation G_c and diffuse radiation D_c on inclined planes (1966-75)

Units : kWh.m^{-2}

		Jan	Feb	Mar	Apr	May	Jun	Jul	Aug	Sep	Oct	Nov	Dec	Mean
Date in month		29	26	29	28	29	21	4	4	4	4	4	4	
10° South	G_c	2.19	3.65	5.74	7.36	8.61	8.72	8.45	7.72	6.17	4.30	2.74	1.72	5.62
	D_c	0.40	0.50	0.65	0.99	0.92	0.91	0.79	1.08	0.84	0.50	0.39	0.30	0.69
30° South	G_c	3.37	5.03	6.86	7.88	8.60	8.58	8.35	8.03	7.10	5.63	4.16	2.88	6.38
	D_c	0.50	0.61	0.77	1.12	1.05	1.03	0.91	1.21	0.98	0.61	0.49	0.37	0.80
Latitude South	G_c	4.25	5.91	7.25	7.58	7.74	7.59	7.42	7.54	7.26	6.38	5.17	3.78	6.49
(51°56')	D_c	0.57	0.71	0.91	1.27	1.21	1.20	1.08	1.35	1.11	0.72	0.58	0.43	0.93
60° South	G_c	4.43	6.03	7.15	7.23	7.19	7.00	6.85	7.12	7.08	6.45	5.38	3.99	6.32
	D_c	0.59	0.74	0.95	1.31	1.28	1.27	1.15	1.39	1.16	0.75	0.60	0.45	0.97
Vertical South	G_c	4.37	5.54	5.72	5.01	4.42	4.16	4.10	4.73	5.39	5.68	5.25	4.09	4.87
	D_c	0.60	0.79	1.07	1.44	1.50	1.49	1.39	1.52	1.26	0.84	0.64	0.46	1.09
Vertical SE/SW	G_c	3.25	4.39	5.25	5.39	5.36	5.18	5.09	5.34	5.34	4.84	4.02	2.99	4.70
	D_c	0.55	0.75	1.05	1.44	1.52	1.52	1.41	1.53	1.25	0.82	0.59	0.41	1.07
Vertical E/W	G_c	1.41	2.44	3.78	4.76	5.60	5.58	5.47	5.11	4.25	3.09	1.98	1.19	3.73
	D_c	0.44	0.66	0.98	1.39	1.52	1.53	1.42	1.52	1.18	0.75	0.50	0.33	1.02
Vertical NE/NW	G_c	0.42	0.86	1.79	2.92	4.01	4.10	3.98	3.41	2.31	1.25	0.59	0.30	2.17
	D_c	0.38	0.57	0.88	1.27	1.45	1.46	1.37	1.40	1.06	0.66	0.43	0.29	0.94
Vertical North	G_c	0.37	0.56	0.84	1.51	2.49	2.62	2.51	1.99	1.11	0.63	0.41	0.28	1.28
	D_c	0.37	0.56	0.84	1.19	1.38	1.40	1.31	1.31	0.99	0.63	0.41	0.28	0.89
INPUT DATA														
Daily sunshine hours		5.8	7.7	9.9	12.3	13.7	14.3	13.5	13.0	10.6	7.8	6.7	5.1	
Angstrom a + b		0.83	0.85	0.85	0.84	0.86	0.86	0.87	0.84	0.84	0.86	0.84	0.82	0.85
Normalisation factor		1.01	1.01	1.02	1.01	1.02	0.99	1.00	1.01	1.04	1.07	1.00	0.99	

CHAPTER 3 - References

1. Bener, P., Frohlich, C. and Valko, P., (1973), Mobile station for automatic measurements of spectral solar and sky radiance and of the fluxes of global and sky radiation on differently orientated planes, Proceedings, Internat. Solar Energy Soc. Conf., Paris, Paper E31, pp.10.

2. Valko, P. and Heimo, A., (1980), The Swiss Mobile Solar Radiation Research System. Chapter IV, pp. 20-27 in the Task IV/2 Handbook, "Validation of the Guidelines for Portable Meteorological Instrument Packages of the Int. Energy Agency Solar R&D Program, Washington.

3. World Meteorological Organisation, (1981), Meteorological aspects of the utilisation of solar radiation as an energy source, WMO Tech. Note No. 172, Geneva, pp.298.

4. Heimo, A., (1984), Diagnosis of the atmosphere through optical measurements performed with a mobile station. Thesis ETH Zurich, No. 7755, 1985, 188p.

5. Brusa, R. and Frohlich, C., (1972), Entwicklung eines neuen Absolutradiometers, WRCD Techn. Note No. 1.

6. Frohlich, C. and Bener, P., (1978), Spektroradiometer zur Bestimmung der atmospharischen Trubung, WRCD Publication No. 555.

7. Schubiger, F., (1983), Ein Numerisches Modell zur Bestimmung der Trubung, der Sichtweite und der Transmissionskoeffizienten der Atmosphare anhand von Breitband-und Spektralstrahlungsmessungen. Working Reports of the Swiss Meteorological Institute No. 115, pp.42 + 4.

8. Valko, P., (1970), On the diffuse irradiance of non horizontal plane surfaces. Proceedings WMO/IUGG Symposium on Radiation including satellite techniques, Bergen, 1968, WMO Techn. Note No. 104, pp.191-195.

9. Valko, P., (1982), Empirical study of the angular distribution of sky radiance and of ground reflected radiation fluxes, Report No. 2, within Project F of the CEC Solar Energy R&D Programme, Reading, pp.13 + 21 figures.

10. Heimo, A. and Valko, P., (1977), First results obtained from the Swiss mobile system for solar radiation measurements. Proceedings, UNESCO/WMO Solar Energy Symposium, Geneva 1976. WMO No. 477, pp.144-152.

11. Valko, P., (1980), Some empirical properties of solar radiation and related parameters, Chapter 8, pp.8.1-8.46 in the Task IV-1 Handbook "An introduction to meteorological measurements and data handling for solar energy applications" of the International Energy Agency Solar R&D Program, Washington.

12. Valko, P. (1982), Empirical study of the angular distribution of sky radiance and of ground reflected radiation fluxes. Report No. 3 within Project F of the CEC Solar Energy R&D Programme, Odeillo, pp.20 + 17 Figures + 2 Appendices (graphs).

13. Valko, P., (1984), Empirical study of the angular distribution of sky radiance and of ground reflected radiation fluxes, Proceedings, Solar World Congress, Perth, 1983, Vol. 4, pp.2183-2188.

14. Valko, P., (1967), Ueber den Zusammenhang zwischen Trubungsfaktor und Trubungskoeffizient, Archiv. Meteor. Geoph. Biokl. B15, pp.359-375.

15. Valko, P. (1984), Empirical study of the angular distribution of sky radiance and of ground reflected radiation fluxes, Report No. 4 within Project F of the CEC Solar Energy R&D Programme.

16. Kittler, R. (1986), Proceedings of C.I.E. inter-sessional conference, "Sunlight in Buildings", Bouwcentrum, Rotterdam, III, pp.273-285.

17. C.I.E. (1973), Standardisation of luminance distribution of clear skies, Publication C.I.E. No. 22, TC-4.2.

18. Liebelt, C. (1978), Leuchtdichte- und Strahdichteverteilung des Himmels, Diss. Lichttechnisches Institut der Universistät Karlsruhe.

19. Steven, M.D. and Unsworth, M.H. (1979), The diffuse irradiance of slopes under cloudless skies, Q.J.R. Met. Soc., 105, 93-602.

20. Verdonschot, J.K.M, Van Den Brink, G.J. and Slob, W.H. (1984), Climatology of solar irradiance on inclined surfaces IV, Parts I, II, III, Commission of the European Communities, Energy, Report No. EUR 9535E EN/I, II, III, CEC Brussels.

21. Slob, W.H. (1983), Private communication of data for selected clear days at Cabauw.

CHAPTER 4

PREDICTION OF SOLAR RADIATION ON SLOPES - OVERCAST DAYS

WITH SPECIAL REFERENCE TO THE EEC REGION

by S. Aydinli and J. Krochmann, Institut für Lichttechnik, Technical University of Berlin.

1. Introduction

The final development of the overcast sky model for the prediction of solar radiation on inclined surfaces on days without sun and its detailed checking against European observations on slopes formed one of the key contractual responsibilities of the Institut fur Lichttechnik, Technical University of Berlin in the second phase of the CEC Inclined Surface Solar Radiation Prediction programme.

The CEC solar radiation data tape compiled during the first CEC solar energy program was used to carry out the detailed checks. It should be stressed, as the next chapter explains in detail, that the overcast sky model forms an integral part of the overall method for predicting mean monthly solar radiation on slopes. Considerable importance must, therefore, be given to the statistical accuracy achieved in modelling solar radiation on slopes under overcast day conditions.

2. Methodology

The overcast sky data on the CEC solar radiation data tape were extracted for 5 different European sites, for which hourly horizontal and inclined surface data were available together with daily sunshine records. Comparisons were made between the precalculated and the measured values of solar radiation on horizontal surfaces. Comparisons were also made between the observed hourly ratios of the slope irradiation/horizontal irradiation and the corresponding predicted values of the ratio. Statistical surveys were prepared of the results giving mean values, minimum values and maximum values together with standard deviations.

3. Defining overcast days by daily observed sunshine

A uniformly overcast sky is characterised according to the Commission Internationale de l'Eclairage solely by its relative luminance distribution (CIE standard overcast sky). In practice it is often necessary to describe an overcast day with an overcast sky as a day when the observed daily sunshine duration S_d equals zero hours.

By adopting this criterion, the following additional sky conditions are also included:

i) partly cloudy skies on calm days with permanently covered sun, but with high radiances on specific areas of cloud.

ii) days with heavy fog or dark thunderclouds, on which the direct beam irradiance does not reach the threshold value of the sunshine recorder (usually 200 W/m^2), a phenomenon most likely to occur on days which have low solar elevations throughout the day, i.e. winter.

Therefore, one must expect the individual measurements of solar irradiances for the overcast days to show great scattering, like the investigations of overcast day illuminances by Krochmann (1) or Nakamura (2).

4. The available observed data

Observed data of hourly horizontal and slope irradiation were available for the stations and inclined surfaces in Table 4.1 for the statistical analysis of the irradiance of overcast skies. Only the values for those days which had no sunshine were used from the observed data.

5. The mathematical basis of the overcast sky model

The predicted irradiance on a horizontal surface ${}_pG_b$ from overcast sky can be found as a function of solar altitude γ and solar distance correction K_d as

$$ {}_pG_b = K_d(2.61 + 182.6 \sin \gamma) \quad W/m^2 \tag{4.1} $$

In practice the variations in K_d are so small compared with the variance of the observed data that the variations in K_d were neglected in the subsequent analysis and K_d was set equal to 1.

For the calculation of overcast sky irradiances on inclined surfaces, the standard CIE relative luminance distribution of overcast sky was adopted for the relative radiance distribution. This enables the value of the ratio, sky slope irradiance/horizontal surface irradiance, to be determined. This ratio depends only on the angle of inclination β, in radians, of the receiving plane.

$$ G_{sb}(\beta,\alpha) = (0.182(1.178(1 + \cos \beta) + (\pi - \beta) \cos \beta + \sin \beta))\,{}_pG_b \tag{4.2} $$

The irradiance on an inclined surface due to reflection from the ground of incident horizontal surface irradiance is given by:

$$ R_b(\beta,\alpha) = \rho_g/2 \, . \, (1 - \cos \beta)\,{}_pG_b \tag{4.3} $$

where ρ_g is albedo of ground. This was assumed generally here to be 0.2. It was also assumed that the ground was unobstructed. However, at Bracknell, the vertical pyranometers are shielded from the ground with a blackened shading system and there a ground albedo of 0.02 was assumed.

Therefore the total predicted irradiance on an inclined surface/horizontal surface irradiance from overcast sky is determined by:

$$ {}_pG_b(\beta,\alpha) = G_{sb}(\beta,\alpha) + R_b(\beta,\alpha) \tag{4.4} $$

Hence the ratio ${}_pG_b(\beta,\alpha)/{}_pG_b$ may be found and compared with the observed ratio $G_b(\beta,\alpha)/G_b$.

Table 4.1 Stations and data used for checking diffuse irradiation models for overcast days defined as days with no bright sunshine.

Stations	Data used	Period	Comments
Carpentras	G	1973 - 1975	Hor.
	D	1973 - 1975	Hor.
	G(180, 90)	11/7/74 - 31/7/75	Vert. South
	G(180, 44)	1973	South tilt = Latitude
Odeillo	G	1/4/71 - 16/6/75	Hor.
	D	1/4/71 - 16/6/75	Hor.
	G(180, 90)	1/4/71 - 16/6/75	Vert. South
Trappes	G	1973 - 1975	Hor.
	G(0, 90)	1/1/75 - 28/2/75	Vert. North
	G(180, 48°46')	1/1/73 - 31/3/74	South tilt = Latitude
	G(135, 50°)	15/3/74 - 30/9/74	SE, Tilt 50°
Valentia	G	1/1/76 - 28/2/78	Hor.
	D	1/1/76 - 28/2/78	Hor.
	G(180, 90)	15/3/74 - 30/9/74	Vert. South
Bracknell	G	1967 - 1976	Hor.
	D	1967 - 1976	Hor.
	G(90, 0)	1967 - 1976	Vert. N.
	G(90, 180)	1967 - 1976	Vert. S.
	G(90, 90)	1967 - 1976	Vert. E.
	G(90, 270)	1967 - 1976	Vert. W.

6. Statistical analysis of extracted observations

Horizontal surfaces

For the statistical analysis of measured results, the observed values of horizontal irradiance were plotted against solar altitude, and the results compared with the predictions of formula (4.1) for different seasons. In addition to the plotted points, the mean observed values of the horizontal irradiance for 5° bands of solar elevations (i.e. 5° - band; 2.5° - 7.5°; 10° - band;7.5° - 12.5°, etc.) were calculated, as well as the maximum and minimum values and the standard deviations in each band. The number of the observed values which were used to determine the average horizontal irradiance for each solar elevation band were also extracted. Only extracted observed values of the diffuse irradiance on the days without sunshine were considered for each station (for Trappes the observed values of the hourly global irradiances on days when $S_d = 0$).

Vertical surfaces

For the inclined surfaces the corresponding statistical analysis was performed to establish the relationship ratio, R, between observed ratios of slope irradiance/horizontal surface irradiance and predicted ratios of these values, estimated using formula 4.4.

7. Discussion of results

The graphical plots which follow present the results graphically and add the associated statistics. The observed values show a great scatter, not only for the horizontal irradiances, but also for the slope irradiance/horizontal irradiance ratios.

The mean values of the horizontal irradiances present a good agreement with equation 4.1, except the data associated with high sun altitudes which are associated with summer conditions. However,the number of the observed values available in these bands is very low. Because of the very great scatter in the observed values for Odeillo, it must be assumed that no correlation exists between solar altitude and overcast day irradiances on horizontal surfaces for the observed data presented for that station.

Turning to inclined surfaces, the mean values of R show good agreement with the mathematical predictions achieved using the CIE standard overcast sky. However, the degree of scatter and also the mean values of R both increase with decrease of solar altitude, i.e. the values are highest in winter and for those observations at low solar elevation at other seasons. Under these conditions the sunshine recorders may not register due to the low direct beam irradiance, so the days are not necessarily overcast. The mean values of R for Trappes for the slope 48°46' - South are very high, especially in winter and spring. This is the consequence of some very high values of observed irradiance when $S_d = 0$.

8. Conclusions

1. There is considerable statistical variability between the mean horizontal surface radiation on overcast days and the observed horizontal surface radiation at any given solar altitude. The mean condition is, however, well described for European conditions by Equation 4.1.

$$G_b = K_d(2.61 + 182.6 \sin\gamma) \quad Wm^{-2} \tag{4.1}$$

2. The mean prediction accuracy in estimating overcast day hourly irradiation on slopes using the Moon and Spencer formulation is good, but again there is considerable statistical variation between prediction and observation from hour to hour, due to the random and fluctuating nature of cloud covers of various types.

3. The two models described in this Chapter form a sound basis for slope radiation modelling concerned with mean values of overcast day irradiation as a function of solar altitude. They, therefore, form a validated basis for the overcast sky part of the CEC inclined surface solar radiation prediction methodology for Europe, which is discussed in detail in the next chapter.

References

1. J. Krochmann, (1963), Uber die Horizontalbeleuchtungsstarke der Tagesbeleuchtung, Lichttechnik 11, p.559-562.

2. H. Nakamura & M. Oki, (1979), Study on the statistic estimation of the horizontal illuminance from unobstructed sky, Journal Light and Visual Environment, 1, p.23-31.

3. S. Aydinli, (1981), Uber die Berechnung der zur Verfugung stehenden Solarenergie und des Tageslichtes, VDI Fortschritt Berichte, Reihe 6, Nr. 79, VDI Verlag GmbH Dusseldorf.

Symbols used in the following statistical tables for overcast days

S.D. Standard deviation

G_b Overcast day horizontal irradiance

R Ratio of observed to predicted value for inclined surfaces, using Moon and Spencer formula.

The heavier circles on the graphs give the mean values for each 5 degree band.

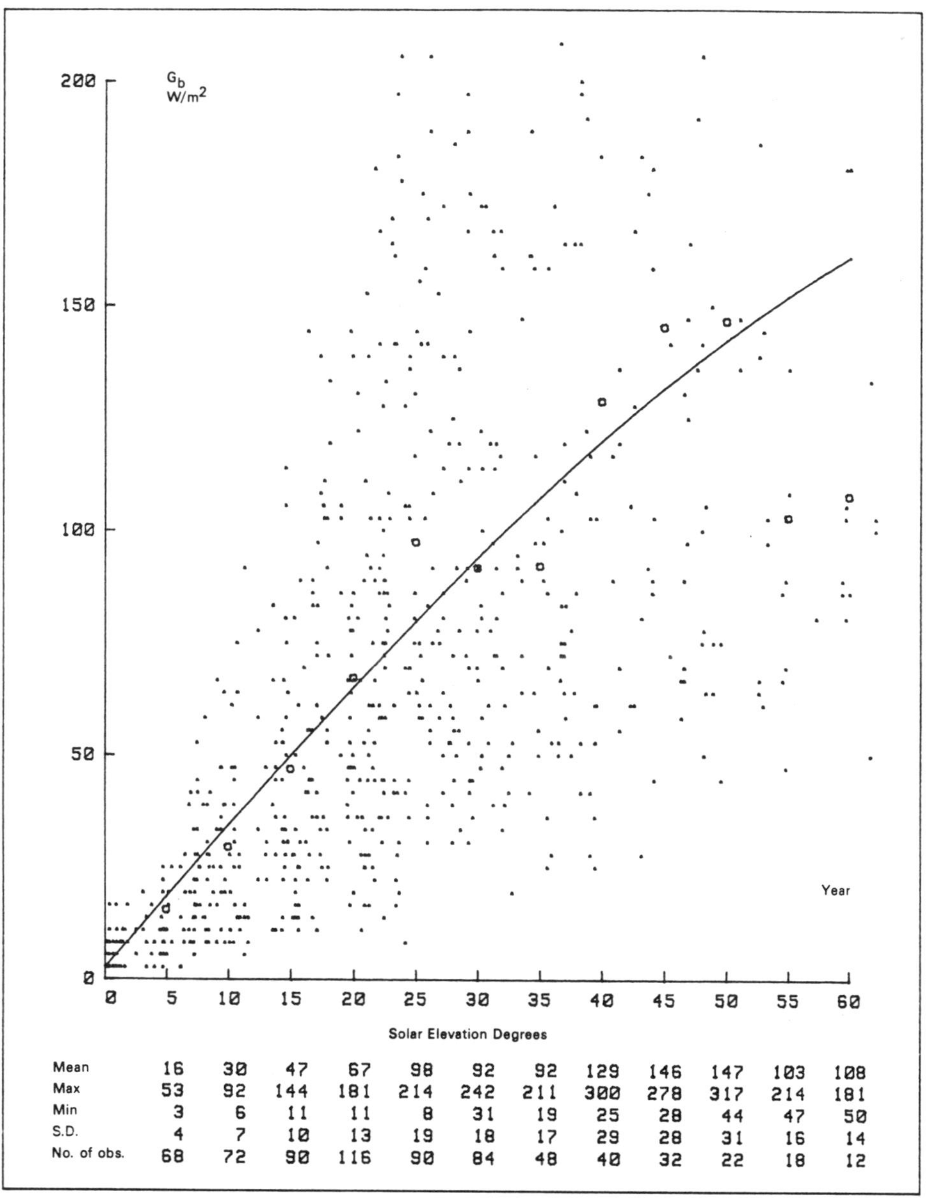

	5	10	15	20	25	30	35	40	45	50	55	60
Mean	16	30	47	67	98	92	92	129	146	147	103	108
Max	53	92	144	181	214	242	211	300	278	317	214	181
Min	3	6	11	11	8	31	19	25	28	44	47	50
S.D.	4	7	10	13	19	18	17	29	28	31	16	14
No. of obs.	68	72	90	116	90	84	48	40	32	22	18	12

Figure 4.1 Statistical distribution of horizontal surface solar radiation from overcast skies.
Station: Carpentras

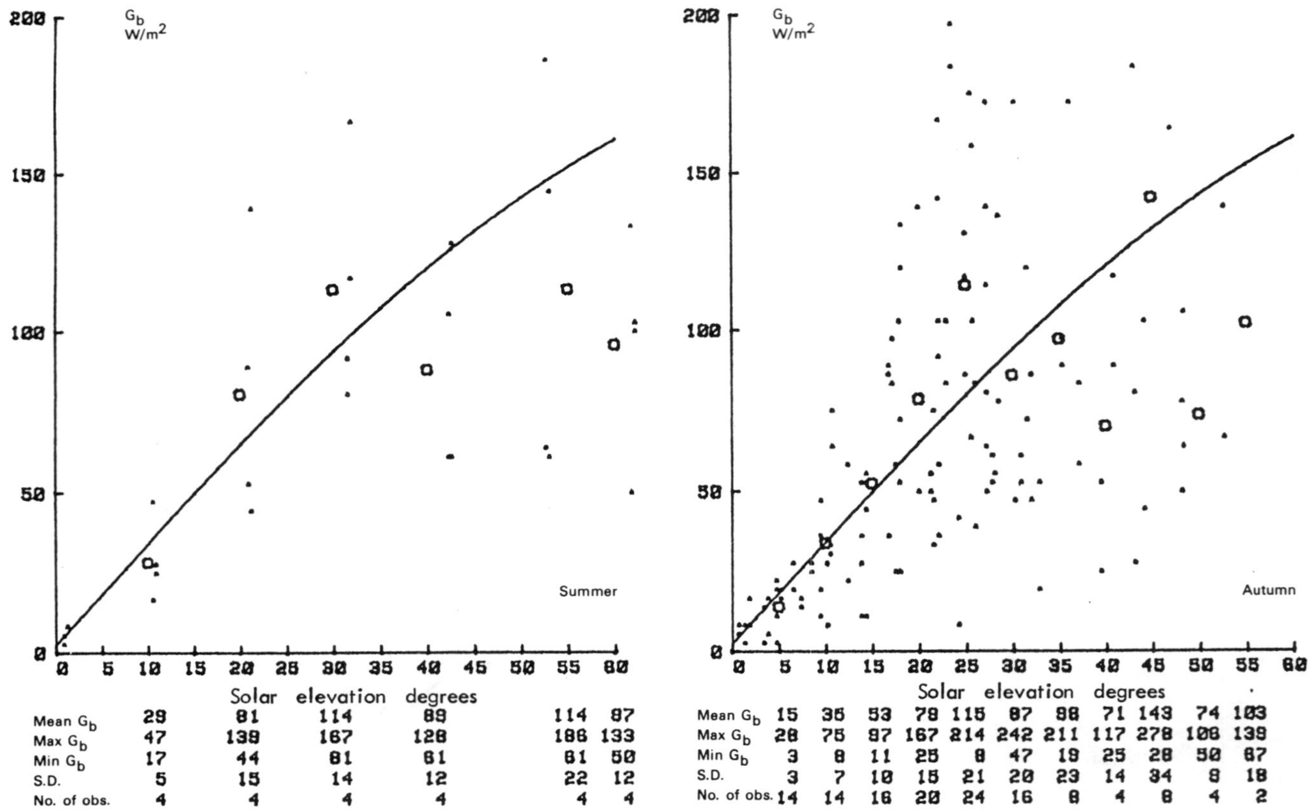

Summer

Solar elevation degrees	10	20	30	40	55	60
Mean G_b	29	81	114	89	114	97
Max G_b	47	139	167	128	186	133
Min G_b	17	44	81	61	61	50
S.D.	5	15	14	12	22	12
No. of obs.	4	4	4	4	4	4

Autumn

Solar elevation degrees	5	10	15	20	25	30	35	40	45	50	55
Mean G_b	15	36	53	78	115	87	98	71	143	74	103
Max G_b	28	75	97	167	214	242	211	117	278	108	139
Min G_b	3	8	11	25	8	47	19	25	28	50	67
S.D.	3	7	10	15	21	20	23	14	34	8	18
No. of obs.	14	14	16	20	24	16	8	4	8	4	2

Figure 4.2a Statistical distribution of horizontal surface solar radiation from overcast skies: Summer and Autumn: Carpentras

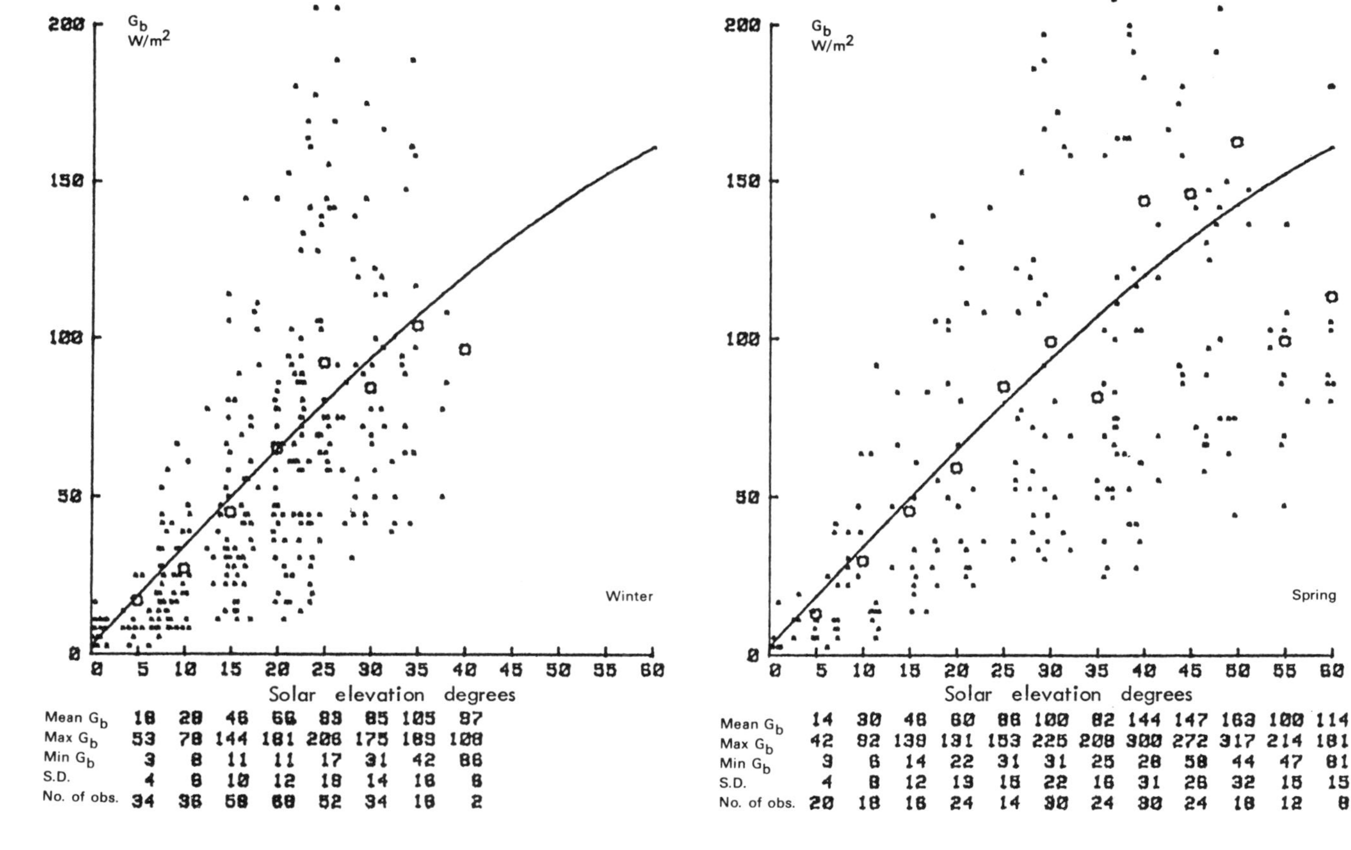

Winter

	0–5	5–10	10–15	15–20	20–25	25–30	30–35	35–40
Mean G_b	18	28	46	66	83	85	105	87
Max G_b	53	78	144	181	206	175	189	108
Min G_b	3	8	11	11	17	31	42	86
S.D.	4	6	10	12	18	14	16	6
No. of obs.	34	36	58	68	52	34	16	2

Spring

	0–5	5–10	10–15	15–20	20–25	25–30	30–35	35–40	40–45	45–50	50–55	55–60
Mean G_b	14	30	48	60	86	100	82	144	147	163	100	114
Max G_b	42	92	139	131	153	225	208	300	272	317	214	181
Min G_b	3	6	14	22	31	31	25	28	58	44	47	81
S.D.	4	8	12	13	15	22	16	31	26	32	15	15
No. of obs.	20	18	16	24	14	30	24	30	24	18	12	8

Figure 4.2b Statistical distribution of horizontal surface solar radiation from overcast skies: Winter and Spring: Carpentras

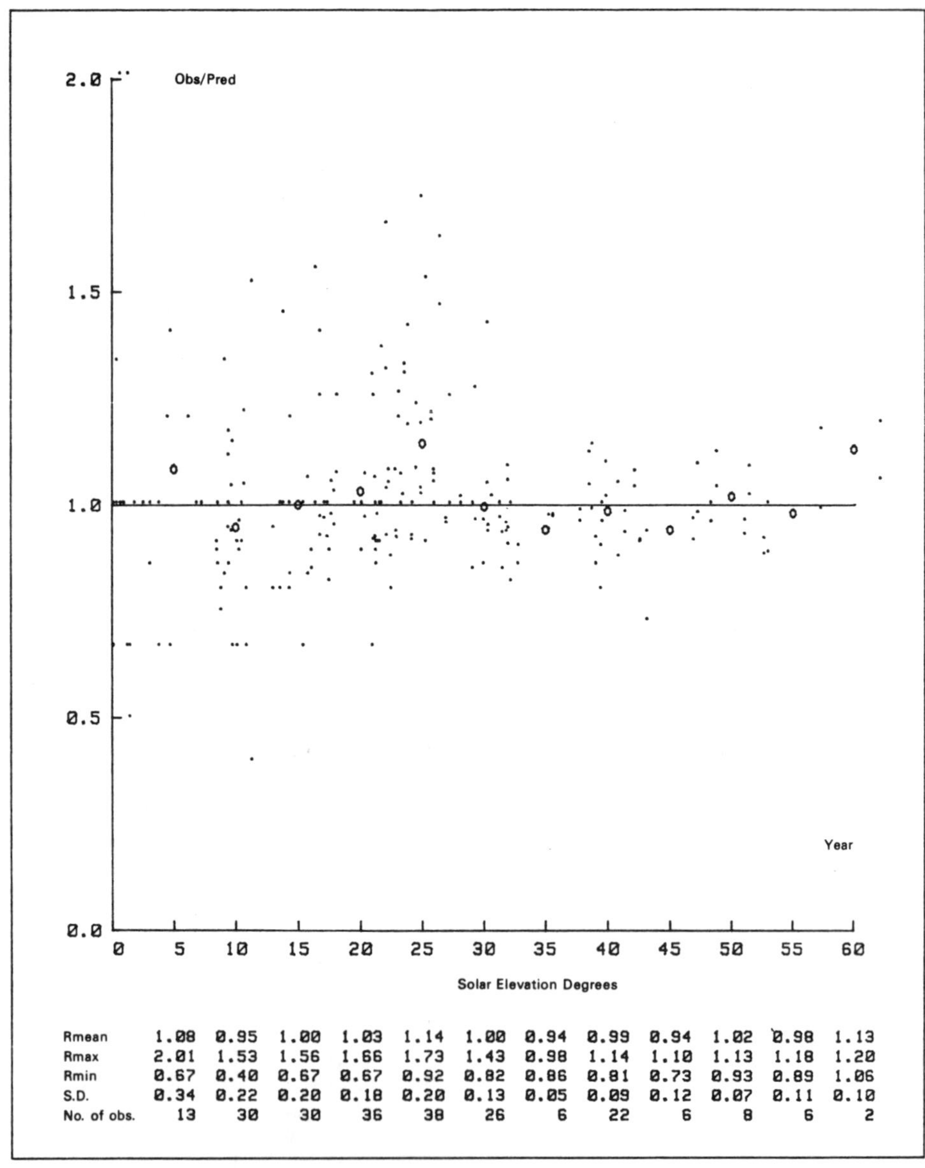

Rmean	1.08	0.95	1.00	1.03	1.14	1.00	0.94	0.99	0.94	1.02	0.98	1.13
Rmax	2.01	1.53	1.56	1.66	1.73	1.43	0.98	1.14	1.10	1.13	1.18	1.20
Rmin	0.67	0.40	0.67	0.67	0.92	0.82	0.86	0.81	0.73	0.93	0.89	1.06
S.D.	0.34	0.22	0.20	0.18	0.20	0.13	0.05	0.09	0.12	0.07	0.11	0.10
No. of obs.	13	30	30	36	38	26	6	22	6	8	6	2

Figure 4.3 Statistical analysis of the ratio of observed radiation on a stated inclined plane to the predicted ratio on the inclined plane for overcast sky conditions.
Station: Carpentras. Plane: Vertical South

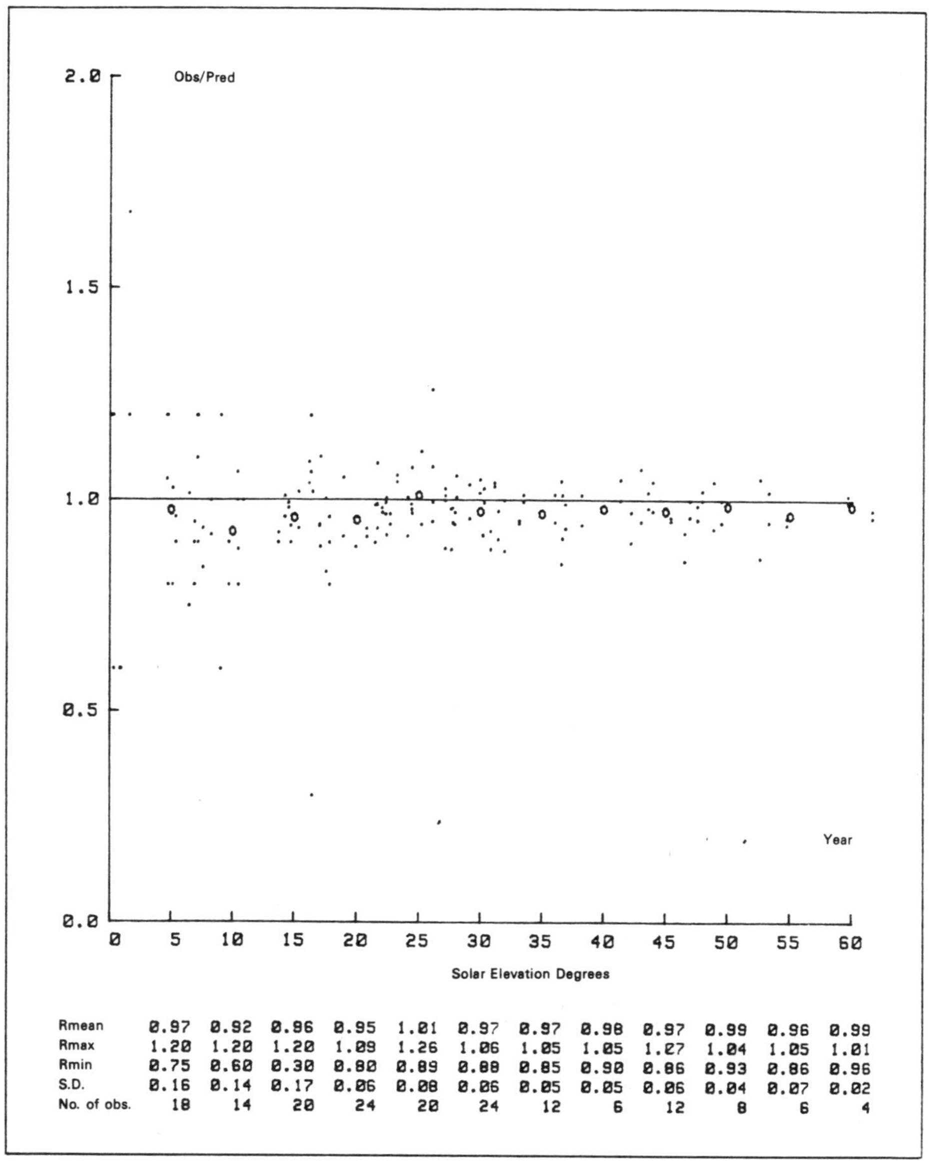

Rmean	0.97	0.92	0.96	0.95	1.01	0.97	0.97	0.98	0.97	0.99	0.96	0.99
Rmax	1.20	1.20	1.20	1.09	1.26	1.06	1.05	1.05	1.27	1.04	1.05	1.01
Rmin	0.75	0.60	0.30	0.80	0.89	0.88	0.85	0.90	0.86	0.93	0.86	0.96
S.D.	0.16	0.14	0.17	0.06	0.08	0.06	0.05	0.05	0.06	0.04	0.07	0.02
No. of obs.	18	14	20	24	20	24	12	6	12	8	6	4

Figure 4.4 Statistical analysis of the ratio of observed radiation on a stated inclined plane to the predicted ratio on the inclined plane for overcast sky conditions.
Station: Carpentras. Plane: 44° South

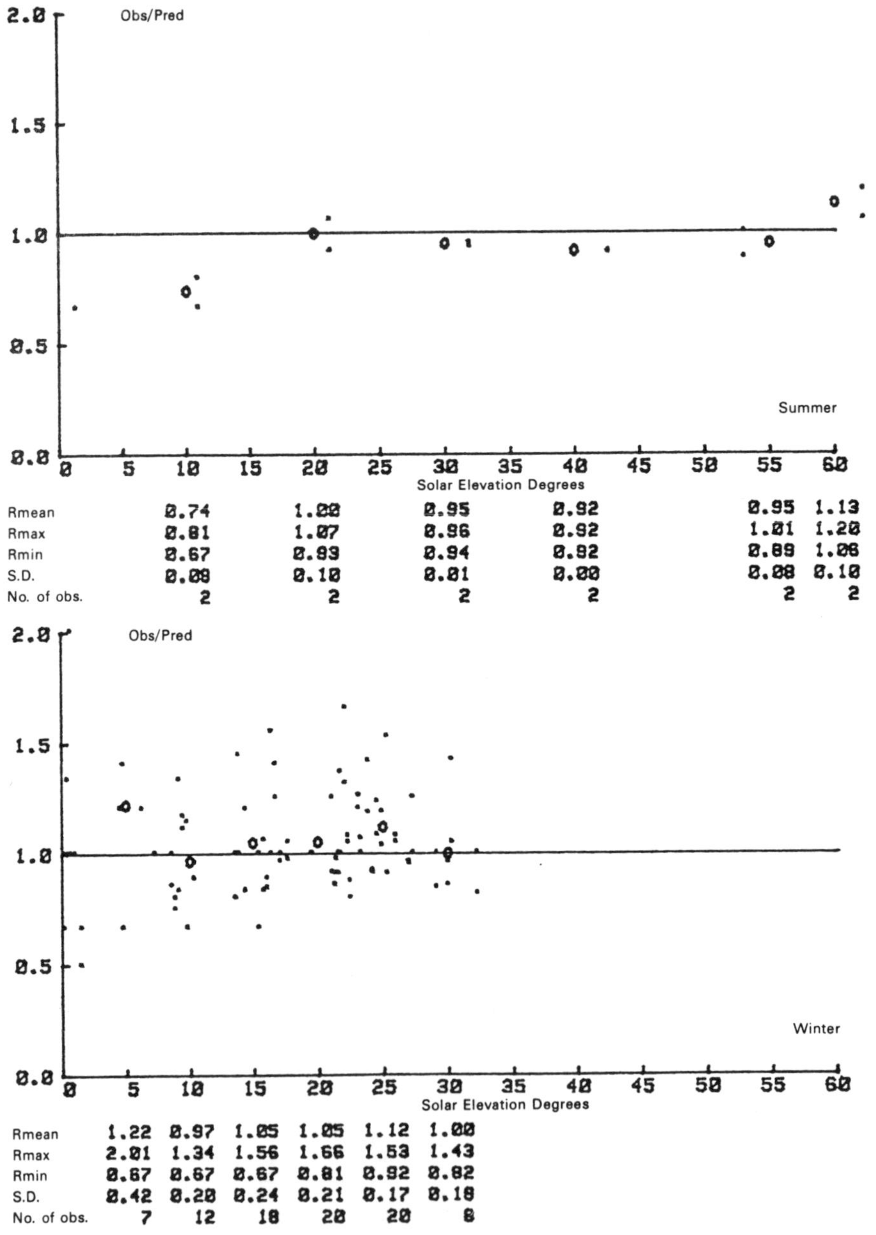

	10	20	30	40	55	60
Rmean	0.74	1.00	0.95	0.92	0.95	1.13
Rmax	0.81	1.07	0.96	0.92	1.01	1.20
Rmin	0.67	0.93	0.94	0.92	0.89	1.06
S.D.	0.09	0.10	0.01	0.00	0.08	0.10
No. of obs.	2	2	2	2	2	2

	5	10	15	20	25	30
Rmean	1.22	0.97	1.05	1.05	1.12	1.00
Rmax	2.01	1.34	1.56	1.66	1.53	1.43
Rmin	0.67	0.67	0.67	0.81	0.92	0.82
S.D.	0.42	0.20	0.24	0.21	0.17	0.18
No. of obs.	7	12	18	20	20	8

Figure 4.5a Statistical analysis of the ratio of observed radiation on a stated inclined plane to the predicted ratio on the inclined plane for overcast sky conditions: Summer and Winter: Carpentras. Vertical South

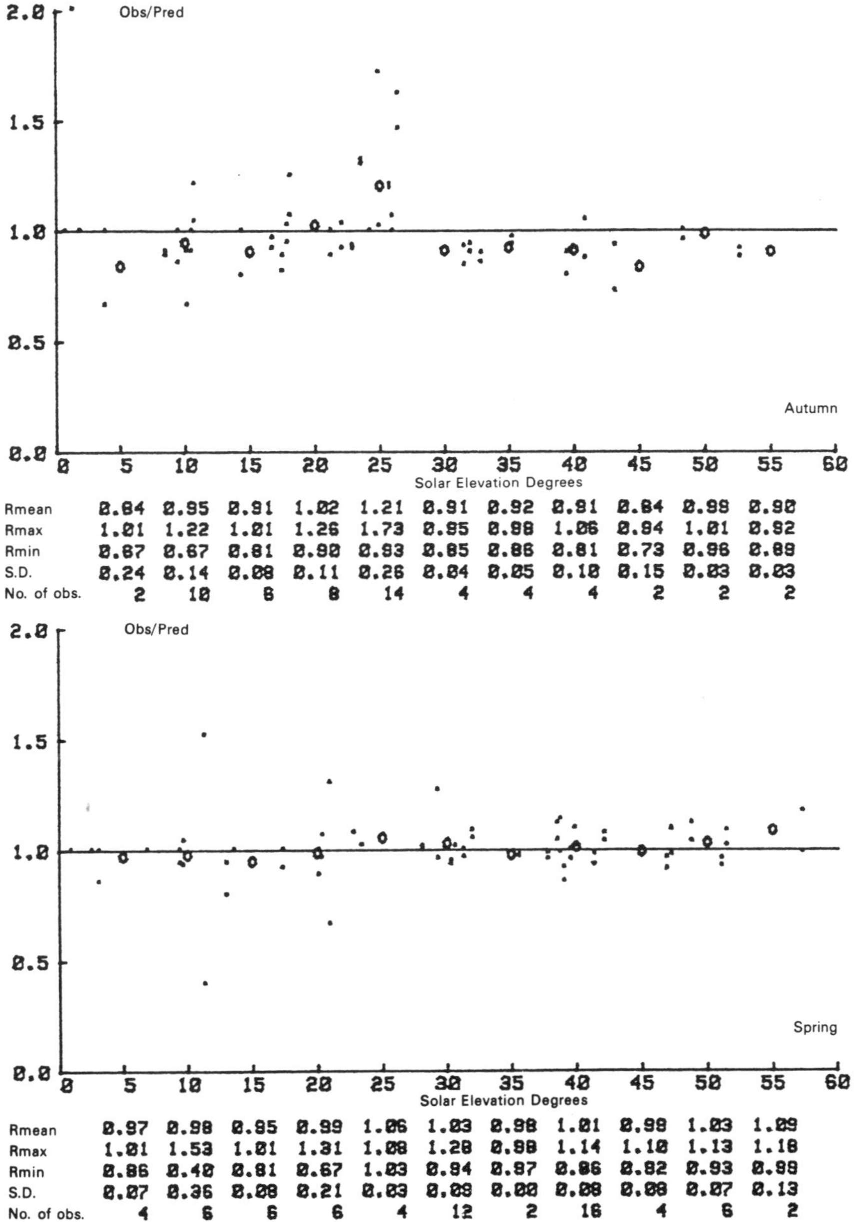

Autumn

	5	10	15	20	25	30	35	40	45	50	55
Rmean	0.84	0.95	0.91	1.02	1.21	0.91	0.92	0.91	0.84	0.98	0.90
Rmax	1.01	1.22	1.01	1.26	1.73	0.95	0.98	1.06	0.94	1.01	0.92
Rmin	0.67	0.67	0.81	0.90	0.93	0.85	0.86	0.81	0.73	0.96	0.89
S.D.	0.24	0.14	0.08	0.11	0.26	0.04	0.05	0.10	0.15	0.03	0.03
No. of obs.	2	10	6	8	14	4	4	4	2	2	2

Spring

	5	10	15	20	25	30	35	40	45	50	55
Rmean	0.97	0.98	0.85	0.99	1.06	1.03	0.98	1.01	0.98	1.03	1.09
Rmax	1.01	1.53	1.01	1.31	1.08	1.28	0.98	1.14	1.10	1.13	1.18
Rmin	0.86	0.40	0.81	0.67	1.03	0.94	0.97	0.86	0.82	0.93	0.99
S.D.	0.07	0.36	0.08	0.21	0.03	0.08	0.00	0.08	0.08	0.07	0.13
No. of obs.	4	6	6	6	4	12	2	16	4	6	2

Figure 4.5b Statistical analysis of the ratio of observed radiation on a stated inclined plane to the predicted ratio on the inclined plane for overcast sky conditions: Autumn and Spring: Carpentras. Vertical South

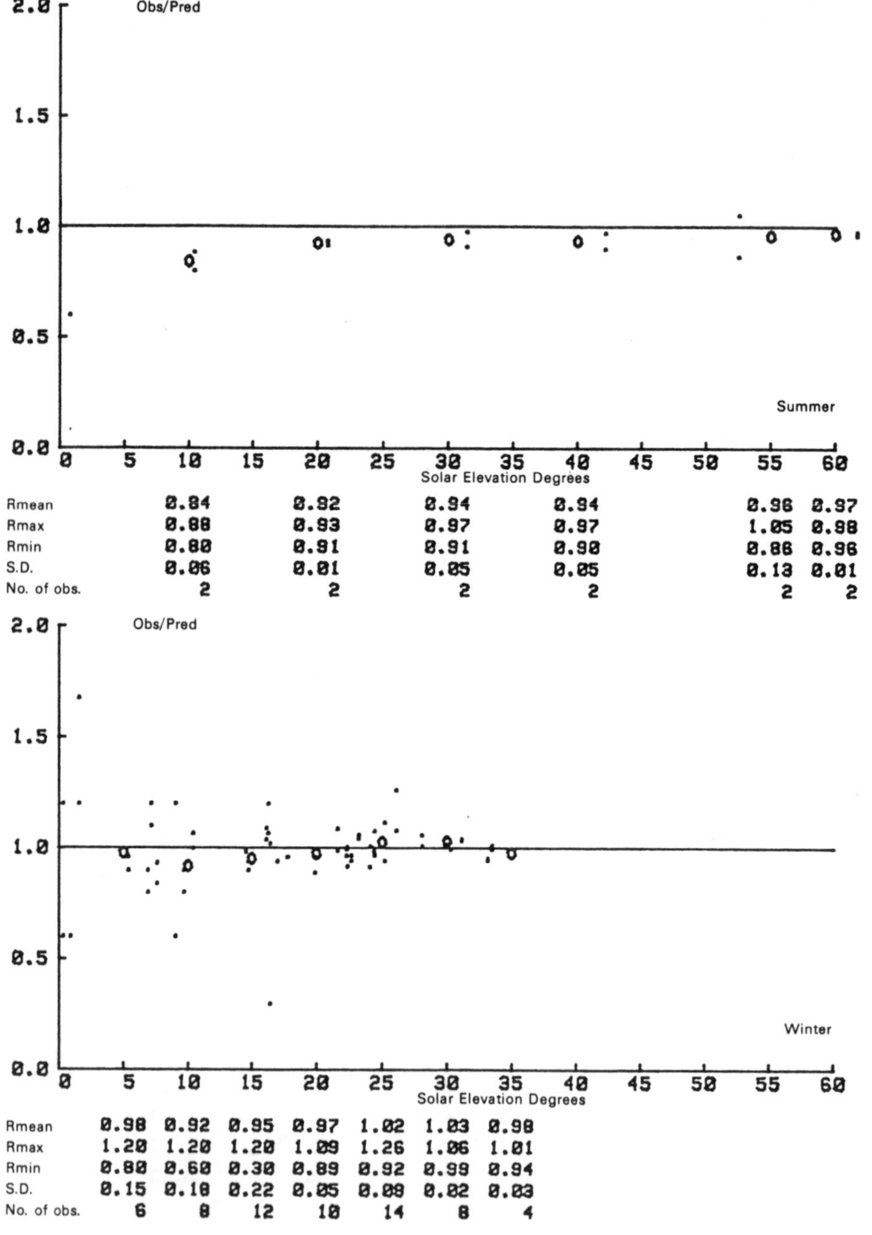

Figure 4.6a Statistical analysis of the ratio of observed radiation on a stated inclined plane to the predicted ratio on the inclined plane for overcast sky conditions: Summer and Winter: Carpentras. 44° South

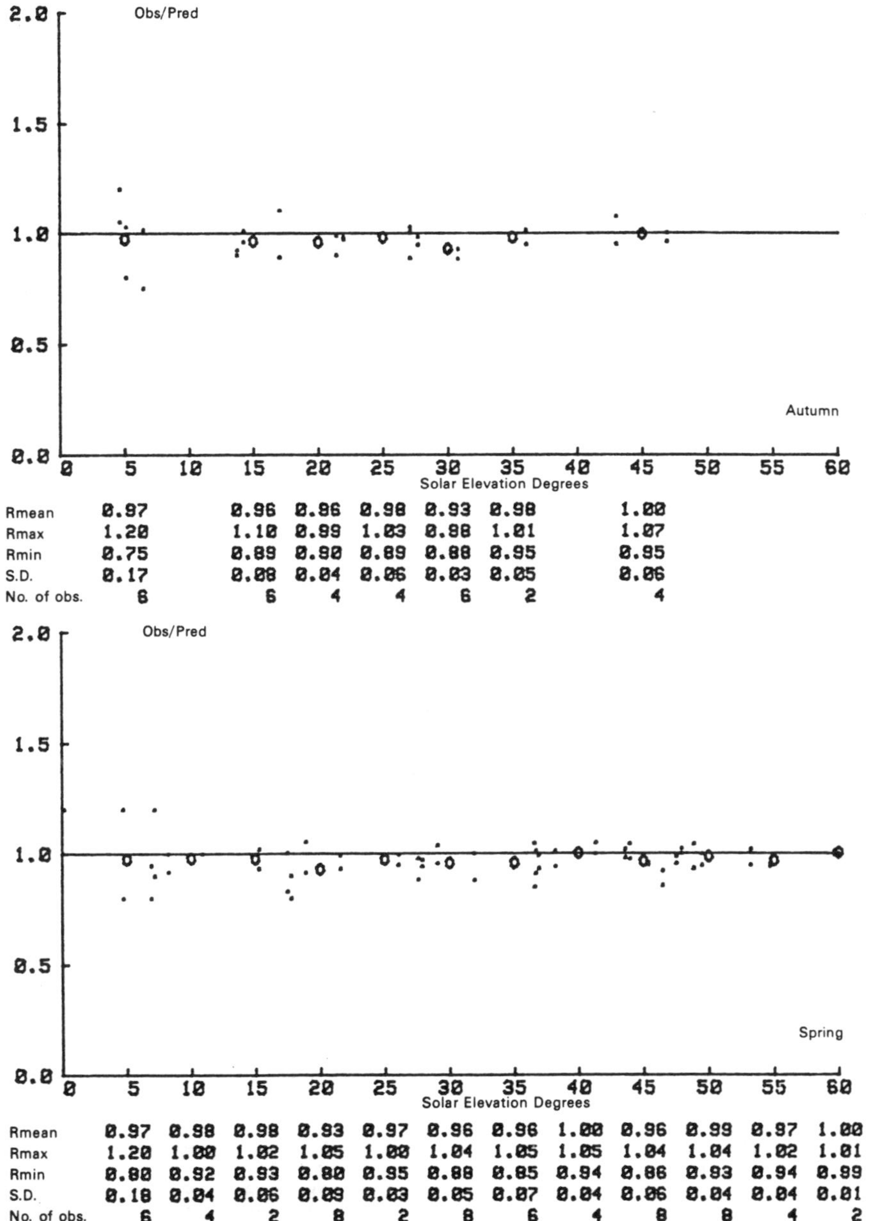

Autumn

	5	10	15	20	25	30	35	40	45	50	55	60
Rmean	0.97		0.96	0.86	0.98	0.93	0.98		1.00			
Rmax	1.20		1.10	0.99	1.03	0.98	1.01		1.07			
Rmin	0.75		0.89	0.80	0.89	0.88	0.95		0.95			
S.D.	0.17		0.08	0.04	0.06	0.03	0.05		0.06			
No. of obs.	6		6	4	4	6	2		4			

Spring

	5	10	15	20	25	30	35	40	45	50	55	60
Rmean	0.97	0.98	0.98	0.93	0.97	0.96	0.96	1.00	0.96	0.99	0.97	1.00
Rmax	1.20	1.00	1.02	1.05	1.00	1.04	1.05	1.05	1.04	1.04	1.02	1.01
Rmin	0.80	0.92	0.93	0.80	0.95	0.88	0.85	0.94	0.86	0.93	0.94	0.99
S.D.	0.18	0.04	0.06	0.09	0.03	0.05	0.07	0.04	0.06	0.04	0.04	0.01
No. of obs.	6	4	2	8	2	8	6	4	8	8	4	2

Figure 4.6b Statistical analysis of the ratio of observed radiation on a stated inclined plane to the predicted ratio on the inclined plane for overcast sky conditions: Autumn and Spring: Carpentras. 44° South

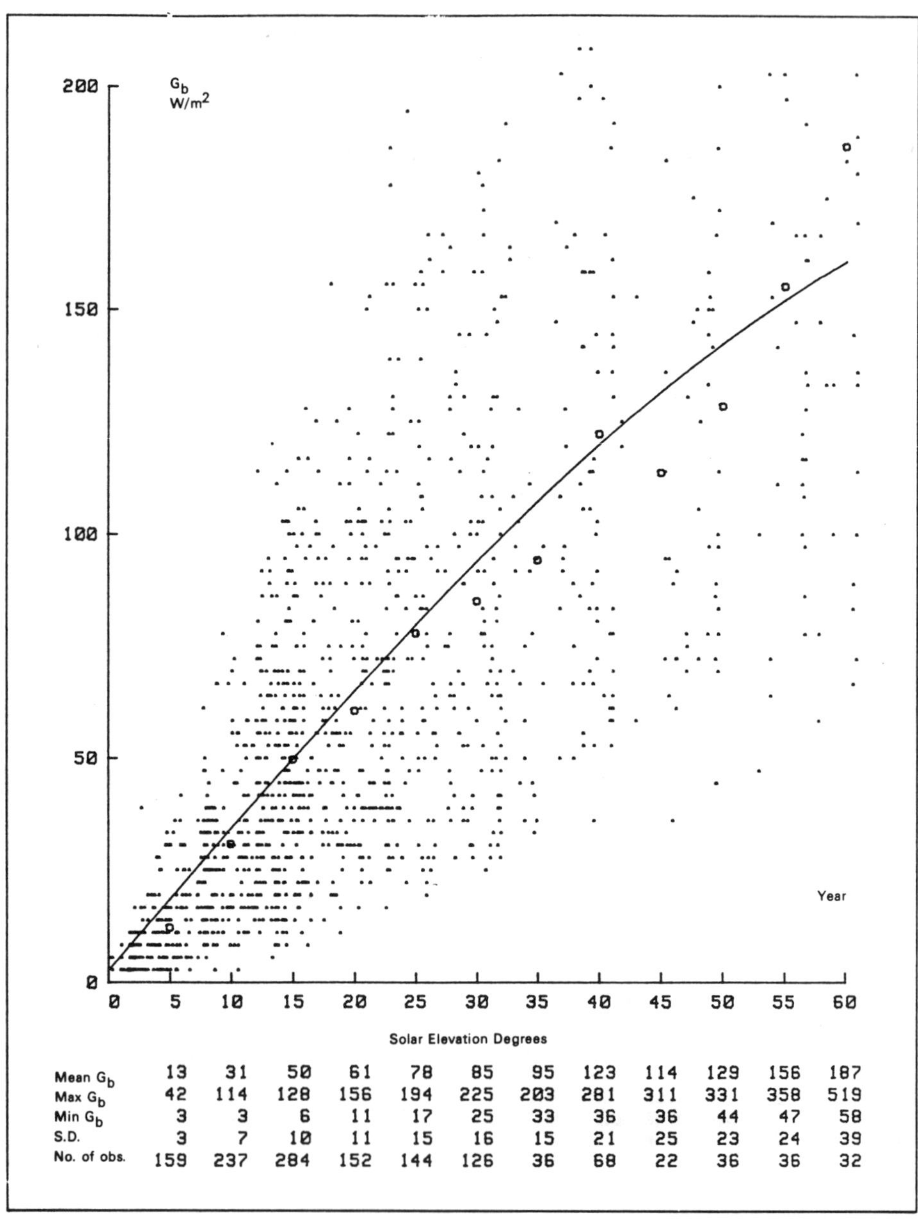

	5	10	15	20	25	30	35	40	45	50	55	60
Mean G_b	13	31	50	61	78	85	95	123	114	129	156	187
Max G_b	42	114	128	156	194	225	203	281	311	331	358	519
Min G_b	3	3	6	11	17	25	33	36	36	44	47	58
S.D.	3	7	10	11	15	16	15	21	25	23	24	39
No. of obs.	159	237	284	152	144	126	36	68	22	36	36	32

Figure 4.7 Statistical distribution of horizontal surface solar radiation from overcast skies.
Station: Valentia

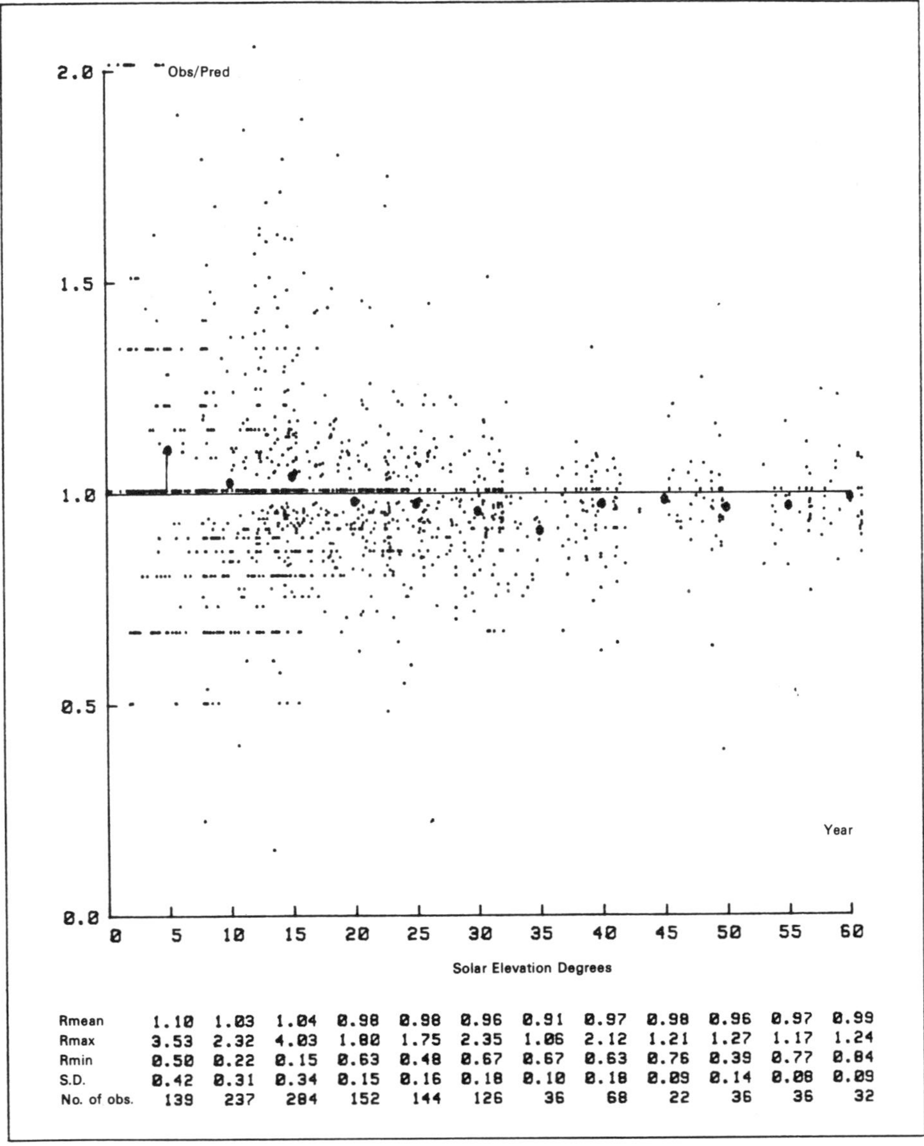

Rmean	1.10	1.03	1.04	0.98	0.98	0.96	0.91	0.97	0.98	0.96	0.97	0.99
Rmax	3.53	2.32	4.03	1.80	1.75	2.35	1.06	2.12	1.21	1.27	1.17	1.24
Rmin	0.50	0.22	0.15	0.63	0.48	0.67	0.67	0.63	0.76	0.39	0.77	0.84
S.D.	0.42	0.31	0.34	0.15	0.16	0.18	0.10	0.18	0.09	0.14	0.08	0.09
No. of obs.	139	237	284	152	144	126	36	68	22	36	36	32

Figure 4.8 Statistical analysis of the ratio of observed radiation on a stated inclined plane to the predicted ratio on the inclined plane for overcast sky conditions.
Station: Valentia. Plane: Vertical South

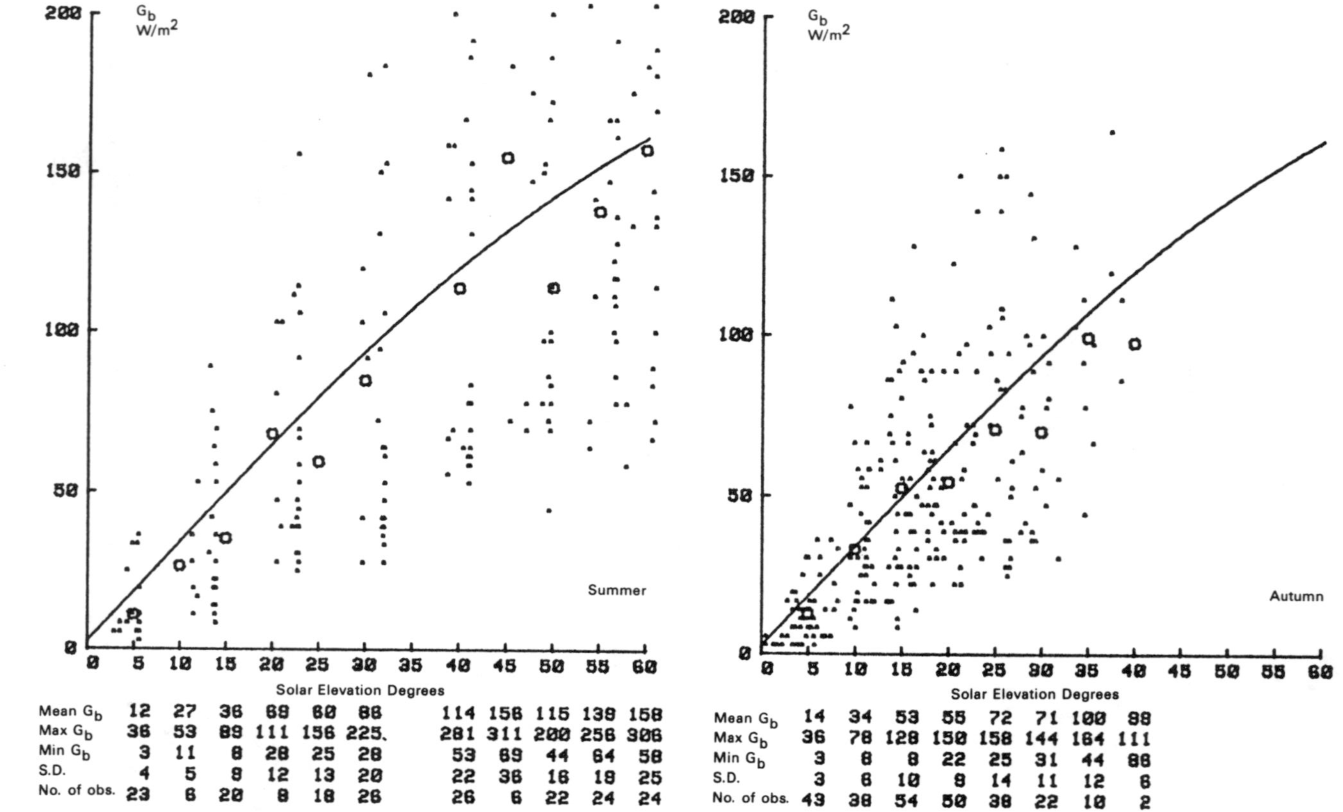

Summer

Solar Elevation Degrees	5	10	15	20	25	30	35	40	45	50	55	60
Mean G_b	12	27	36	68	60	86		114	156	115	138	158
Max G_b	36	53	89	111	156	225.		281	311	200	256	306
Min G_b	3	11	8	28	25	28		53	69	44	64	58
S.D.	4	5	8	12	13	20		22	36	16	18	25
No. of obs.	23	6	20	8	18	26		26	6	22	24	24

Autumn

Solar Elevation Degrees	5	10	15	20	25	30	35	40
Mean G_b	14	34	53	55	72	71	100	98
Max G_b	36	78	128	150	158	144	164	111
Min G_b	3	8	8	22	25	31	44	86
S.D.	3	6	10	9	14	11	12	6
No. of obs.	43	38	54	50	38	22	10	2

Figure 4.9a Statistical distribution of horizontal surface solar radiation from overcast skies: Summer and Autumn: Valentia

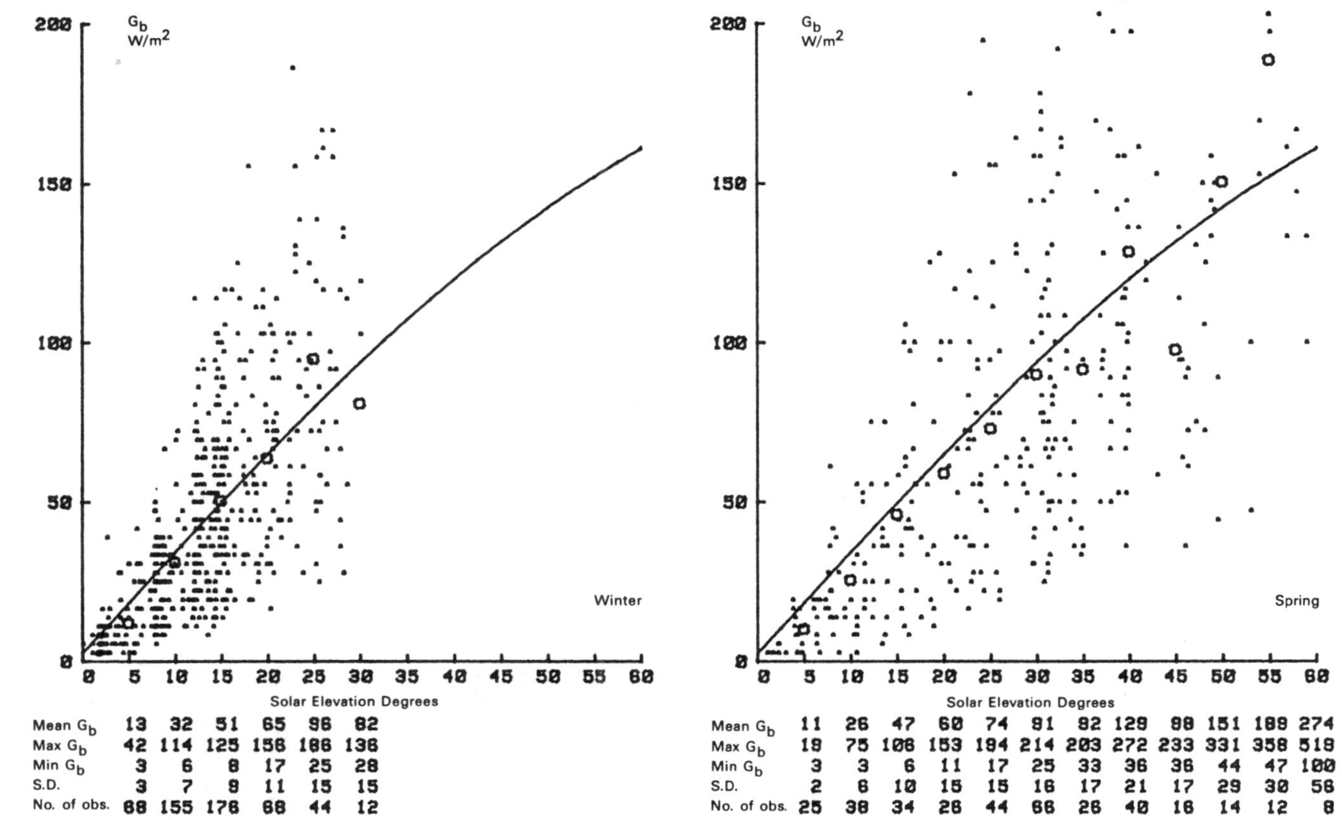

Figure 4.9b Statistical distribution of horizontal surface solar radiation from overcast skies: Winter and Spring: Valentia

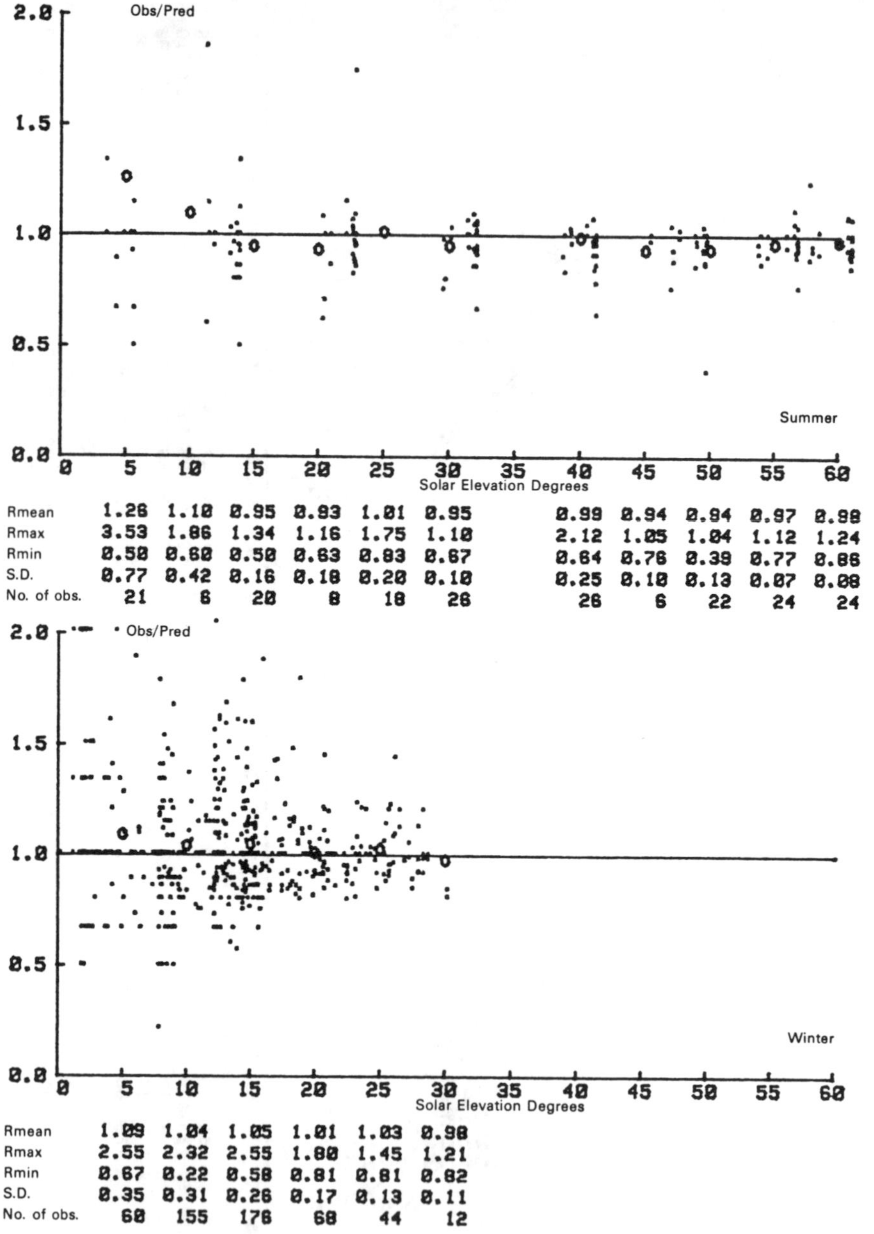

	5	10	15	20	25	30	35	40	45	50	55	60
Rmean	1.26	1.10	0.95	0.93	1.01	0.95		0.99	0.94	0.94	0.97	0.98
Rmax	3.53	1.86	1.34	1.16	1.75	1.10		2.12	1.05	1.04	1.12	1.24
Rmin	0.50	0.60	0.50	0.63	0.83	0.67		0.64	0.76	0.39	0.77	0.86
S.D.	0.77	0.42	0.16	0.18	0.20	0.10		0.25	0.10	0.13	0.07	0.08
No. of obs.	21	6	20	8	18	26		26	6	22	24	24

	5	10	15	20	25	30
Rmean	1.09	1.04	1.05	1.01	1.03	0.98
Rmax	2.55	2.32	2.55	1.80	1.45	1.21
Rmin	0.67	0.22	0.58	0.81	0.81	0.82
S.D.	0.35	0.31	0.26	0.17	0.13	0.11
No. of obs.	60	155	176	68	44	12

Figure 4.10a Statistical analysis of the ratio of observed radiation on a stated inclined plane to the predicted ratio on the inclined plane for overcast sky conditions: Summer and Winter: Valentia. Vertical South

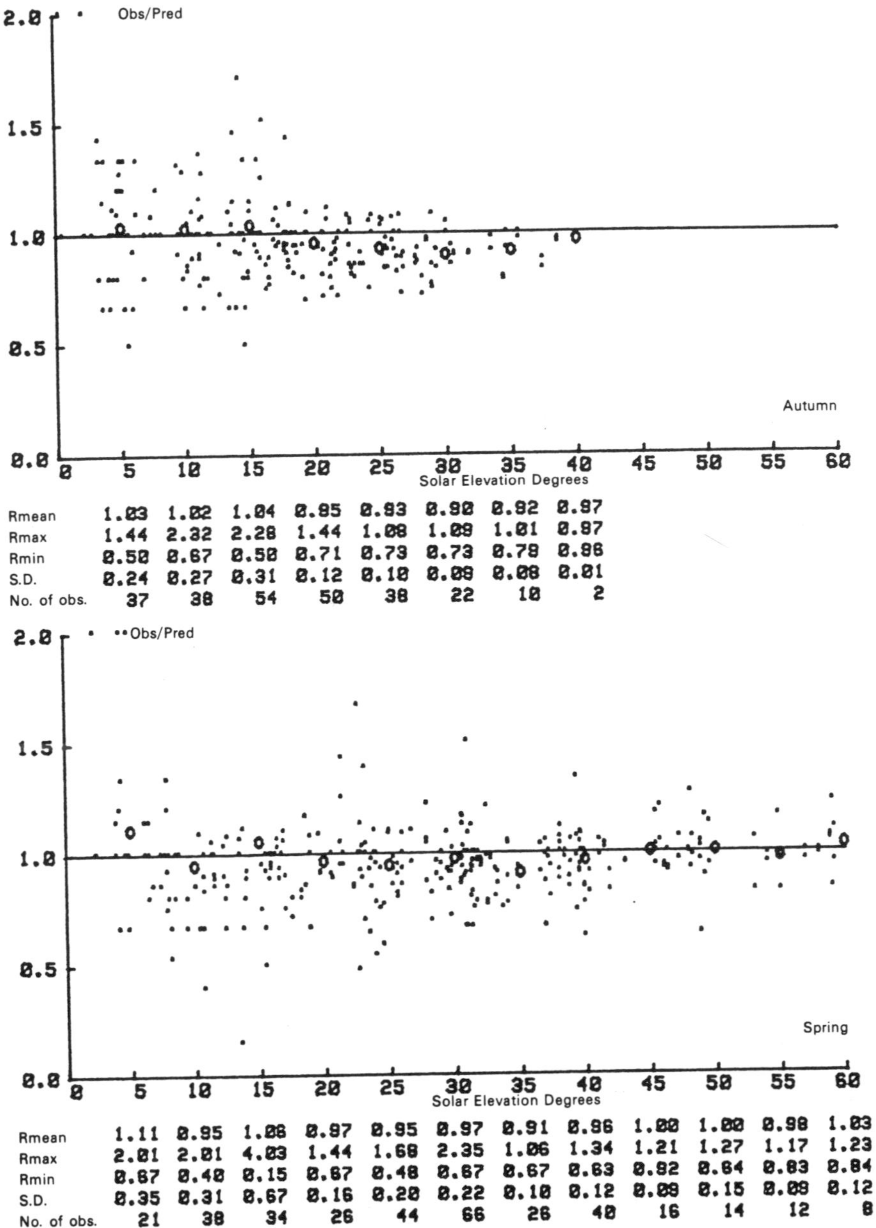

Rmean	1.03	1.02	1.04	0.85	0.83	0.90	0.82	0.97
Rmax	1.44	2.32	2.28	1.44	1.08	1.09	1.01	0.97
Rmin	0.50	0.67	0.50	0.71	0.73	0.73	0.78	0.96
S.D.	0.24	0.27	0.31	0.12	0.10	0.09	0.08	0.01
No. of obs.	37	38	54	50	38	22	10	2

Rmean	1.11	0.95	1.06	0.97	0.95	0.97	0.91	0.96	1.00	1.00	0.98	1.03
Rmax	2.01	2.01	4.03	1.44	1.68	2.35	1.06	1.34	1.21	1.27	1.17	1.23
Rmin	0.67	0.40	0.15	0.67	0.48	0.67	0.67	0.63	0.82	0.64	0.83	0.84
S.D.	0.35	0.31	0.67	0.16	0.20	0.22	0.10	0.12	0.09	0.15	0.09	0.12
No. of obs.	21	38	34	26	44	66	26	40	16	14	12	8

Figure 4.10b Statistical analysis of the ratio of observed radiation on a stated inclined plane to the predicted ratio on the inclined plane for overcast sky conditions: Autumn and Spring: Valentia. Vertical South

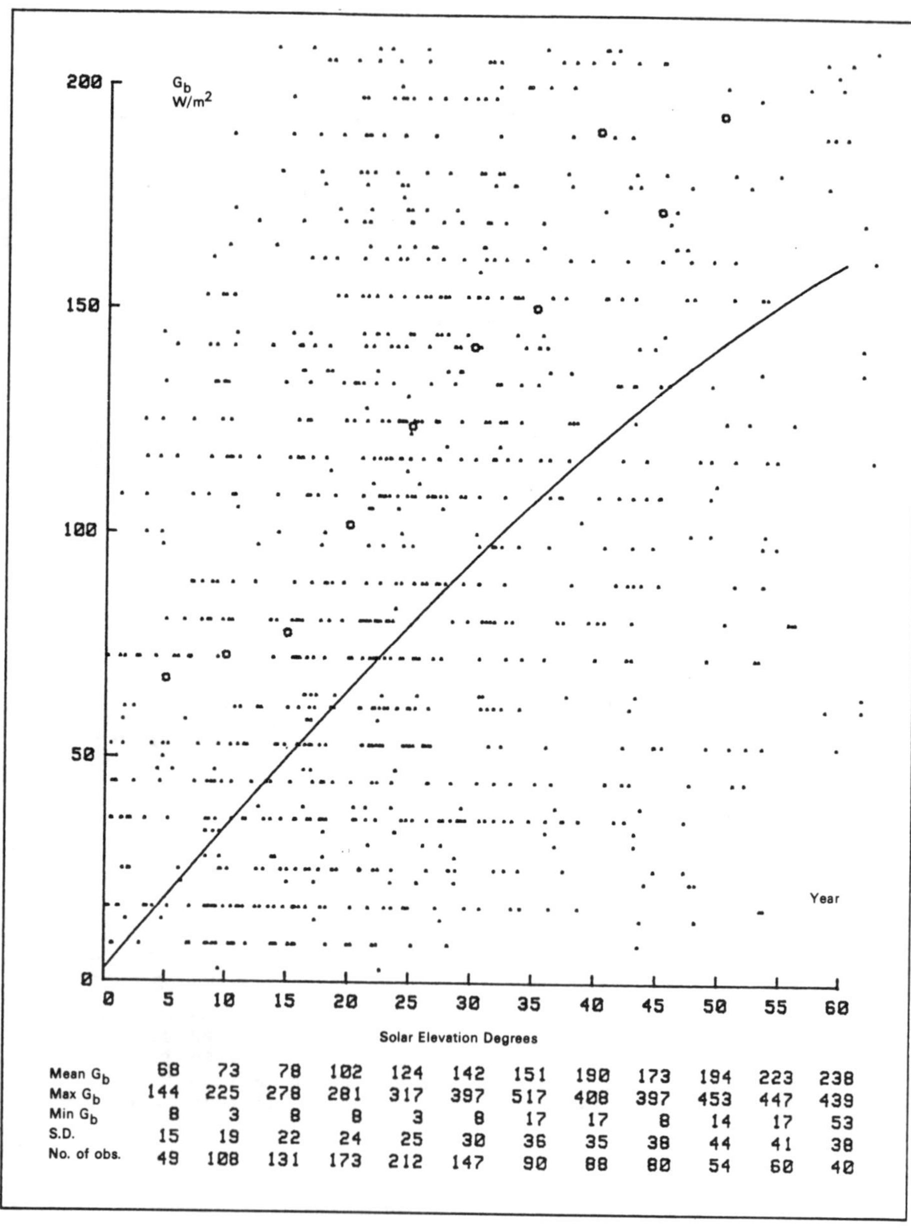

Mean G_b	68	73	78	102	124	142	151	190	173	194	223	238
Max G_b	144	225	278	281	317	397	517	408	397	453	447	439
Min G_b	8	3	8	8	3	8	17	17	8	14	17	53
S.D.	15	19	22	24	25	30	36	35	38	44	41	38
No. of obs.	49	108	131	173	212	147	90	88	80	54	60	40

Figure 4.11 Statistical distribution of horizontal surface solar radiation from overcast skies.
Station: Odeillo

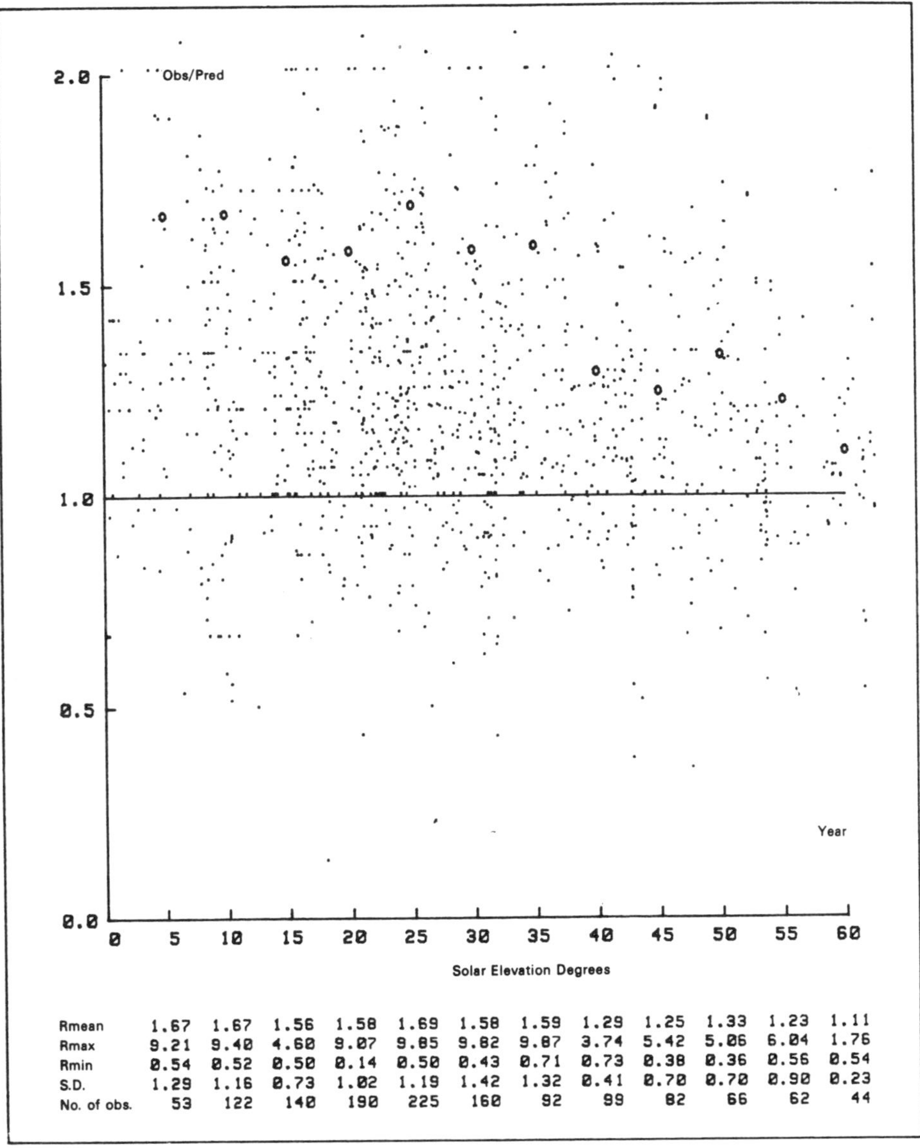

Rmean	1.67	1.67	1.56	1.58	1.69	1.58	1.59	1.29	1.25	1.33	1.23	1.11
Rmax	9.21	9.40	4.60	9.07	9.85	9.82	9.87	3.74	5.42	5.06	6.04	1.76
Rmin	0.54	0.52	0.50	0.14	0.50	0.43	0.71	0.73	0.38	0.36	0.56	0.54
S.D.	1.29	1.16	0.73	1.02	1.19	1.42	1.32	0.41	0.70	0.70	0.90	0.23
No. of obs.	53	122	140	190	225	160	92	99	82	66	62	44

Figure 4.12 Statistical analysis of the ratio of observed radiation on a stated inclined plane to the predicted ratio on the inclined plane for overcast sky conditions.
Station: Odeillo. Plane: Vertical South

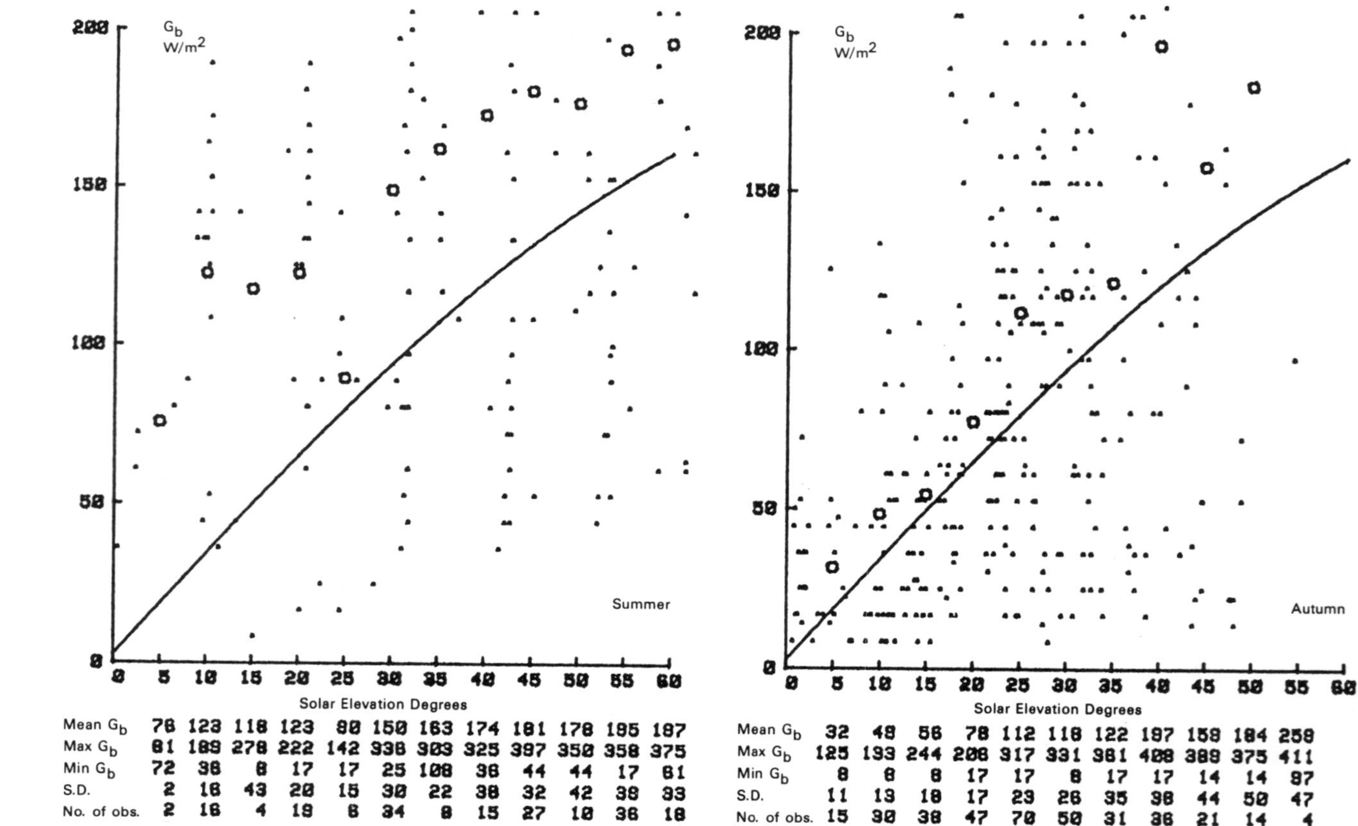

Summer

Solar Elevation Degrees	5	10	15	20	25	30	35	40	45	50	55	60
Mean G_b	78	123	118	123	80	150	163	174	181	178	185	187
Max G_b	81	189	278	222	142	338	303	325	397	350	358	375
Min G_b	72	36	8	17	17	25	108	36	44	44	17	61
S.D.	2	16	43	20	15	30	22	38	32	42	38	33
No. of obs.	2	16	4	19	8	34	8	15	27	10	36	18

Autumn

Solar Elevation Degrees	5	10	15	20	25	30	35	40	45	50	55
Mean G_b	32	48	56	78	112	118	122	197	158	184	258
Max G_b	125	193	244	208	317	331	381	408	388	375	411
Min G_b	8	8	8	17	17	8	17	17	14	14	97
S.D.	11	13	18	17	23	26	35	38	44	50	47
No. of obs.	15	30	38	47	70	50	31	36	21	14	4

Figure 4.13a Statistical distribution of horizontal surface solar radiation from overcast skies: Summer and Autumn: Odeillo

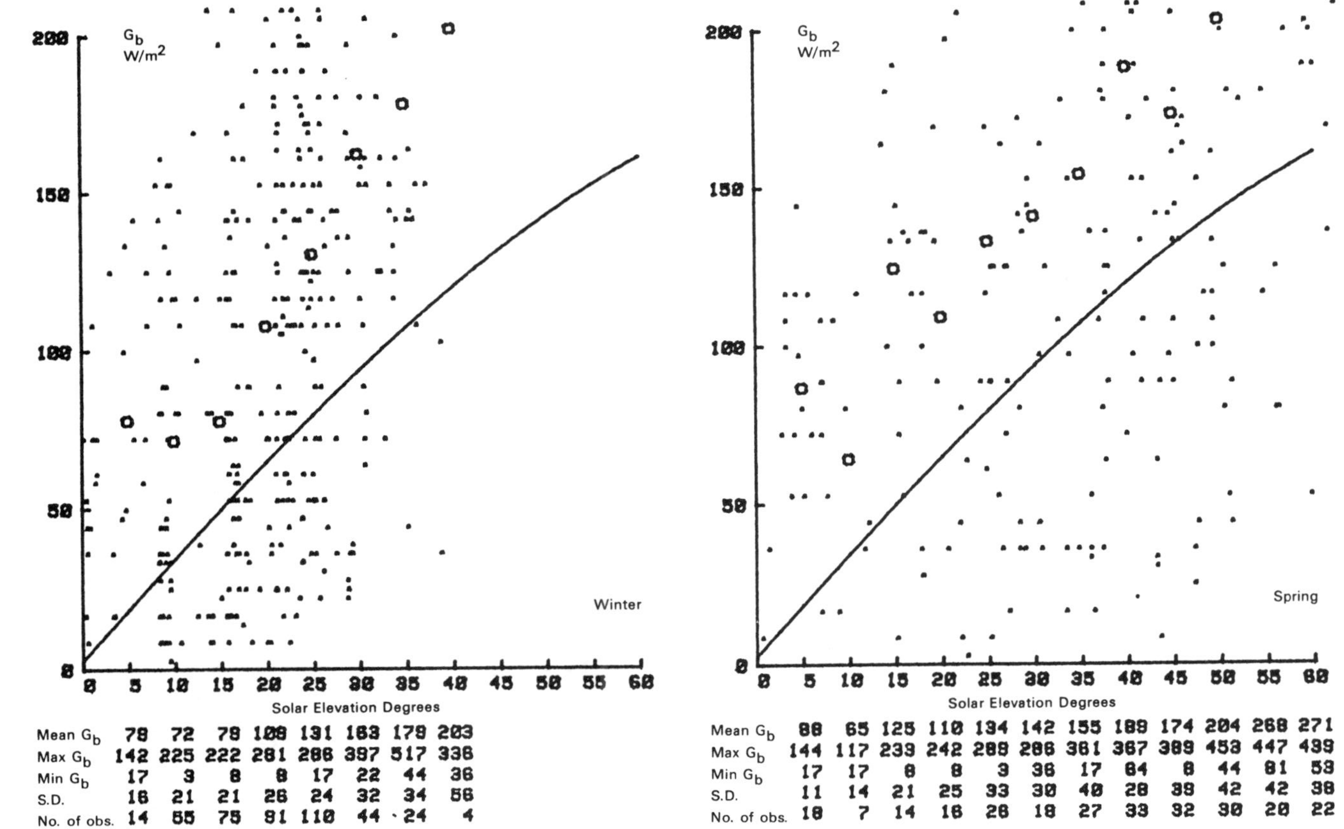

Winter

Mean G_b	78	72	78	108	131	163	178	203
Max G_b	142	225	222	281	286	387	517	336
Min G_b	17	3	8	8	17	22	44	36
S.D.	16	21	21	26	24	32	34	56
No. of obs.	14	55	75	81	110	44	·24	4

Spring

Mean G_b	88	65	125	110	134	142	155	188	174	204	268	271
Max G_b	144	117	233	242	289	286	361	367	389	453	447	439
Min G_b	17	17	8	8	3	36	17	64	8	44	81	53
S.D.	11	14	21	25	33	30	40	28	38	42	42	38
No. of obs.	18	7	14	16	26	18	27	33	32	30	20	22

Figure 4.13b Statistical distribution of horizontal surface solar radiation from overcast skies: Winter and Spring: Odeillo

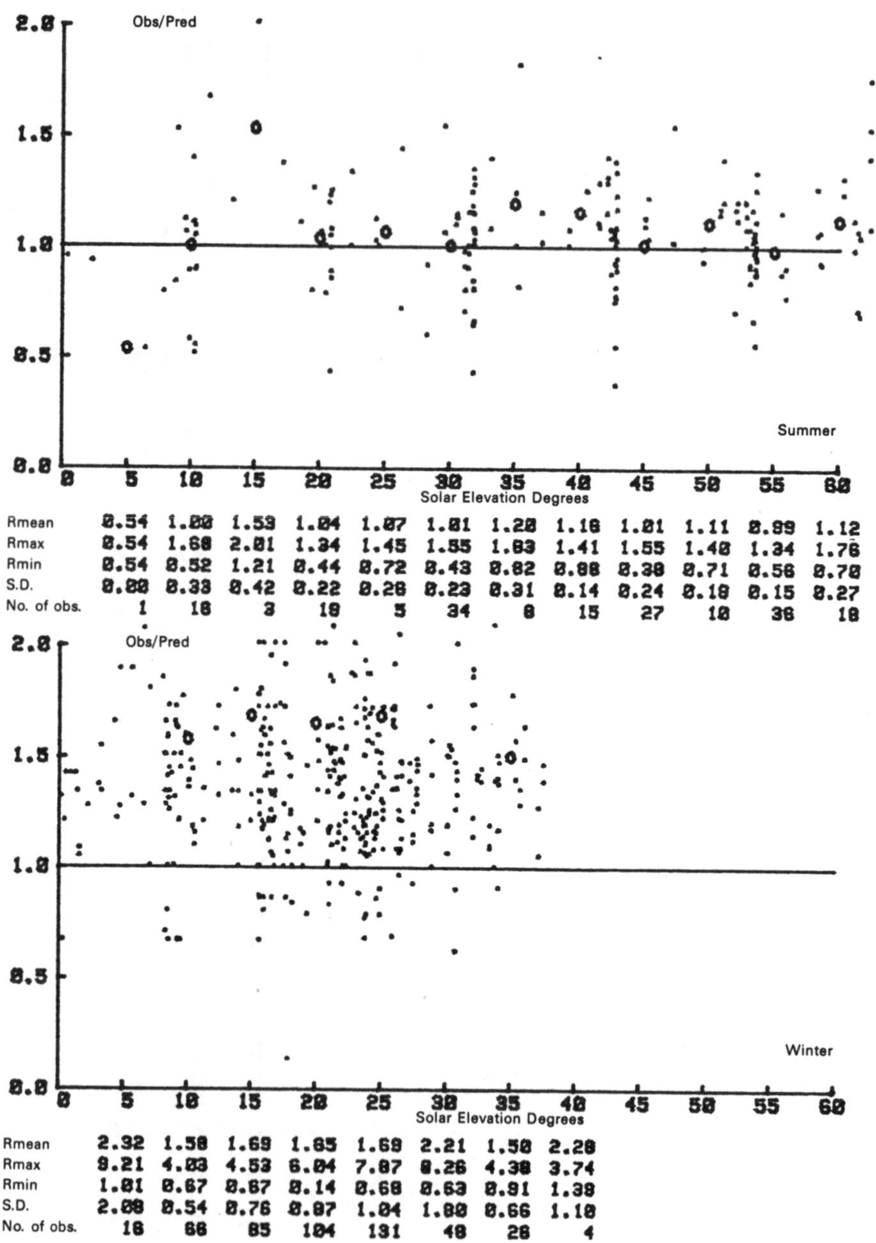

Summer:

	5	10	15	20	25	30	35	40	45	50	55	60
Rmean	0.54	1.00	1.53	1.04	1.07	1.01	1.20	1.16	1.01	1.11	0.89	1.12
Rmax	0.54	1.68	2.01	1.34	1.45	1.55	1.83	1.41	1.55	1.40	1.34	1.76
Rmin	0.54	0.52	1.21	0.44	0.72	0.43	0.82	0.88	0.38	0.71	0.56	0.70
S.D.	0.00	0.33	0.42	0.22	0.28	0.23	0.31	0.14	0.24	0.18	0.15	0.27
No. of obs.	1	18	3	19	5	34	8	15	27	10	36	18

Winter:

	5	10	15	20	25	30	35	40
Rmean	2.32	1.58	1.69	1.65	1.68	2.21	1.50	2.20
Rmax	9.21	4.03	4.53	6.04	7.87	8.26	4.38	3.74
Rmin	1.01	0.67	0.67	0.14	0.68	0.63	0.91	1.38
S.D.	2.08	0.54	0.76	0.87	1.04	1.80	0.66	1.10
No. of obs.	18	66	85	104	131	48	28	4

Figure 4.14a Statistical analysis of the ratio of observed radiation on a stated inclined plane to the predicted ratio on the inclined plane for overcast sky conditions: Summer and Winter: Odeillo. Vertical South

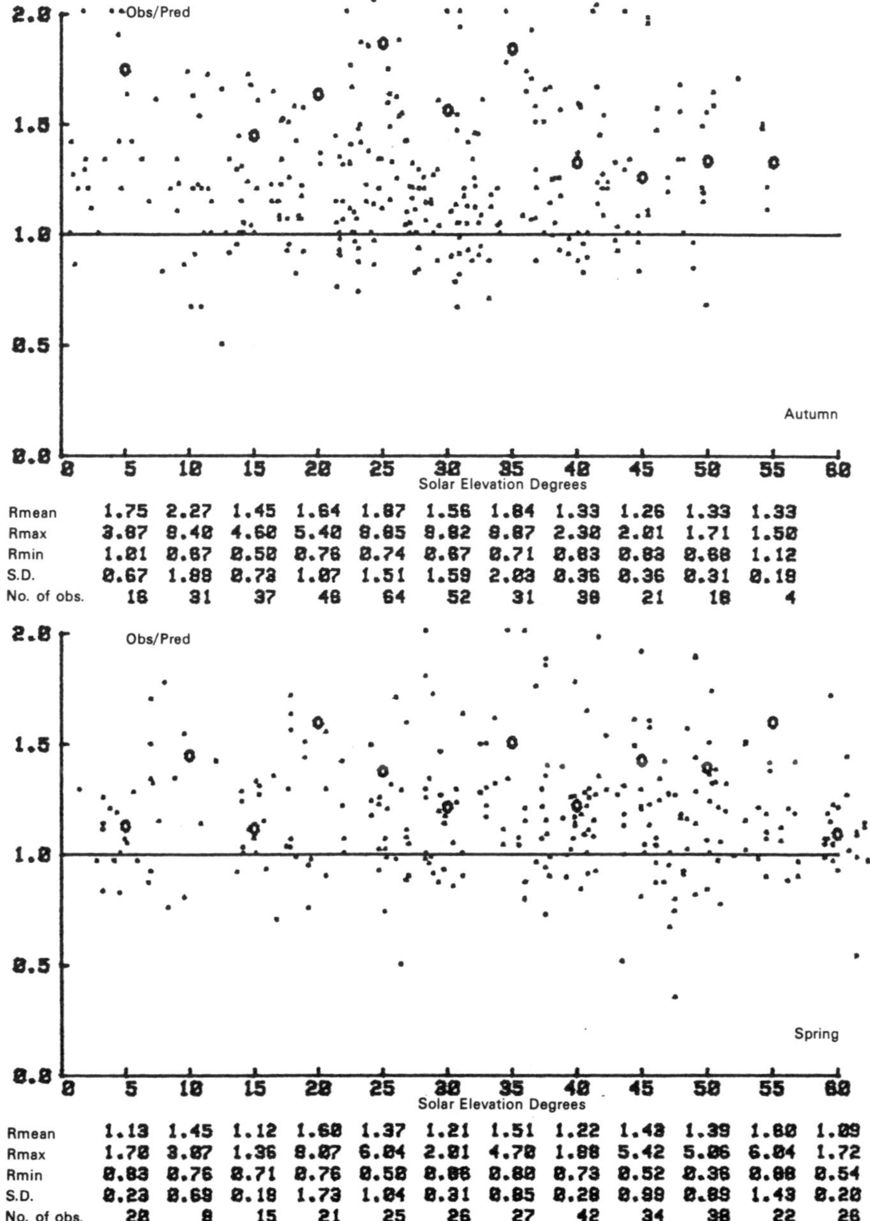

	5	10	15	20	25	30	35	40	45	50	55
Rmean	1.75	2.27	1.45	1.64	1.87	1.56	1.84	1.33	1.26	1.33	1.33
Rmax	3.87	8.40	4.60	5.40	8.85	8.82	8.87	2.30	2.01	1.71	1.50
Rmin	1.01	0.67	0.50	0.76	0.74	0.67	0.71	0.83	0.83	0.68	1.12
S.D.	0.67	1.88	0.73	1.07	1.51	1.59	2.03	0.36	0.36	0.31	0.18
No. of obs.	18	31	37	48	64	52	31	38	21	18	4

	5	10	15	20	25	30	35	40	45	50	55	60
Rmean	1.13	1.45	1.12	1.60	1.37	1.21	1.51	1.22	1.43	1.39	1.60	1.09
Rmax	1.70	3.87	1.36	8.07	6.04	2.81	4.70	1.88	5.42	5.06	6.04	1.72
Rmin	0.83	0.76	0.71	0.76	0.58	0.86	0.80	0.73	0.52	0.36	0.88	0.54
S.D.	0.23	0.69	0.18	1.73	1.04	0.31	0.85	0.28	0.99	0.89	1.43	0.20
No. of obs.	28	8	15	21	25	26	27	42	34	38	22	26

Figure 4.14b Statistical analysis of the ratio of observed radiation on a stated inclined plane to the predicted ratio on the inclined plane for overcast sky conditions: Autumn and Spring: Odeillo. Vertical South

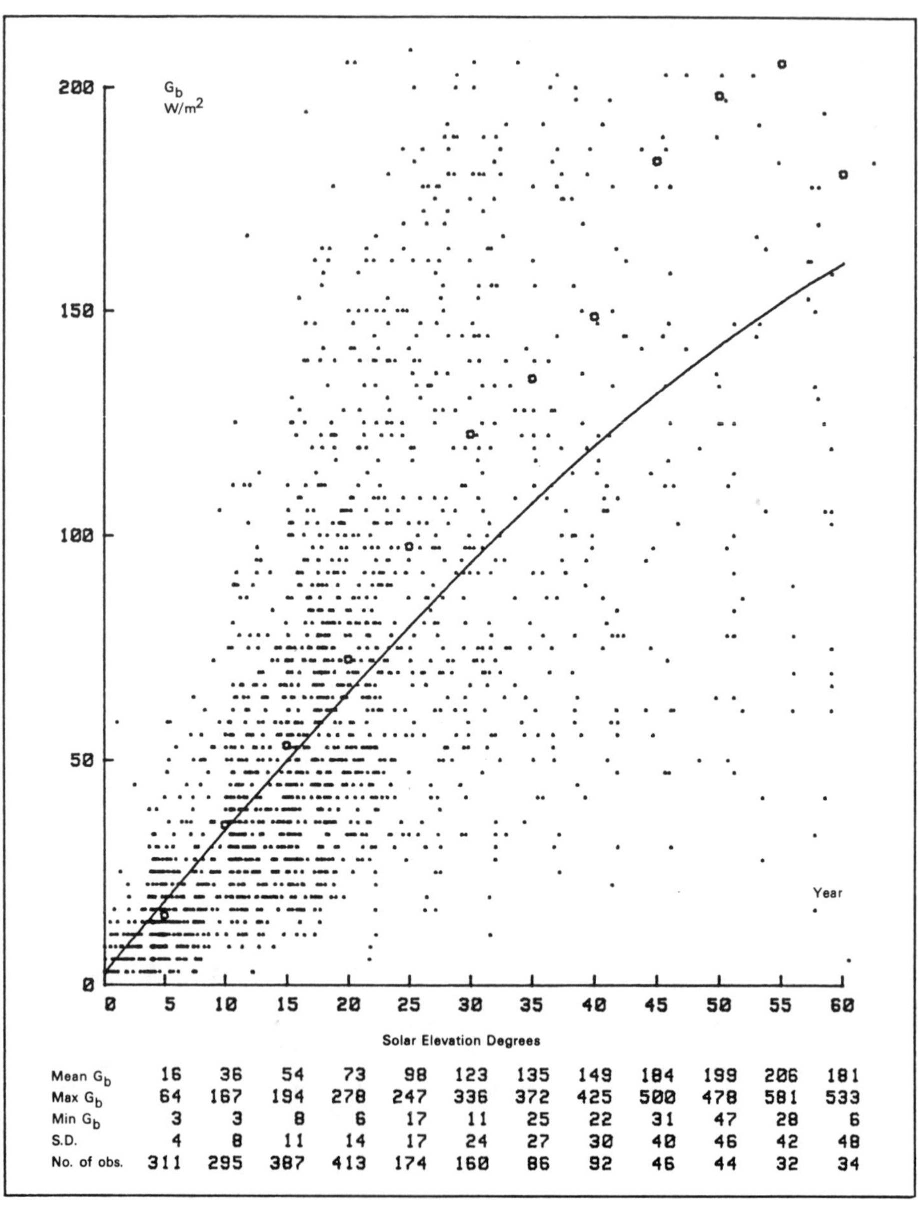

	0	5	10	15	20	25	30	35	40	45	50	55
Mean G_b	16	36	54	73	98	123	135	149	184	199	206	181
Max G_b	64	167	194	278	247	336	372	425	500	478	581	533
Min G_b	3	3	8	6	17	11	25	22	31	47	28	6
S.D.	4	8	11	14	17	24	27	30	40	46	42	48
No. of obs.	311	295	387	413	174	160	86	92	46	44	32	34

Figure 4.15 Statistical distribution of horizontal surface solar radiation from overcast skies.
Station: Trappes

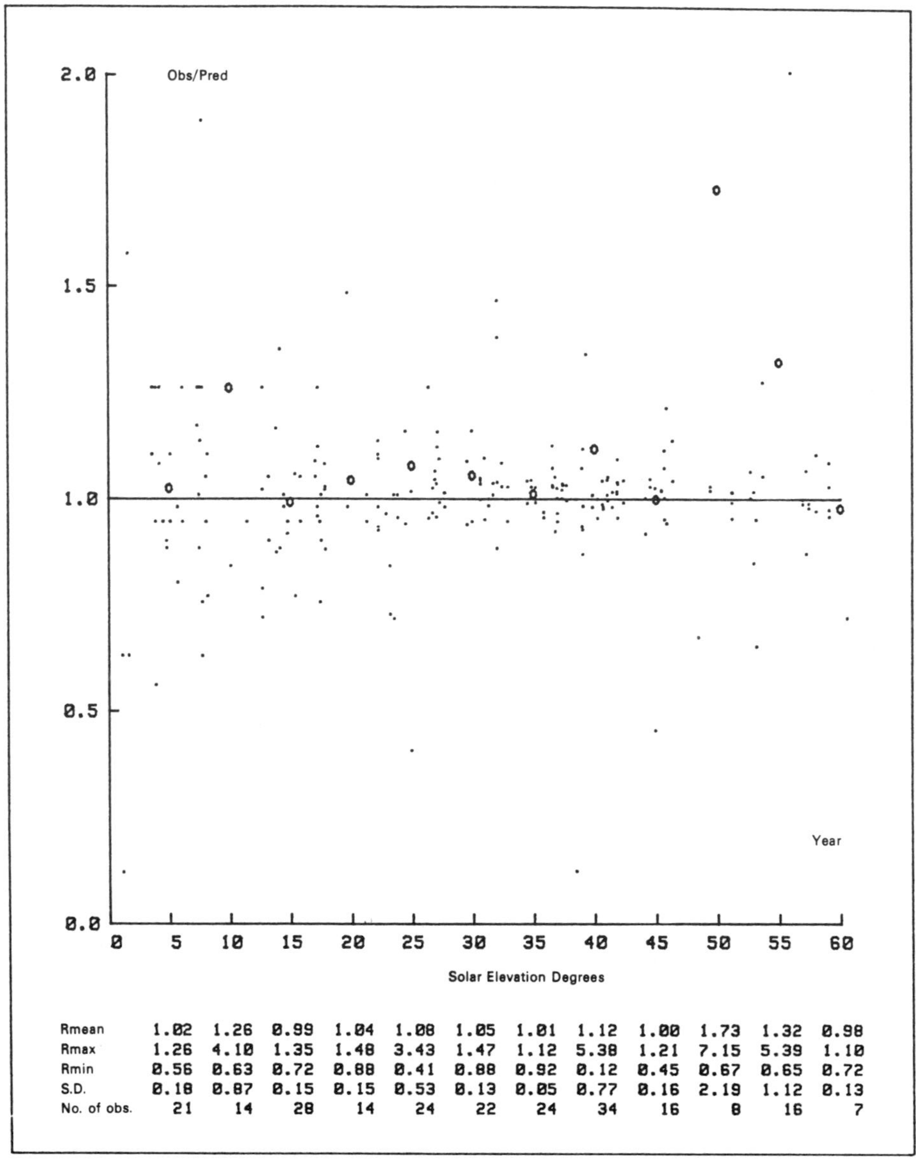

Rmean	1.02	1.26	0.99	1.04	1.08	1.05	1.01	1.12	1.00	1.73	1.32	0.98
Rmax	1.26	4.10	1.35	1.48	3.43	1.47	1.12	5.38	1.21	7.15	5.39	1.10
Rmin	0.56	0.63	0.72	0.88	0.41	0.88	0.92	0.12	0.45	0.67	0.65	0.72
S.D.	0.18	0.87	0.15	0.15	0.53	0.13	0.05	0.77	0.16	2.19	1.12	0.13
No. of obs.	21	14	28	14	24	22	24	34	16	8	16	7

Figure 4.16 Statistical analysis of the ratio of observed radiation on a stated inclined plane to the predicted ratio on the inclined plane for overcast sky conditions.
Station: Trappes. Plane: 50° South East

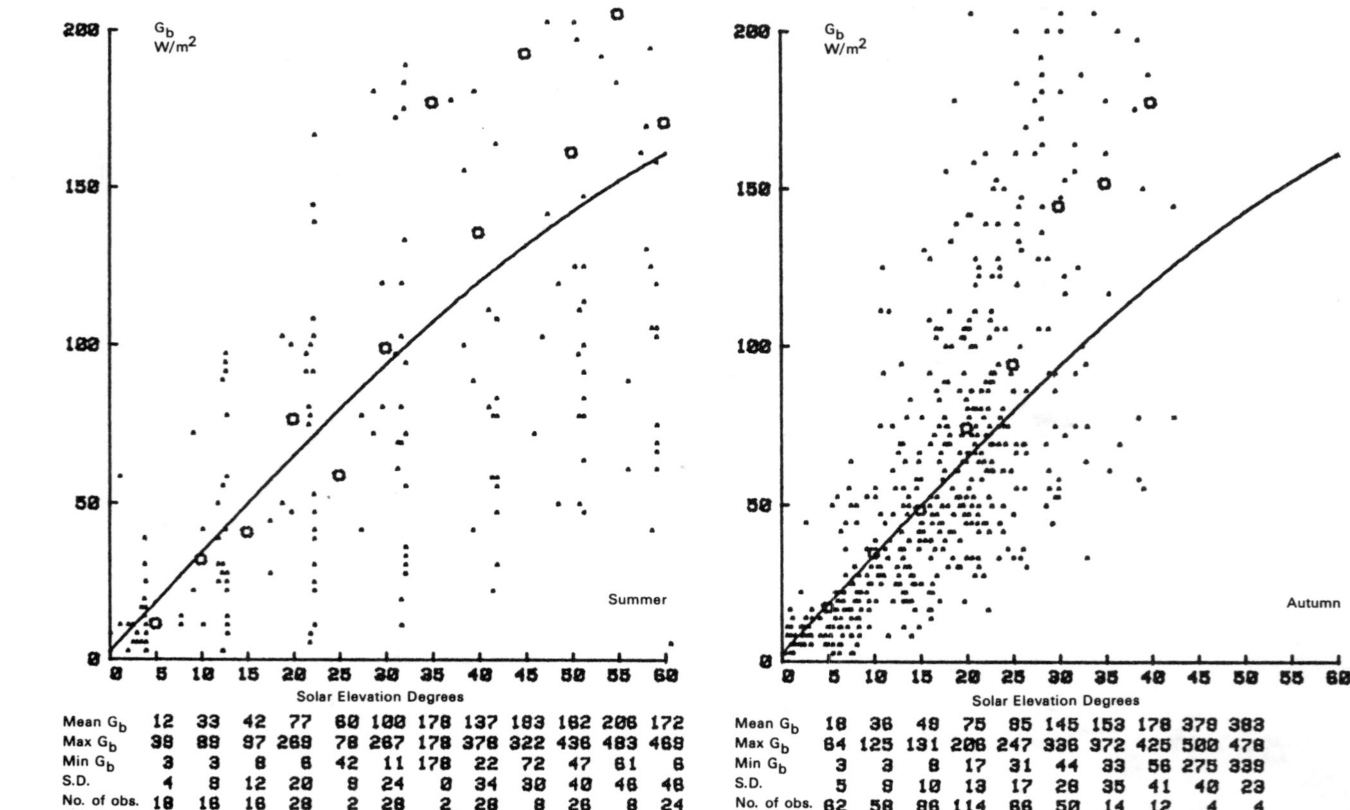

	5	10	15	20	25	30	35	40	45	50	55	60
Mean G_b	12	33	42	77	60	100	170	137	183	162	206	172
Max G_b	38	88	97	268	78	267	178	378	322	436	483	468
Min G_b	3	3	8	6	42	11	178	22	72	47	61	6
S.D.	4	8	12	20	8	24	0	34	30	40	46	46
No. of obs.	18	16	16	28	2	28	2	28	8	26	8	24

	5	10	15	20	25	30	35	40	45	50
Mean G_b	18	36	48	75	85	145	153	178	378	383
Max G_b	84	125	131	206	247	336	372	425	500	478
Min G_b	3	3	8	17	31	44	33	56	275	339
S.D.	5	8	10	13	17	28	35	41	40	23
No. of obs.	62	58	86	114	66	50	14	12	4	4

Figure 4.17a Statistical distribution of horizontal surface solar radiation from overcast skies: Summer and Autumn: Trappes

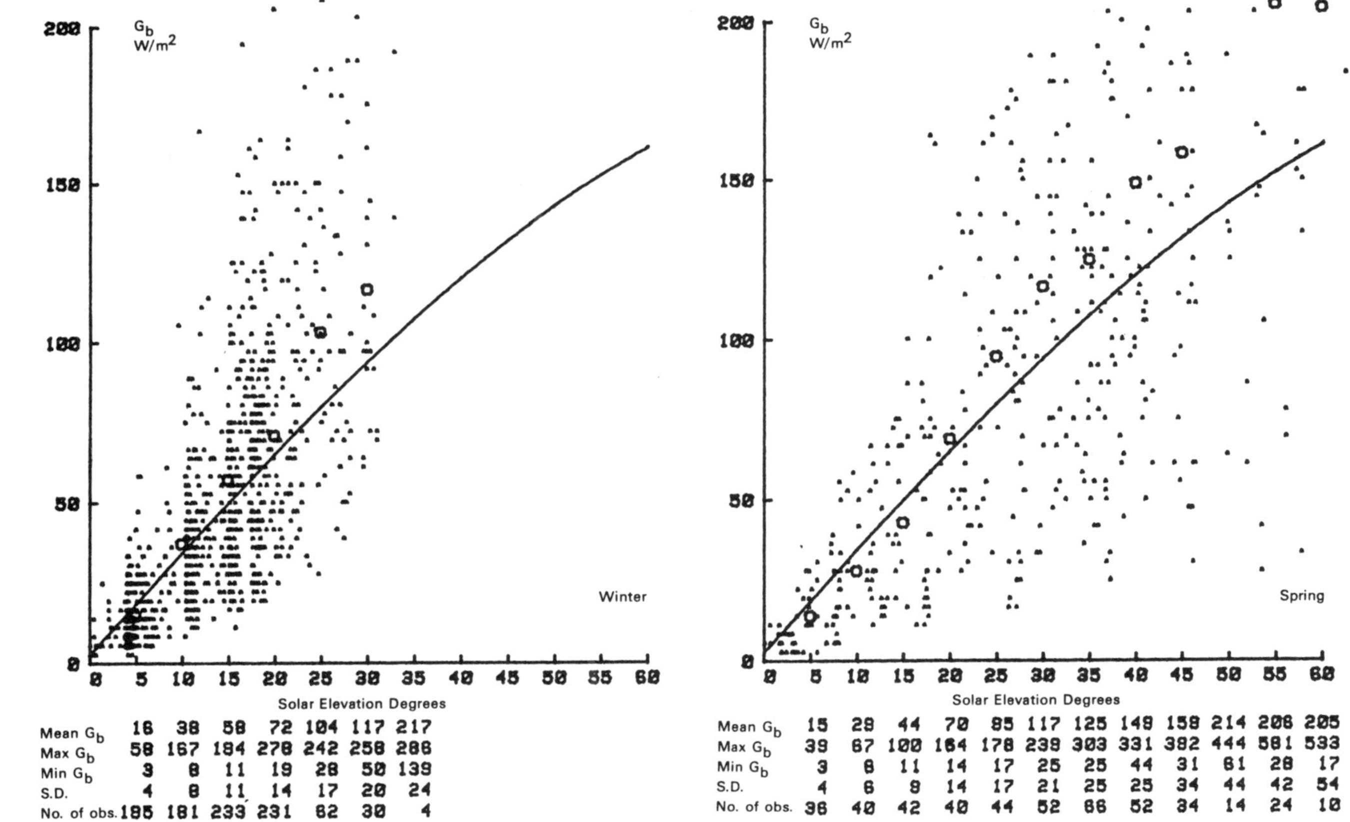

Winter:

Solar Elevation Degrees	5	10	15	20	25	30	35
Mean G_b	16	38	58	72	104	117	217
Max G_b	58	167	184	278	242	258	286
Min G_b	3	8	11	19	28	50	139
S.D.	4	8	11	14	17	20	24
No. of obs.	185	181	233	231	82	30	4

Spring:

Solar Elevation Degrees	5	10	15	20	25	30	35	40	45	50	55	60
Mean G_b	15	28	44	70	85	117	125	149	158	214	206	205
Max G_b	39	67	100	164	178	238	303	331	382	444	581	533
Min G_b	3	8	11	14	17	25	25	44	31	61	28	17
S.D.	4	6	9	14	17	21	25	25	34	44	42	54
No. of obs.	36	40	42	40	44	52	66	52	34	14	24	10

Figure 4.17b Statistical distribution of horizontal surface solar radiation from overcast skies: Winter and Spring: Trappes

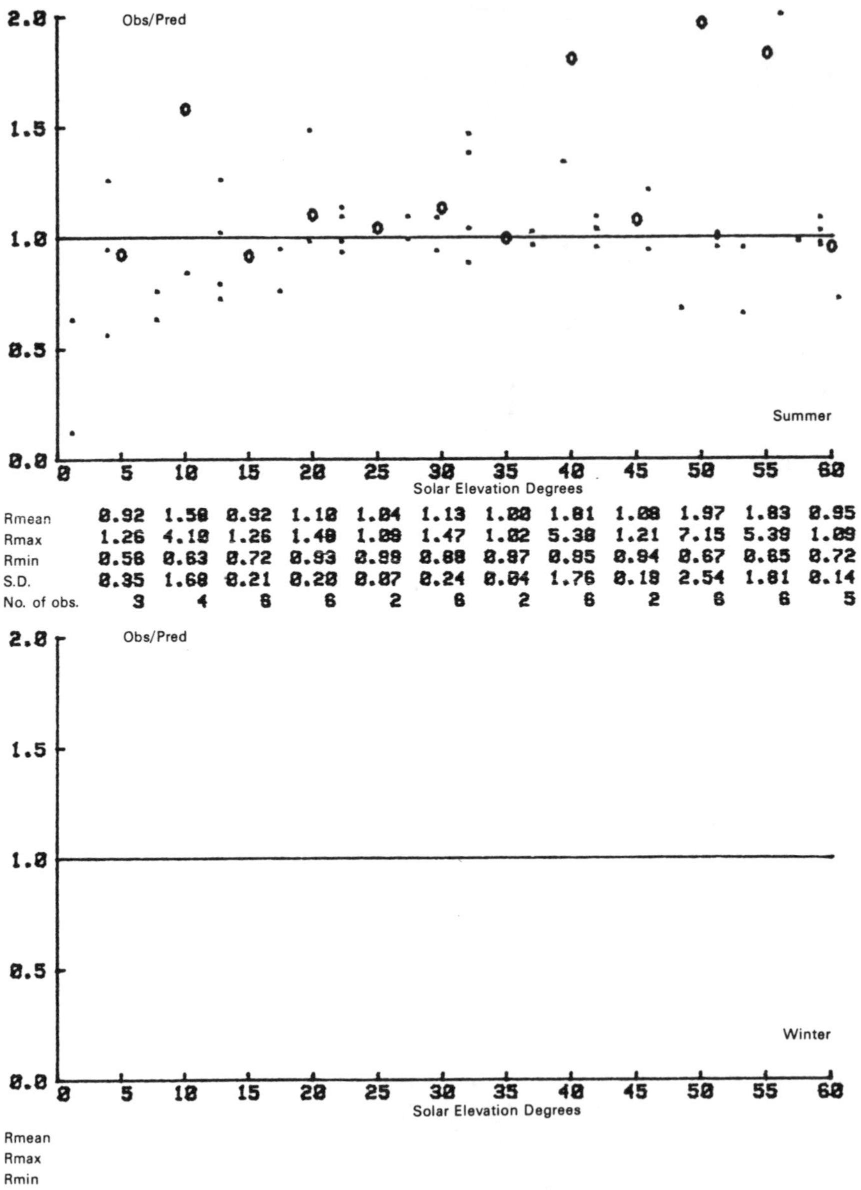

	5	10	15	20	25	30	35	40	45	50	55	60
Rmean	0.92	1.58	0.92	1.10	1.04	1.13	1.00	1.81	1.08	1.97	1.83	0.95
Rmax	1.26	4.18	1.26	1.48	1.09	1.47	1.02	5.38	1.21	7.15	5.39	1.09
Rmin	0.58	0.63	0.72	0.93	0.99	0.88	0.97	0.95	0.94	0.67	0.65	0.72
S.D.	0.35	1.68	0.21	0.20	0.07	0.24	0.04	1.76	0.18	2.54	1.81	0.14
No. of obs.	3	4	8	6	2	8	2	8	2	8	6	5

Figure 4.18a Statistical analysis of the ratio of observed radiation on a stated inclined plane to the predicted ratio on the inclined plane for overcast sky conditions: Summer and Winter: Trappes. 50° South East

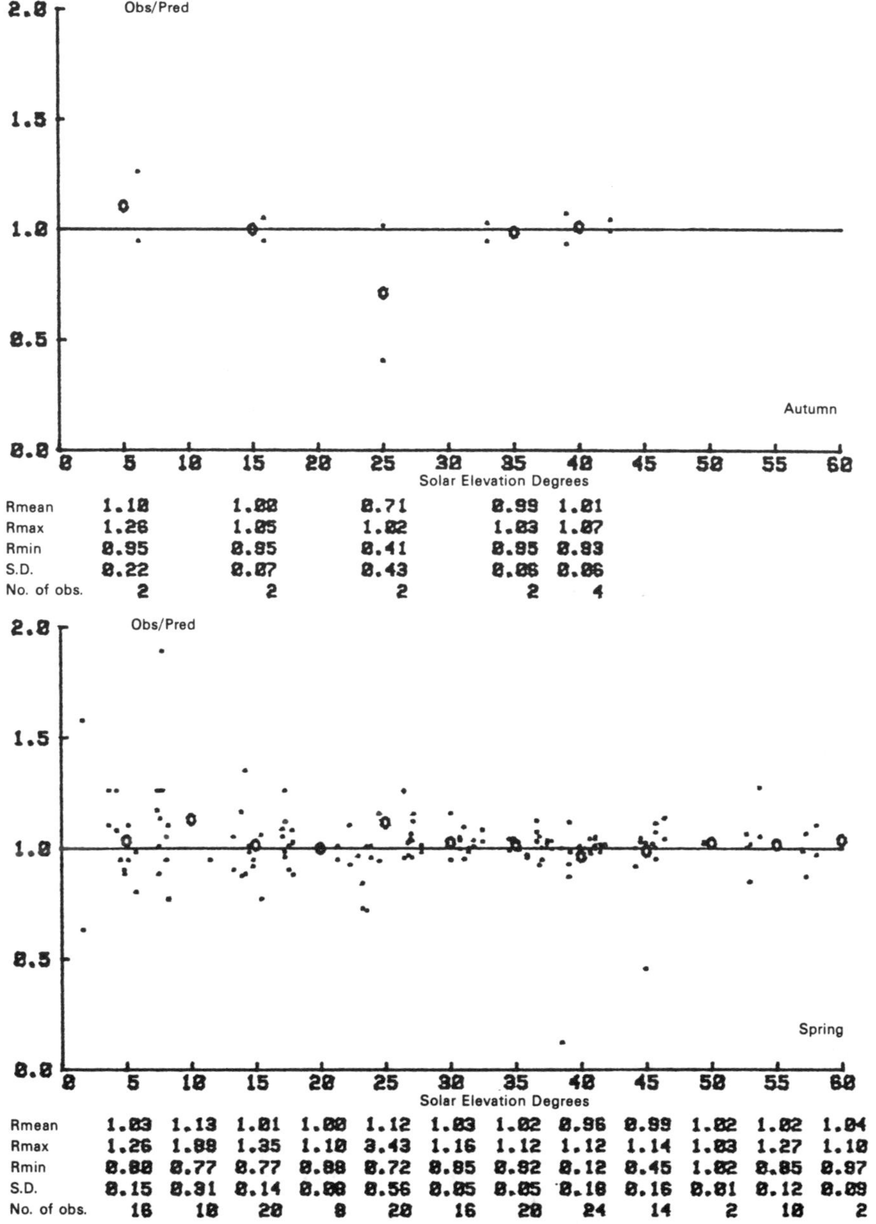

	5	15	25	35	40
Rmean	1.10	1.00	0.71	0.99	1.01
Rmax	1.26	1.05	1.02	1.03	1.07
Rmin	0.95	0.95	0.41	0.95	0.93
S.D.	0.22	0.07	0.43	0.06	0.06
No. of obs.	2	2	2	2	4

	5	10	15	20	25	30	35	40	45	50	55	60
Rmean	1.03	1.13	1.01	1.00	1.12	1.03	1.02	0.96	0.99	1.02	1.02	1.04
Rmax	1.26	1.88	1.35	1.10	3.43	1.16	1.12	1.12	1.14	1.03	1.27	1.10
Rmin	0.80	0.77	0.77	0.88	0.72	0.85	0.92	0.12	0.45	1.02	0.85	0.97
S.D.	0.15	0.31	0.14	0.08	0.56	0.05	0.05	0.18	0.16	0.01	0.12	0.09
No. of obs.	16	10	20	8	20	16	20	24	14	2	10	2

Figure 4.18b Statistical analysis of the ratio of observed radiation on a stated inclined plane to the predicted ratio on the inclined plane for overcast sky conditions: Autumn and Spring: Trappes. 50° South East

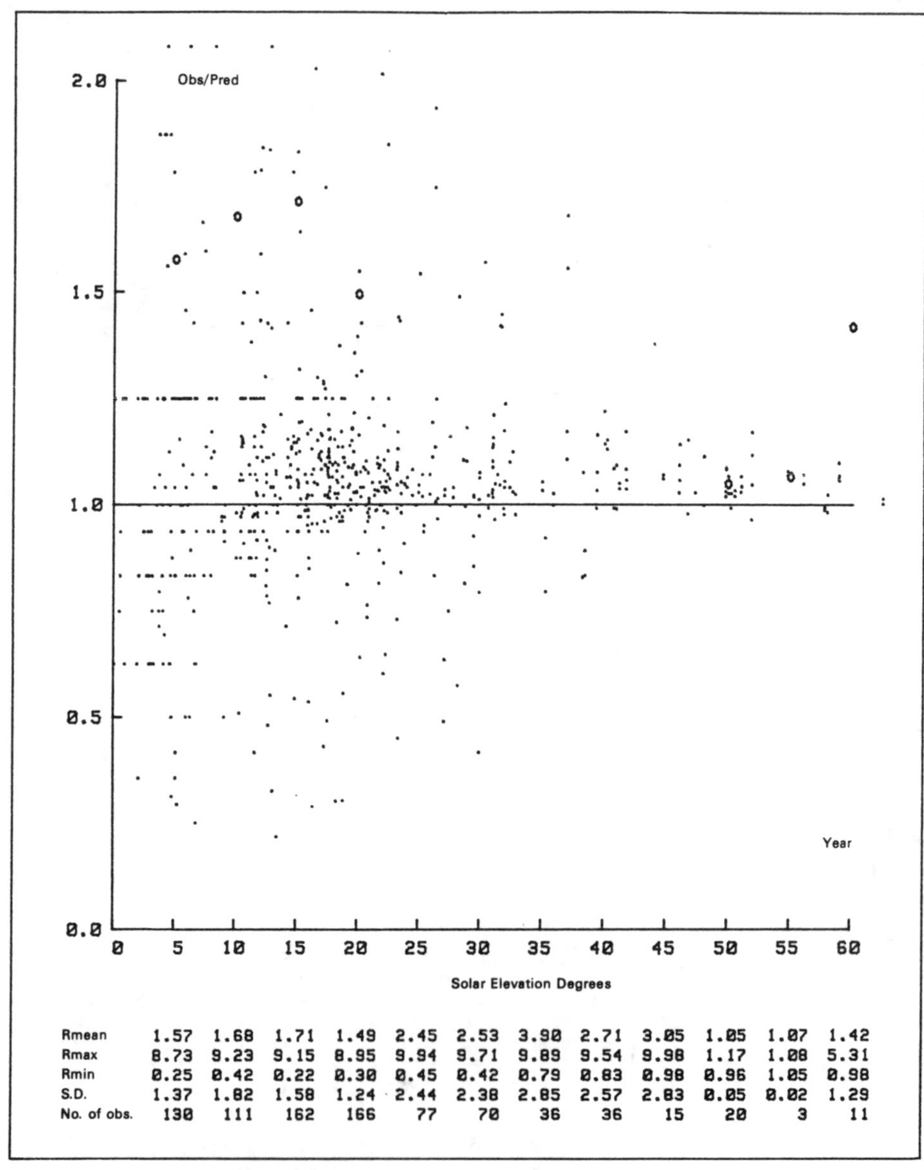

Rmean	1.57	1.68	1.71	1.49	2.45	2.53	3.90	2.71	3.05	1.05	1.07	1.42
Rmax	8.73	9.23	9.15	8.95	9.94	9.71	9.89	9.54	9.98	1.17	1.08	5.31
Rmin	0.25	0.42	0.22	0.30	0.45	0.42	0.79	0.83	0.98	0.96	1.05	0.98
S.D.	1.37	1.82	1.58	1.24	2.44	2.38	2.85	2.57	2.83	0.05	0.02	1.29
No. of obs.	130	111	162	166	77	70	36	36	15	20	3	11

Figure 4.19 Statistical analysis of the ratio of observed radiation on a stated inclined plane to the predicted ratio on the inclined plane for overcast sky conditions.
Station: Trappes. Plane: 48° 46' South

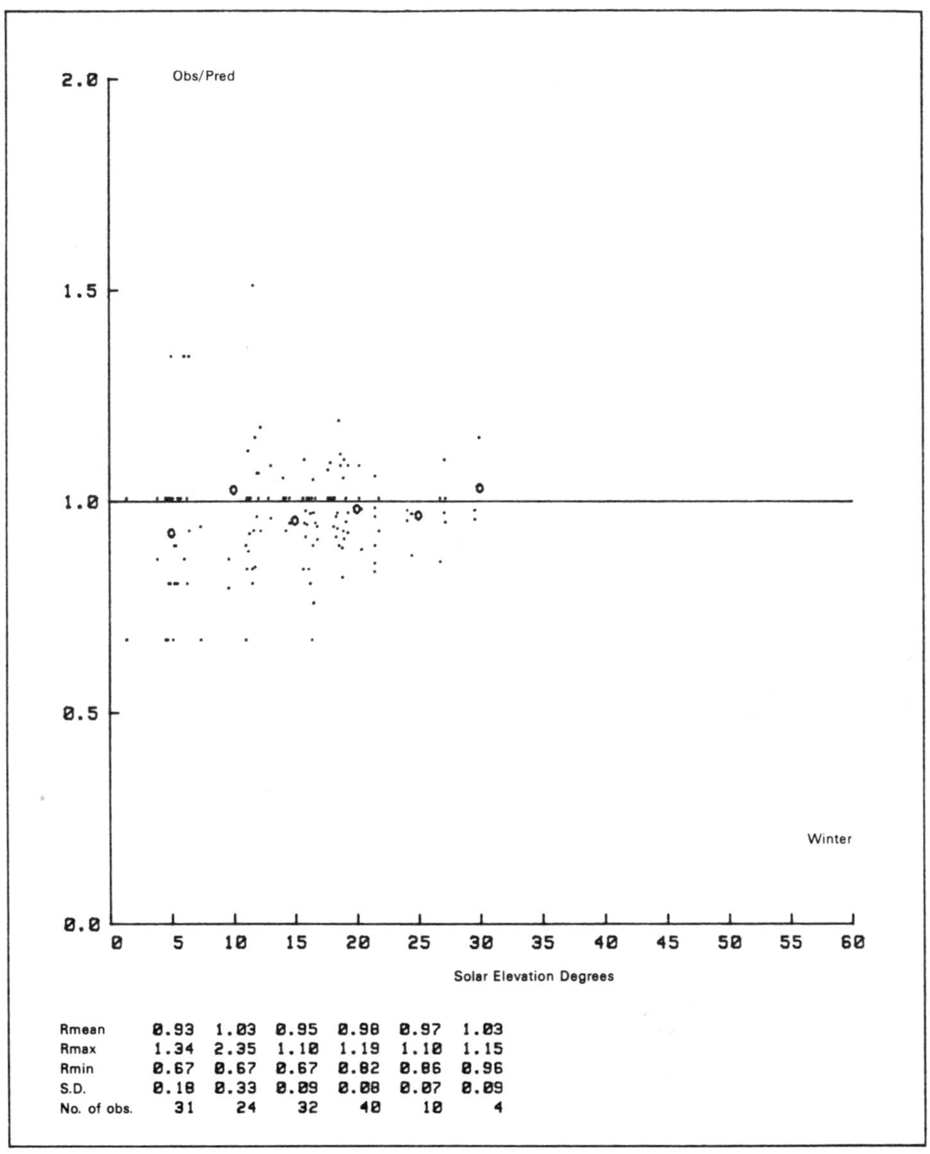

Figure 4.20 Statistical analysis of the ratio of observed radiation on a stated inclined plane to the predicted ratio on the inclined plane for overcast sky conditions.
Station: Trappes. Plane: Vertical North

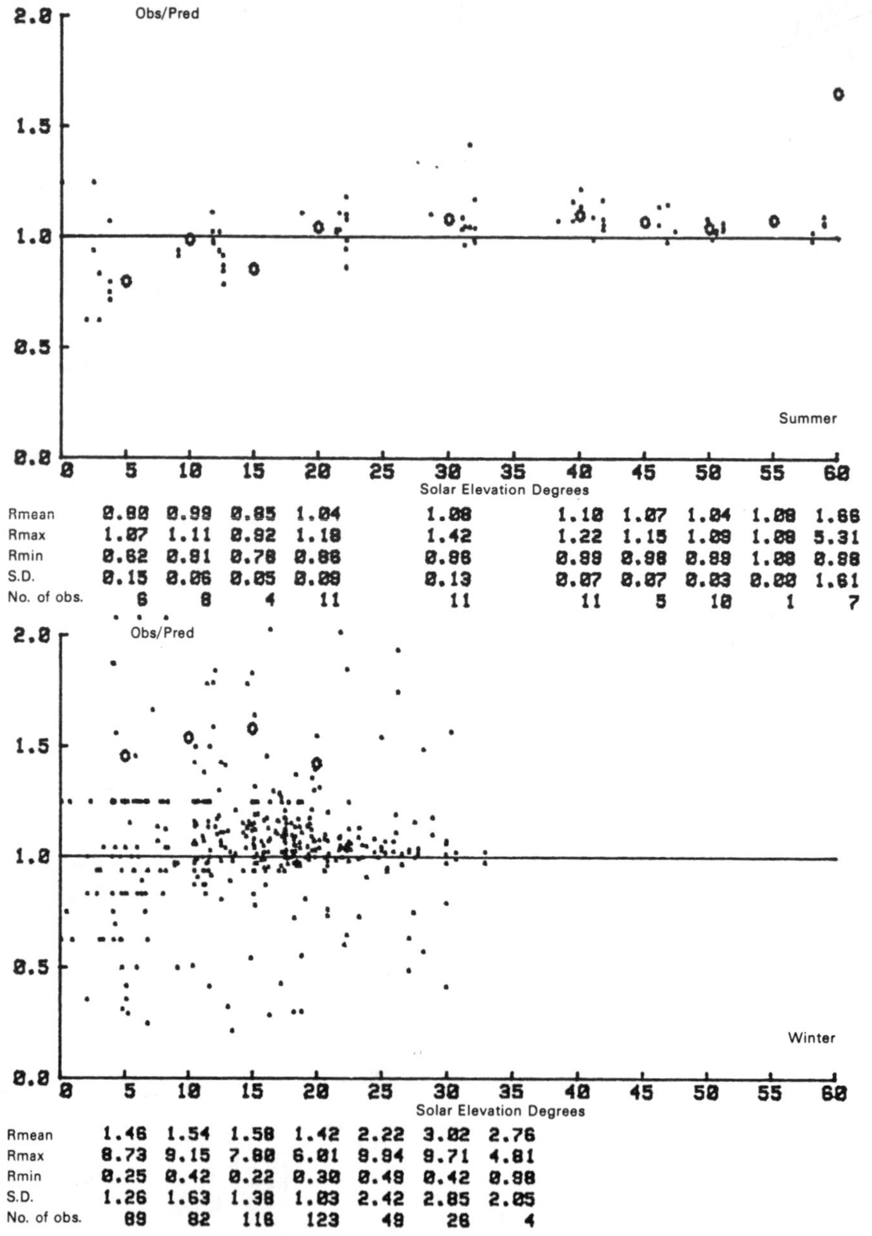

Summer

	5	10	15	20	25	30	35	40	45	50	55	60
Rmean	0.80	0.99	0.85	1.04		1.00		1.10	1.07	1.04	1.00	1.66
Rmax	1.07	1.11	0.92	1.18		1.42		1.22	1.15	1.09	1.00	5.31
Rmin	0.62	0.91	0.78	0.86		0.96		0.99	0.98	0.99	1.00	0.98
S.D.	0.15	0.06	0.05	0.09		0.13		0.07	0.07	0.03	0.00	1.61
No. of obs.	6	8	4	11		11		11	5	10	1	7

Winter

	5	10	15	20	25	30	35
Rmean	1.46	1.54	1.58	1.42	2.22	3.02	2.76
Rmax	8.73	9.15	7.80	6.01	9.94	9.71	4.81
Rmin	0.25	0.42	0.22	0.30	0.49	0.42	0.98
S.D.	1.26	1.63	1.38	1.03	2.42	2.85	2.05
No. of obs.	89	82	116	123	49	26	4

Figure 4.21a Statistical analysis of the ratio of observed radiation on a stated inclined plane to the predicted ratio on the inclined plane for overcast sky conditions: Summer and Winter: Trappes. 48° 46' South

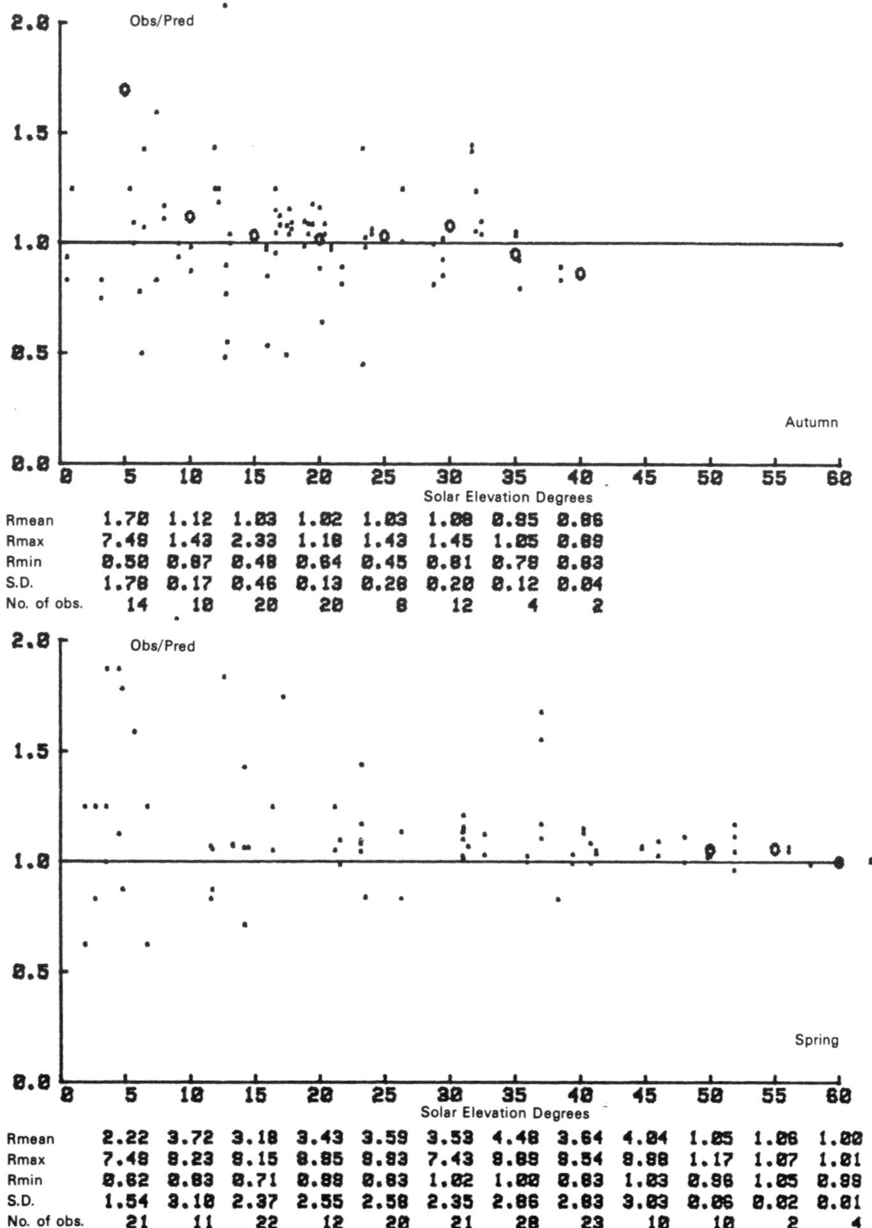

Rmean	1.70	1.12	1.03	1.02	1.03	1.08	0.85	0.86
Rmax	7.48	1.43	2.33	1.18	1.43	1.45	1.05	0.89
Rmin	0.50	0.87	0.48	0.64	0.45	0.81	0.78	0.83
S.D.	1.78	0.17	0.46	0.13	0.28	0.20	0.12	0.04
No. of obs.	14	10	20	20	8	12	4	2

Rmean	2.22	3.72	3.18	3.43	3.59	3.53	4.48	3.64	4.04	1.05	1.06	1.00
Rmax	7.48	8.23	8.15	8.85	8.83	7.43	8.89	8.54	8.88	1.17	1.07	1.01
Rmin	0.62	0.83	0.71	0.88	0.83	1.02	1.00	0.83	1.03	0.96	1.05	0.98
S.D.	1.54	3.10	2.37	2.55	2.58	2.35	2.86	2.83	3.03	0.06	0.02	0.01
No. of obs.	21	11	22	12	20	21	28	23	10	10	2	4

Figure 4.21b Statistical analysis of the ratio of observed radiation on a stated inclined plane to the predicted ratio on the inclined plane for overcast sky conditions: Autumn and Spring: Trappes. 48° 46' South

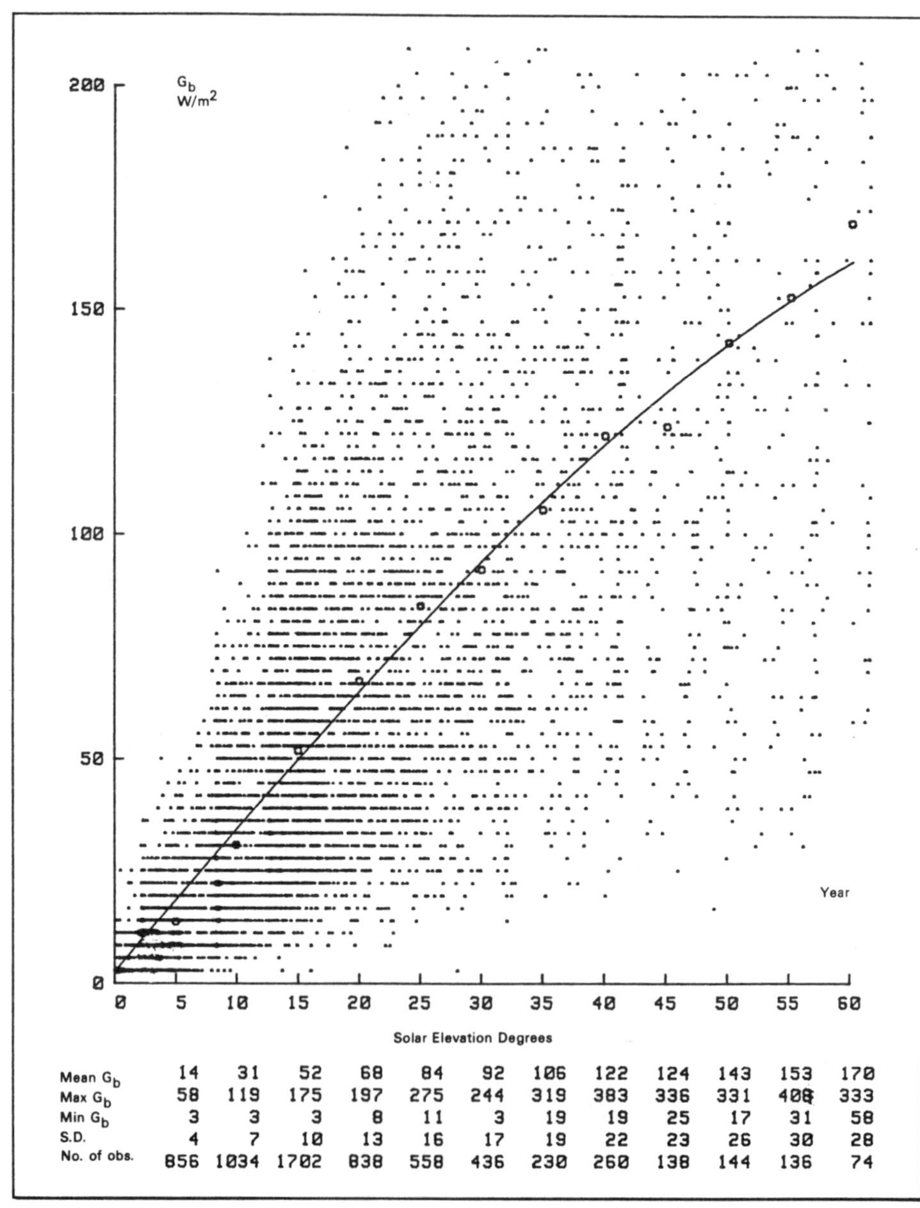

Mean G_b	14	31	52	68	84	92	106	122	124	143	153	170
Max G_b	58	119	175	197	275	244	319	383	336	331	408	333
Min G_b	3	3	3	8	11	3	19	19	25	17	31	58
S.D.	4	7	10	13	16	17	19	22	23	26	30	28
No. of obs.	856	1034	1702	838	558	436	230	260	138	144	136	74

Figure 4.22 Statistical distribution of horizontal surface solar radiation from overcast skies.
Station: Bracknell

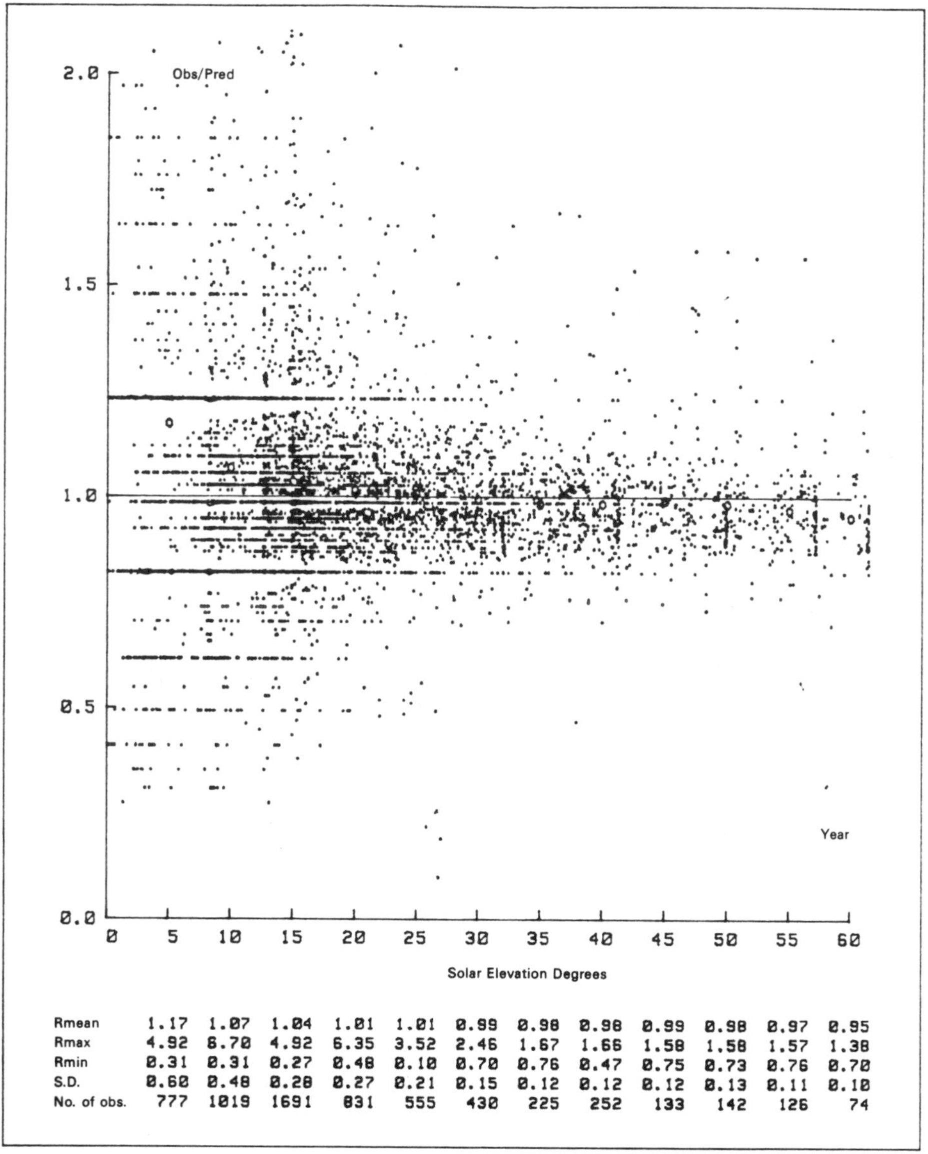

Rmean	1.17	1.07	1.04	1.01	1.01	0.99	0.98	0.98	0.99	0.98	0.97	0.95
Rmax	4.92	6.70	4.92	6.35	3.52	2.46	1.67	1.66	1.58	1.58	1.57	1.38
Rmin	0.31	0.31	0.27	0.48	0.10	0.70	0.76	0.47	0.75	0.73	0.76	0.70
S.D.	0.60	0.48	0.28	0.27	0.21	0.15	0.12	0.12	0.12	0.13	0.11	0.10
No. of obs.	777	1019	1691	831	555	430	225	252	133	142	126	74

Figure 4.23 Statistical analysis of the ratio of observed radiation on a stated inclined plane to the predicted ratio on the inclined plane for overcast sky conditions.
Station: Bracknell. Plane: Vertical South

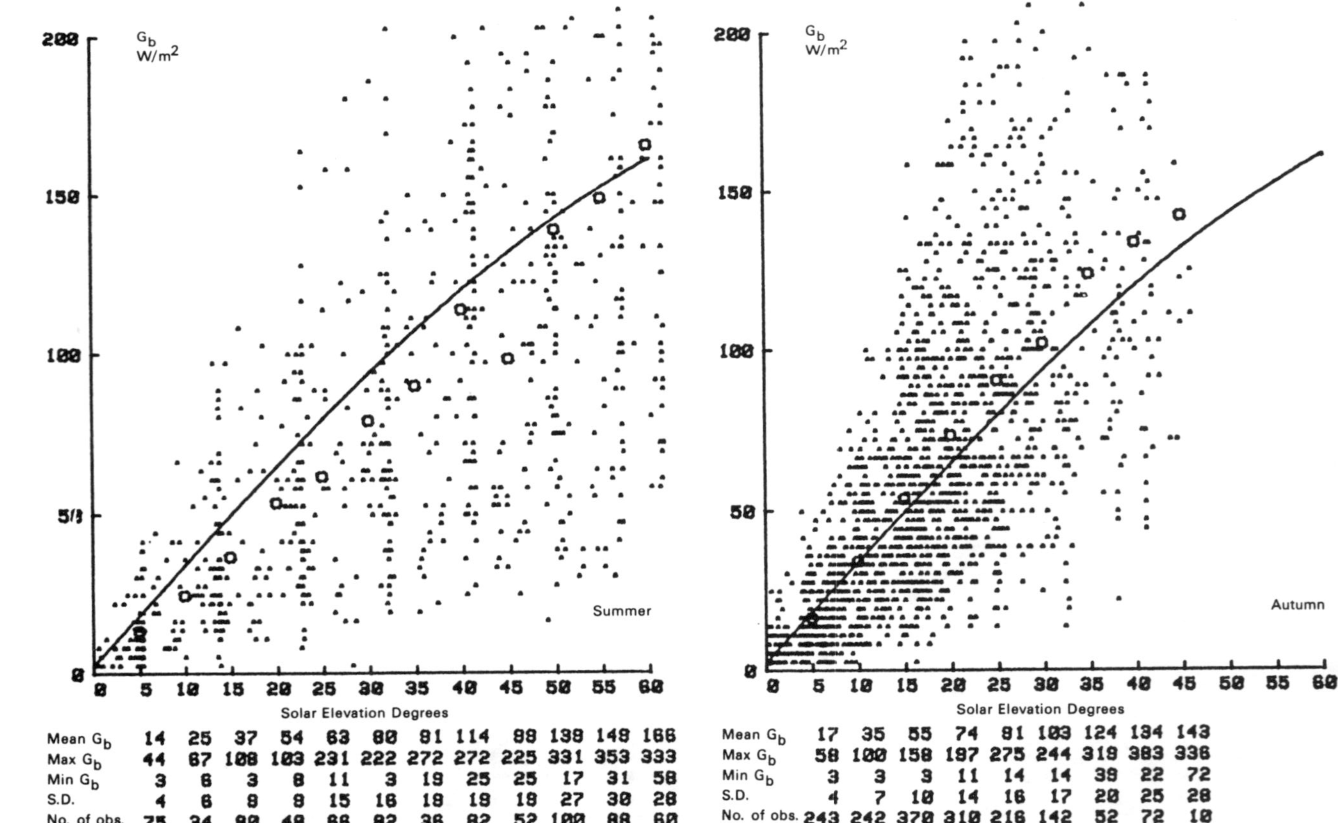

Mean G_b	14	25	37	54	63	80	81	114	88	138	148	166
Max G_b	44	67	108	103	231	222	272	272	225	331	353	333
Min G_b	3	6	3	8	11	3	19	25	25	17	31	58
S.D.	4	6	8	8	15	16	18	19	18	27	30	28
No. of obs.	75	34	80	48	66	82	36	82	52	100	88	60

Mean G_b	17	35	55	74	81	103	124	134	143
Max G_b	58	100	158	197	275	244	319	383	336
Min G_b	3	3	3	11	14	14	39	22	72
S.D.	4	7	10	14	16	17	20	25	28
No. of obs.	243	242	370	310	216	142	52	72	10

Figure 4.24a Statistical distribution of horizontal surface solar radiation from overcast skies: Summer and Autumn: Bracknell

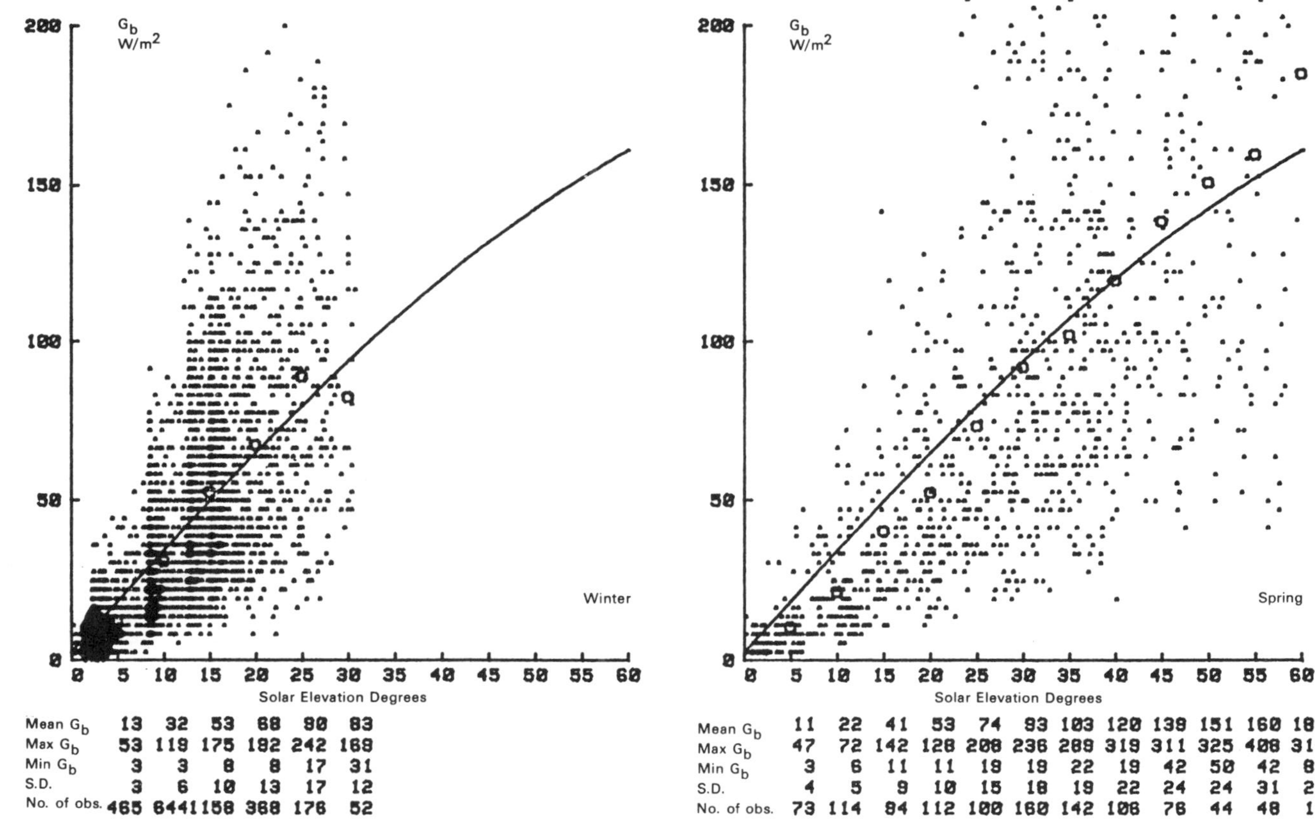

Mean G_b	13	32	53	68	90	83
Max G_b	53	119	175	192	242	169
Min G_b	3	3	8	8	17	31
S.D.	3	6	10	13	17	12
No. of obs.	465	644	1158	368	176	52

Mean G_b	11	22	41	53	74	93	103	120	138	151	160	186
Max G_b	47	72	142	128	208	236	289	319	311	325	408	314
Min G_b	3	6	11	11	19	19	22	19	42	50	42	83
S.D.	4	5	9	10	15	18	19	22	24	24	31	28
No. of obs.	73	114	84	112	100	160	142	106	76	44	48	14

Figure 4.24b Statistical distribution of horizontal surface solar radiation from overcast skies: Winter and Spring: Bracknell

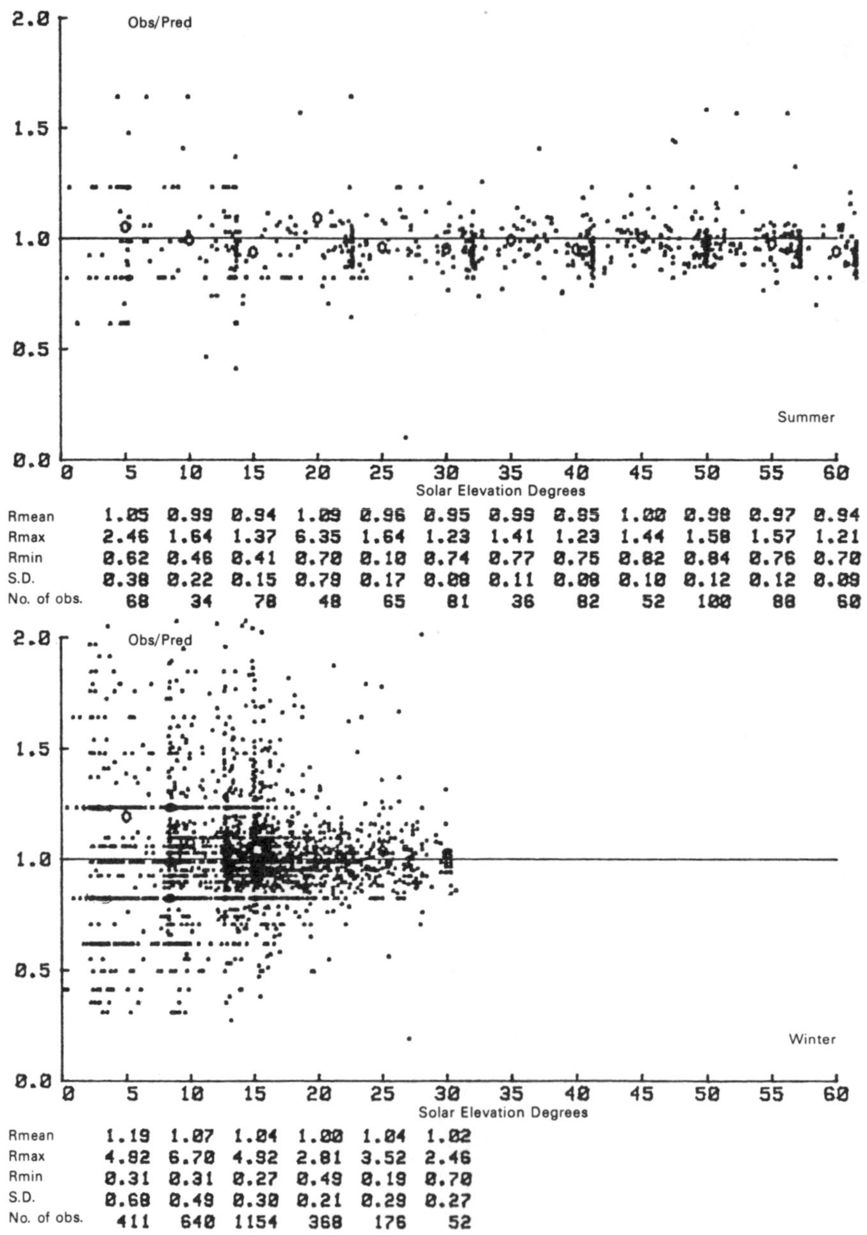

Rmean	1.05	0.99	0.94	1.09	0.96	0.95	0.99	0.95	1.00	0.98	0.97	0.94
Rmax	2.46	1.64	1.37	6.35	1.64	1.23	1.41	1.23	1.44	1.58	1.57	1.21
Rmin	0.62	0.46	0.41	0.70	0.10	0.74	0.77	0.75	0.82	0.84	0.76	0.70
S.D.	0.38	0.22	0.15	0.79	0.17	0.08	0.11	0.08	0.10	0.12	0.12	0.09
No. of obs.	68	34	78	48	65	81	36	82	52	100	88	60

Rmean	1.19	1.07	1.04	1.00	1.04	1.02
Rmax	4.82	6.70	4.92	2.81	3.52	2.46
Rmin	0.31	0.31	0.27	0.49	0.19	0.70
S.D.	0.68	0.49	0.30	0.21	0.29	0.27
No. of obs.	411	640	1154	368	176	52

Figure 4.25a Statistical analysis of the ratio of observed radiation on a stated inclined plane to the predicted ratio on the inclined plane for overcast sky conditions: Summer and Winter: Bracknell. Vertical South

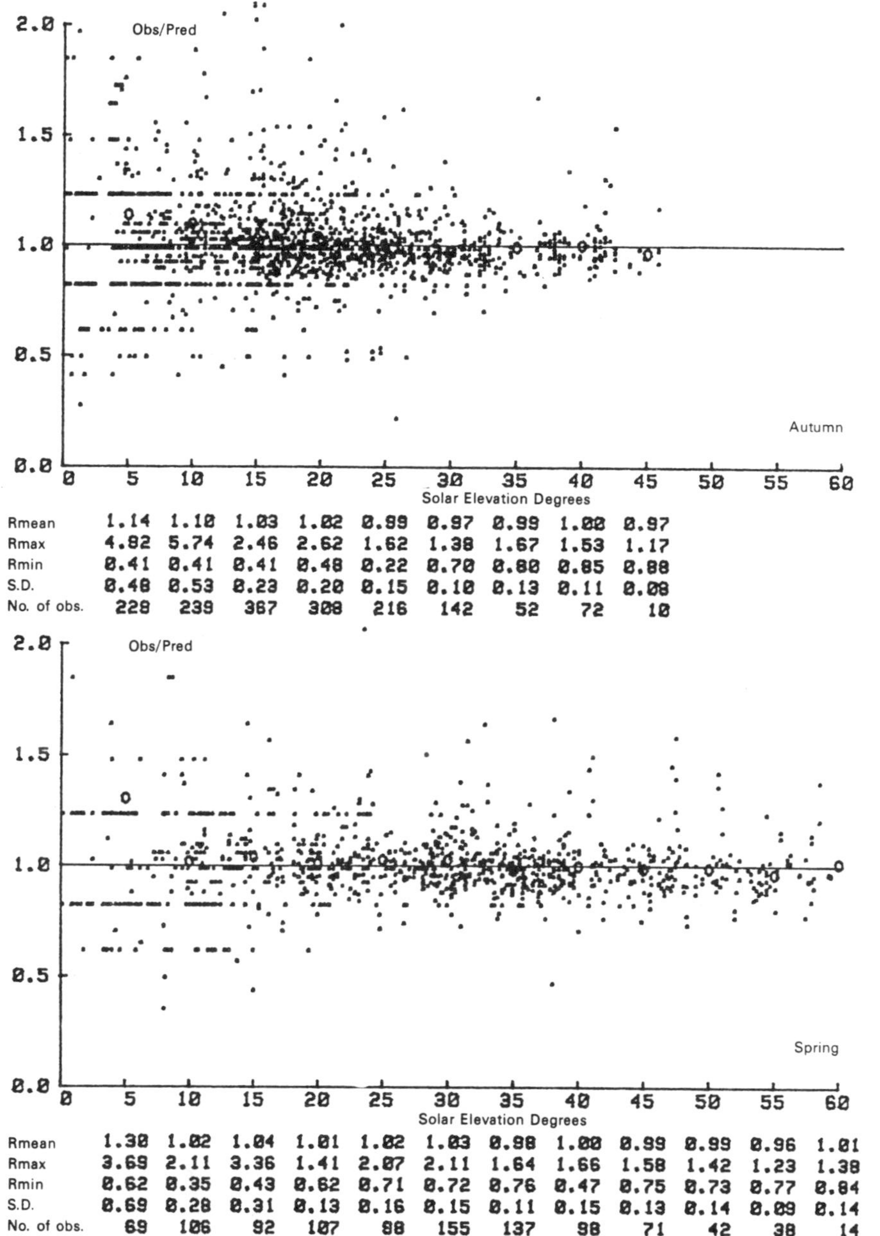

Rmean	1.14	1.10	1.03	1.02	0.99	0.97	0.99	1.00	0.97
Rmax	4.82	5.74	2.46	2.62	1.62	1.38	1.67	1.53	1.17
Rmin	0.41	0.41	0.41	0.48	0.22	0.70	0.80	0.85	0.88
S.D.	0.48	0.53	0.23	0.20	0.15	0.10	0.13	0.11	0.08
No. of obs.	228	239	367	308	216	142	52	72	10

Rmean	1.30	1.02	1.04	1.01	1.02	1.03	0.98	1.00	0.99	0.99	0.96	1.01
Rmax	3.69	2.11	3.36	1.41	2.07	2.11	1.64	1.66	1.58	1.42	1.23	1.38
Rmin	0.62	0.35	0.43	0.62	0.71	0.72	0.76	0.47	0.75	0.73	0.77	0.84
S.D.	0.69	0.28	0.31	0.13	0.16	0.15	0.11	0.15	0.13	0.14	0.09	0.14
No. of obs.	69	106	92	107	88	155	137	98	71	42	38	14

Figure 4.25b Statistical analysis of the ratio of observed radiation on a stated inclined plane to the predicted ratio on the inclined plane for overcast sky conditions: Autumn and Spring: Bracknell. Vertical South

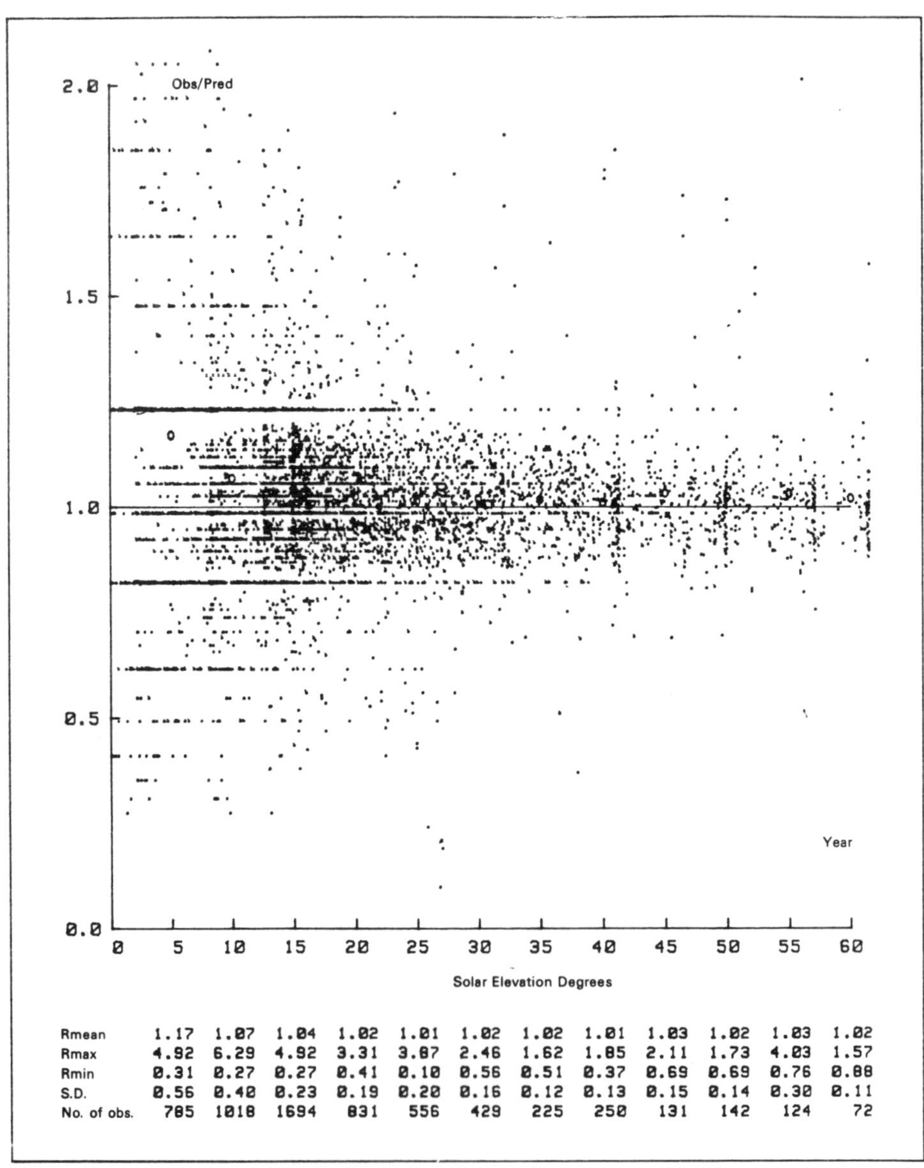

Rmean	1.17	1.07	1.04	1.02	1.01	1.02	1.02	1.01	1.03	1.02	1.03	1.02
Rmax	4.92	6.29	4.92	3.31	3.87	2.46	1.62	1.85	2.11	1.73	4.03	1.57
Rmin	0.31	0.27	0.27	0.41	0.10	0.56	0.51	0.37	0.69	0.69	0.76	0.88
S.D.	0.56	0.40	0.23	0.19	0.20	0.16	0.12	0.13	0.15	0.14	0.30	0.11
No. of obs.	785	1018	1694	831	556	429	225	250	131	142	124	72

Figure 4.26 Statistical analysis of the ratio of observed radiation on a stated inclined plane to the predicted ratio on the inclined plane for overcast sky conditions.
Station: Bracknell. Plane: Vertical East

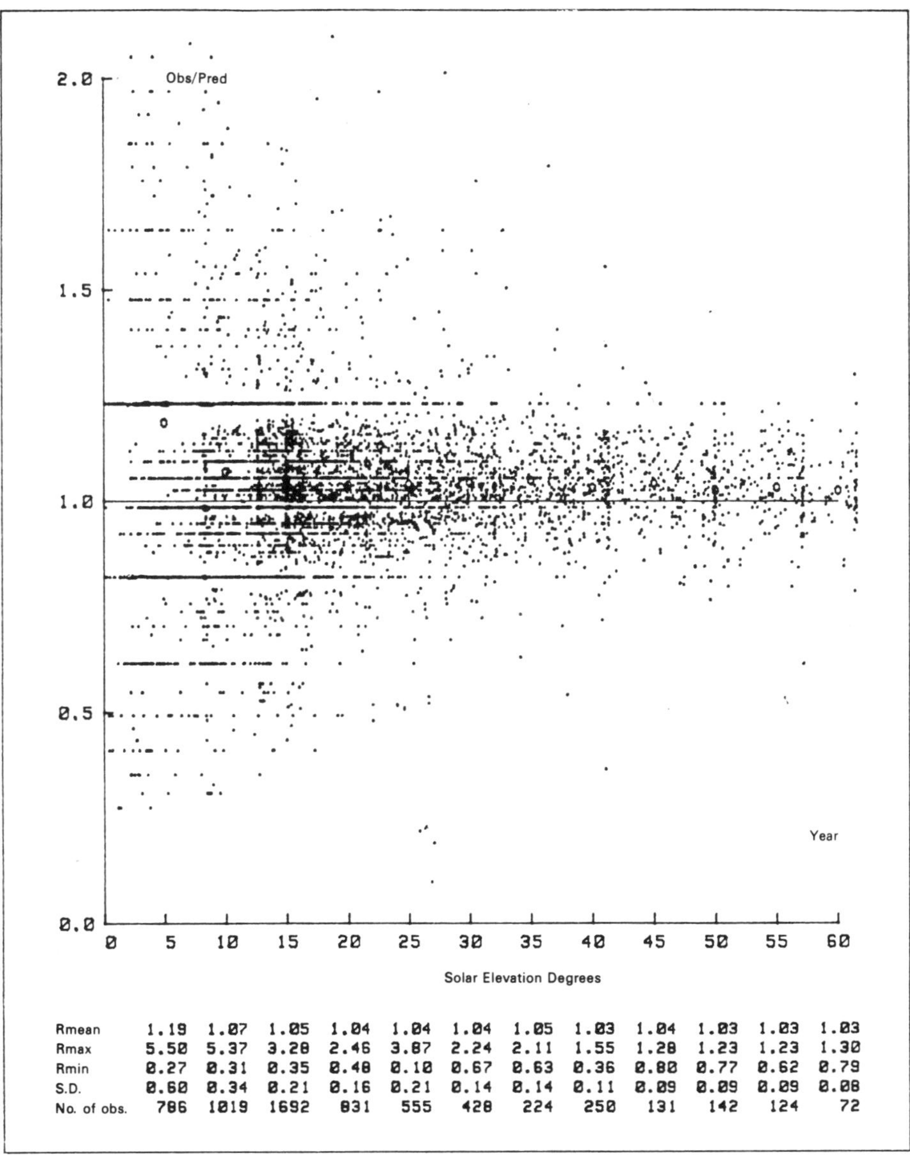

	5	10	15	20	25	30	35	40	45	50	55	60
Rmean	1.19	1.07	1.05	1.04	1.04	1.04	1.05	1.03	1.04	1.03	1.03	1.03
Rmax	5.50	5.37	3.28	2.46	3.87	2.24	2.11	1.55	1.28	1.23	1.23	1.30
Rmin	0.27	0.31	0.35	0.48	0.10	0.67	0.63	0.36	0.80	0.77	0.62	0.79
S.D.	0.60	0.34	0.21	0.16	0.21	0.14	0.14	0.11	0.09	0.09	0.09	0.08
No. of obs.	786	1019	1692	831	555	428	224	250	131	142	124	72

Figure 4.27 Statistical analysis of the ratio of observed radiation on a stated inclined plane to the predicted ratio on the inclined plane for overcast sky conditions.
Station: Bracknell. Plane: Vertical West

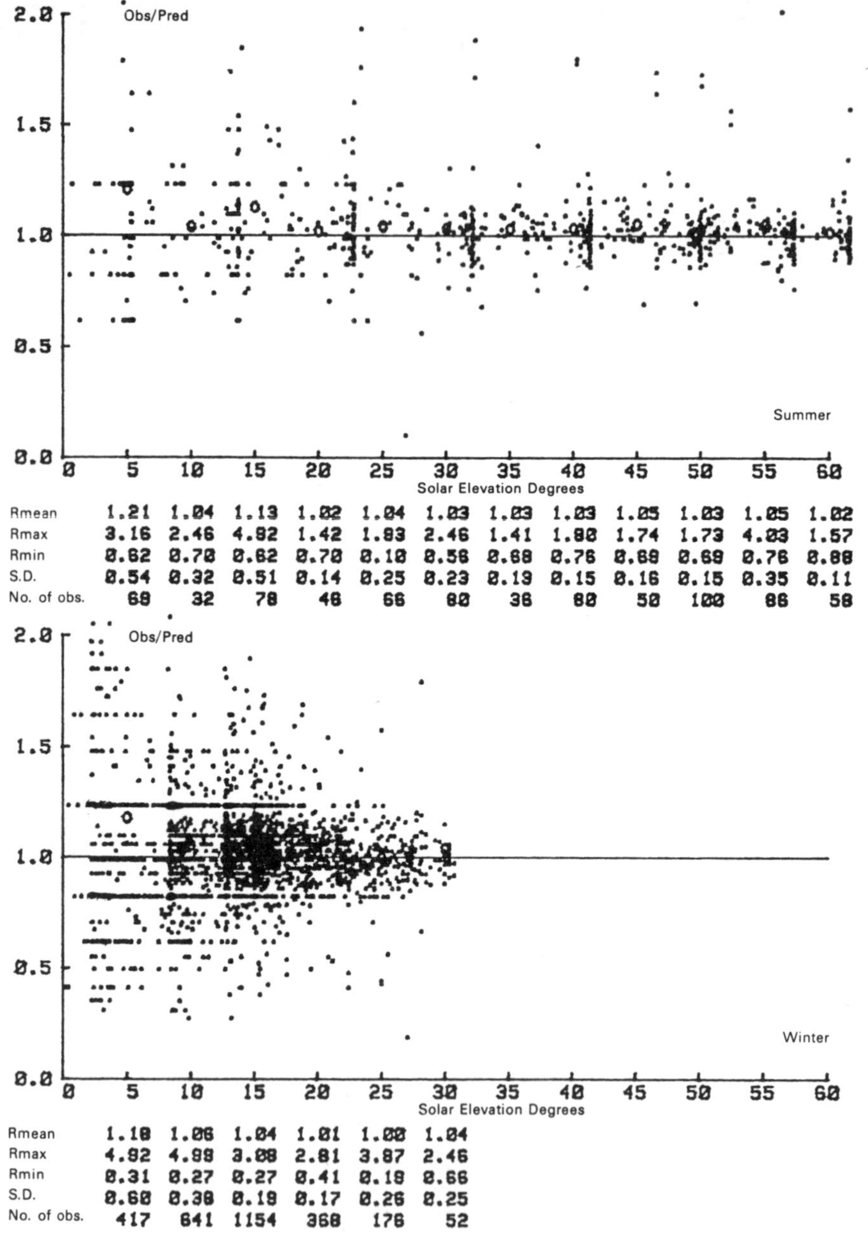

	5	10	15	20	25	30	35	40	45	50	55	60
Rmean	1.21	1.04	1.13	1.02	1.04	1.03	1.03	1.03	1.05	1.03	1.05	1.02
Rmax	3.16	2.46	4.82	1.42	1.83	2.46	1.41	1.80	1.74	1.73	4.03	1.57
Rmin	0.62	0.70	0.62	0.70	0.10	0.56	0.68	0.76	0.69	0.69	0.76	0.88
S.D.	0.54	0.32	0.51	0.14	0.25	0.23	0.13	0.15	0.16	0.15	0.35	0.11
No. of obs.	69	32	78	48	66	80	36	80	50	100	86	58

	5	10	15	20	25	30
Rmean	1.18	1.06	1.04	1.01	1.00	1.04
Rmax	4.92	4.98	3.08	2.81	3.87	2.46
Rmin	0.31	0.27	0.27	0.41	0.18	0.66
S.D.	0.60	0.38	0.18	0.17	0.26	0.25
No. of obs.	417	641	1154	368	176	52

Figure 4.28a Statistical analysis of the ratio of observed radiation on a stated inclined plane to the predicted ratio on the inclined plane for overcast sky conditions: Summer and Winter: Bracknell. Vertical East

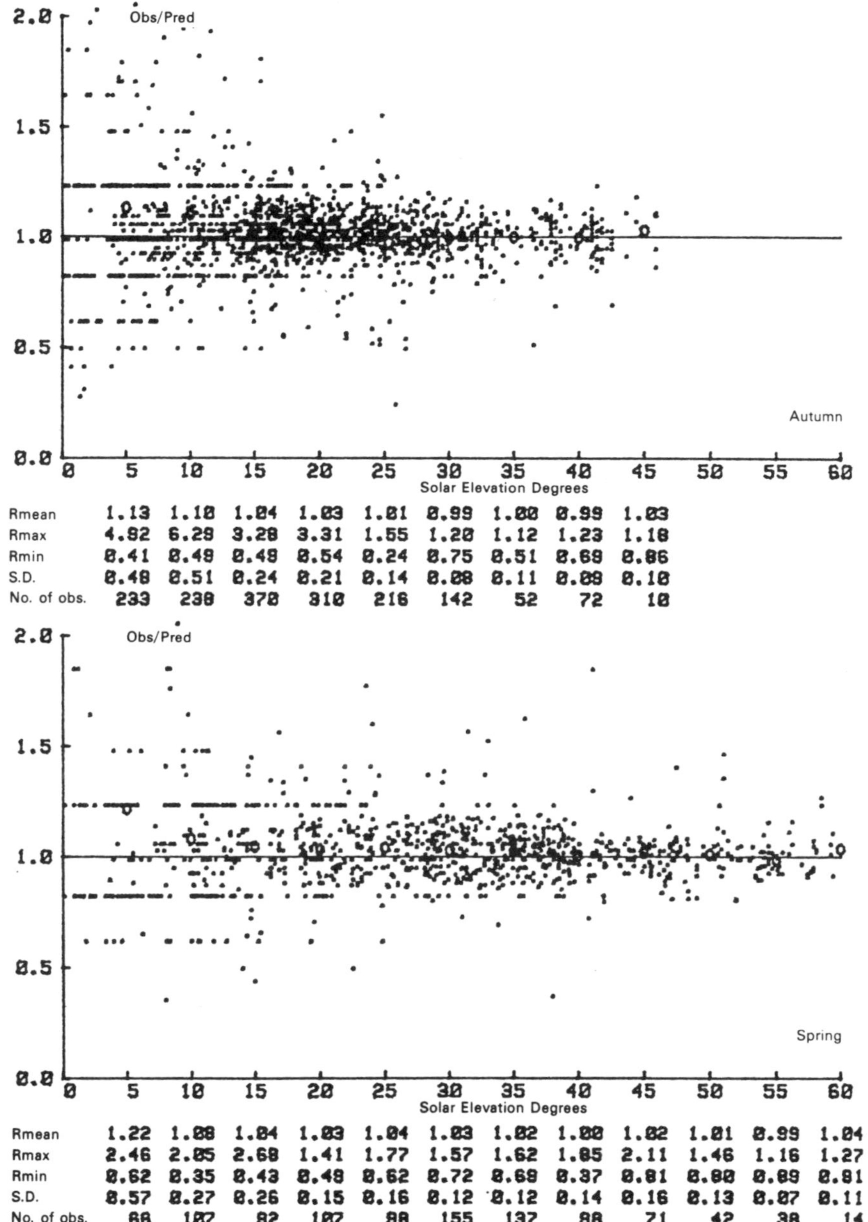

Rmean	1.13	1.10	1.04	1.03	1.01	0.99	1.00	0.99	1.03
Rmax	4.92	6.29	3.28	3.31	1.55	1.20	1.12	1.23	1.18
Rmin	0.41	0.49	0.48	0.54	0.24	0.75	0.51	0.69	0.86
S.D.	0.48	0.51	0.24	0.21	0.14	0.08	0.11	0.09	0.10
No. of obs.	233	238	370	310	216	142	52	72	10

Rmean	1.22	1.08	1.04	1.03	1.04	1.03	1.02	1.00	1.02	1.01	0.99	1.04
Rmax	2.46	2.05	2.68	1.41	1.77	1.57	1.62	1.85	2.11	1.46	1.16	1.27
Rmin	0.62	0.35	0.43	0.48	0.62	0.72	0.69	0.37	0.81	0.80	0.89	0.81
S.D.	0.57	0.27	0.26	0.15	0.16	0.12	0.12	0.14	0.16	0.13	0.07	0.11
No. of obs.	66	107	82	107	88	155	137	88	71	42	38	14

Figure 4.28b Statistical analysis of the ratio of observed radiation on a stated inclined plane to the predicted ratio on the inclined plane for overcast sky conditions: Autumn and Spring: Bracknell. Vertical East

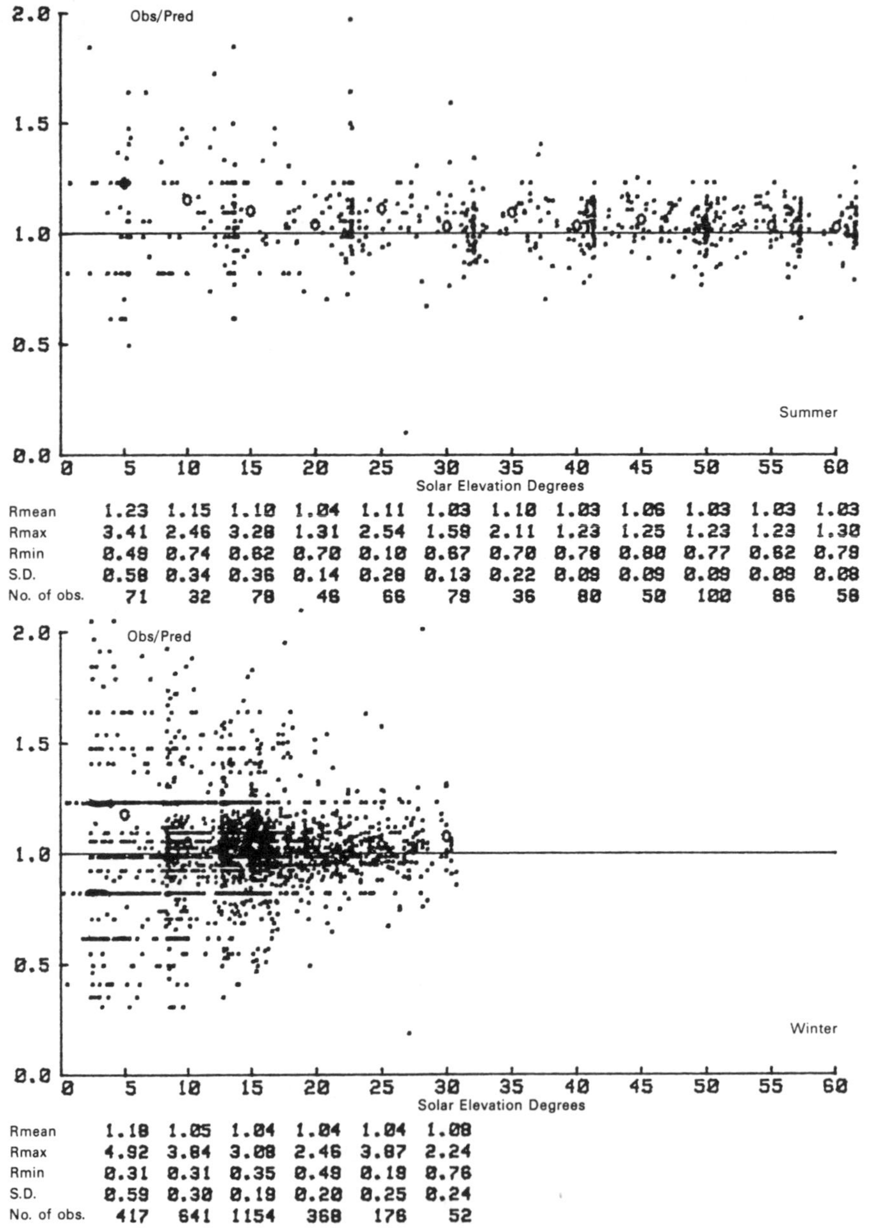

Rmean	1.23	1.15	1.10	1.04	1.11	1.03	1.10	1.03	1.06	1.03	1.03	1.03
Rmax	3.41	2.46	3.28	1.31	2.54	1.58	2.11	1.23	1.25	1.23	1.23	1.30
Rmin	0.49	0.74	0.62	0.70	0.10	0.67	0.70	0.78	0.80	0.77	0.62	0.79
S.D.	0.58	0.34	0.36	0.14	0.28	0.13	0.22	0.09	0.09	0.09	0.09	0.08
No. of obs.	71	32	78	46	66	79	36	80	50	100	86	58

Rmean	1.18	1.05	1.04	1.04	1.04	1.08
Rmax	4.92	3.84	3.08	2.46	3.87	2.24
Rmin	0.31	0.31	0.35	0.49	0.19	0.76
S.D.	0.59	0.30	0.19	0.20	0.25	0.24
No. of obs.	417	641	1154	368	176	52

Figure 4.29a Statistical analysis of the ratio of observed radiation on a stated inclined plane to the predicted ratio on the inclined plane for overcast sky conditions: Summer and Winter: Bracknell. Vertical West

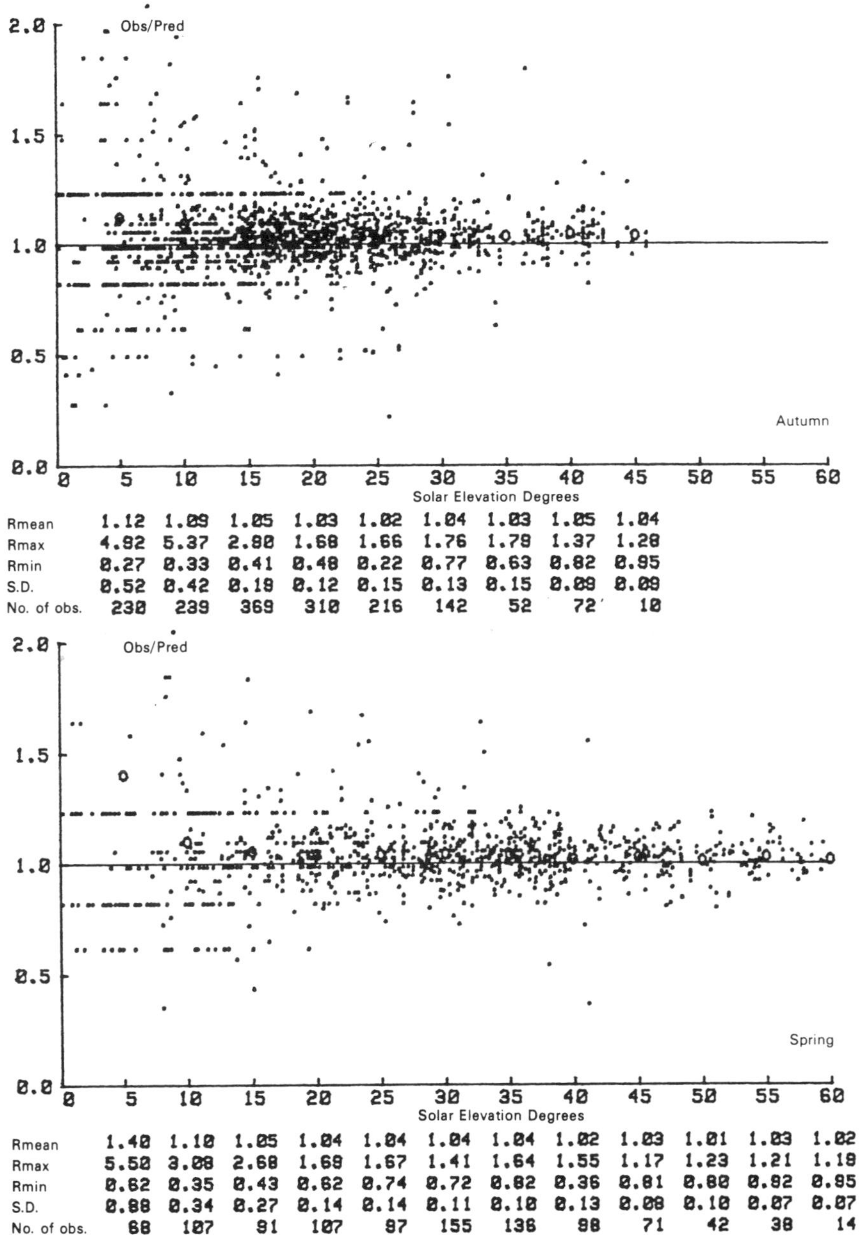

	0–5	5–10	10–15	15–20	20–25	25–30	30–35	35–40	40–45
Rmean	1.12	1.09	1.05	1.03	1.02	1.04	1.03	1.05	1.04
Rmax	4.92	5.37	2.80	1.68	1.66	1.76	1.79	1.37	1.28
Rmin	0.27	0.33	0.41	0.48	0.22	0.77	0.63	0.82	0.95
S.D.	0.52	0.42	0.19	0.12	0.15	0.13	0.15	0.09	0.09
No. of obs.	230	239	369	310	216	142	52	72	10

	0–5	5–10	10–15	15–20	20–25	25–30	30–35	35–40	40–45	45–50	50–55	55–60
Rmean	1.40	1.10	1.05	1.04	1.04	1.04	1.04	1.02	1.03	1.01	1.03	1.02
Rmax	5.50	3.08	2.68	1.69	1.67	1.41	1.64	1.55	1.17	1.23	1.21	1.18
Rmin	0.62	0.35	0.43	0.62	0.74	0.72	0.82	0.36	0.81	0.80	0.82	0.95
S.D.	0.88	0.34	0.27	0.14	0.14	0.11	0.10	0.13	0.08	0.10	0.07	0.07
No. of obs.	68	107	91	107	97	155	136	98	71	42	38	14

Figure 4.29b Statistical analysis of the ratio of observed radiation on a stated inclined plane to the predicted ratio on the inclined plane for overcast sky conditions: Autumn and Spring: Bracknell. Vertical West

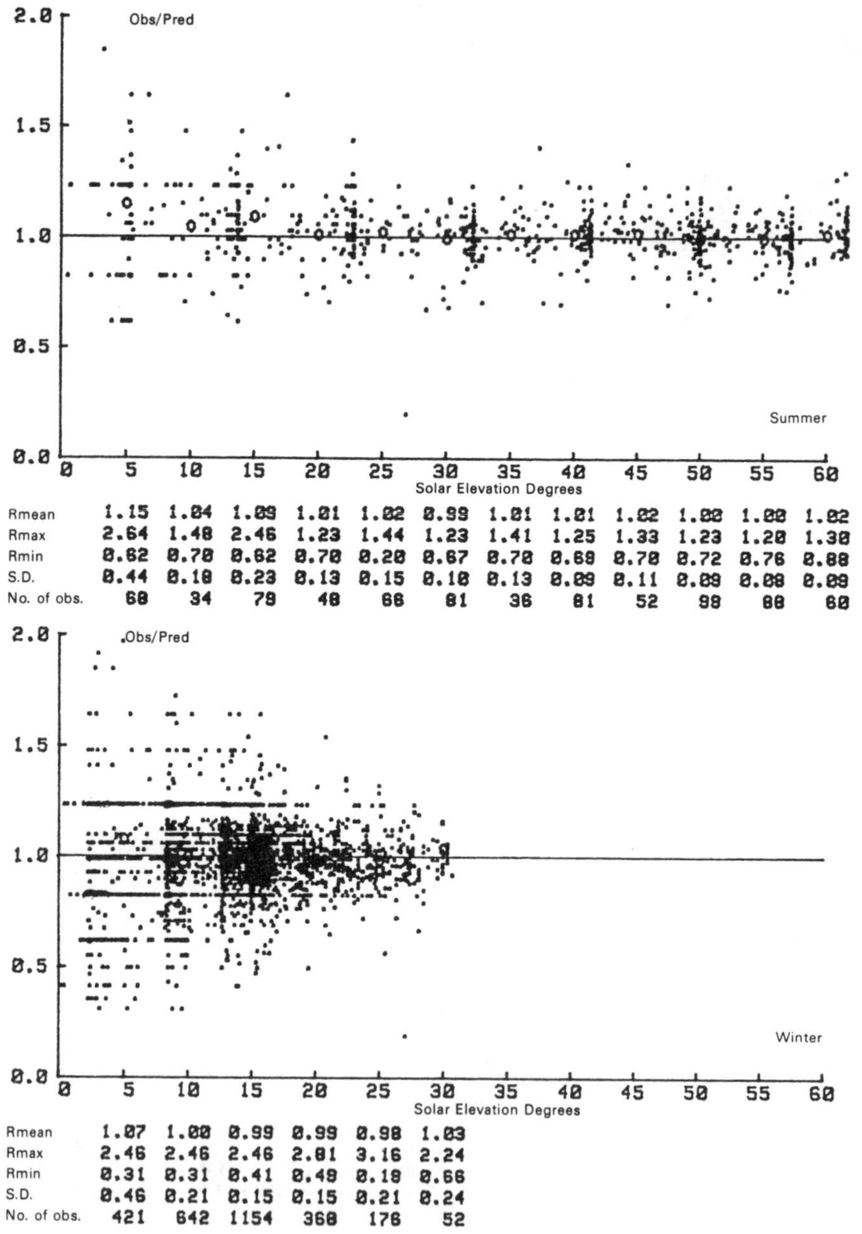

Rmean	1.15	1.04	1.09	1.01	1.02	0.99	1.01	1.01	1.02	1.00	1.00	1.02
Rmax	2.64	1.48	2.46	1.23	1.44	1.23	1.41	1.25	1.33	1.23	1.20	1.30
Rmin	0.62	0.70	0.62	0.70	0.20	0.67	0.70	0.68	0.70	0.72	0.76	0.88
S.D.	0.44	0.10	0.23	0.13	0.15	0.10	0.13	0.09	0.11	0.09	0.08	0.09
No. of obs.	68	34	79	48	66	81	36	81	52	98	88	60

Rmean	1.07	1.00	0.99	0.99	0.98	1.03
Rmax	2.46	2.46	2.46	2.81	3.16	2.24
Rmin	0.31	0.31	0.41	0.49	0.18	0.66
S.D.	0.46	0.21	0.15	0.15	0.21	0.24
No. of obs.	421	642	1154	368	176	52

Figure 4.30a Statistical analysis of the ratio of observed radiation on a stated inclined plane to the predicted ratio on the inclined plane for overcast sky conditions. Summer and Winter: Bracknell. Vertical North

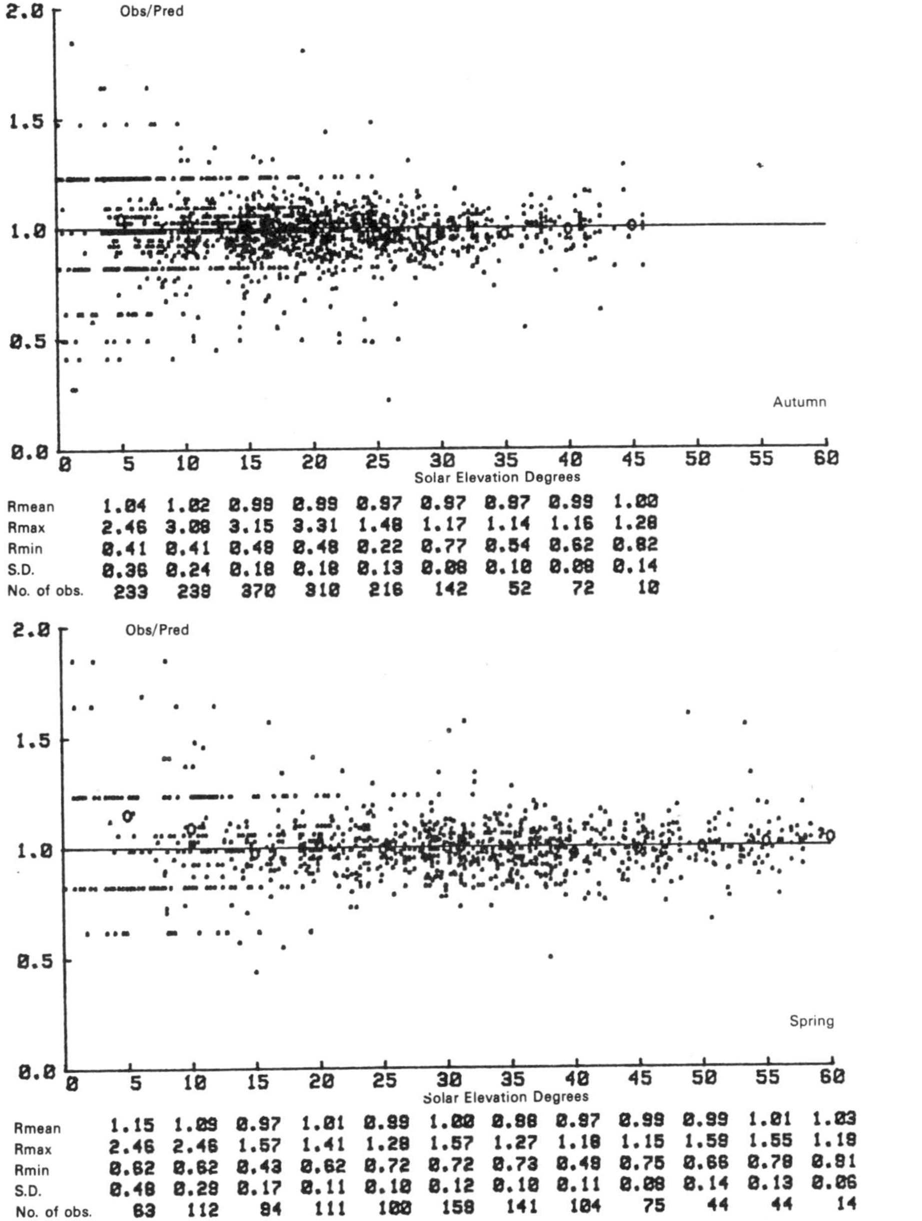

Rmean	1.04	1.02	0.99	0.99	0.97	0.97	0.97	0.99	1.00
Rmax	2.46	3.08	3.15	3.31	1.48	1.17	1.14	1.16	1.28
Rmin	0.41	0.41	0.48	0.48	0.22	0.77	0.54	0.62	0.82
S.D.	0.36	0.24	0.18	0.18	0.13	0.08	0.10	0.08	0.14
No. of obs.	233	239	370	310	216	142	52	72	10

Rmean	1.15	1.09	0.97	1.01	0.99	1.00	0.98	0.97	0.99	0.99	1.01	1.03
Rmax	2.46	2.46	1.57	1.41	1.28	1.57	1.27	1.18	1.15	1.59	1.55	1.19
Rmin	0.62	0.62	0.43	0.62	0.72	0.72	0.73	0.49	0.75	0.66	0.78	0.81
S.D.	0.48	0.28	0.17	0.11	0.10	0.12	0.10	0.11	0.08	0.14	0.13	0.06
No. of obs.	63	112	94	111	100	159	141	104	75	44	44	14

Figure 4.30b Statistical analysis of the ratio of observed radiation on a stated inclined plane to the predicted ratio on the inclined plane for overcast sky conditions: Autumn and Spring: Bracknell. Vertical North

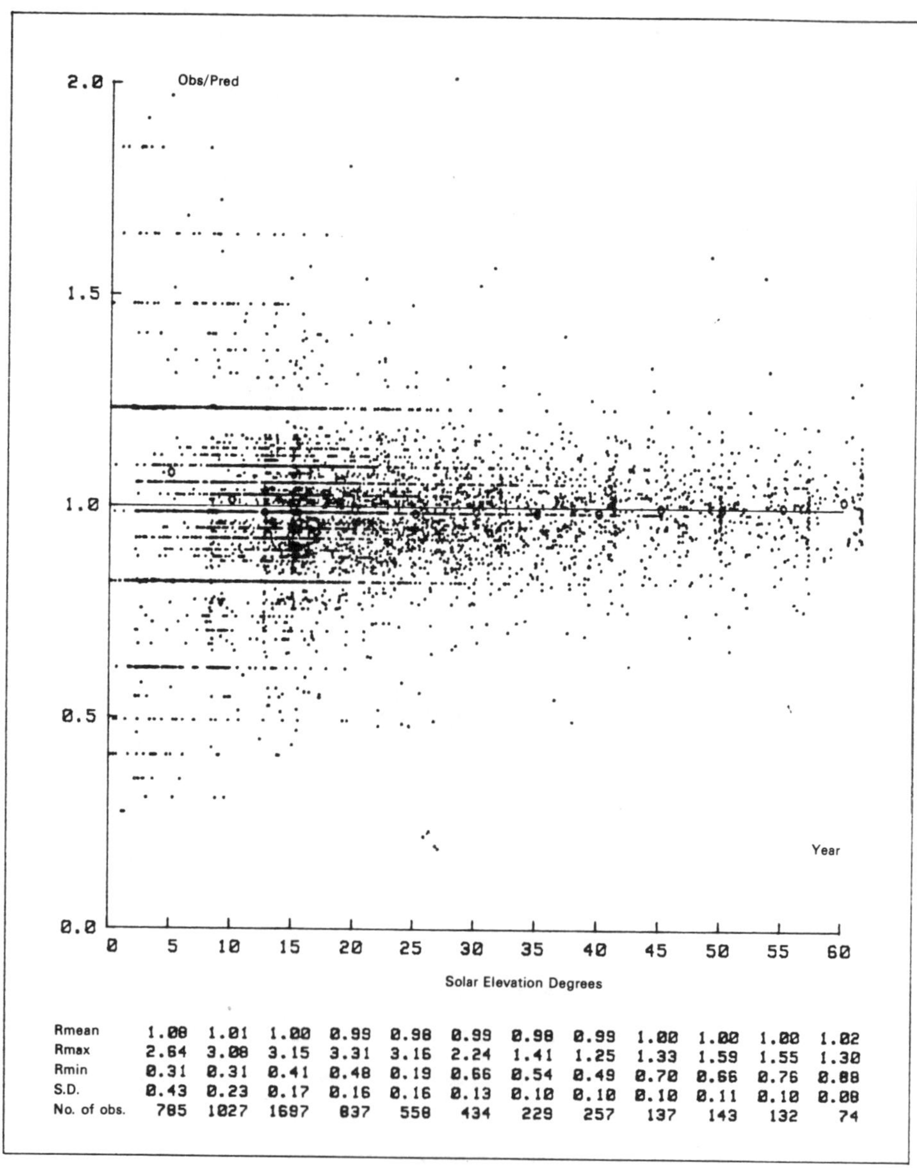

Rmean	1.08	1.01	1.00	0.99	0.98	0.99	0.98	0.99	1.00	1.00	1.00	1.02
Rmax	2.64	3.08	3.15	3.31	3.16	2.24	1.41	1.25	1.33	1.59	1.55	1.30
Rmin	0.31	0.31	0.41	0.48	0.19	0.66	0.54	0.49	0.70	0.66	0.76	0.88
S.D.	0.43	0.23	0.17	0.16	0.16	0.13	0.10	0.10	0.10	0.11	0.10	0.08
No. of obs.	785	1027	1697	837	558	434	229	257	137	143	132	74

Figure 4.31 Statistical analysis of the ratio of observed radiation on a stated inclined plane to the predicted ratio on the inclined plane for overcast sky conditions.
Station: Bracknell. Plane: Vertical North

CHAPTER 5

THE ESTIMATION OF MONTHLY MEAN HOURLY IRRADIANCE AND DAILY IRRADIATION ON INCLINED PLANES

Abstract

Part I of this Chapter, first of all, discusses the Berlin method for assessing the monthly mean direct beam irradiance. It then proceeds to discuss the new University of Sheffield model actually used to produce the monthly mean hourly values of the beam irradiance used in the development of the daily slope irradiation tables and maps published in the CEC European Solar Radiation Atlas, Vol. II, Inclined Surfaces.

Part II develops the corresponding diffuse irradiance model. The Berlin method is first explained. Then a method suitable for high latitudes is given. This leads on to the discussion of the development of the final model used to produce estimates of hourly diffuse irradiance on horizontal planes used in the production of the inclined plane atlas.

Part III discusses how the methods for estimating the diffuse irradiance from cloudless skies and from overcast skies described in Chapters 3 and 4 are combined to estimate the monthly mean diffuse sky irradiance on slopes. The assembly of a model for computing monthly mean daily irradiation on inclined planes is then described. Checks on the model against observation are given. The final computation process used to produce the monthly mean values of G_m and D_m on inclined planes in the CEC European Solar Radiation Atlas, Vol. II, is explained.

CHAPTER 5 - PART I

PREDICTION OF MONTHLY MEAN HOURLY VALUES OF THE DIRECT BEAM IRRADIANCE

1. Introduction

The monthly mean hourly values of the direct beam irradiance are, except in the sunniest climates, substantially below the clear day values due to the interception of the beam by clouds. The monthly mean relative duration of bright sunshine, σ_m, provides an indication of the average proportion of the month during which the sun is likely to be shining ($\sigma_m = S_m/S_o$ where S_m is the monthly mean daily bright sunshine, and S_o is the monthly mean daily astronomical daylength).

If the sunshine recorder were a perfect on-off switch device, if the sunshine was evenly distributed across each day, and if the atmospheric clarity always remained the same throughout the month, one could estimate the mean direct beam irradiance normal to the beam I_m as

$$I_m = \sigma_m I_c \qquad Wm^{-2} \tag{5.1}$$

Unfortunately none of these three assumptions holds true. Sunshine recorders, which produce burn traces on suitable standard cards, are not perfect on-off switch recording devices. They have burn thresholds which vary from type of instrument to instrument, and also with weather conditions, e.g. high humidity versus low humidity. A range of beam intensity thresholds, needed to start a record, is encountered according to instrument type and weather. The threshold is typically around 150-250 W/m^2. With a higher sun, when there is scattered cloud, the burn record tends to coalesce, and more sunshine is recorded than should be. There may be, on average, proportionally more sun at one time of day than another, for example the mornings may be sunnier than the afternoons.

Furthermore the atmospheric clarity on partially cloudy days is often markedly different from that on clear days. The precipitable water vapour on cloudless days is usually lower than the monthly average, because reduced atmospheric water vapour implies less risk of cloud. Changes of the turbidity coefficient are more complex to assess. Weather with broken clouds that bring rain, may wash the particulates out of the atmosphere and produce relatively clear conditions. Cloudy weather is often windier than clear weather, so aerosols are more widely dispersed and the urban influence becomes less. Inversions which trap the pollutants are more likely to occur in cloudless anticyclonic weather. This project attempted to seek mean relationships rather than relationships applicable to specific days, for which more detailed input data are required, concerning cloud type, levels of pollution and so on.

2. Choice of solar declination for monthly mean irradiance models

No computations can be carried out until the solar geometry has been established. The first decision needed to be made to establish the representative solar geometry for any month is to select the mean solar declination for that month. The mean monthly solar declination was calculated by taking the mean of the daily values of the noon declination for each day in each month, using a 365 day year. The equivalent date in the month was then established by finding the day whose declination was closest to the monthly mean declination. The values of the monthly mean declination used in the CEC studies are given in Table 6.2 together with the associated equivalent dates in each month. Thus, following the procedures set out in Appendix 1, it is simple to establish the representative mean values of the various solar angles for each month from the latitude of the site and the declination. All calculations in this chapter were carried out in solar time.

3. Choice of effective daylength

Equation 5.1 was based on the use of the astronomical daylength to compute σ_m. Some allowance can be made for effects of card burn threshold on the daily relative sunshine duration by adopting a specified minimum solar elevation at which it is assumed the sunshine recorder card will start to burn. This value is then used to calculate the effective daylength. It is possible to make this threshold dependent on the Linke Turbidity Factor (1)(2), but this makes the calculations more complicated. The Sheffield University method for estimating monthly mean solar radiation adopts a 4 degree solar elevation as the basis for calculating the effective daylength, S_{04}, for each month. The relevant algorithm is given in Appendix 1. Monthly mean values of S_{04} for latitudes 75°S-75°N are given in Chapter 6, Tables 6.7(a) to 6.7(c). The 4 degree sunrise/sunset day relative sunshine duration σ_{4m} is given by:

$$\sigma_{4m} = S_m/S_{04}$$

where S_m is the monthly mean daily sunshine.

4. Correction function f_5 for adjusting beam interpolation

One may then attempt pragmatically to adjust Equation 5.1 by introducing a pragmatic correction factor f_5 to allow for the various effects discussed. One may thus rewrite equation 5.1 as

$$I_m = f_5 \, \sigma_{4m} \, I_c \tag{5.2}$$

where f_5 is a pragmatic modifying factor to adjust the relationship between I_m, I_c and σ_{4m}. This relationship function has to be derived from a study of observed beam irradiances on horizontal surfaces at different types of site. For practical applications within the framework of the European studies discussed here, f_5 to be useful must have its magnitude defined as some function of the available input parameters in the CEC European Solar Radiation Atlas, Vol. I, Global radiation on horizontal surfaces (3).

However, as the new Sheffield University model for predicting monthly mean hourly values of the beam irradiance is developed from the Berlin Technical University model, it is necessary to discuss the Berlin model in some detail first.

5. The Berlin method for estimating monthly mean hourly beam irradiance from the Berlin clear sky and overcast sky models

5.1 Choice of clear sky Linke Turbidity Factors - Berlin method (4)

The Berlin computational method involves a non-linear interpolation, hour by hour, between the clear sky predicted beam irradiances, and the overcast sky predicted beam irradiances, based on the use of the monthly mean hourly duration of bright sunshine in each hour.

The clear day predictions have to be made first, which require an estimate of the clear day Linke Turbidity Factor. The Berlin method makes use of the Linke Turbidity Factors recommended by Steinhauser (5) which are given in Table 2.13. A subjective choice has first to be made between the four site types in Table 2.13. The Linke Turbidity Factor was not adjusted for solar altitude, as is the case in the Sheffield University method eventually adopted. Once the Linke Turbidity Factor has been decided, the beam irradiance for cloudless skies can be established for each hour from Equation 2.14, and the corresponding cloudless sky diffuse irradiance on horizontal surfaces calculated from Equation 2.28. The differences between the original Berlin clear sky diffuse irradiance model and the new Sheffield EC clear sky diffuse irradiance model have already been discussed in detail in Chapter 2.

The overcast day values are established using the methodology given in Chapter 4. However, one other input is needed, the monthly mean hourly sunshine, in order to proceed to interpolation for monthly mean conditions.

5.2 Monthly mean hourly sunshine

The monthly mean hourly sunshine is not normally available for most solar radiation stations, so the research group at the Institut für Lichttechnik, Berlin, had to develop a method for estimating this quantity from values of the monthly mean daily bright sunshine. Using data from 13 German stations, a relationship was developed between monthly mean hourly relative duration of bright sunshine, σ_h and monthly mean daily relative duration of bright sunshine σ_d (6). This method is described in detail in Appendix 6. It was adopted as a standard procedure in the Berlin method for estimating σ_h from S_m, the monthly mean daily bright sunshine.

5.3 Berlin method for estimates of monthly mean beam irradiance from σ_h

The remaining problem was to relate the values of f_5 to σ_h. Studies by Aydinli (7) at the Institut für Lichttechnik, confirmed, like many other studies, that the monthly mean direct beam irradiance cannot be estimated from the clear day values, simply using the concept of hourly sunshine probability for three reasons:- i) the observed atmospheric turbidity tends to be greater on partially clouded days than on clear days, ii) sunshine recorders tend to overestimate the sunshine available with intermittent sun due to the burn patches coalescing, iii) sunshine recorders also tend to underestimate the sunshine available at low solar elevations in winter due to their burn threshold. Allowances need to be made for these effects.

The calculation of the monthly mean direct beam irradiance from the clear sky model in the Berlin method thus involves two corrections:

1. A simple basic proportional correction to allow for the actual hourly relative sunshine duration σ_h in any particular month at each specific hour of day.

2. A second pragmatically derived correction to allow for the variations in atmospheric turbidity, card burn etc. associated with differing amounts of hourly relative sunshine duration σ_h. This second correction factor they designated as R_b rather than f_5.

The correction factor R_b was derived empirically from 10 years of irradiation observations at Hamburg by Aydinli. Its form is illustrated graphically in Figure 5.1. It will be noted that the function has a value of 1 for clear sky conditions and that the value falls to a minimum of 0.75 when the hourly sunshine probability is around 40%-50%, and then rises again at low values of σ_h. The precise reasons for this rise are not clear, but, as the EC method is developed to cover monthly mean values of sunshine probability, there are, in fact, little data observed in the EEC area where the noon hourly sunshine probability is less than 20%, and these values are for mid-winter at high latitudes, when pyranometers become unreliable observing instruments, so the bottom end of the curve remains tentative. However, pragmatically this correction has been shown to give reasonably good daily predictions across the Europe area north of the Alps and also for Singapore and other lower latitude sites.

The average direct beam irradiance $I_m(\beta, \alpha)$ on any slope is thus calculated by the Berlin method from the corresponding hourly values of the clear sky direct beam irradiance as

$$I_m(\beta, \alpha) = \sigma_h \cdot R_b \cdot I_c(\beta, \alpha) \quad Wm^{-2} \tag{5.3}$$

where σ_h is the monthly mean probable hourly sunshine at a given time of day

R_b is the polynomial function of σ_h given below, refer Equation 5.4

$I_c(\beta,\alpha)$ is the clear sky direct beam irradiance on the inclined planes calculated at that time using the Linke Turbidity Factor for that month, Wm^{-2}.

R_b is defined by:

$$R_b = ((-3.341\,\sigma_h + 6.924)\,\sigma_h - 4.066)\,\sigma_h + 1.48 \tag{5.4}$$

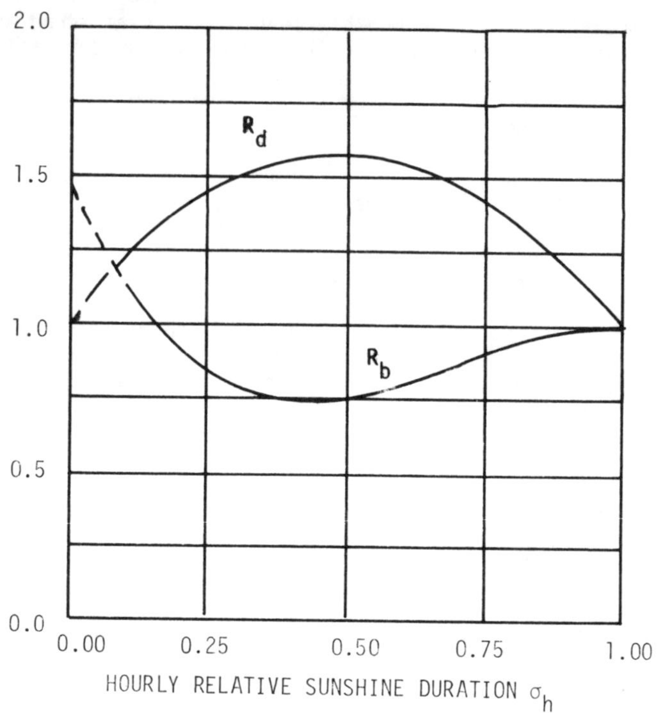

Figure 5.1 Additional correction functions for interpolated average direct beam, R_b, and diffuse irradiances, R_d, as a function of monthly mean hourly relative sunshine duration. Berlin model.

Table 5.1 Constants to use with Equations 5.8 and 5.9.

j \ i	0	1	2
0	-0.42907	0.098539	-0.011429
1	6.1280	-0.46284	0.0050835
2	-8.5179	0.64546	-0.0067794
3	3.8176	-0.28110	0.0028390

Notes: The practical limits of σ_{4m} as set by observed sunshine data are between about 0.1 and 0.7, but the function has been adjusted to converge on the clear sky values when σ_{4m} = 1.0.

6. The Sheffield University method for estimation of monthly mean hourly beam irradiance

6.1 Adoption of 4 degree sunrise/sunset to define effective daylength

Having checked the Berlin beam model at the hourly level, it appeared certain key difficulties could not be satisfactorily resolved. There was no way in which to scientifically resolve the sunshine recorder threshold problem at low solar elevations which results from the practical limitations of sunshine recorders. While no bright sunshine may be recorded at very low solar elevations, the monthly mean beam irradiance close to sunrise and sunset is certainly not zero. While the impact of setting the direct beam irradiance at sunrise to zero on the global irradiance on horizontal surfaces is very small, the impact on surfaces facing the rising and setting sun in contrast is quite appreciable, i.e. up to 50-100 W/m^2 error.

The new approach substitutes the daily value of the monthly mean relative sunshine σ_{4m}, calculated using the 4° elevation sunrise/sunset daylength for the hourly radiation for the hourly value σ_h.

6.2 Practical development of model

It was then assumed that the monthly mean direct beam irradiance on a horizontal surface could be calculated from the clear day values by a formula of the form

$$I_m(0,0) = I_c(0,0) \times \sigma_{4m} \times f(\sigma_{4m}, \gamma) \qquad (5.5)$$

The unknown daily variation of sunshine availability was allowed for by making the correction function $f_5 = f(\sigma_{4m}, \gamma)$ dependent on solar altitude as well as on 4 degree sunrise/sunset relative sunshine duration.

Monthly mean hourly observed data of global and diffuse radiation on a horizontal surface for the period 1969 to 1973 from Hamburg were used to develop the new model. This data set was selected for two important reasons. It was known firstly that the calibration standards at the Hamburg Observatory were high and secondly that the shadow band corrections applied to the diffuse pyranometer observations had been carefully considered and proper account taken of non-uniform sky radiance effects.

The hourly values of $I_m(0,0)$ were set equal to $G_m - D_m$. As the Berlin interpolation formula had shown good results at the daily level, it was accepted as a reasonable first approximation, substituting σ_{4m}, for σ_h i.e.

$$R_b = ((-3.341\ \sigma_{4m} + 6.924)\ \sigma_{4m} - 4.066)\ \sigma_{4m} + 1.48$$

Then, selecting appropriate Linke Turbidity Factors for each month for urban areas from Steinhauser's values in Table 2.13, hourly values of $I_c(0,0)$ could be estimated using the monthly mean declination, and next, by applying the modified interpolation Berlin formula, values of $I_m(0,0)$ could be estimated and compared with the observed Hamburg values. Then the mean hourly errors in prediction of $I_m(0,0)$ were taken out against solar altitude, on a 10° solar altitude band basis, and a polynomial derived to correct the implicit solar altitude errors, due to the various factors already discussed. The new solar altitude dependent model was then reiterated against the data and the errors re-examined, this time against σ_{4m}. The process was repeated until a convergence was achieved which made the hourly beam irradiance errors acceptably small. Finally, to develop the final algorithm, certain adjustments were made, so the curves passed through 1.0 when $\sigma_{4m} = 1$. This point, of course, lies well outside the observed range of values.

Cross comparisons were next made with observed hourly beam irradiance data for horizontal surfaces from 13 other stations in Europe, where hourly data sets of comparable quality to Hamburg were available, though sometimes with less carefully considered corrections to D_m for shadow band effects. These stations are listed in Table 2.20.

The final algorithm derived took the following form:

$$I_m(\beta, \alpha) = f_5 \, \sigma_{4m} \, I_c(\beta, \alpha) \qquad Wm^{-2} \tag{5.7}$$

where f_5 is computed by the following procedure:

If $\gamma < 45$ then

$$f_5 = \sum_{j=0}^{3} \left(\sum_{i=0}^{2} (a_{ij} \gamma^i) \, \sigma_{4m}{}^j \right) \tag{5.8}$$

otherwise

$$f_5 = \sum_{j=0}^{3} \left(\sum_{i=0}^{2} (a_{ij} \times 45^i) \, \sigma_{4m}{}^j \right) \tag{5.9}$$

where the values of a_{ij} are in Table 5.1

The objective method used to find the Linke Turbidity Factor with which to estimate I_c is discussed in detail later.

The form of the new interpolation function is given in Figure 5.2. It will be noted that the function has values below one for typical values of σ_{4m} around 40% to 50% but rises again at very low values of σ_{4m} at higher solar altitudes, thus taking account of the two important defects of sunshine recorders as radiation measurement instruments, the tendency to overburn in partially clouded skies with widely scattered cloud, especially important in windy conditions, and the failure of the sunshine recorder to burn at low solar altitudes, so the mean record underestimates, for months of low solar altitude, the actual monthly mean sunshine observed due to the threshold effect. This effect is especially important in winter. Because of the practical limitations of sunshine records, it is not possible to answer the question, "What is the actual relation between the Linke Turbidity Factor on clear days and on average days ?", because a composite correction involving both issues results from this analytic approach based on the use of sunshine records.

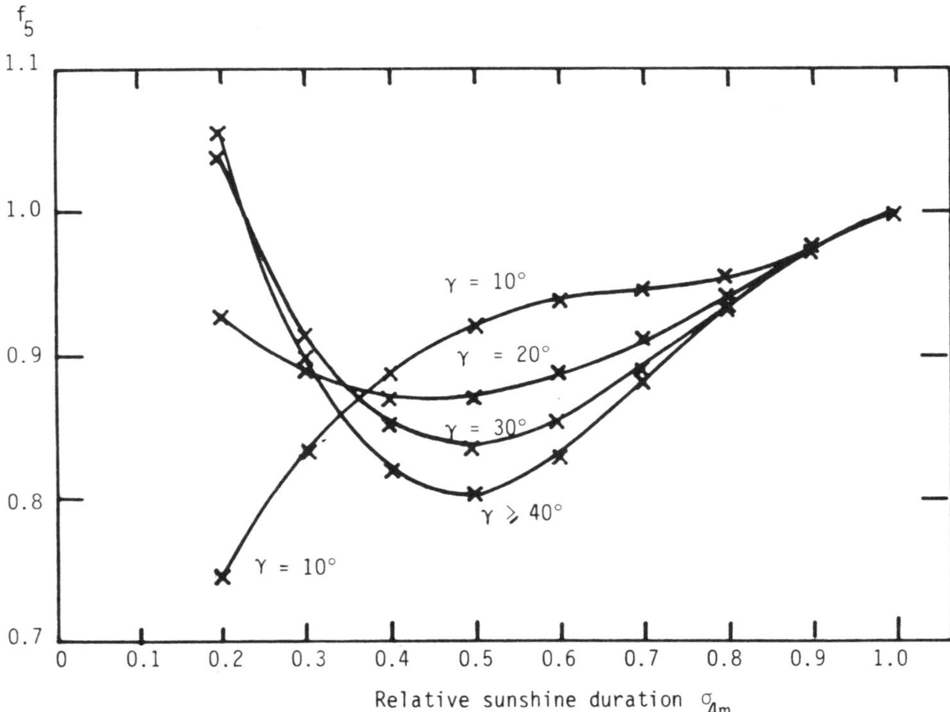

Figure 5.2 Correction function f_5 used to compute the monthly mean direct beam irradiance from the clear sky irradiance and the 4° sunrise/sunset relative sunshine duration σ_{4m} as a function of solar altitude .

6.3 The accuracy of prediction of monthly mean hourly values of $I_m(0,0)$ using the Sheffield University model

A daily beam irradiation prediction model was then developed following the principles already set out at length in Chapter 2. The daily beam irradiation values were obtained by numerically summing the hourly predicted values computed at each mid hour between sunrise and sunset, using for these mean beam value predictions, the monthly mean values of the declination to estimate the solar geometry.

Monthly mean hourly data from the following sites for the indicated periods were then used to check the monthly mean beam predictions

Hamburg	1964/73	Cambridge	1966/71	Bracknell	1966/75
Lerwick	1966/75	Aberporth	1968/75	Jersey	1966/75
Eskdalemuir	1966/75	Valentia	1964/74	Paris/Trappes	1965/75
Aldergrove	1969/75	London W.C.	1967/75	Locarno Monti	1958/71
Kew	1966/75	Carpentras	1968/75	Uccle	Not known

The monthly mean daily beam irradiation on the horizontal surface $I_m(0,0)$ was computed as $G_m - D_m$, using the daily irradiation totals. The model was then operated iteratively to find, for each month for each station, the daily value of the air mass 2 Linke Turbidity Factor needed to match the mean monthly daily beam irradiation output to the observed daily beam values on horizontal surfaces. The derived values of the AM2 Linke Turbidity Factor are given in Table 5.2. Using these derived values of T_L, the hourly values of $I_m(0,0)$ were computed for each month for each size using the model and compared with the hourly observed values ($G_m - D_m$). The errors were then divided into solar altitude bands 0-10°, 10°-20°, etc. and means and standard deviations extracted for each band. The results of this comparison are given in Table 5.3.

Hamburg, the site for which the model was developed, is placed first in Table 5.3, followed by the other stations in descending order of latitude. The values for Uccle are placed at the end, as this was a supplementary study made after the completion of the main project. The prediction errors (observed/predicted) were separately assessed for morning and afternoon. In each case the standard deviations of the errors are given. As the model was specifically adjusted to Hamburg, the mean errors at Hamburg are small. However, the model does not distinguish between the sunshine available in the morning and in the afternoon, so there are differences of up to 10% between morning and afternoon errors. The mean of the morning and afternoon values for Hamburg however is very close to 1.00. In general the mean beam prediction errors are less than ±5%. There are some systematic differences between morning and afternoon. At inland sites the morning beam values tend to be underpredicted by typically about 2-5% while the afternoon beam values tend to be overpredicted by typically about 2-5%. At coastal sites the morning beam values tend to be overpredicted by about 5% and the afternoon beam values underpredicted by about 5-8%. At inland sites convective cloud forming in the afternoon tends to reduce afternoon sunshine. In contrast at coastal sites, morning sea fog takes time to disperse, also the formation of convective cloud is less likely with off shore winds due to sea breezes, so there is more afternoon sunshine very close to the coast.

Table 5.2 Values of Angstrom (a + b) from European Solar Radiation Atlas, Volume I, (Second Edition) and derived values of air mass 2 Linke Turbidity Factor for fifteen European stations.

	Jan		Feb		Mar		Apr		May		Jun		Jul		Aug		Sep		Oct		Nov		Dec	
	a+b	T_L	a+b	T_L	a+b	T_L	a+b	T_L	a+b	T_L	a+b	T_L	a+b	T_L	a+b	T_L	a+b	T_L	a+b	T_L	a+b	T_L	a+b	T_L
Lerwick	.79	.93	.80	2.4	.84	2.8	.82	3.2	.87	3.5	.84	3.1	.84	3.9	.81	3.6	.82	3.3	.80	2.8	.78	2.0	.80	1.6
Eskdalemuir	.78	1.6	.80	2.5	.81	3.2	.80	3.7	.81	4.4	.79	4.1	.79	4.4	.77	4.4	.78	3.6	.79	3.0	.77	2.3	.75	1.8
Aldergrove	.79	1.7	.77	3.0	.79	3.2	.74	3.9	.78	4.1	.78	3.3	.78	4.3	.78	4.6	.77	3.3	.74	3.4	.77	2.5	.75	2.0
Hamburg	.71	3.6	.77	3.7	.77	4.1	.77	4.3	.77	4.9	.74	5.1	.72	5.6	.72	4.9	.74	4.4	.75	3.9	.76	3.2	.72	2.8
Cambridge	.79	2.7	.78	3.2	.79	3.4	.76	4.4	.76	4.3	.76	4.3	.77	4.4	.76	4.5	.77	3.6	.77	3.2	.74	2.0	.80	2.3
Aberporth	.73	2.8	.75	3.0	.80	2.7	.79	3.0	.79	3.2	.80	3.0	.81	3.3	.76	3.7	.76	3.1	.76	3.0	.73	3.1	.72	2.7
Valentia	.83	2.1	.85	2.6	.85	2.6	.84	3.0	.86	3.1	.86	3.3	.87	3.5	.84	3.3	.84	3.0	.86	2.9	.84	2.5	.82	2.3
London	.66	3.0	.67	3.8	.68	4.2	.71	5.1	.70	5.3	.72	4.3	.70	4.8	.68	5.0	.70	4.1	.67	3.6	.64	3.5	.65	3.1
Kew	.70	3.1	.71	3.3	.73	3.9	.73	4.2	.72	5.0	.72	4.4	.71	4.7	.69	4.8	.71	4.0	.71	3.4	.68	3.3	.71	2.8
Bracknell	.76	2.4	.75	3.5	.78	3.5	.74	4.4	.74	4.9	.74	4.1	.73	4.4	.73	4.5	.73	3.9	.72	3.3	.72	2.9	.72	2.4
Uccle	.70	3.9	.73	4.1	.73	4.5	.75	4.5	.76	4.5	.75	4.6	.74	4.8	.71	4.6	.72	4.1	.70	3.9	.71	3.8	.66	3.5
Jersey	.80	3.3	.77	3.5	.80	3.1	.78	3.5	.78	3.8	.76	3.7	.76	4.2	.75	4.4	.77	3.6	.74	3.2	.75	3.6	.76	3.1
Trappes	.78	2.7	.75	3.6	.76	3.4	.77	3.8	.77	4.0	.76	3.6	.76	4.2	.73	4.1	.74	3.5	.73	3.2	.74	3.3	.77	2.9
Locarno-Monti	.78	2.2	.79	2.5	.83	3.2	.82	3.7	.85	2.8	.84	2.7	.81	3.2	.81	3.2	.82	2.9	.79	2.5	.78	2.3	.73	2.2
Carpentras	.70	3.3	.74	3.0	.78	3.1	.80	3.2	.79	3.3	.79	3.5	.77	3.9	.76	3.9	.77	3.4	.74	3.5	.70	3.3	.69	3.3

Table 5.3 Mean errors (observed/predicted) in the estimation of monthly mean hourly direct beam horizontal surface irradiance using proposed model as a function of solar altitude.

Station		Solar Altitude Band - Degrees					
		10-20	20-30	30-40	40-50	50-60	60-70
Hamburg	A.M. Error	1.004	1.023	1.026	1.036	1.009	-
	A.M. S.D.	.057	.047	.037	.046	.026	-
	P.M. Error	1.048	.993	.982	.961	.961	-
	P.M. S.D.	.082	.062	.036	.034	.040	-
Lerwick	A.M. Error	.937	.996	.994	1.000	1.008	-
	A.M. S.D.	.122	.068	.067	.066	.044	-
	P.M. Error	.909	.950	1.041	1.080	1.057	-
	P.M. S.D.	.116	.050	.070	.045	.047	-
Eskdalemuir	A.M. Error	1.099	1.074	1.049	1.001	.942	-
	A.M. S.D.	.074	.078	.060	.039	.017	-
	P.M. Error	.999	.991	.965	.957	.975	-
	P.M. S.D.	.102	.047	.051	.042	.027	-
Aldergrove	A.M. Error	.982	1.002	.972	.964	.962	-
	A.M. S.D.	.099	.077	.062	.045	.039	-
	P.M. Error	.971	.971	.963	.966	.967	-
	P.M. S.D.	.116	.064	.053	.036	.080	-
Cambridge	A.M. Error	.994	1.062	1.079	1.039	1.009	.970
	A.M. S.D.	.114	.083	.044	.073	.047	-
	P.M. Error	.954	.980	.981	.945	.954	.970
	P.M. S.D.	.128	.060	.031	.024	.035	-
Aberporth	A.M. Error	.908	.909	.915	.971	.995	.989
	A.M. S.D.	.179	.116	.095	.054	.032	-
	P.M. Error	.961	1.014	1.080	1.093	1.079	1.079
	P.M. S.D.	.074	.046	.039	.034	.049	-
Valentia	A.M. Error	.792	.846	.861	.950	1.011	1.022
	A.M. S.D.	.180	.149	.133	.075	.041	-
	P.M. Error	.956	1.036	1.110	1.137	1.153	1.101
	P.M. S.D.	.116	.051	.029	.054	.037	-
London W.C.	A.M. Error	1.103	1.057	1.049	1.005	.977	.977
	A.M. S.D.	.162	.063	.052	.057	.031	-
	P.M. Error	1.045	.962	.972	.963	.947	.977
	P.M. S.D.	.161	.069	.039	.032	.034	-
Kew	A.M. Error	1.036	1.040	1.053	1.022	.992	.95
	A.M. S.D.	.072	.055	.051	.056	.019	-
	P.M. Error	1.007	.981	.982	.971	.945	.928
	P.M. S.D.	.096	.048	.051	.038	.026	-
Bracknell	A.M. Error	1.065	1.041	1.055	1.021	.987	.945
	A.M. S.D.	.139	.076	.059	.056	.024	-
	P.M. Error	.997	.978	.986	.978	.937	.909
	P.M. S.D.	.091	.045	.047	.043	.038	-

Table 5.3 continued Mean errors (observed/predicted) in the estimation of monthly mean hourly direct beam horizontal surface irradiance using proposed model as a function of solar altitude.

Station		Solar Altitude Band - Degrees 10-20	20-30	30-40	40-50	50-60	60-70
Jersey	A.M. Error	.856	.838	.907	.924	.977	1.014
	A.M. S.D.	.157	.116	.093	.077	.036	0.027
	P.M. Error	.978	1.043	1.116	1.141	1.074	1.053
	P.M. S.D.	.074	.066	.056	.049	.047	0.025
Paris/Trappes	A.M. Error	.961	.994	1.03	1.042	1.038	1.013
	A.M. S.D.	.083	.053	.037	.034	.018	.010
	P.M. Error	.925	.958	.993	1.000	.995	.979
	P.M. S.D.	.045	.037	.039	.029	.021	.028
Locarno-Monti	A.M. Error	.809	.949	1.010	1.069	1.089	1.100
	A.M. S.D.	.125	.097	.059	.040	.027	0.006
	P.M. Error	.795	.951	.982	1.014	1.042	1.076
	P.M. S.D.	.187	.142	.090	.057	.028	0.006
Carpentras	A.M. Error	.889	.971	1.026	1.042	1.026	1.035
	A.M. S.D.	.062	.042	.030	.011	.031	.020
	P.M. Error	.895	.981	.993	1.009	1.014	1.019
	P.M. S.D.	.089	.074	.056	.037	.018	.021
Uccle	A.M. Error	.997	1.006	1.030	1.060	1.024	1.015
	A.M. S.D.	.095	.025	.029	.028	.019	-
	P.M. Error	.987	.969	.967	.975	.972	.995
	P.M. S.D.	.002	.044	.031	.032	.021	-

Table 5.4 Number of pairs of consecutive hourly observations of overcast conditions for Hamburg, 1964-73 classified by cloud type. Source: Kasten & Czeplak.

Season	Cloud type	Solar elevation (degrees) 10	20	30	40	50	60
Year	Ci, Cc, Cs	41	49	45	18	20	8
	Ac, As	307	231	200	161	134	52
	Sc, Cu	515	334	232	180	146	56
	St	892	504	254	146	53	17
	Ns	26	27	15	15	12	6
Number of obs.		1781	1145	746	520	365	139

The prediction results at Valentia are the least satisfactory, but there is a hill to the south east of the site which may affect the morning observations at low solar elevations. There is also some obstruction by mountains at Locarno Monti which again may explain low values of the error ratio at low solar altitudes at that site.

There was no way of taking account of differences between morning and afternoon sunshine in producing the standard tables for the inclined surface atlas from the data actually available, so the sunshine was assumed to be symmetrical about noon, and the beam predictions are identical in the CEC European Solar Radiation Atlas, Vol. II, Inclined Surfaces, for east and west facing surfaces.

In general, considering the limitations of the techniques, the agreement between prediction and observation of the beam irradiation seems very satisfactory.

7. Dogniaux's method for estimating I_m from I_c

An alternative approach has been suggested by Dogniaux (8) based on observations at Uccle, Belgium. The relationship proposed is:

$$I_m = \sigma_h^n \, I_c \tag{5.10}$$

where σ_h is the monthly mean hourly sunshine duration

and where $n = 1 + 0.36\ \sigma_h$.

Rearranging to match the Berlin formulation we obtain:

$$I_m = R_b' \, \sigma_h \, I_c \tag{5.11}$$

where $R_b' = (\sigma_h)^{0.36\ \sigma_h}$

which may be compared with the Berlin formulation given in Equation 5.4.

$$R_b = ((-3.341\ \sigma_h + 6.924)\ \sigma_h - 4.066)\ \sigma_h + 1.48$$

The Dogniaux function has a value of $R_b' = 1$ at $\sigma_h = 0$, whereas the Berlin formulation gives a value around 1.5. However, close examination of Dogniaux's plotted points shows the Dogniaux formula underestimates by 10%-20% at low values of σ_h, so the simplified form of the relationship is achieved at the expense of some loss of accuracy. However, the Berlin formulation, while better at very low values of σ_h, underestimates the observed ratio of I_m/I_c at Uccle, by about 15% in the critical range of σ_h = 0.4 - 0.5 where most of the observations actually lie. As there are no monthly mean values of σ_h above 0.525 at Uccle, the form of the curve between 0.525 and 1.0 is unchecked in the Dogniaux relationship.

8. Conclusions on the prediction of direct beam irradiance

The development of a new validated model has been described for estimating monthly mean hourly values of the beam irradiance on any slope. The accuracy of prediction has been shown to be satisfactory for sites where observed hourly values of G_m and D_m are available, but as no systematic figures were available of sunshine during the morning and afternoon, the model cannot distinguish variations between easterly and westerly orientations. The practical use of the model requires knowledge of the mean clear day Linke Turbidity Factor for each month. This can be estimated subjectively, but, as part II of this chapter explains, it is possible to develop an objective methodology for estimating T_L from observation of global radiation alone.

CHAPTER 5 - PART II

ESTIMATION OF MONTHLY MEAN DIFFUSE IRRADIATION ON HORIZONTAL SURFACES

9. Introduction

Computation of the solar radiation on inclined planes requires that the diffuse irradiation component from the sky can be split away from the global radiation. As the monthly mean diffuse irradiation provides typically over 50% of the monthly horizontal surface global irradiation in most of Northern Europe, the accurate estimation of monthly mean hourly diffuse irradiance on horizontal surfaces is a very important first stage in solar radiation modelling of solar radiation on inclined planes in the European region. Considerable effort, therefore, had to be dedicated in this programme to the problem of modelling of monthly mean diffuse irradiation.

Diffuse irradiation measurements are usually presented at four levels of time aggregation:

1. Means of instantaneous irradiance values averaged over individual hours of observation.

2. Daily values of diffuse irradiation on specific days.

3. Means of diffuse irradiation values for specific hours averaged over stated periods, usually one month.

4. Means of daily values averaged over stated periods, again usually one month.

The values normally used in simulation studies of solar radiation impacts are of the first type, while values used in general climatological studies of solar radiation are of types 3 and 4.

As diffuse irradiation measurements on horizontal surfaces were not available for the majority of European sites, suitable methodologies for estimating diffuse irradiance systematically from measurements of global radiation and sunshine had to be developed as a key part of this project for estimating monthly mean values of daily slope irradiation.

This section of this chapter describes the evolution of the methodologies used to predict hourly mean values of the diffuse irradiance on horizontal surfaces using, as input data, the monthly mean daily observed data in the CEC European Solar Radiation Atlas, Vol. I, Global Radiation on Horizontal Surfaces (3). The inputs eventually used are monthly mean daily global solar radiation, monthly mean daily bright sunshine, and monthly mean values of the sum of the Angstrom regression coefficients (a + b). The development of such methodologies proved very difficult.

10. Relationships between diffuse irradiation, sunshine and cloud type

Chapter 2 discussed at some length the estimation of the diffuse irradiance from cloudless skies, i.e. when the relative duration of bright sunshine is equal to one. It showed the key factor affecting clear day diffuse irradiance was the atmospheric turbidity. Chapter 4 discussed the prediction of overcast day irradiance adopting a standard solar altitude dependent model for the whole European region. The high variance present in the individual hourly observations was commented upon in that Chapter.

If one considers individual overcast hours, there is a wide range of variation. Statistically the observed overcast day data will be widely distributed about the mean, as Chapter 4 has demonstrated. The variance at any given solar altitude depends most on cloud type/types (there may be more than one layer of cloud).

Kasten and Czeplak (9) discuss the issues of radiation on overcast days in some detail in relation to Hamburg, drawing on work from various other sources. The overcast day data for Hamburg for the period 1964-73 were analysed in 5 cloud types:

Cirrus = Ci, Cc, Cs,
Altus = Ac, As,
Cumulus = Sc, Cu,
Stratus = St.
Nimbostratus = Ns.

Relationships between cloud type, overcast day irradiance on horizontal surfaces were taken out as a function of the sine of the solar altitude. At a 30 degree solar elevation, the annual mean irradiance values for overcast days varied from about 275 Wm^{-2} for Cirrus to less than 100 Wm^{-2} for Stratus and Nimbostratus clouds. The authors then expressed the overcast day irradiance values as a function of the clear day irradiance value, i.e. G_b/G_c. They found the increase in the ratio with solar altitude was rather small, so that the transmittances of Cirrus, Altus, Cumulus, Stratus and Nimbostratus could be generalised as 0.61, 0.27, 0.25, 0.18 and 0.16 respectively. Table 5.4 gives the frequency of the numbers of observations of overcast skies related to cloud type.

A more detailed seasonal breakdown is given in Kasten's original paper. It will be noted that Stratus dominates at lower solar elevations, while, at the higher solar elevations found in summer, Stratocumulus and Cumulus dominate with Altocumulus and Altostratus also being common. It follows that higher typical overcast day diffuse irradiances at any given solar altitude are to be expected in summer than in winter at Hamburg. It is, therefore, not possible to predict individual hourly values of the overcast day diffuse irradiance without knowledge of cloud type, though as Chapter 4 indicated, it is possible to produce relatively sound monthly mean estimates for overcast days.

If one now considers days of intermediate amounts of sunshine between $S_d = 0$ and $S_d = 1$, one must firstly note that the hourly values of S_h may vary very widely across one day, so it is not possible to estimate the hourly diffuse irradiance on specific days without knowledge of the hourly sunshine in that day as well as of cloud type. There is considerable reflection of solar radiation from the side walls of clouds, so more diffuse radiation is received on partially clouded days at any given solar altitude than on clear or cloudy days. Given a data set, it is obviously possible to study the mean relationship between monthly mean hourly sunshine S_h and the monthly mean hourly diffuse horizontal irradiation D_m as a function of solar altitude. Any such data set in Northern Europe will comprise a large number of overcast hours, a few clear hours and quite a number of hours of intermediate sunshine, all with varying transmittances due to the combined influence of atmospheric turbidity and cloud type. The prediction of individual hourly values of the diffuse irradiance for specific days is thus a very difficult problem.

Different mathematical relationship functions are found according to whether one seeks relationship functions between:

1. Individual hourly diffuse radiation observations and individual hourly sunshine values or individual hourly cloudiness observations,

2. Daily diffuse radiation observations and daily sunshine for individual days,

3. Monthly mean hourly values of the diffuse radiation and other monthly mean parameters like monthly mean daily global radiation, monthly mean daily bright sunshine,

4. Monthly mean daily values of the diffuse radiation and other monthly mean daily parameters in monthly mean daily sunshine or cloudiness.

This CEC project has been mainly concerned with relationship functions of the third type, i.e. the production of monthly mean estimates of hourly diffuse irradiance, from which the estimated values of the daily diffuse irradiation are obtained. The mean values of D_m obtained from the observations in Northern Europe are the consequence of combining the many overcast hour values, the often few clear hour values, and the typically considerably higher hourly diffuse values for quite a number of days of intermediate sunshine. The centroid of the points will thus be below the curve for individual days or hours plotted as a function of relative duration of bright sunshine. Thus the methodologies which follow are not suitable for simulation studies where relationship functions of the first and second type are needed.

11. Predicting monthly mean diffuse hourly irradiance on any surface from monthly mean sunshine - Berlin method

A large number of studies have shown that the diffuse irradiance is greater under partially overcast skies than under clear or totally overcast conditions, for example Page (10) and Kasten and Czeplak (11). The diffuse irradiance cannot, therefore, be obtained by simple linear interpolation between the clear and overcast sky conditions. The Berlin group developed a suitable diffuse weighting function using monthly mean hourly sunshine duration as the basis of interpolation. Two basic corrections have to be applied using this method:

1. A simple linear weighting of the clear sky and overcast sky diffuse values on any particular slope in proportion to hourly relative sunshine duration at that particular hour, i.e.

$$(\sigma_h\, D_c(\beta,\ \alpha) + (1 - \sigma_h)\, D_b(\beta,\ \alpha)) \tag{5.11}$$

where σ_h is the hourly sunshine duration.

2. The application of a pragmatic correction factor R_d to allow for the increased amounts of diffuse radiation when the sky is partially clouded. Thus:

$$D_m(\beta,\ \alpha) = (\sigma_h\, D_c(\beta,\ \alpha) + (1 - \sigma_h)\, D_b(\beta,\ \alpha))\ .\ R_d \tag{5.12}$$

where σ_h is the hourly relative sunshine duration for a specific hour in a specific month

$D_c(\beta,\ \alpha)$ is the calculated clear sky value of the diffuse irradiance on the appropriate surface for that locality, Wm^{-2}.

$D_b(\beta,\ \alpha)$ is the overcast sky value of the diffuse irradiance on the same surface, Wm^{-2}, calculated using the methods of Chapter 4.

R_d is a polynomial function of σ_h given below:

$$R_d = ((0.444\ \sigma_h - 2.98)\sigma_h + 2.54)\ \ \sigma_h + 1.0 \tag{5.13}$$

The values of R_d were determined by Aydinli (12), using principally data from Hamburg. The predictions using this formula were then checked successfully against observations from other European countries and for sites at lower latitudes. The form of this function is illustrated in Figure 5.2. It will be noted that, on scientific grounds, the function must pass through 1 at $\sigma_h = 1$ and at $\sigma_h = 0$. The peak value of correction factor is about 1.645 and this occurs around 50% possible sunshine, when the combined effects of direct reflections from and multiple reflections between clouds are greatest.

12. Estimation of monthly mean hourly diffuse irradiation

An alternative approach in the prediction of monthly mean diffuse sky irradiation is by using the daily diffuse values. The most widely used approaches for the prediction of daily diffuse solar radiation in the past have been the techniques developed by Liu and Jordan (13) and independently by Page (14). The Page method was based on the use of the linear dimensionless equation:

$$\bar{D}_d/\bar{G}_d = c + d\ \bar{G}_d/\bar{G}_o \tag{5.14}$$

where $\bar{D}_d$ is the monthly mean daily diffuse irradiation on a horizontal surface, which should be properly corrected for shading band effects,

$\bar{G}_d$ is the monthly mean daily global irradiation on a horizontal surface

$\bar{G}_o$ is the monthly mean daily extraterrestrial irradiance on a horizontal surface

and c and d are climatologically determined regression coefficients.

Liu and Jordan suggested that their relationship values which were not linear were universally applicable, while Page maintained the values of c and d were strongly climatologically dependent. The mean values for a widely scattered set of stations across the world found by Page were c = 1.00 and d = 1.13.

An alternative formulation is in terms of monthly mean daily sunshine $\bar{S}_d$:

$$\bar{D}_d/\bar{G}_d = c' + d'(\bar{S}_d/\bar{S}_o) \qquad (5.15)$$

where $\bar{S}_o$ is the monthly mean astronomical duration of sunshine.

Table 5.5 gives the values of c and d found by Page (15) using monthly mean values of $\bar{G}_d/\bar{G}_o$ as a basis of the regression. Table 5.6 gives the similar values using $\bar{S}_d/\bar{S}_o$ as the basis of the regression equation for a range of European stations. The daylength used allowed for refraction in the atmosphere, and so is slightly different from the daylength defined in Equation 5.3. The data confirm Page's earlier hypothesis that it is not satisfactory to treat c and d as universal constants. The data set, however, is not entirely self consistent because of the different shading band techniques used at different sites. However, in order to determine c and d, one must have values of $\bar{D}_d$ and $\bar{G}_d$ or estimate them.

13. Estimating monthly mean hourly diffuse horizontal irradiance from the monthly mean daily diffuse irradiation

In order to estimate the hourly diffuse irradiance from daily values, one has to make certain assumptions about the nature of the inter-relationship of the irradiance at any hour with the daily irradiation.

Liu and Jordan (16) suggested that the ratio of diffuse irradiance at any hour to the daily diffuse irradiation on a horizontal surface at the earth's surface was nearly identical to the ratio of the irradiance at that hour on a horizontal surface outside the atmosphere to the daily extraterrestrial irradiance on that day. This extraterrestrial ratio which they called r_o, may be estimated from fundamental considerations as

$$r_o = \frac{\pi}{24}\left(\frac{\cos\omega - \cos\omega_s}{\sin\omega_s - \omega_s\cos\omega_s}\right) \qquad (5.16)$$

where ω is the actual hour angle, radians

ω_s is the hour angle at sunset, radians.

ω_s may be derived from the astronomical daylength S_o {refer Appendix 1, algorithm 4, and Chapter 6, Tables 6.6(a)-6.6(c)} as $\frac{15 \times S_o}{2} \times \frac{\pi}{180}$.

Table 5.5 Experimentally derived values of c and d in the regression equation $\bar{D}_d/\bar{G}_d = c + d\ \bar{G}_d/\bar{G}_o$ with predicted values of $\bar{D}_d/\bar{G}_d$ for $\bar{G}_d/\bar{G}_o$ = 0.2 and 0.5 for a number of European stations. $\bar{G}_d/\bar{G}_o$ = 0.2 is very broadly representative of winter conditions and $\bar{G}_d/\bar{G}_o$ = 0.5 of summer conditions.

Station	Period	Lat. N	Value of c	Value of d	Pred $\bar{D}_d/\bar{G}_d$		Corr. Coeff.	Method of obtaining diffuse
					$\bar{G}_d/\bar{G}_o$ 0.20	$\bar{G}_d/\bar{G}_o$ 0.50	r	
Bergen	(1968-72)	60°24'	1.224	-1.706	.883	.371	-0.95	Occulting disc
Lerwick	(1965-70)	60°08'	1.078	-1.140	.850	.508	-0.96	Shading ring
Hamburg	(1964-73)	53°38'	1.043	-1.037	.836	.525	-0.96	Shading ring
Cambridge	(1965-70)	52°13'	0.937	-0.841	.769	.516	-0.91	Shading ring
Aberporth	(1965-70)	52°08'	1.064	-1.140	.835	.494	-0.97	Shading ring
De Bilt	(1971-75)*	52°06'	1.119	-1.227	.874	.506	-0.96	Global-pyrheliometer resolved components
Valentia	(1964-74)	51°56'	0.958	-0.851	.788	.533	-0.94	Shading ring
London Weather Centre	(1965-70)	51°31'	0.990	-1.103	.769	.439	-0.98	Shading ring
Kew	(1965-70)	51°28'	0.980	-1.026	.775	.467	-0.97	Shading ring
Bracknell	(1965-70)	51°21'	0.995	-0.990	.797	.500	-0.95	Shading ring
Uccle	(1951-65)	50°49'	0.999	-1.006	.798	.496	-0.98	Occulting disc
Trappes	(1971-75)	48°46'	1.004	-1.050	.794	.479	-0.97	Shading ring
Hohenpeizenburg	(1976)	47°48'	0.891	-0.900	.711	.441	-0.85	Shading ring
Locarno-Monti	(1958-71)	46°10'	0.626	-0.395	.547	.429	-0.76	Shading ring
Locarno-Monti	(1964)	46°10'	0.886	-0.897	.707	.438	-0.88	Shading ring
Carpentras	(1973-75)	44°05'	0.934	-0.985	.737	.441	-0.96	Shading ring
Odeillo	(Apr 71-Mar 75)	42°29'	0.642	-0.508	.541	.388	-0.15	Shading ring (uncorrected)

* Incomplete data

Table 5.6 Experimentally derived values of c' and d' in the regression equation $\bar{D}_d/\bar{G}_d = c' + d'(\bar{S}_d/\bar{S}_o)$ for a number of European stations plus predicted values of $\bar{D}_d/\bar{G}_d$ for $\bar{S}_d/\bar{S}_o = 0.2$ and $(\bar{S}_d/\bar{S}_o) = 0.5$. $\bar{S}_d/\bar{S}_o$ is the percentage possible sunshine defined on the Smithsonian Tables basis.

Station	Period	Lat. N	Value of c'	Value of d'	Pred $\bar{D}_d/\bar{G}_d$		Corr. Coeff.	Method of obtaining diffuse
					$\bar{S}_d/\bar{S}_o$ 0.20	$\bar{S}_d/\bar{S}_o$ 0.50	r	
Bergen	(1968-72)	60°24'	0.928	-1.153	.699	.353	-0.97	Occulting disc
Lerwick	(1965-70)	60°08'	0.902	-1.023	.697	.390	-0.98	Shading ring
Eskdalemuir	(1965-70)	55°19'	0.767	-0.531	.661	.502	-0.87	Shading ring
Hamburg	(1964-73)	53°38'	0.867	-0.652	.737	.541	-0.96	Shading ring
Cambridge	(1965-70)	52°13'	0.796	-0.582	.680	.505	-0.93	Shading ring
Aberporth	(1965-70)	52°08'	0.883	-0.896	.703	.435	-0.97	Shading ring
De Bilt	(1971-75)*	52°06'	0.896	-0.731	.750	.531	-0.97	Global-pyrheliometer resolved components
Valentia	(Mar 76-Feb 78)	51°56'	0.748	-0.582	.632	.457	-0.89	Shading ring
Valentia	(1964-74)	51°56'	0.839	-0.779	.679	.439	-0.97	Shading ring
London Weather Centre	(1965-70)	51°31'	0.849	-0.723	.704	.488	-0.98	Shading ring
Kew	(1965-70)	51°28'	0.882	-0.896	.703	.435	-0.97	Shading ring
Bracknell	(1967-75)	51°21'	0.802	-0.599	.682	.503	-0.94	Shading ring
Uccle	(1951-65)	50°49'	0.871	-0.731	.725	.506	-0.99	Occulting disc
Locarno-Monti	(1961-68)	46°10'	0.636	-0.407	.555	.432	-0.92	Shading ring
Carpentras	(1973-75)	44°05'	0.788	-0.631	.662	.472	-0.98	Shading ring
Odeillo	(Apr 71-Mar 75)	42°29'	0.788	-0.804	.628	.386	-0.86	Shading ring (uncorrected)

* Incomplete data

This approach enables a universal chart to be produced giving r_o as function of daylength/sunset hour angle, and time elapsed from solar noon which appears in many standard texts. Equation 5.16 implies a cosine relationship between D_m and the hour angle for a specific day.

Thus, rearranging equation 5.16, one may write

$$r_o = a \cos \omega + b \qquad (5.17)$$

where

$$a = \frac{\pi}{24(\sin \omega_s - \omega_s \cos \omega_s)}$$

$$b = \frac{-\pi \cos \omega_s}{24(\sin \omega_s - \omega_s \cos \omega_s)}$$

As

$$\sin \gamma = \cos \phi \cos \delta \cos \omega + \sin \phi \sin \delta$$

it follows that as $\cos \omega = \dfrac{\sin \gamma - \sin \phi \sin \delta}{\cos \phi \cos \delta}$,

we may redefine equation 5.17 as follows:

$$r_o = a' \sin \gamma + b'$$

where

$$a' = \frac{\pi}{24(\sin \omega_s - \omega_s \cos \omega_s) \cos \phi \cos \delta}$$

$$b' = \frac{-\pi . (\cos \omega_s - \tan \phi \tan \delta)}{24(\sin \omega_s - \omega_s \cos \omega_s)} = 0$$

The Liu and Jordan formula therefore implies that the monthly mean hourly diffuse solar radiation should be a sine function of solar altitude γ. However, Page (17), by actually studying the observed monthly data, showed a linear relationship with solar altitude γ was more appropriate in most cases for Northern Europe.

$$D_m = (2 + a_o \gamma) K_d \qquad (5.19)$$

where a_o is the slope for a particular month. The value of 2 Wm^{-2} represents the diffuse irradiance at sunrise and sunset and K_d, the correction to mean solar distance. It was, therefore, decided to carry out further detailed studies to establish the typical relationship functions found in Europe using Equation 5.19.

14. Studies of the relationship between monthly mean hourly diffuse solar radiation and solar altitude

Seven sites for which both monthly mean hourly global and diffuse radiation were available were used the initial stages of this study. These sites were Aberporth in Wales, Lerwick in Scotland, Hamburg in Germany, Kew in England, Valentia in West Ireland, Trappes on the outskirts of Paris, and Carpentras in S.E. France, the sunniest of the sites. Correlations were sought between monthly mean observed hourly values of D_m and solar altitude γ using equation 5.19.

In the majority of cases very good correlations were achieved, especially at higher latitudes in the less sunny climates, Table 5.7. In some months, in the sunnier climates, the correlations were less good, as the curves took on a curvilinear form, more in line with the Liu and Jordan assumptions. When the values were curvilinear, an x is entered in the table. The curvilinear form seemed primarily associated with conditions where the monthly mean relative sunshine duration calculated on an astronomical sunrise basis σ_m was above 40%. Figure 5.3 plots the observed data for Hamburg and Carpentras, and shows the difference between the high latitude relatively overcast site and the lower latitude sunnier site. The linear nature of the winter relationship and the curvilinear nature of the summer relationship is evident for Carpentras. In general the values of a_o at midwinter are lower than those found for the summer, a fact in line with Kasten and Czeplak's review of cloud type influence on diffuse irradiation in Hamburg already discussed in paragraph 10.

Using a subjective assessment of atmospheric clarity, attempts were made to relate the value of a_o to σ_m. The tentative results are plotted in Figure 5.4. This shows that, at high values of σ_m, clear sites have less diffuse solar radiation than sites with more pollution, a result entirely in line with the results of Chapter 2, but at low values of σ_m the clear sites have more diffuse radiation, a result not in line with the results of Chapter 4 for overcast days. An independent study of overcast days by Page and Colquhoun (18) for the UK had, however, shown that the daily transmittance on overcast days did depend on the clarity of the climate and the greater the pollution, the less diffuse radiation was observed on overcast days. This result is also in line with practical observation. Anyone who has lived in a highly polluted urban environment is aware of the darkening effects of pollution. The curves in Figure 5.4 reveal a crossover point around σ_m = 45%.

15. Practical diffuse irradiance model for latitudes 48°N and above

While it was clear that the results from this study were not adequate to cover the whole European region, the good prediction accuracy achieved at high latitudes and the simple nature of the formula involved indicated that it was worth developing the method into a usable practical model, so that given observed or predicted values of the daily irradiation, the hourly irradiances on horizontal surfaces could be easily computed for sites in Northern Europe.

The daily integral of equation 5.19 has to be evaluated to do this, i.e.

$$\int_{SR}^{SS} K_d(2 + a_o\ \gamma)\ d\omega$$

The integral has to be evaluated for hour angles between sunrise and sunset, setting

$$\gamma = \sin^{-1}(\cos\phi\ \cos\delta\ \cos\omega - \sin\phi\ \sin\delta)$$

Table 5.7 Linear relationships for the prediction of hourly diffuse solar radiation as a function of solar altitude for 7 European sites. Table includes values of monthly mean daily relative duration of bright sunshine σ_m, expressed as a percentage, a_o, the regression constant in the formula $D_m = K_d(2 + a_o \gamma)$ and the correlation coefficient r for each month.

		J	F	M	A	M	J	J	A	S	O	N	D
Lerwick	σ_m	12.7	18.3	21.7	28.6	26.1	27.1	23.1	24.0	22.1	17.0	12.4	9.0
	a_o	4.89	4.97	4.79	5.11	5.23	5.20	5.50	5.28	4.91	4.55	4.32	4.21
	r	.980	.995	.998	.999	.999	.999	.998	.998	.999	.996	.995	.984
Valentia	σ_m	18.1	26.7	30.7	36.6	34.8	31.5	29.3	32.1	30.9	24.0	20.1	15.6
	a_o	4.98	5.41	5.65	5.49	5.63	5.50	5.58	5.32	5.40	5.04	5.02	4.85
	r	.999	.999	.998	.999	.999	.998	.999	.999	.999	.997	.997	.998
Aberporth	σ_m	20.1	25.9	31.8	35.0	40.1	42.5	34.2	36.6	36.0	28.5	22.1	21.3
	a_o	5.10	5.29	5.49	5.34	x	x	5.53	5.37	5.17	5.21	4.99	4.69
	r	.998	.995	.998	.999	x	x	.998	.997	.998	.998	.996	.993
Hamburg	σ_m	16.1	21.5	30.5	35.3	42.3	47.0	41.5	47.6	40.1	32.0	20.8	14.8
	a_o	5.13	5.53	5.58	5.36	5.49	5.50	5.54	5.48	5.35	5.14	4.61	4.61
	r	.996	.998	.999	.999	.999	.999	.998	.999	.998	.998	.997	.996
Kew	σ_m	18.5	24.8	32.6	35.1	41.6	44.0	39.5	40.9	39.2	33.2	27.6	18.3
	a_o	4.49	4.65	4.66	4.75	4.99	4.80	5.03	5.02	4.98	4.70	4.67	4.38
	r	N.A.	N.A.	N.A.	N.A.	N.A.	N.A.	N.A.	N.A.	N.A.	N.A.	N.A.	N.A.
Trappes	σ_m	17.4	37.0	34.1	39.9	42.0	50.1	47.9	54.1	45.2	32.8	22.3	20.3
	a_o	4.93	5.26	5.15	5.20	x	x	x	5.09	4.99	5.11	4.84	4.78
	r	.999	.999	.999	.998	x	x	x	.998	.999	.998	.996	.998
Carpentras	σ_m	49.7	60.5	52.9	61.5	61.9	70.4	80.4	75.8	62.0	69.1	50.0	57.1
	a_o	5.01	4.52	4.87	x	x	x	x	x	x	x	4.94	5.05
	r	.998	.998	.999	x	x	x	x	x	x	x	.919	.995

x implies a curvilinear form of the relationship was found

N.A. Not available.

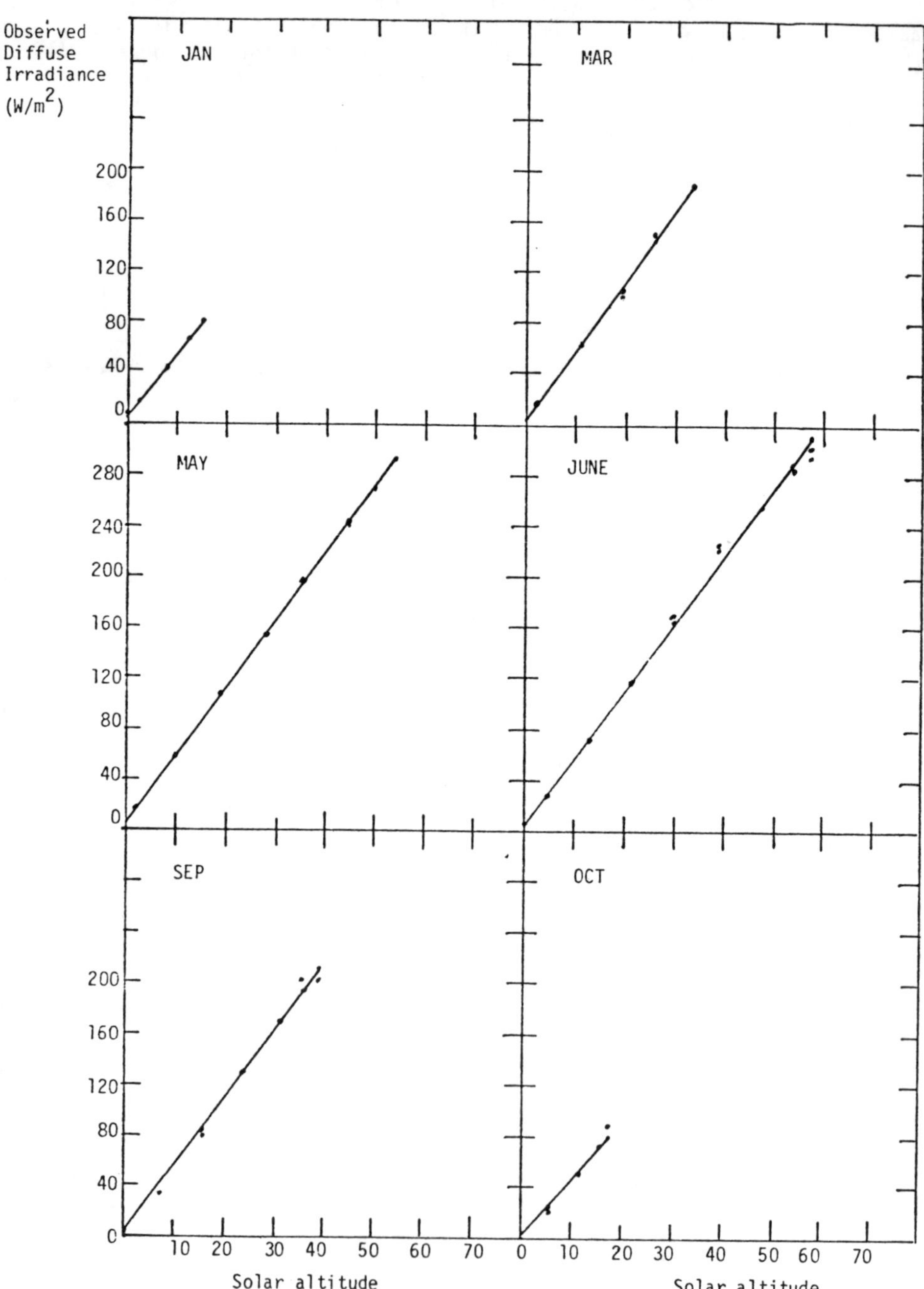

Figure 5.3a Relationship between monthly mean hourly diffuse irradiance and solar altitude for Hamburg.

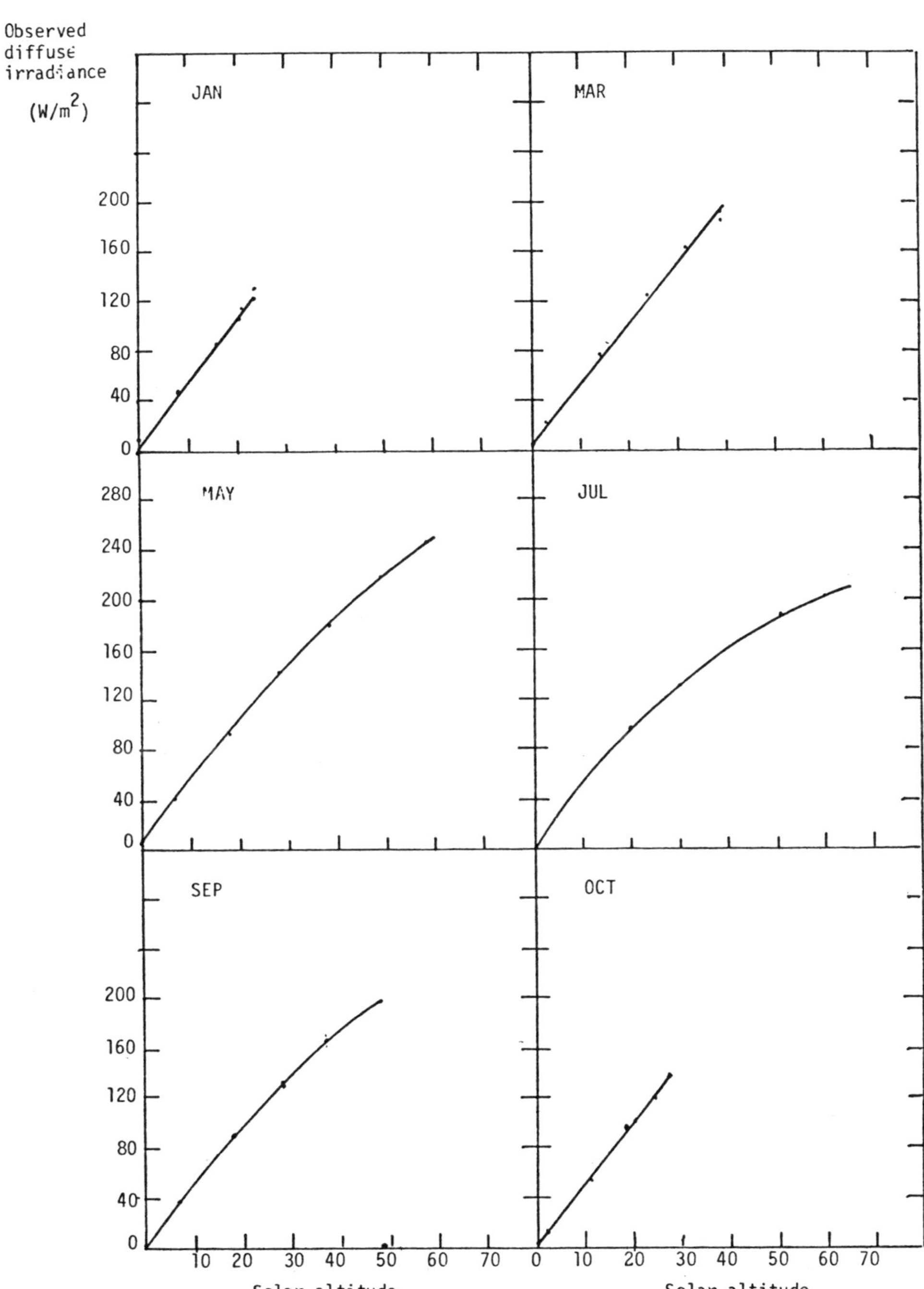

Figure 5.3b Relationship between monthly mean hourly diffuse irradiance and solar altitude for Carpentras.

As no simple way of performing the integration could be identified, systematic numerical integration was adopted to compute the daily values of the diffuse horizontal radiation from values of the diffuse irradiance computed at 15 minute intervals between sunrise and sunset for the middle day of each month. The monthly integrations were performed for latitudes 0-60°N setting $a_o = 5.0$, its representative mean value. Then polynomial equations were sought to relate the values of

$$\int_{SR}^{SS} 5.0 \sin \gamma \, d\omega$$ to the latitude of the site for each month.

These polynomials are given in Table 5.8. The number of terms in the polynomial was selected to provide a minimum accuracy of 1.0% in the evaluation of the integral.

It then becomes simple to derive the actual value of a_o for any site for any month from the monthly mean daily values of the horizontal surface irradiation D_{dm}, because

$$D_{dm} = 2K_d \, \omega_s + K_d(a_o/5.0) \cdot 5.0 \int \gamma \, d\omega \tag{5.20}$$

Hence
$$a_o = 5.0(D_{dm} - 2K_d \, \omega_s)/(K_d \cdot 5.0 \int \gamma \, d\omega) \tag{5.21}$$

where ω_s is the hour angle at sunset as in the Liu and Jordan method

K_d is the correction to m.s.d.

The value of $5.0 \int \gamma \, d\omega$ is easily obtained using the equations in Table 5.8. Once a_o is determined, it is possible to estimate the monthly mean diffuse irradiance at any solar altitude simply from equation 5.19. Proceeding in this way, monthly values of a_o were then evaluated for the same 15 sites whose details are given in Table 2.20.

The resulting values of a_o found in this way are given in Table 5.9, together with the associated values of the sum of the Angstrom (a + b) regression constants. The monthly mean hourly predicted values were then compared with the observed values categorising the whole year's data into 10° bands of solar altitude. The results are given in Table 5.10. The mean errors for each solar altitude band are acceptably small and the standard deviations are typically ±2-3% except for the lowest solar altitude band treated when they rose to 4-5%.

16. Derivation of an objective method for estimating the diffuse irradiance from the values of the Angstrom regression coefficients

Plotting the derived values of a_o against the Angstrom (a + b) showed links between the two. The data for Carpentras and Locarno Monti were clearly different from the rest of the data set for reasons already discussed. The data for Hamburg were also higher than the rest of the data set, as the diffuse corrections made in Hamburg for non-isotropic shadow band effects were substantial, while most of the other stations applied no extra corrections, thus accounting for a difference of about 5% in monthly mean values of diffuse irradiation. The values for Eskdalemuir were also low, but Eskdalemuir is in a valley and is therefore a relatively

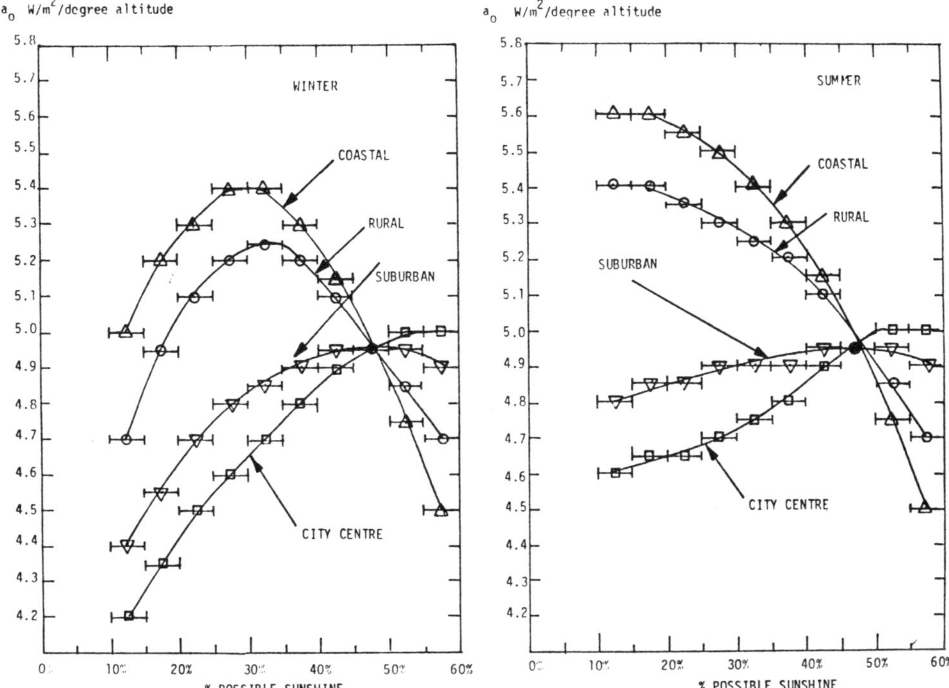

Figure 5.4 Values of a_o in the diffuse irradiance formula $D_m = K_d(2 + a_o\ \gamma)$ classified by site type and plotted against relative astronomical duration of bright sunshine. Winter left hand side, summer right hand side.

Table 5.8 Polynomials to calculate monthly values of $\int_{SR}^{SS} 5.0\,\gamma\,d\omega$ as a function of latitude on the 15th day of each month, except February when 14th day is evaluated.

$$\int_{SR}^{SS} 5.0\ \gamma\, d\omega = A_0 + A_1\phi + A_2\phi^2 + A_3\phi^3 + A_4\phi^4 + A_5\phi^5$$

where ϕ = latitude, computed for latitudes 0° - 60°N.

JANUARY

$a_0 = 2.36385$
$a_1 = -3.33949 \times 10^{-2}$
$a_2 = -2.75084 \times 10^{-4}$
$a_3 = 7.95097 \times 10^{-6}$
$a_4 = -1.26973 \times 10^{-7}$.
$a_5 = 9.53668 \times 10^{-10}$.

FEBRUARY

$a_0 = 2.54274$
$a_1 = -2.62335 \times 10^{-2}$
$a_2 = -2.61513 \times 10^{-4}$
$a_3 = 2.07640 \times 10^{-6}$

MARCH

$a_0 = 2.69938$
$a_1 = -1.06521 \times 10^{-2}$.
$a_2 = -4.49611 \times 10^{-4}$.
$a_3 = 2.63846 \times 10^{-6}$.

APRIL

$a_0 = 2.60885$
$a_1 = 2.23491 \times 10^{-2}$
$a_2 = -1.05281 \times 10^{-3}$
$a_3 = 7.37671 \times 10^{-6}$

MAY

$a_0 = 2.40176$
$a_1 = 4.34484 \times 10^{-2}$
$a_2 = -1.14394 \times 10^{-3}$
$a_3 = 6.81221 \times 10^{-6}$

JUNE

$a_0 = 2.27848$
$a_1 = 5.10228 \times 10^{-2}$.
$a_2 = -1.09595 \times 10^{-3}$.
$a_3 = 6.13401 \times 10^{-6}$.

JULY

$a_0 = 2.32973$
$a_1 = 4.82164 \times 10^{-2}$
$a_2 = -1.12456 \times 10^{-3}$
$a_3 = 6.46078 \times 10^{-6}$

AUGUST

$a_0 = 2.52522$
$a_1 = 3.20199 \times 10^{-2}$
$a_2 = -1.08120 \times 10^{-3}$
$a_3 = 6.71171 \times 10^{-6}$

SEPTEMBER

$a_0 = 2.70000$
$a_1 = 8.47400 \times 10^{-4}$.
$a_2 = -6.33798 \times 10^{-4}$.
$a_3 = 3.82796 \times 10^{-6}$.

OCTOBER

$a_0 = 2.62640$
$a_1 = -2.05576 \times 10^{-2}$
$a_2 = -3.23961 \times 10^{-4}$
$a_3 = 2.12458 \times 10^{-6}$

NOVEMBER

$a_0 = 2.43114$
$a_1 = -3.20146 \times 10^{-2}$
$a_2 = -1.52007 \times 10^{-4}$
$a_3 = 6.25928 \times 10^{-7}$
$a_4 = 1.37121 \times 10^{-8}$

DECEMBER

$a_0 = 2.31314$
$a_1 = -3.60901 \times 10^{-2}$.
$a_2 = -7.92760 \times 10^{-5}$.
$a_3 = -3.31241 \times 10^{-6}$.
$a_4 = 2.12382 \times 10^{-7}$.
$a_5 = -3.84721 \times 10^{-9}$.
$a_6 = 2.60800 \times 10^{-11}$.

The minimum number of terms to evaluate to a 1% accuracy have been selected in each case.

Table 5.9 Values of a_o in formula $D_m = (2 + a_o \gamma) K_d$ Wm^{-2} for prediction of hourly values of monthly mean diffuse solar radiation on horizontal planes for 15 European sites and also the sum of Angstrom regression coefficients (a + b).
Units: a_o Wm^{-2} per degree solar altitude.

		J	F	M	A	M	J	J	A	S	O	N	D
Hamburg**	a_o	5.46	5.48	5.48	5.30	5.48	5.47	5.55	5.54	5.42	5.17	4.62	4.57
	(a+b)	.71	.77	.77	.77	.77	.74	.72	.72	.74	.75	.76	.72
Lerwick*	a_o	5.25	5.02	5.36	5.63	5.27	5.12	5.47	5.36	5.19	4.94	4.73	4.42
	(a+b)	.79	.80	.84	.82	.87	.84	.84	.81	.82	.80	.78	.80
Eskdalemuir*	a_o	4.31	5.08	5.17	5.15	4.93	4.97	4.84	5.15	4.97	4.59	4.53	3.98
	(a+b)	.78	.80	.81	.80	.81	.79	.79	.77	.78	.79	.77	.75
Aldergrove*	a_o	4.84	5.18	5.28	5.45	5.33	5.22	5.30	5.52	5.36	4.89	4.84	4.56
	(a+b)	.79	.77	.79	.74	.78	.78	.78	.78	.77	.74	.77	.75
Cambridge*	a_o	5.22	5.20	5.19	5.08	5.32	5.26	5.22	5.02	5.20	4.99	5.14	4.68
	(a+b)	.79	.78	.79	.76	.76	.76	.77	.76	.77	.77	.74	.80
Aberporth*	a_o	5.55	5.38	5.38	5.48	5.12	4.98	5.32	5.37	5.37	5.15	5.16	5.10
	(a+b)	.73	.75	.80	.79	.79	.80	.81	.76	.76	.76	.73	.72
Valentia**	a_o	5.37	5.41	5.58	5.46	5.61	5.48	5.55	5.37	5.45	5.09	5.10	4.90
	(a+b)	.83	.85	.85	.84	.86	.86	.87	.84	.84	.86	.84	.82
London W.C.*	a_o	4.41	4.45	4.49	4.78	4.78	4.56	4.80	4.73	4.70	4.46	4.40	4.04
	(a+b)	.66	.67	.68	.71	.70	.72	.70	.68	.70	.67	.64	.65
Kew*	a_o	4.77	4.76	4.81	4.80	5.00	4.84	5.03	5.05	5.04	4.74	4.80	4.41
	(a+b)	.70	.71	.73	.73	.72	.72	.71	.69	.71	.71	.68	.71
Bracknell*	a_o	4.95	5.12	5.01	5.15	5.31	4.95	5.22	5.16	5.30	4.91	4.82	4.59
	(a+b)	.76	.75	.78	.74	.74	.75	.73	.73	.73	.72	.72	.72
Jersey*	a_o	5.23	5.21	5.08	5.26	5.05	4.75	4.94	5.13	5.02	4.90	4.74	4.42
	(a+b)	.80	.77	.80	.78	.78	.76	.76	.75	.77	.74	.75	.76
Uccle	a_o	4.87	5.07	5.15	5.16	5.26	5.06	5.22	5.39	5.35	5.12	4.70	4.40
	(a+b)	.70	.73	.73	.75	.76	.75	.74	.71	.72	.70	.71	.66
Paris/Trappes*	a_o	5.06	5.15	5.04	5.13	5.01	4.94	4.93	5.10	5.01	5.06	4.82	4.63
	(a+b)	.78	.75	.76	.77	.77	.76	.76	.73	.74	.73	.74	.77
Carpentras	a_o	5.15	4.57	4.82	4.53	4.48	4.12	3.78	4.22	4.43	4.66	4.96	5.01
	(a+b)	.70	.74	.78	.80	.79	.79	.77	.76	.77	.74	.70	.69
Locarno Monti	a_o	5.12	4.97	4.85	4.86	4.74	4.51	4.64	4.60	4.61	4.37	4.28	4.74
	(a+b)	.78	.79	.83	.82	.85	.84	.81	.81	.82	.79	.78	.73

* Isotropic shading band corrections known to be applied.
** Special local corrections applied for shading band effects.

Table 5.10 Mean errors (observed/predicted) in the estimation of monthly mean hourly diffuse horizontal surface irradiation using the Page linear model $D_m = (2 + a_o \gamma) K_d$ for 15 European sites.

Station		Solar Altitude Band - Degrees					
		10-20	20-30	30-40	40-50	50-60	60-70
Hamburg	A.M. Error	.983	1.019	1.024	.997	.962	-
	A.M. S.D.	.038	.015	.022	.024	.014	-
	P.M. Error	.993	1.013	1.012	.999	.981	-
	P.M. S.D.	.039	.017	.023	.025	.015	-
Lerwick	A.M. Error	.935	.987	1.027	1.021	1.018	-
	A.M. S.D.	.064	.039	.015	.005	.011	-
	P.M. Error	.946	.997	1.017	1.024	1.029	-
	P.M. S.D.	.038	.027	.013	.015	.022	-
Eskdalemuir	A.M. Error	.958	1.014	1.014	1.005	1.006	-
	A.M. S.D.	.050	.032	.015	.014	.015	-
	P.M. Error	.955	.992	1.009	1.011	.995	-
	P.M. S.D.	.035	.023	.022	.017	.006	-
Aldergrove	A.M. Error	.980	1.004	1.037	1.036	1.003	-
	A.M. S.D.	.048	.017	.021	.037	.018	-
	P.M. Error	.979	1.008	1.015	1.011	.999	-
	P.M. S.D.	.038	.035	.013	.016	.021	-
Cambridge	A.M. Error	.974	1.010	1.018	.99	.968	.954
	A.M. S.D.	.032	.025	.024	.025	.003	-
	P.M. Error	1.004	1.012	1.017	1.001	.978	.973
	P.M. S.D.	.030	.026	.033	.018	.020	-
Aberporth	A.M. Error	.987	1.008	1.024	1.011	.98	.949
	A.M. S.D.	.035	.030	.023	.018	.023	-
	P.M. Error	1.012	1.016	1.011	.984	.952	.922
	P.M. S.D.	.044	.034	.023	.016	.012	-
Valentia	A.M. Error	.941	.980	1.012	1.006	.993	.968
	A.M. S.D.	.052	.029	.019	.014	.016	-
	P.M. Error	1.010	1.037	1.030	1.009	.975	.95
	P.M. S.D.	.033	.02	.021	.015	.012	-
London Weather Centre	A.M. Error	.994	1.021	1.032	1.008	.960	.934
	A.M. S.D.	.048	.032	.021	.025	.016	-
	P.M. Error	.993	1.011	1.021	.996	.965	.945
	P.M. S.D.	.043	.034	.031	.021	.015	-
Kew	A.M. Error	.988	1.018	1.03	1.007	.974	.950
	A.M. S.D.	.034	.027	.018	.026	.017	-
	P.M. Error	.989	.994	1.005	.996	.971	.950
	P.M. S.D.	.032	.023	.020	.017	.013	-
Bracknell	A.M. Error	.978	1.012	1.013	.997	.955	.935
	A.M. S.D.	.031	.025	.015	.019	.013	-
	P.M. Error	1.013	1.013	1.018	1.005	.969	.938
	P.M. S.D.	.049	.043	.031	.028	.016	-

Table 5.10 continued Mean errors (observed/predicted) in the estimation of monthly mean hourly diffuse horizontal surface irradiation using the Page linear model $D_m = (2 + a_o \gamma) K_d$ for 15 European sites.

Station		Solar Altitude Band - Degrees					
		10-20	20-30	30-40	40-50	50-60	60-70
Jersey	A.M. Error	1.001	1.044	1.049	1.052	.991	.93
	A.M. S.D.	.041	.034	.044	.039	.025	.014
	P.M. Error	1.010	1.020	.987	.963	.954	.922
	P.M. S.D.	.035	.022	.021	.014	.016	0.005
Uccle	A.M. Error	.977	1.024	1.029	1.005	.970	.944
	A.M. S.D.	.041	.017	.018	.020	.015	-
	P.M. Error	.983	1.018	1.028	1.001	.969	.954
	P.M. S.D.	.035	.022	.019	.021	.014	-
Paris/Trappes	A.M. Error	1.005	1.026	1.024	1.013	.972	.938
	A.M. S.D.	.043	.032	.032	.027	.023	.008
	P.M. Error	.997	1.024	1.027	1.013	.975	.950
	P.M. S.D.	.046	.047	.039	.023	.023	.008
Locarno-Monti	A.M. Error	.981	1.002	1.001	.985	.959	.931
	A.M. S.D.	.034	.034	.040	.037	.021	.016
	P.M. Error	1.020	1.038	1.045	1.021	.990	.958
	P.M. S.D.	.031	.041	.043	.041	.029	.022
Carpentras	A.M. Error	1.095	1.044	1.022	.984	.937	.874
	A.M. S.D.	.101	.071	.061	.042	.030	.030
	P.M. Error	1.104	1.040	1.029	1.005	.946	.897
	P.M. S.D.	.111	.086	.097	.063	.037	.025

obstructed site. The values for Hamburg, Carpentras, Locarno Monti and Eskdalemuir were excluded from the data set. Linear regressions were then sought between a_o and $(a + b)$ for each month for the remaining 11 sites of the form

$$a_o = a_o' + a_1'(a + b)$$

The results are given in Table 5.11. The monthly values of the intercept and slope are plotted in Figure 5.5. With the exception of May, the intercepts and slopes plot as smooth curves. Ignoring May, it was possible to develop two annual Fourier functions which enables a_o to be objectively estimated from the monthly mean values of Angstrom $(a + b)$ on the Horizontal Surface Atlas so the diffuse irradiance in any month at any solar altitude at latitudes above about 48°N can be very simply evaluated using the two Fourier functions below to find a_o which is then substituted into Equation 5.19. The limitation to European latitudes above 48°N is stressed.

$$a_o' = b + \sum_{i=1}^{2} (c_i \cos(iM\, 2\pi/12) + d_i \sin(iM\, 2\pi/12)) \qquad (5.22)$$

where M is the month number (e.g. January: M = 1)

$b = 2.19$, $c_1 = 0.431$, $c_2 = 0.0150$, $d_1 = -1.09$ and $d_2 = 0.0317$

$$a_1' = b + \sum_{i=1}^{2} (c_i \cos(iM\, 2\pi/12) + d_i \sin(iM\, 2\pi/12)) \qquad (5.23)$$

where M is as above

$b = 3.74$, $c_1 = -0.718$, $c_2 = -0.206$, $d_1 = 1.42$ and $d_2 = 0.0535$

17. Development of the final EC monthly mean diffuse radiation model

Consideration of the facts already discussed, taking note of the Berlin method, both its strengths and weaknesses, and a detailed examination of Kasten and Czeplak's studies of relationships between D_m and cloudiness for Hamburg, led to abandonment of the use of the overcast day values of G_b in the processes of interpolation for the monthly mean diffuse model. Kasten's work shows that, to a first approximation, the relationship between D_m and cloud amount is linear between about 1 okta out to about 6 oktas, thereafter the diffuse radiation decreased fairly sharply with further increase of cloud amount.

It was decided to establish a base line value of D_m for interpolation purposes, called $D_{.25}$ set up at $\sigma_{4m} = .25$ to replace the value of D_b in the Berlin method. This base line reference diffuse irradiance level was set on the basis of the diffuse studies reported in the previous paragraphs as

$$D_{.25} = K_d(2 + 5.3\gamma) \quad Wm^{-2} \qquad (5.24)$$

Table 5.11 Linear regression of coefficient a_o in Page's monthly mean diffuse irradiance equation $D_m = (2 + a_o \gamma)K_d$ against sum of monthly Angstrom regression coefficients $(a + b)$ using data from Lerwick, Aldergrove, Cambridge, Aberporth, London Weather Centre, Kew, Bracknell, Jersey, Paris (Trappes), Uccle and Valentia (11 sites).

$$a_o = a_1 + a_2(a + b)$$

Month	Intercept a_1	Slope a_2	Correlation Coefficient r	Significance level
J	1.96	4.08	0.675	(5%)
F	1.61	4.59	0.787	(1%)
M	0.99	5.32	0.893	(0.2%)
A	1.01	5.49	0.781	(1%)
M	2.96	2.87	0.666	(5%)
J	1.81	4.15	0.728	(2%)
J	2.36	3.66	0.826	(0.2%)
A	2.94	3.02	0.648	(5%)
S	3.25	2.55	0.512	(20%)
O	3.38	2.08	0.547	(10%)
N	3.01	2.49	0.585	(10%)
D	2.70	2.50	0.499	(20%)

Table 5.12 Constants used in formula for calculating additional correction functions f_6 for estimating monthly mean diffuse irradiance on horizontal surfaces.

i	b_i	c_i	d_i
0	0.5212	0.83202	7.12889
1	2.429	0.011619	-14.9747
2	-3.383	-1.8832×10^{-4}	9.10674
3	1.432	9.8559×10^{-7}	-

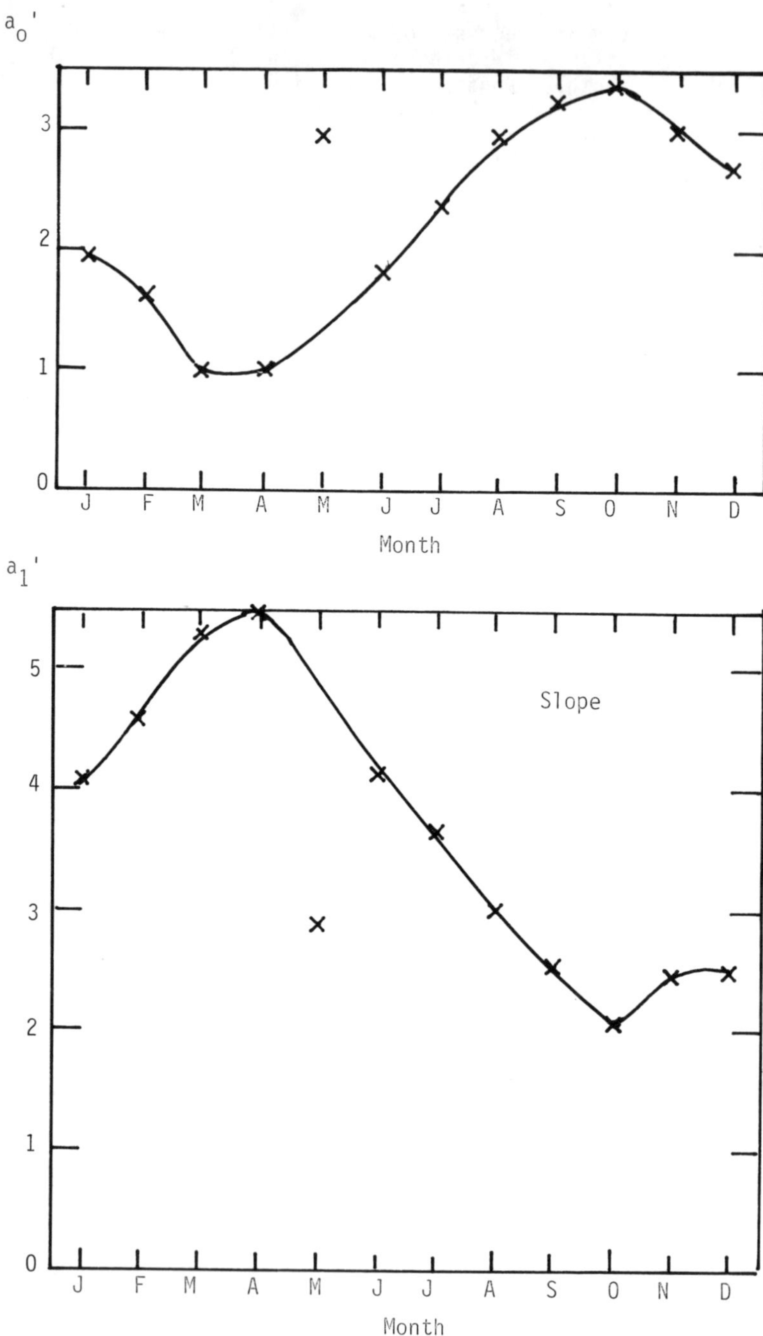

Figure 5.5 Monthly values of a_0' and a_1' derived, using equation 5.21, from the sum of the Ångstrom regression coefficients (a + b) plotted against month. Ignoring May, it is possible to plot a smooth curve and fit it by a Fourier series.

A first estimate of the actual value of D_m could then be obtained from the value of σ_{4m} by linear interpolation using the value of D_c to establish the value for $\sigma_{4m} = 1$, computed using the methodologies of Chapter 2. Calling this first estimate ${}_uD_m$, then

$$ {}_uD_m = \sigma_{4m} D_c + (1 - \sigma_{4m})(D_{.25} - 0.25 \sigma_{4m} D_c)/.75 \tag{5.25} $$

Figure 5.6 gives a graphical explanation of the basic interpolation process.

Then, using the observed monthly mean hourly Hamburg diffuse irradiation data in association with the observed monthly mean daily values of sunshine, S_m, the hourly errors in predicting monthly mean hourly irradiation resulting from the use of equation 5.25 were established against the actual Hamburg hourly observations. The diffuse correction function needed for Hamburg f_6' was found to involve both a function of solar altitude γ, as well as a function of relative sunshine duration σ_{4m}. By successive numerical approximation the correction function was derived from the Hamburg observations as

$$ f_6' = (\sum_{i=0}^{3} b_i \sigma_{4m}^{\ i})(\sum_{i=0}^{3} c_i \gamma^i) \tag{5.26} $$

where the values of b_i and c_i are given in Table 5.12.

A daily global irradiation model was then constructed from the direct beam and diffuse models to predict monthly mean hourly global irradiation on horizontal surfaces, following the detailed principles set out in Chapter 2 for the corresponding clear day models. The mean daily air mass 2 turbidity values were then estimated from the daily difference between $G_m - D_m$ by reiteration as explained in Part I of this Chapter, using the data from the standard 14 stations. (The corresponding data for Uccle became available at a later stage.) These daily air mass 2 Linke Turbidity Factors for monthly mean conditions were then used to establish the values of D_c in equation 5.25 using the methods given in Chapter 2, and hence daily estimated values of D_m were established for each month for the 14 sites, correcting by Equation 5.26. These predicted monthly mean daily values were then compared with the observed monthly mean daily values. The monthly mean and annual mean daily errors in the estimation of D_m were then established for each of the 14 selected sites. Examination of these annual errors in the monthly mean daily diffuse predictions showed that the errors fell in one direction for the very clear sites, and in another direction for the less clear sites.

Using the annual mean values of the Angstrom $(\overline{a + b})$ from the European Solar Radiation Atlas, Vol. I, Global radiation on horizontal surfaces, correlations were then sought between the annual mean diffuse errors and the annual mean values of the Angstrom regression coefficients, $(\overline{a + b})$, for the 14 sites. A final annual correction function was thus established as $1/\sum_{i=0}^{2} d_i(\overline{a + b})^i$. The values of d_i found are given in Table 5.12.

Thus the final correction function adopted for the diffuse interpolation correction function f_6 was given by

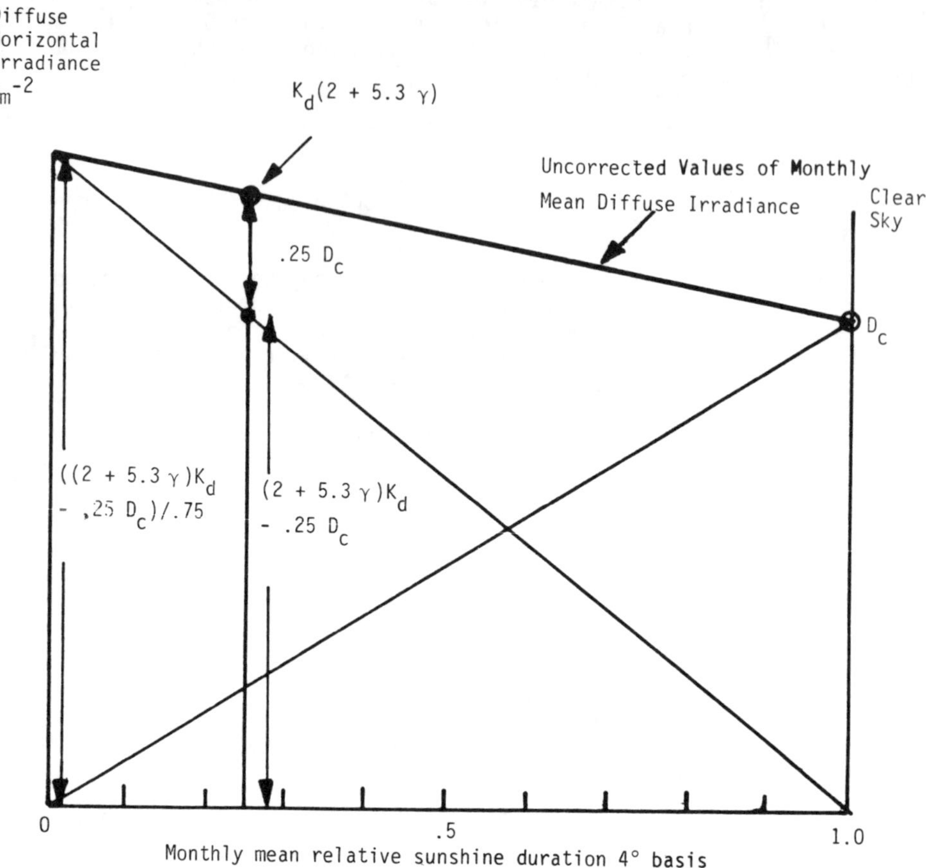

Figure 5.6 Process of linear interpolation used to obtain first estimate of monthly mean diffuse irradiance on horizontal surfaces ${}_{u}D_{m}$ from the monthly mean relative duration of sunshine on 4 degree sunrise basis. D_c is obtained from clear day model.

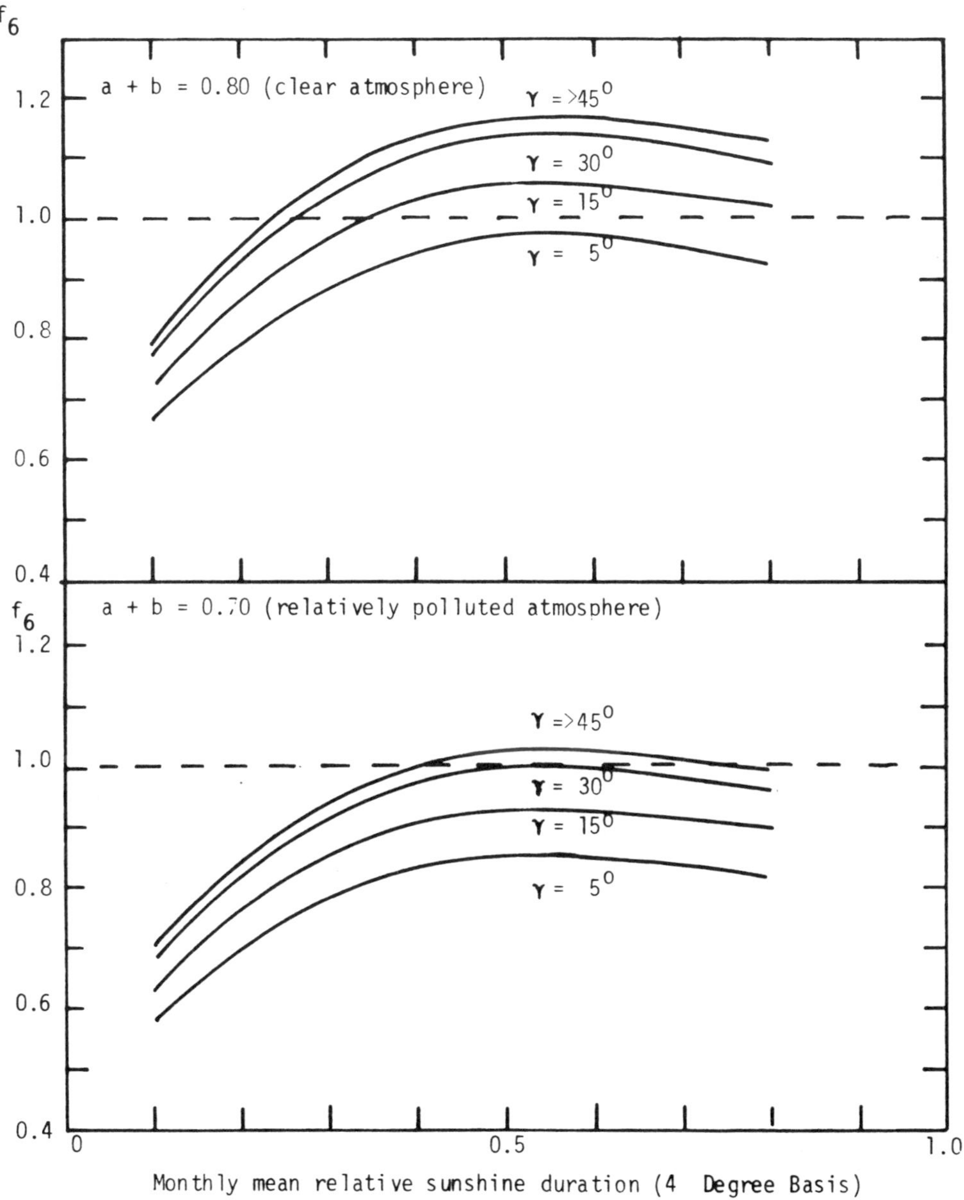

Figure 5.7 Relationship between the diffuse correction function f_6 and 4 degree sunrise/sunset relative sunshine duration as a function of solar altitude γ. Top diagram for a relatively clear atmosphere. Bottom diagram for a relatively polluted atmosphere.

$$f_6 = (\sum_{i=0}^{3} b_i \sigma_{4m}{}^i)(\sum_{i=0}^{3} c_i \gamma^i)/\sum_{i=0}^{i=2} d_i(\overline{a + b})^i \quad (5.27)$$

The form of this function is illustrated by Figure 5.7 for two values of $(\overline{a + b})$, while Table 6.13 of Chapter 6 provides detailed values of the function for values of $(\overline{a + b}) = .74$ with a correction table for other values of $\overline{a + b}$ at the bottom.

The lowest values of $\overline{a + b}$ are around 0.66 for central London and the highest values for very clear maritime sites exceed 0.84. The range of correction factors to allow for the overall annual clarity of the atmosphere thus ranges from 0.853 to 1.060. This finding, of course, confirms the findings reported in the previous section.

Assembling the new global irradiance model, it was then possible to estimate for the 14 reference stations monthly mean hourly values of the diffuse irradiance, and, assuming these hourly values of the irradiance to be representative of the monthly mean hourly values of the diffuse irradiation to estimate the errors of the final model against hourly observations of D_m. The errors for individual hours were then classified by 10 degree bands of solar altitude dividing the data into morning hours and afternoon hours. The means and standard deviations were then extracted for each solar altitude band for each station separately for mornings and afternoons.

The results of this analysis are set out in Table 5.13. Very good agreement was achieved between prediction and observation. This diffuse irradiance model was then adopted as the standard method for estimating monthly mean hourly horizontal surface diffuse irradiation from the observed monthly mean sunshine data, which were available for about 100 sites in the CEC European Solar Radiation Atlas, Vol. I. However, for sites where the diffuse radiation observations were not made, the problem of determining the correct value of air mass 2 Linke Turbidity Factor still remained to be resolved.

18. Development of an objective method for assessing the Linke Turbidity Factors for the prediction of monthly mean hourly irradiances

The difficulties of using subjective approaches to the assessment of the Linke Turbidity Factor were explained in Chapter 2. When the new monthly mean model was first developed, a subjective approach was first adopted. However, considerable difficulties soon emerged. It was at this stage that other collaborators in the project suggested the value of seeking a truly objective approach. The project action leader, M. Dogniaux, had met with considerable success in finding a relationship between the sum of the Angstrom regression coefficients and the atmospheric transparency index (19) and he encouraged the Sheffield University Research Group to re-examine their methodology to see if they could make it objective rather than subjective, using the sum of the Angstrom Regression Coefficients as the basis.

The procedure adopted, using the data for the 14 European sites where reliable hourly data on monthly mean values of the global and diffuse irradiation were available to the project, was first of all to determine by

Table 5.13 Mean errors (observed/predicted) in the estimation of monthly mean hourly diffuse horizontal surface irradiation using the proposed model as a function of solar altitude band for 15 European sites.

Station		Solar Altitude Band - Degrees					
		10-20	20-30	30-40	40-50	50-60	60-70
Hamburg	A.M. Error	.978	1.001	1.009	1.003	.991	-
	A.M. S.D.	.031	.021	.016	.016	.013	-
	P.M. Error	.987	.996	.998	1.005	.991	-
	P.M. S.D.	.029	.018	.018	.020	.015	-
Lerwick	A.M. Error	.953	.989	1.017	1.010	1.011	-
	A.M. S.D.	.047	.029	.012	.004	.003	-
	P.M. Error	.965	.999	1.007	1.013	1.021	-
	P.M. S.D.	.022	.018	.009	.014	.014	-
Eskdalemuir	A.M. Error	.973	1.011	1.002	.999	1.008	-
	A.M. S.D.	.036	.025	.015	.014	.013	-
	P.M. Error	.970	.990	.998	1.006	.997	-
	P.M. S.D.	.021	.018	.021	.016	.010	-
Aldergrove	A.M. Error	.987	.998	1.027	1.035	1.010	-
	A.M. S.D.	.041	.017	.019	.033	.018	-
	P.M. Error	.986	1.002	1.005	1.010	1.007	-
	P.M. S.D.	.031	.030	.013	.023	.027	-
Cambridge	A.M. Error	.974	1.001	1.007	.988	.984	.986
	A.M. S.D.	.025	.019	.020	.021	.009	-
	P.M. Error	1.004	1.003	1.006	1.000	.995	1.006
	P.M. S.D.	.024	.021	.030	.017	.015	-
Aberporth	A.M. Error	.985	1.001	1.014	1.011	.995	.978
	A.M. S.D.	.030	.025	.019	.014	.023	-
	P.M. Error	1.010	1.008	1.002	.985	.967	.950
	P.M. S.D.	.030	.029	.018	.015	.010	-
Valentia	A.M. Error	.953	.979	1.005	1.001	.994	.973
	A.M. S.D.	.039	.023	.018	.011	.013	-
	P.M. Error	1.023	1.036	1.022	1.003	.977	.955
	P.M. S.D.	.029	.026	.021	.013	.011	-
London Weather Centre	A.M. Error	.990	1.007	1.018	1.010	.983	.979
	A.M. S.D.	.034	.023	.023	.020	.014	-
	P.M. Error	.989	.997	1.006	.997	.988	.991
	P.M. S.D.	.030	.021	.022	.018	.010	-
Kew	A.M. Error	.982	1.003	1.016	1.009	.999	.994
	A.M. S.D.	.028	.019	.017	.018	.017	-
	P.M. Error	.983	.980	.992	.998	.996	.994
	P.M. S.D.	.024	.016	.013	.014	.010	-
Bracknell	A.M. Error	.975	1.000	1.000	.998	.975	.975
	A.M. S.D.	.026	.023	.015	.018	.012	-
	P.M. Error	1.009	1.000	1.004	1.006	.989	.978
	P.M. S.D.	.034	.030	.022	.024	.012	-

Table 5.13 continued — Mean errors (observed/predicted) in the estimation of monthly mean hourly diffuse horizontal surface irradiation using the proposed model as a function of solar altitude band for 15 European sites.

Station		Solar Altitude Band - Degrees					
		10-20	20-30	30-40	40-50	50-60	60-70
Jersey	A.M. Error	.971	1.013	1.033	1.059	1.032	.998
	A.M. S.D.	.043	.015	.024	.021	.021	.015
	P.M. Error	.980	.991	.973	.970	.995	.989
	P.M. S.D.	.036	.027	.018	.012	.021	.006
Uccle	A.M. Error	.971	1.008	1.016	1.009	.995	.980
	A.M. S.D.	.037	.016	.003	.014	.011	-
	P.M. Error	.977	1.001	1.015	1.005	.994	.990
	P.M. S.D.	.023	.013	.010	.013	.012	-
Paris/Trappes	A.M. Error	.986	1.007	1.007	1.015	.997	.987
	A.M. S.D.	.034	.019	.020	.017	.018	-
	P.M. Error	.978	1.005	1.010	1.014	1.000	1.000
	P.M. S.D.	.024	.029	.022	.012	.021	-
Locarno-Monti	A.M. Error	.927	.978	.988	.996	.997	.978
	A.M. S.D.	.055	.024	.024	.025	.022	.002
	P.M. Error	.965	1.014	1.032	1.032	1.029	1.006
	P.M. S.D.	.077	.027	.031	.025	.015	.012
Carpentras	A.M. Error	.982	1.008	.994	.993	1.001	.987
	A.M. S.D.	.023	.017	.022	.008	.013	.028
	P.M. Error	.990	1.003	.999	1.014	1.010	1.013
	P.M. S.D.	.031	.025	.036	.026	.022	.022

Table 5.14 Monthly values of the constant a_o in the formula $T_L = a_o - a_o(a + b)$ for assessing objectively the monthly mean air mass 2 Linke Turbidity Factor from the sum of the Angstrom regression coefficients (a + b), derived using data for 15 European stations.

Month	a_o	Correlation Coefficient
Jan	10.66	-.755
Feb	13.23	-.692
Mar	15.42	-.842
Apr	17.05	-.837
May	18.55	-.822
Jun	16.94	-.786
Jul	18.16	-.783
Aug	16.96	-.832
Sep	14.94	-.810
Oct	12.85	-.693
Nov	11.34	-.684
Dec	9.49	-.730

successive reiteration the air mass 2 Linke Turbidity Factor needed to match the model daily beam predictions for each month to the observed monthly mean daily values of $(G_m - D_m)$. A linear relationship between T_L and (a + b) was assumed. It was also noted that the Linke Turbidity Factor must be equal to zero, when (a + b) = 1, then it followed that (1,0) was a constrained point on the curve, thus

$$T_L = a_o - a_o(a + b) \tag{5.28}$$

Table 5.14 gives the monthly values of a_o derived using the monthly data for the 14 stations with the associated regression coefficients. Comparing Table 5.14 with Table 2.26, it will be noted that the values of a_o found are different in both cases, implying different values for the clarity for cloudless days. However, the methodology discussed here is primarily a methodology for estimating monthly mean values of G_m, $I_m(0,0)$ and D_m, while Chapter 2 provides the basic methodology for estimating G_{max} $I_{max}(0,0)$ and D_{max}. The precise values of the cloudless day values are not significant if the resulting mean monthly mean values are correct.

CHAPTER 5 - PART III

MODELLING GLOBAL, BEAM AND DIFFUSE IRRADIATION ON SLOPES

19. Introduction

In general there is inadequate knowledge concerning the radiance distribution of partially clouded skies, so the estimation of monthly mean hourly diffuse radiation on slopes from the hourly values of the corresponding hourly horizontal surface values is subject to some uncertainty. There were no observations of monthly mean diffuse irradiation alone available for checking any assumptions made. The estimation of the direct beam contribution is however a relatively trivial trigonometric problem. By combining the direct beam predictions for inclined planes with the diffuse predictions, one obtains the global predictions, and there were observations available for global radiation on inclined planes. The validation of the model developed had therefore to be based on verification of the global model. There is another uncertainty for inclined planes, in contrast with horizontal planes, which is the influence of the ground albedo. In checking predictive models for inclined planes, a value of the ground albedo has to be estimated. In the studies that follow, and in the standard EC computational process, the ground albedo was set at a constant value of 0.2 unless otherwise stated.

20. Direct beam component on inclined planes of monthly mean hourly irradiance model

On trigonometrical principles it follows that

$$I_m(\beta,\alpha) = I_m(0,0) \times \cos \nu/\sin\gamma \qquad (5.29)$$

where ν is the angle of incidence

γ is the solar altitude.

It is thus simple to establish the direct beam contribution from the horizontal values using the underlying solar geometry.

21. Basic assumptions - monthly mean diffuse model for inclined planes

In the absence of systematised studies of the properties of partially clouded skies, the most appropriate assumption seemed to be to combine the clear day and overcast day models in some appropriate way. It was assumed that the diffuse irradiance on inclined planes had three components:

i) An overcast / sky partially clouded sky component whose radiance distribution followed the principles set out in Chapter 4.

ii) A clear sky component, whose radiance distribution followed the principles set out in Chapter 3,

iii) A ground reflected component related to the monthly mean global irradiance on the horizontal plane and the albedo.

The 4 degree monthly mean daily relative sunshine duration was first used to split the monthly mean hourly diffuse irradiance into two components using the following formula

$$ {}_{pc}D_m = D_m - \sigma_{4m} D_c \qquad (5.30) $$

where ${}_{pc}D_m$ is the monthly mean diffuse irradiance assumed to be associated with overcast / partially cloudy sky conditions

and $\sigma_{4m} D_c$ is the monthly mean diffuse irradiance assumed to be associated with cloudless sky conditions.

The value of the diffuse sky slope irradiance attributed to the overcast/partially clouded sky can then be established using the standard Moon and Spencer methodology, as discussed in Chapter 4. Thus, introducing the subscript s to denote sky alone, and a subscript u to indicate a final correction has yet to be applied as explained below

$$ {}_{upc}D_{ms}(\beta,\alpha) = f_4 \; {}_{pc}D_m \qquad (5.31) $$

where f_4 is the ratio of the inclined surface diffuse irradiance to the horizontal surface diffuse irradiance for overcast days

As Chapter 4 has shown, the value of f_4 for a plane of tilt β is given by

$$ f_4 = 0.1819(1.178(1 + \cos\beta) + \frac{\pi}{180}(180 - \beta)\cos\beta + \sin\beta) \qquad (5.32) $$

where the tilt β is in degrees.

Values of f_4 are tabulated in Chapter 6, Table 6.1.

The corresponding monthly mean blue sky component on an inclined plane, ${}_{uc}D_{ms}(\beta, \alpha)$ was set equal to $f_2 \; \sigma_{4m} D_c$ following the principles set out in Chapter 3, where f_2 is the conversion factor for a plane of given tilt and orientation with respect to the sun for the cloudless non-isotropic sky at that Linke Turbidity Factor. Detailed values of f_2 are given in Chapter 6, Tables 6.14a-j.

These two diffuse sky components were then added together and a further correction applied. This further correction was developed, using detailed monthly mean observations of vertical slope irradiation at Bracknell, to match the monthly mean predictions better to the observations. This correction is only applied when the wall solar azimuth angle lies between ±45° and ±135°. Thus, adding the two sky components

$$ {}_{u}D_{ms} = {}_{uc}D_{ms}(\beta,\alpha) + {}_{upc}D_{ms}(\beta,\alpha) \qquad (5.33) $$

Then, if α_s is between 45° and 135° or between 225° and 315°.

$$ D_{ms}(\beta, \alpha) = {}_{u}D_{ms}(\beta, \alpha)(1 + \sin\beta \; \sin 2(\alpha_s - 45)(.19 - .19 \sin\gamma)) \qquad (5.34a) $$

Otherwise no correction is applied, and

$$D_{ms}(\beta,\alpha) = {}_uD_{ms}(\beta,\alpha) \qquad Wm^{-2} \qquad (5.34b)$$

Finally the ground component $R_m(\beta, \alpha)$ has to be estimated. This is estimated from the monthly mean global irradiance on the horizontal plane G_m and the ground albedo ρ_s as

$$R_m(\beta,\alpha) = f_3 \; G_m \qquad Wm^{-2} \qquad (5.35)$$

where $f_3 = 0.5 \; \rho_s(1 - \cos \beta)$.

The monthly mean global irradiance on the inclined plane $G_m(\beta,\alpha)$ is then obtained as the sum of the three components

$$G_m(\beta,\alpha) = I_m(\beta,\alpha) + D_{ms}(\beta,\alpha) + R_m(\beta,\alpha) \qquad (5.36)$$

The monthly mean total diffuse irradiance from ground and sky on the inclined plane is given by

$$D_m(\beta,\alpha) = D_{ms}(\beta,\alpha) + R_m(\beta,\alpha) \qquad (5.37)$$

<u>The daily irradiation model used to predict daily irradiation on inclined planes</u>

A daily irradiation model was then assembled from its three components. In the production of the CEC standard tables for inclined surfaces, numerical values of the monthly mean hourly irradiance on horizontal surfaces are calculated on the half hour at hourly intervals between sunrise and sunset using the monthly mean declination. By adding the three components over the day, values of the predicted monthly mean daily global irradiation on horizontal surfaces are achieved. These values are then compared with the observed horizontal values month by month, and the hourly predictions normalised, so that the sum of the predicted hourly beam and diffuse irradiances is set equal to the monthly mean daily global irradiation. These normalised hourly values of $I_m(0,0)$ and D_m are then used to compute the corresponding slope values. The detailed step by step methodology is set out in Chapter 6. In preparing the CEC European Solar Radiation Atlas, Vol. II, Inclined Surfaces, normalisation factors for the monthly mean daily values are systematically entered on the monthly mean outputs.

Table 5.15 provides an analysis of the normalisation factors needed to match predicted and observed data in the preparation of the CEC Atlas. Appendix 8 contains a more detailed discussion of some normalisation factors.

<u>Checking the monthly mean daily slope irradiation model against observation</u>

Data for checking the model came from slope observations made at a number of European sites. The results of these daily checks are presented in Table 5.16. The acccuracy of prediction is typically of the order of

Table 5.15 Statistical summary of modelling prediction errors in the estimation of monthly mean daily global solar radiation on horizontal surfaces using the recommended computational method.

The normalisation factor is the observed monthly mean global daily solar radiation divided by the predicted monthly mean global daily solar radiation.

Basis of table. Observed and predicted data for 96 sites in Europe as presented in the Inclined Surface Atlas Tables for which monthly values of Angstrom (a + b) were available.

Month	Jan	Feb	Mar	Apr	May	Jun	Jul	Aug	Sep	Oct	Nov	Dec
Mean normalisation factor	1.017	1.013	1.028	1.025	1.026	1.018	1.023	1.031	1.029	1.008	0.994	1.006
Standard deviation	0.060	0.053	0.045	0.046	0.047	0.045	0.045	0.042	0.043	0.052	0.059	0.066
Normalisation bands	Frequency of occurrence											
0.80-0.81	0	0	0	0	0	0	0	0	0	0	0	0
0.82-0.83	0	0	0	0	0	0	0	0	0	0	1	0
0.84-0.85	1	0	0	0	0	1	1	0	0	0	2	1
0.86-0.87	1	0	0	0	0	0	1	0	0	0	2	2
0.88-0.89	0	2	0	0	2	1	0	0	1	2	0	4
0.90-0.91	2	3	0	1	1	1	1	3	1	1	3	4
0.92-0.93	4	2	1	1	1	2	3	1	2	7	8	2
0.94-0.95	5	2	5	4	1	3	0	1	1	5	5	6
0.96-0.97	10	11	6	12	7	9	3	3	4	11	9	6
0.98-0.99	10	17	13	10	9	4	9	8	6	9	14	13
1.00-1.01	14	16	13	10	16	19	15	14	16	18	15	16
1.02-1.03	11	8	18	14	13	19	18	16	21	13	11	12
1.04-1.05	15	16	15	16	16	18	26	23	17	12	13	11
1.06-1.07	8	8	5	14	19	14	13	16	17	6	8	6
1.08-1.09	6	4	13	9	8	4	3	9	6	9	2	7
1.10-1.11	4	4	5	4	3	1	3	2	3	3	3	1
1.12-1.13	2	3	2	0	0	0	0	0	0	0	0	2
1.14-1.15	1	0	0	1	0	0	0	0	1	0	0	1
1.16-1.17	1	0	0	0	0	0	0	0	0	0	0	1
1.18-1.19	1	0	0	0	0	0	0	0	0	0	0	1
TOTAL	96	96	96	96	96	96	96	96	96	96	96	96

Table 5.16 Prediction errors (100 x (predicted/observed) - 100) %, using the final CEC monthly mean irradiation model for slope predictions to predict monthly mean daily slope irradiation.

	J	F	M	A	M	J	J	A	S	O	N	D	Annual mean error	S.D.
Bracknell														
South vert	6.2	5.4	3.0	0.1	2.0	2.6	4.0	1.8	1.1	4.0	4.7	4.6	3.3	1.8
East vert	3.2	2.5	2.2	-3.8	-5.6	-5.7	-1.9	0.2	-0.9	2.1	1.1	12.1	0.46	4.6
West vert	9.5	10.5	13.3	1.5	0.6	1.3	-1.5	1.3	7.4	3.0	14.0	10.5	5.95	5.26
Valentia														
South vert	2.2	-0.3	-0.5	-3.4	-2.1	-1.5	-4.1	-4.3	-1.9	0.6	1.2	1.2	-1.16	2.0
Carpentras Aug 74/ Jul 75														
South vert	0.6	1.	-0.4	0.9	1.1	1.8	6.0	4.2	-2.5	-2.1	-2.1	2.3	0.93	2.4
Carpentras 1973														
South tilt to equal latitude	3.2	0.4	1.8	1.1	3.0	3.0	3.8	5.2	5.0	1.2	-5.2	0.1	1.82	2.71
Trappes 1973														
South vert	-9.9	-8.3	2.9	-4.9	-1.3	-1.8	-1.7	-2.0	-0.9	-6.0	-2.8	-4.6	-3.92	2.76
Odeillo 1971/75														
South vert	13.7	52	-1.5	-7.5	-3.9	-6.5	-5.4	-6.8	-3.5	3.0	12.9	8.7	0.72	7.44
Hamburg 1952/54														
South vert	-	-0.7	2.1	0.5	-6.8	-5.7	-5.5	2.4	-1.5	-3.0	6.2	2.6	-1.28	3.64

5%. Bigger differences emerge in winter, when the problem of obstruction exerts a greater impact. No systematic errors of the model could be identified. The model was, therefore, adopted to produce the monthly mean inclined surface tables for the CEC European Solar Radiation Atlas, Vol. II, Inclined Surfaces, using an albedo of 0.20 throughout. The daily diffuse irradiation tabulated in the Atlas is the sum of the sky and ground components. Table 5.16 provides an example of a monthly mean irradiation Table from the European Solar Radiation Atlas, Vol. II, with its associated heading. The Atlas Tables, printed to a bigger format, combine the clear day and monthly mean values on a single page. D_m in Table 5.16 refers to $D_m(\beta, \alpha)$ in the terminology of this book and G_m to $G_m(\beta,\alpha)$. The monthly mean daily beam irradiation on any slope may be found as $G_m - D_m$. The input data are given at the bottom of the Table. The normalization factor ratio observed/predicted is also given. All horizontal data were normalized before being used to predict slope irradiation. The units used are kWh m^{-2} per day. These units were chosen for engineering convenience in place of the standard SI unit, the megajoule.

Conclusion

An objective validated method for estimating the monthly mean hourly and daily irradiation on inclined planes has been developed that uses, as inputs, monthly mean observed data in the CEC European Solar Radiation Atlas, Vol. I, Global Radiation on Horizontal Surfaces. The model has shortcomings, but the methodology used had to accept the limitations of input data available. With more refined inputs account could be taken of:

i) Actual ground albedo, particularly important with snow covered ground,

ii) Site obstruction, particularly important in hilly and mountainous terrain,

iii) Differences between morning and afternoon sunshine.

The practical methodology however enabled the 102 site tables to be prepared objectively without debate about probable atmospheric properties. These tables were then used to prepare the European maps in the CEC European Solar Radiation Atlas, Vol. II, Inclined Surfaces. Chapter 6 sets out a step by step method for performing the calculations. Colloquial mainframe and microcomputer programs have been developed in the Department of Building Science, University of Sheffield, to facilitate easy application of the model.

Table 5.17 Sample table of monthly mean slope irradiation from the CEC European Solar Radiation Atlas, Vol. II. This is the upper part of the table. The lower part of the table is illustrated in Table 3.12 in Chapter 3. G_m represents $G_m(\beta,\alpha)$ and D_m represents $D_m(\beta,\alpha)$ in the Atlas.

\+ +

IRELAND VALENTIA

Latitude : 51°56'N Longitude : 10°15'W Altitude : 20 m

Estimated monthly means of daily global radiation G_m and diffuse radiation D_m on inclined planes(1966-75)

Units : $kWh.m^{-2}$

		Jan	Feb	Mar	Apr	May	Jun	Jul	Aug	Sep	Oct	Nov	Dec	Mean
10° South	G_m	0.80	1.70	2.91	4.19	5.23	5.37	4.89	4.33	3.27	1.85	1.02	0.65	3.02
	D_m	0.47	0.84	1.47	2.16	2.71	2.91	2.73	2.33	1.69	0.97	0.57	0.39	1.61
30° South	G_m	1.01	2.10	3.27	4.38	5.17	5.19	4.76	4.40	3.55	2.17	1.28	0.83	3.18
	D_m	0.45	0.82	1.43	2.11	2.63	2.80	2.63	2.27	1.64	0.93	0.54	0.37	1.56
Latitude South	G_m	1.14	2.31	3.32	4.15	4.65	4.55	4.21	4.06	3.49	2.28	1.43	0.95	3.05
(51°56')	D_m	0.40	0.75	1.31	1.93	2.38	2.52	2.36	2.06	1.50	0.84	0.49	0.33	1.41
60° South	G_m	1.15	2.32	3.25	3.96	4.34	4.21	3.91	3.84	3.38	2.26	1.44	0.97	2.92
	D_m	0.38	0.71	1.25	1.84	2.27	2.39	2.23	1.96	1.43	0.80	0.46	0.31	1.34
Vertical South	G_m	1.07	2.07	2.63	2.86	2.89	2.71	2.56	2.67	2.61	1.93	1.32	0.91	2.18
	D_m	0.29	0.57	0.99	1.46	1.78	1.85	1.72	1.53	1.12	0.62	0.36	0.24	1.05
Vertical SE/SW	G_m	0.85	1.69	2.37	2.89	3.15	3.00	2.79	2.76	2.44	1.62	1.05	0.71	2.11
	D_m	0.29	0.57	0.99	1.44	1.76	1.83	1.70	1.51	1.12	0.63	0.36	0.24	1.04
Vertical E/W	G_m	0.47	1.05	1.78	2.54	3.12	3.11	2.82	2.54	1.95	1.08	0.59	0.38	1.79
	D_m	0.29	0.55	0.96	1.41	1.75	1.82	1.69	1.48	1.08	0.61	0.36	0.24	1.02
Vertical NE/NW	G_m	0.28	0.58	1.14	1.85	2.53	2.64	2.36	1.95	1.31	0.66	0.34	0.23	1.33
	D_m	0.28	0.52	0.92	1.35	1.71	1.80	1.67	1.44	1.05	0.59	0.34	0.22	0.99
Vertical North	G_m	0.26	0.50	0.89	1.39	2.01	2.18	1.95	1.51	1.02	0.57	0.32	0.22	1.07
	D_m	0.26	0.50	0.89	1.33	1.71	1.81	1.68	1.42	1.02	0.57	0.32	0.22	0.98
INPUT DATA														
Daily sunshine hours		1.3	2.6	3.6	5.0	5.7	5.4	4.7	4.8	3.9	2.4	1.6	1.1	
Angstrom a + b		0.83	0.85	0.85	0.84	0.86	0.86	0.87	0.84	0.84	0.86	0.84	0.82	0.85
Normalisation factor		1.01	0.98	1.02	1.02	1.03	1.00	0.98	1.00	1.00	0.94	1.00	1.04	

CHAPTER 5 - References

1. Page, J.K. and Rodgers, G.G., (1979), Mathematical basis of programs for computing average hourly irradiance and daily irradiation on sloping surfaces, Volume 1 of final report for Contract 292-77-UK in CEC Solar Energy Programme, Project F, Action 3.2, Internal Report of Department of Building Science, University of Sheffield.

2. Page, J.K., Rodgers, G.G. and Souster, C.G., (1980), Systematic design assessment techniques for solar buildings, presented to the Royal Society Discussion Meeting on Solar Energy, Royal Society, London, Nov. 1978, Phil. Trans. R. Soc. London A295, 379-401.

3. Commission of the European Communities (1984), European Solar Radiation Atlas, Volume 1, 2nd Edition, Global Radiation on Horizontal Surfaces, TUV, Verlag.

4. Krochmann, J., (1979), Calculation method and its comparison with measuring results, Final report for contract 290-77-ESD in CEC Solar Energy programme, Project F, Action 3.2, Internal Report of Institut für Lichttechnik, Berlin.

5. Steinhauser, F., (1934), Die mittlere Trübung der Luft an verschiedenen Orten, Gerlands Beiträge zur Geophysik, Vol.42,pp.110-121.

6. Krochmann, J., (1974), Uber die Sonnenscheinwahrscheinlichkeit in der Bundesrepublik Deutschland Lichttechnik 25, Nr. 10, 11.

7. Aydinli, S., (1981), Uber die Berechnung der zur Verfugung stehenden Solarenergie und des Tageslichtes, Fortschr.-Ber. VDI-Z, Vol. 6, No. 79.

8. Dogniaux, R., (1984), Private communication.

9. Kasten, F. and Czeplak, G., (1980), Solar radiation and terrestrial radiation dependent on the amount and type of cloud, Solar Energy, Vol. 24, 177-189.

10. Page, J.K., (1979), Methods for the estimation of solar radiation on vertical and inclined surfaces, in Solar Energy Conversion, An introductory course, Ed. Dixon, A.E. and Leslie, T.D., Pergamon Press, p.37-99.

11. Kasten, F. and Czeplak, G., (1980), Solar and terrestrial radiation dependent on the amount and type of cloud, Solar Energy, Vol.24,177-189.

12. Aydinli, S., (1980), The calculation of the available solar radiation and daylight, in Proc. Symp. in Daylight, Berlin, 1980, Commission Internationale de l'eclairage, published by J. Krochmann, Institut für Lichttechnik, Berlin.

13. Liu, B.Y.H. and Jordan, R.C., (1960), The inter-relationship and characteristic distribution of direct diffuse and total solar radiation, Solar Energy, Vol. 4, 1-19.

14. Page, J.K., (1964), The estimation of monthly mean values of daily total short wave radiation on vertical and inclined surfaces from sunshine records for latitudes 40°N-40°S, Proc. of UN Conference on New Sources of Energy, Rome, Vol. 4, p.378.

15. Page, J.K. and Rodgers, G.G., ibid., reference 1.

16. Liu, B.Y.H. and Jordan, R.C., (1977), Availability of solar energy for flat plate solar collectors, in Applications of Solar Energy for Heating and Cooling of Buildings, American Society of Heating, Refrigeration and Air Conditioning Engineers, ASHRAE GRP 170 p.V3.

17. Page J.K., (1978), Geographical variations in the climatic factors influencing solar building design, Opening lecture presented to the North East London Polytechnic, UNESCO International Conference, Solar Technology for Building, RIBA, London. 25-29 July, 1977. (Ed. C. Stambolis), RIBA Publications, London, Vol. 1, 38-70.

18. Page, J.K. and Colquhoun, I., (1980), Report on data for overcast days processed from UK Meteorological Office Summaries, 1972-75, BRE Contract Report No. F 3/2/159 (unpublished).

19. Dogniaux, R. and Lemoine, M., (1983), Classification of radiation sites in terms of different indices of atmospheric transparency, in Solar Energy R&D in the European Community, Series F, Volume 2, Solar Radiation Data, (Ed. W. Palz), D. Reidel Publishing Company, p.94-107.

CHAPTER 6

INSTRUCTIONS FOR THE MANUAL CALCULATION OF HOURLY AND DAILY VALUES OF CLEAR DAY AND MONTHLY MEAN SOLAR IRRADIATION ON HORIZONTAL AND INCLINED SURFACES USING THE METHODOLOGY ADOPTED TO PRODUCE THE CEC INCLINED SURFACE RADIATION ATLAS

6.1 INTRODUCTION

These calculations should be carried out using local apparent (solar) time. The computational procedures used for estimating mean monthly values of solar radiation are described first. The modifications in the procedure needed to obtain estimates of maximum monthly values of solar radiation are then indicated in the second section of the Chapter. Conversion procedures to convert from local mean time to local apparent time are described in Appendix 1. A worked example is provided to demonstrate the calculations in detail.

6.2 REQUIRED INPUTS

Essential inputs

1. Station latitude and height in metres. (It is convenient to also record the longitude, though it is not used in the calculations.)

2. Monthly mean daily duration of bright sunshine in hours for each of the months in the year.

3. Ground albedo - the recommended value is 0.2. This value was used in the preparation of the inclined surface tables and maps in the CEC Atlas.

4. Slope inclination β and orientation α of each slope to be considered. The orientation is measured as the bearing of the normal to the surface. The slope inclination is measured from the horizontal plane, i.e. vertical surface = 90°.

Desirable inputs

5. Monthly mean values of the sum of the Angstrom regression coefficients (a + b) and the annual mean value of (a + b). These values are available for over one hundred European sites in the systematic tables provided in the CEC European Solar Radiation Atlas, Vol. II, Inclined Surfaces.

If the inputs discussed above under 5 are not available, one of the procedures discussed in Section 6.4 below must be adopted instead, as it is essential that an adequate description of the atmospheric clarity be provided for the site in question. In the absence of observed data, this may have to be achieved by judgement of atmospheric conditions for the specific site.

6. Monthly mean values of the daily horizontal global irradiation in $Whm^{-2}\ d^{-1}$. If observations are available, the horizontal predictions can be normalized to match the observations. In the absence of such data normalization is not possible. Normalization is not an essential process. However, it was used throughout the CEC European Solar Radiation Atlas, Vol. II, Inclined Surfaces, to produce the published tables.

6.3 NOTATION USED IN THE COMPUTATION OF MONTHLY MEAN VALUES OF SOLAR IRRADIANCE AND DAILY SOLAR IRRADIATION

In reading the procedures below note the meaning of the following key subscripts:

Subscript c	Cloudless day value
Subscript m	Monthly mean value
Subscript b	Overcast day value
Subscript s	Component from sky alone.

If no indication is given of the slope inclination, the irradiance symbol normally refers to the horizontal surface, i.e. $D_m(\beta,\alpha)$ is the monthly mean diffuse irradiance on a plane of inclination β and orientation α, D_m is the monthly mean diffuse irradiance on a horizontal surface. There is one exception to this rule. The direct beam intensity I is expressed normal to the beam and I(0,0) represents the horizontal surface beam irradiance.

6.4 DETAILED COMPUTATIONAL PROCEDURES

The sequence of calculation is important otherwise the right inputs for the next step may not be available at the right moment. Figure 6.1 and Figure 6.2 summarise the procedures to be followed to obtain mean values of hourly irradiance and daily irradiation on horizontal and inclined planes and the loops in the calculation method involved. The steps are numbered consecutively.

PROCEDURES CARRIED OUT AT A GENERAL LEVEL, INDEPENDENT OF TIME OF YEAR

1. CORRECTION FOR STATION HEIGHT

Calculate station height correction to relative air mass p/p_o.

$$p/p_o = 1.0 - z/10000 \qquad \text{dimensionless}$$

where z is station height (metres above sea level).

A better formula is $p/p_o = e^{-z/z_h}$ where z_h =scale height of the atmosphere = 8 Km. The first formula was used in the production of the inclined surface solar radiation Atlas.

GENERAL PROCEDURES FOR INCLINED PLANES

Note: Steps 2 and 3 can usefully be deferred until the inclined surface calculations are commenced.

1. Station height correction p/p_0

LOOP OVER EACH MONTH FOR HORIZONTAL SURFACES

4.1/4.2 Select solar declination δ Table 6.2 and correction to m.s.d. K_d

4.3 Select extraterrestrial irradiance at normal incidence I_{oj} Table 6.3

4.4 Compute air mass 2 Linke Turbidity Factor T_L

4.5 Select astronomical daylength, Table 6.6

4.6 Find times of sunrise and sunset t_r and t_s

4.7 Select 4° SR/SS daylength, Table 6.7 and compute daily relative sunshine duration 4° sunrise basis σ_{4m}

5.1 Establish set of times for hourly calculations considering the
5.2 symmetry of the solar geometry about noon

LOOP OVER EACH HOUR

6.1/6.2 Establish solar geometry for mid point of hour

7 Compute:
7.1 Relative air mass, m
7.2 Rayleigh optical thickness, δ_R
7.3 Linke Turbidity Factor corrected for solar altitude, $T_L(\gamma)$
7.4 Clear sky direct beam irradiance at normal incidence, I_c
7.5 Clear sky direct beam irradiance on a horizontal surface, $I_c(0,0)$

8. Estimate clear sky horizontal surface diffuse irradiance D_c using atmospheric transmittance after absorption alone, Table 6.8, and in addition scattering correction function, Table 6.9.

9. Estimate clear sky hourly global radiation on a horizontal surface G_c.

10. Estimate overcast sky horizontal surface irradiance $G_b = D_b$

11. Select hourly mean beam correction functions, Table 6.10, and estimate the mean direct beam irradiance on horizontal $I_m(0,0)$

12. Estimate the two components of mean sky diffuse irradiance on horizontal D_m

13. Add (11) and (12) to obtain mean global irradiance on horizontal G_m

14. Add to obtain daily beam, diffuse & global irradiation on horizontal surface for clear, mean* and overcast conditions for each month.

* Normalisation of hourly values of irradiance should be carried out at this stage. The normalisation is applied to the hourly values of $I_m(0,0)$, D_m, and G_m.

Figure 6.1 Summary of calculation method for estimation of the hourly components of daily solar radiation on inclined planes for monthly mean conditions using the EC method. Work in solar time.

Stage I Horizontal surface calculations. Reference numbers refer to steps described in text which follows.

LOOP OVER EACH ANGLE OF SLOPE

2. Select overcast sky inclined/horizontal surface conversion ratio f_4 Table 6.1

LOOP OVER EACH ANGLE OF SLOPE AND FOR EACH ALBEDO

3. Find ratio of ground reflected irradiance on an inclined surface to global irradiance on a horizontal surface f_3

LOOP OVER EACH ANGLE OF SLOPE

LOOP OVER EACH MONTH

LOOP OVER EACH HOUR

LOOP OVER EACH ORIENTATION

15.1/15.2 Establish inclined surface solar geometry

16.1 Calculate clear sky beam irradiance on plane $I_c(\beta,\alpha)$

16.2.1 Calculate ratio f_2 for estimating clear sky diffuse irradiance on plane for given solar altitude and wall solar azimuth angle Tables 6.14(A) - (J)

16.2.2 Calculate clear sky diffuse irradiance on plane $D_{cs}(\beta,\alpha)$

16.3 Calculate clear sky ground reflected irradiance on plane $R_c(\beta,\alpha)$

16.4 Add components to obtain clear sky diffuse global irradiance on slope $D_c(\beta,\alpha)$ and $G_c(\beta,\alpha)$

17.1 Calculate from D_b and f_4 hourly overcast sky inclined plane sky diffuse irradiance $D_{bs}(\beta,\alpha)$

17.2 Calculate from D_b and f_3 hourly overcast sky inclined plane ground reflected diffuse irradiance $R_b(\beta,\alpha)$

17.3 Sum 17.1 and 17.2 to obtain global (= diffuse) overcast sky irradiance on the plane $G_b(\beta,\alpha)$ and $D_b(\beta,\alpha)$

LOOP OVER EACH ORIENTATION

18.1 Calculate monthly mean direct beam irradiance $I_m(\beta,\alpha)$

18.2 Calculate monthly mean diffuse sky irradiance $D_{ms}(\beta,\alpha)$

18.3 Calculate monthly mean diffuse irradiance reflected from ground $R_m(\beta,\alpha)$

18.4 Add to get monthly mean overall diffuse irradiance $D_m(\beta,\alpha)$
18.5 and global irradiance $G_m(\beta,\alpha)$

19. Obtain daily integrals of three components of the solar radiation and global radiation by summation of hourly values on each plane month by month

Figure 6.2 Summary of calculation method for estimation of hourly components of daily solar radiation on inclined planes for monthly mean conditions using the EC method. Work in solar time. Stage II Inclined plane calculations. Reference numbers refer to steps described in text which follows.

2. OVERCAST SKY HORIZONTAL TO INCLINED SURFACE CONVERSION RATIO f_4.

Find the ratio of inclined surface overcast sky irradiance to horizontal surface overcast sky irradiance f_4 (dimensionless) for each angle of slope (degrees) using Table 6.1

3. RATIO OF GROUND REFLECTED DIFFUSE IRRADIANCE ON AN INCLINED PLANE TO HORIZONTAL SURFACE GLOBAL IRRADIANCE f_3.

Calculate the ratio f_3 (dimensionless) for each angle of slope (β) and for each ground albedo (ρ_s) being used

$$f_3 = 0.5\ \rho_s(1 - \cos \beta) \qquad \text{dimensionless}$$

where ρ_s is ground albedo

β is angle of inclination of the plane measured from the horizontal plane (Vertical = 90°)

The tables in the CEC European Solar Radiation Atlas, Vol. II, Inclined Surfaces were produced using a value of 0.2 for the ground albedo.

PROCEDURES CARRIED OUT AT A MONTHLY LEVEL TO COMPUTE MONTHLY MEAN TABLES

4. ESTABLISH VALUES FOR MONTHLY BASIC DATA APPLICABLE TO ALL SLOPES FOR EACH MONTH TO BE STUDIED

4.1 Select representative values of monthly mean solar declination δ (degrees) using Table 6.2. Note Table 6.2 also gives the dates associated with these monthly mean values of the solar declination

4.2 Tabulate monthly values of correction to mean solar distance for mean conditions K_d (dimensionless) using Table 6.2.

4.3 Select monthly values of extraterrestrial irradiance at normal incidence I_{oj} (Wm^{-2}) using Table 6.3. Use the column for monthly mean conditions.

4.4 Estimate representative monthly values of the air mass 2 Linke Turbidity Factor T_L.

Follow procedure 4.4.1 or procedure 4.4.2 or procedure 4.4.3 according to latitude

4.4.1 Northern Hemisphere above 25°N - Climates with a well defined winter and summer.

4.4.1.1 If station is one for which numerical tables exist in the CEC European Solar Radiation Atlas, Volume I, select monthly values of Angstrom (a + b) from the published tables. Then use Table 6.4 to estimate the air mass 2 Linke Turbidity Factor.

Table 6.1 Overcast sky inclined surface to horizontal surface diffuse irradiance ratio for different angles of slope, f_4.

Angle of slope β	Ratio f_4	Angle of slope β	Ratio f_4	Angle of slope β	Ratio f_4
2	1.000	32	0.891	62	0.652
4	0.999	34	0.878	64	0.633
6	0.996	36	0.865	66	0.615
8	0.993	38	0.851	68	0.597
10	0.989	40	0.836	70	0.578
12	0.984	42	0.821	72	0.560
14	0.978	44	0.806	74	0.541
16	0.971	46	0.790	76	0.523
18	0.964	48	0.774	78	0.504
20	0.956	50	0.757	80	0.486
22	0.947	52	0.740	82	0.468
24	0.937	54	0.723	84	0.450
26	0.927	56	0.705	86	0.432
28	0.915	58	0.688	88	0.414
30	0.904	60	0.670	90	0.396

Table 6.2 Recommended values of solar declination (δ) with representative dates and associated day numbers (J) for use in the calculation of monthly mean and mean monthly maximum levels of solar radiation together with values of the correction factor to mean solar distance K_d.

	Computation of monthly means of global radiation				Computation of mean monthly maxima of global radiation							
					Northern hemisphere				Southern hemisphere			
Month	Day	Day No. (365 day year)	δ* deg.	K_d	Day	Day No. (366 day year)	δ** deg.	K_d	Day	Day No. (366 day year)	δ** deg.	K_d
Jan	17	17	-20.71	1.032	29	29	-18.16	1.030	4	4	22.80	1.033
Feb	15	46	-12.81	1.025	26	57	- 9.04	1.020	4	35	16.48	1.028
Mar	16	75	- 1.80	1.011	29	89	3.43	1.003	4	64	6.39	1.017
Apr	15	105	9.77	0.994	28	119	14.19	0.986	4	95	-5.74	1.000
May	15	135	18.83	0.978	29	150	21.64	0.973	4	125	-16.00	0.983
Jun	11	162	23.07	0.969	21	173	23.43	0.967	4	156	-22.45	0.971
Jul	17	198	21.16	0.967	4	186	22.87	0.967	29	211	-18.73	0.970
Aug	16	228	13.65	0.975	4	217	17.22	0.971	29	242	- 9.73	0.981
Sep	16	259	2.89	0.990	4	248	7.15	0.984	28	272	2.06	0.997
Oct	16	289	-8.72	1.007	4	278	-4.39	1.001	29	303	13.50	1.015
Nov	15	319	-18.37	1.022	4	309	-15.42	1.018	28	333	21.32	1.028
Dec	11	345	-22.99	1.031	4	339	-22.25	1.029	22	357	23.46	1.033

* The values of monthly mean solar declination are the average of the individual daily values. For use in the southern hemisphere the sign should be reversed. The day is the day on which this declination occurs.

** The representative values of solar declination used to estimate the mean monthly maxima of global radiation were obtained using Dogniaux's Algorithm which is based on a 366 day year, adopting the representative dates shown in the Table.

Table 6.3 Values of I_{oj} for the monthly mean representative dates and monthly mean maximum representative days. Units: Wm^{-2}.

Month	Day	Monthly mean	Monthly maximum estimates N. Hemisphere		Monthly maximum estimates S. Hemisphere	
		I_{oj}	Day	I_{oj}	Day	I_{oj}
Jan	17	1411	29	1408	4	1413
Feb	15	1401	26	1394	4	1406
Mar	16	1382	29	1372	4	1390
Apr	15	1358	28	1349	4	1367
May	15	1337	29	1330	4	1345
Jun	11	1325	21	1323	4	1327
Jul	17	1322	4	1321	29	1325
Aug	16	1333	4	1327	29	1341
Sep	16	1353	4	1345	28	1362
Oct	16	1377	4	1367	29	1386
Nov	15	1397	4	1390	28	1404
Dec	11	1409	4	1407	22	1412

4.4.1.2 If station has an available observed record of daily global radiation and also an observed record of daily sunshine hours and is not in Atlas, first correct global radiation to WRR Scale and then derive Angstrom regression equations on a month by month basis using the daily observed data, assuming a solar constant of 1367 Wm^{-2}. It is essential that the values of a and b in the regression equations are derived on a monthly basis, as monthly mean values of Angstrom (a + b) are needed. Then use Table 6.4 to estimate the air mass 2 Linke Turbidity Factor.

4.4.1.3 If station is in area covered geographically by CEC Solar Radiation Atlas, Volume I, 2nd Edition, and procedures 4.3.1.1 and 4.3.1.2 cannot be followed then

either

1. estimate monthly values of Angstrom (a + b) by comparison with tabulated data for nearby stations with the same broad atmospheric characteristics and use these values of Angstrom (a + b) to estimate air mass 2 Linke Turbidity Factors, month by month

or

2. use Table 6.5 to estimate suitable monthly representative values of air mass 2 Linke Turbidity Factors directly.

4.4.1.4 If station lies outside area covered geographically by CEC European Solar Radiation Atlas, Vol I, 2nd Edition.

either

1. Adopt the method proposed by Dogniaux and Lemoine, Solar Energy R&D in the European Community, Series F, Vol. 2, Solar Radiation Data, D. Reidel Publishing Company, p.94-107, to estimate the Linke Turbidity Factor. Refer Appendix 2 for summary.

or

2. use Table 6.5 to estimate suitable monthly representative air mass 2 Linke Turbidity Factors taking account of local atmospheric conditions to make amendments if appropriate

or

3. use the WMO related method briefly outlined in 4.4.2 and given in detail in Appendix 2.

Table 6.4 Monthly values of the constant f_m used in the estimation of representative air mass 2 Linke Turbidity Factors T_L for monthly mean conditions using monthly values of Angstrom (a + b) derived from daily observations on a monthly basis

$$T_L = f_m - f_m(a + b)$$

based on observed global and diffuse radiation data for 13 European stations.

Northern Hemisphere above 25°N	Jan	Feb	Mar	Apr	May	Jun	Jul	Aug	Sep	Oct	Nov	Dec
f_m	10.66	13.23	15.42	17.05	18.55	16.94	18.16	16.96	14.94	12.85	11.34	9.49
Southern Hemisphere above 25°S*	Jul	Aug	Sep	Oct	Nov	Dec	Jan	Feb	Mar	Apr	May	June

* Values for the Southern Hemisphere are tentative.

Table 6.5 Suggested monthly mean values of air mass 2 Linke Turbidity Factor for use in the absence of observed information on values of Angstrom a + b for sites above latitude 25°N and above latitude 25°S.

Latitude 25°N and above	Jan	Feb	Mar	Apr	May	Jun	Jul	Aug	Sep	Oct	Nov	Dec
Very clear atmosphere	1.8	1.9	2.0	2.0	2.2	2.3	2.8	2.5	2.2	2.1	2.0	1.8
Mountain - very clear atmosphere	2.2	2.3	2.4	2.4	2.4	2.5	3.0	2.7	2.5	2.4	2.3	2.2
Clear maritime atmosphere	2.4	2.5	2.7	3.4	3.4	3.4	3.9	3.7	2.9	2.7	2.5	2.4
Urban atmosphere	3.0	3.0	3.0	3.7	3.7	3.8	4.8	4.1	3.2	3.0	3.0	3.0
Very large city or industrial atmosphere	4.1	4.1	3.9	4.6	5.0	5.0	6.0	5.5	4.5	3.6	3.8	3.9
Latitude 25°S and above*	Jul	Aug	Sep	Oct	Nov	Dec	Jan	Feb	Mar	Apr	May	Jun

* Results for the Southern Hemisphere are tentative.

4.4.2 Latitudes between 25°N and 25°S

Seasons tend to be divided into wet seasons and dry seasons. Within continental land masses, there is usually a lot of dust during the dry season, and the turbidities may become very high. The rain in wet seasons tends to lay the dust and clear the atmosphere and so the clarity may vary very greatly according to season. Island climates in the oceans enjoy a clearer atmosphere as far as dust is concerned, but the high humidity often leads on to considerable cloudiness. Appendix 2 includes an outline of a method for estimating T_L based on WMO experience. Further tests of the method are desirable to verify the procedures for this belt.

4.4.3 Southern Hemisphere - Climates with a well defined winter and summer usually south of latitude 25°S

Proceed as above, but in using either Table 6.4 or Table 6.5, use indicated transformation of months to allow for the fact Southern Hemisphere winter is centred on June/July.

4.5 Find the monthly mean astronomical daylength S_o (hours) either using Table 6.6 (a), (b) or (c). Adopt linear interpolation between degrees of latitude. Alternatively proceed by computation (Refer 4.7 below).

4.6 Calculate the times of sunrise (t_r) and sunset (t_s) in local apparent time from the astronomical daylength.

$$t_r = 12 - S_o/2 \quad \text{hours}$$

$$t_s = 12 + S_o/2 \quad \text{hours}$$

4.7 Find the monthly mean daily relative sunshine duration based on 4 degree angle of sunrise and sunset σ_{4m}

Calculate monthly mean daily relative sunshine duration on a four degree sunrise/sunset basis σ_{4m} as follows

$$\sigma_{4m} = S_m/S_{04} \quad \text{dimensionless}$$

where S_m = monthly mean observed hours of bright sunshine

S_{04} = the daylength in hours calculated on the assumption that the sunshine recorder starts to burn when the sun has an elevation above 4°

S_{04} may be calculated as follows:

$$S_{04} = \frac{1}{7.5} \cos^{-1}[(\sin 4° - \sin \phi \sin \delta)/(\cos \phi \cos \delta)] \quad \text{hours}$$

where ϕ is the latitude (degrees)

δ is the declination on the representative date of the month (degrees). (Refer stage 4.1)

Table 6.6a Astronomical daylength S_0 (hours), latitudes 26°-75° North.

Latitude	Jan 17	Feb 15	Mar 16	Apr 15	May 15	Jun 11	Jul 17	Aug 16	Sep 16	Oct 16	Nov 16	Dec 11
75 N	0.00	4.26	11.10	17.33	24.00	24.00	24.00	20.67	13.45	7.34	0.00	0.00
74 N	0.00	5.00	11.16	16.92	24.00	24.00	24.00	19.72	13.35	7.69	0.00	0.00
73 N	0.00	5.59	11.21	16.57	24.00	24.00	24.00	19.01	13.27	7.99	0.00	0.00
72 N	0.00	6.08	11.26	16.27	24.00	24.00	24.00	18.45	13.19	8.24	0.00	0.00
71 N	0.00	6.49	11.30	16.00	22.94	24.00	24.00	17.98	13.12	8.47	2.04	0.00
70 N	0.00	6.85	11.34	15.76	21.27	24.00	24.00	17.58	13.06	8.68	3.22	0.00
69 N	1.33	7.16	11.37	15.55	20.36	24.00	24.00	17.23	13.01	8.86	4.01	0.00
68 N	2.75	7.43	11.41	15.36	19.68	24.00	21.78	16.93	12.96	9.03	4.63	0.00
67 N	3.61	7.68	11.43	15.19	19.13	24.00	20.77	16.65	12.91	9.18	5.14	0.24
66 N	4.25	7.91	11.46	15.03	18.67	21.74	20.05	16.41	12.87	9.31	5.57	2.35
65 N	4.78	8.11	11.48	14.89	18.27	20.80	19.48	16.18	12.83	9.44	5.95	3.27
64 N	5.22	8.29	11.51	14.76	17.91	20.11	19.00	15.98	12.79	9.56	6.28	3.94
63 N	5.61	8.47	11.53	14.63	17.60	19.56	18.59	15.80	12.76	9.66	6.58	4.48
62 N	5.96	8.62	11.55	14.52	17.32	19.10	18.23	15.62	12.73	9.76	6.85	4.94
61 N	6.27	8.77	11.57	14.41	17.06	18.69	17.91	15.46	12.70	9.86	7.09	5.34
60 N	6.55	8.91	11.58	14.31	16.83	18.34	17.61	15.32	12.67	9.95	7.32	5.69
59 N	6.80	9.04	11.60	14.22	16.61	18.02	17.35	15.18	12.64	10.03	7.53	6.01
58 N	7.04	9.15	11.62	14.13	16.41	17.73	17.10	15.05	12.62	10.11	7.72	6.30
57 N	7.25	9.27	11.63	14.05	16.22	17.46	16.88	14.93	12.59	10.18	7.90	6.56
56 N	7.45	9.37	11.64	13.97	16.05	17.22	16.67	14.81	12.57	10.25	8.07	6.80
55 N	7.64	9.47	11.66	13.90	15.89	17.00	16.47	14.71	12.55	10.31	8.23	7.03
54 N	7.82	9.57	11.67	13.83	15.73	16.79	16.29	14.60	12.53	10.38	8.37	7.24
53 N	7.98	9.66	11.68	13.76	15.59	16.59	16.12	14.51	12.51	10.43	8.51	7.43
52 N	8.14	9.74	11.69	13.70	15.45	16.40	15.96	14.41	12.49	10.49	8.65	7.61
51 N	8.29	9.83	11.70	13.64	15.32	16.23	15.81	14.33	12.48	10.54	8.77	7.79
50 N	8.43	9.90	11.71	13.58	15.20	16.07	15.66	14.24	12.46	10.60	8.89	7.95
49 N	8.56	9.98	11.72	13.52	15.08	15.91	15.53	14.16	12.44	10.64	9.01	8.10
48 N	8.69	10.05	11.73	13.47	14.97	15.76	15.39	14.09	12.43	10.69	9.11	8.25
47 N	8.81	10.12	11.74	13.42	14.86	15.62	15.27	14.01	12.41	10.74	9.22	8.39
46 N	8.93	10.18	11.75	13.37	14.76	15.49	15.15	13.94	12.40	10.78	9.32	8.53
45 N	9.04	10.25	11.76	13.32	14.66	15.36	15.04	13.87	12.39	10.82	9.41	8.65
44 N	9.14	10.31	11.77	13.28	14.56	15.24	14.93	13.81	12.37	10.86	9.51	8.78
43 N	9.25	10.37	11.78	13.23	14.47	15.12	14.82	13.75	12.36	10.90	9.59	8.89
42 N	9.35	10.42	11.78	13.19	14.38	15.01	14.72	13.68	12.35	10.94	9.68	9.01
41 N	9.44	10.48	11.79	13.15	14.30	14.90	14.62	13.62	12.34	10.98	9.76	9.11
40 N	9.53	10.53	11.80	13.11	14.22	14.79	14.53	13.57	12.32	11.01	9.84	9.22
39 N	9.62	10.59	11.81	13.07	14.14	14.69	14.44	13.51	12.31	11.05	9.92	9.32
38 N	9.71	10.64	11.81	13.03	14.06	14.59	14.35	13.46	12.30	11.08	10.00	9.42
37 N	9.79	10.68	11.82	12.99	13.99	14.50	14.26	13.41	12.29	11.12	10.07	9.51
36 N	9.87	10.73	11.83	12.96	13.91	14.40	14.18	13.35	12.28	11.15	10.14	9.61
35 N	9.95	10.78	11.83	12.92	13.84	14.31	14.10	13.31	12.27	11.18	10.21	9.70
34 N	10.03	10.82	11.84	12.89	13.77	14.23	14.02	13.26	12.26	11.21	10.27	9.78
33 N	10.10	10.87	11.84	12.86	13.71	14.14	13.94	13.21	12.25	11.24	10.34	9.87
32 N	10.18	10.91	11.85	12.82	13.64	14.06	13.87	13.16	12.24	11.27	10.40	9.95
31 N	10.25	10.95	11.86	12.79	13.58	13.98	13.79	13.12	12.23	11.29	10.47	10.03
30 N	10.32	10.99	11.86	12.76	13.51	13.90	13.72	13.07	12.22	11.32	10.53	10.11
29 N	10.39	11.03	11.87	12.73	13.45	13.82	13.65	13.03	12.21	11.35	10.59	10.19
28 N	10.45	11.07	11.87	12.70	13.39	13.75	13.58	12.99	12.21	11.38	10.64	10.26
27 N	10.52	11.11	11.88	12.67	13.33	13.67	13.52	12.95	12.20	11.40	10.70	10.34
26 N	10.58	11.15	11.88	12.64	13.28	13.60	13.45	12.91	12.19	11.43	10.76	10.41

Table 6.6b Astronomical daylength S_0 (hours), latitudes 25° North - 25° South.

Latitude	Jan 17	Feb 15	Mar 16	Apr 15	May 15	Jun 11	Jul 17	Aug 16	Sep 16	Oct 16	Nov 16	Dec 11
25 N	10.65	11.19	11.89	12.61	13.22	13.53	13.39	12.87	12.18	11.45	10.81	10.48
24 N	10.71	11.23	11.89	12.59	13.16	13.46	13.32	12.83	12.17	11.48	10.87	10.55
23 N	10.77	11.26	11.90	12.56	13.11	13.39	13.26	12.79	12.16	11.50	10.92	10.62
22 N	10.83	11.30	11.90	12.53	13.06	13.32	13.20	12.75	12.16	11.53	10.97	10.68
21 N	10.89	11.33	11.91	12.51	13.00	13.25	13.14	12.71	12.15	11.55	11.02	10.75
20 N	10.95	11.37	11.91	12.48	12.95	13.19	13.08	12.68	12.14	11.57	11.07	10.82
19 N	11.00	11.40	11.92	12.45	12.90	13.12	13.02	12.64	12.13	11.60	11.12	10.88
18 N	11.06	11.44	11.92	12.43	12.85	13.06	12.96	12.60	12.13	11.62	11.17	10.94
17 N	11.11	11.47	11.93	12.40	12.80	13.00	12.91	12.57	12.12	11.64	11.22	11.01
16 N	11.17	11.50	11.93	12.38	12.75	12.94	12.85	12.53	12.11	11.66	11.27	11.07
15 N	11.22	11.53	11.94	12.35	12.70	12.87	12.79	12.50	12.10	11.69	11.32	11.13
14 N	11.28	11.57	11.94	12.33	12.65	12.81	12.74	12.46	12.10	11.71	11.37	11.19
13 N	11.33	11.60	11.94	12.30	12.60	12.75	12.68	12.43	12.09	11.73	11.41	11.25
12 N	11.39	11.63	11.95	12.28	12.55	12.69	12.63	12.39	12.08	11.75	11.46	11.31
11 N	11.44	11.66	11.95	12.26	12.51	12.63	12.58	12.36	12.07	11.77	11.51	11.37
10 N	11.49	11.69	11.96	12.23	12.46	12.57	12.52	12.33	12.07	11.79	11.55	11.43
9 N	11.54	11.72	11.96	12.21	12.41	12.52	12.47	12.29	12.06	11.81	11.60	11.49
8 N	11.59	11.76	11.97	12.18	12.37	12.46	12.42	12.26	12.05	11.84	11.64	11.54
7 N	11.65	11.79	11.97	12.16	12.32	12.40	12.36	12.23	12.05	11.86	11.69	11.60
6 N	11.70	11.82	11.97	12.14	12.27	12.34	12.31	12.20	12.04	11.88	11.73	11.66
5 N	11.75	11.85	11.98	12.12	12.23	12.28	12.26	12.16	12.03	11.90	11.78	11.72
4 N	11.80	11.88	11.98	12.09	12.18	12.23	12.21	12.13	12.03	11.92	11.82	11.77
3 N	11.85	11.91	11.99	12.07	12.14	12.17	12.15	12.10	12.02	11.94	11.87	11.83
2 N	11.90	11.94	11.99	12.05	12.09	12.11	12.10	12.06	12.01	11.96	11.91	11.89
1 N	11.95	11.97	12.00	12.02	12.05	12.06	12.05	12.03	12.01	11.98	11.96	11.94
0	12.00	12.00	12.00	12.00	12.00	12.00	12.00	12.00	12.00	12.00	12.00	12.00
1 S	12.05	12.03	12.00	11.98	11.95	11.94	11.95	11.97	11.99	12.02	12.04	12.06
2 S	12.10	12.06	12.01	11.95	11.91	11.89	11.90	11.94	11.99	12.04	12.09	12.11
3 S	12.15	12.09	12.01	11.93	11.86	11.83	11.85	11.90	11.98	12.06	12.13	12.17
4 S	12.20	12.12	12.02	11.91	11.82	11.77	11.79	11.87	11.97	12.08	12.18	12.23
5 S	12.25	12.15	12.02	11.88	11.77	11.72	11.74	11.84	11.97	12.10	12.22	12.28
6 S	12.30	12.18	12.03	11.86	11.73	11.66	11.69	11.80	11.96	12.12	12.27	12.34
7 S	12.35	12.21	12.03	11.84	11.68	11.60	11.64	11.77	11.95	12.14	12.31	12.40
8 S	12.41	12.24	12.03	11.82	11.63	11.54	11.58	11.74	11.95	12.16	12.36	12.46
9 S	12.46	12.28	12.04	11.79	11.59	11.48	11.53	11.71	11.94	12.19	12.40	12.51
10 S	12.51	12.31	12.04	11.77	11.54	11.43	11.48	11.67	11.93	12.21	12.45	12.57
11 S	12.56	12.34	12.05	11.74	11.49	11.37	11.42	11.64	11.93	12.23	12.49	12.63
12 S	12.61	12.37	12.05	11.72	11.45	11.31	11.37	11.61	11.92	12.25	12.54	12.69
13 S	12.67	12.40	12.06	11.70	11.40	11.25	11.32	11.57	11.91	12.27	12.59	12.75
14 S	12.72	12.43	12.06	11.67	11.35	11.19	11.26	11.54	11.90	12.29	12.63	12.81
15 S	12.78	12.47	12.06	11.65	11.30	11.13	11.21	11.50	11.90	12.31	12.68	12.87
16 S	12.83	12.50	12.07	11.62	11.25	11.06	11.15	11.47	11.89	12.34	12.73	12.93
17 S	12.88	12.53	12.07	11.60	11.20	11.00	11.09	11.43	11.88	12.36	12.78	12.99
18 S	12.94	12.56	12.08	11.57	11.15	10.94	11.04	11.40	11.87	12.38	12.83	13.06
19 S	13.00	12.60	12.08	11.55	11.10	10.88	10.98	11.36	11.87	12.40	12.88	13.12
20 S	13.05	12.63	12.09	11.52	11.05	10.81	10.92	11.32	11.86	12.43	12.93	13.18
21 S	13.11	12.67	12.09	11.49	11.00	10.75	10.86	11.29	11.85	12.45	12.98	13.25
22 S	13.17	12.70	12.10	11.47	10.94	10.68	10.80	11.25	11.84	12.47	13.03	13.32
23 S	13.23	12.74	12.10	11.44	10.89	10.61	10.74	11.21	11.84	12.50	13.08	13.38
24 S	13.29	12.77	12.11	11.41	10.84	10.54	10.68	11.17	11.83	12.52	13.13	13.45
25 S	13.35	12.81	12.11	11.39	10.78	10.47	10.61	11.13	11.82	12.55	13.19	13.52

Table 6.6c Astronomical daylength S_0 (hours), latitudes 26° - 75° South.

Latitude	Date Jan 17	Feb 15	Mar 16	Apr 15	May 15	Jun 11	Jul 17	Aug 16	Sep 16	Oct 16	Nov 16	Dec 11
26 S	13.42	12.85	12.12	11.36	10.72	10.40	10.55	11.09	11.81	12.57	13.24	13.59
27 S	13.48	12.89	12.12	11.33	10.67	10.33	10.48	11.05	11.80	12.60	13.30	13.66
28 S	13.55	12.93	12.13	11.30	10.61	10.25	10.42	11.01	11.79	12.62	13.36	13.74
29 S	13.61	12.97	12.13	11.27	10.55	10.18	10.35	10.97	11.79	12.65	13.41	13.81
30 S	13.68	13.01	12.14	11.24	10.49	10.10	10.28	10.93	11.78	12.68	13.47	13.89
31 S	13.75	13.05	12.14	11.21	10.42	10.02	10.21	10.88	11.77	12.71	13.53	13.97
32 S	13.82	13.09	12.15	11.18	10.36	9.94	10.13	10.84	11.76	12.73	13.60	14.05
33 S	13.90	13.13	12.16	11.14	10.29	9.86	10.06	10.79	11.75	12.76	13.66	14.13
34 S	13.97	13.18	12.16	11.11	10.23	9.77	9.98	10.74	11.74	12.79	13.73	14.22
35 S	14.05	13.22	12.17	11.08	10.16	9.69	9.90	10.69	11.73	12.82	13.79	14.30
36 S	14.13	13.27	12.17	11.04	10.09	9.60	9.82	10.64	11.72	12.85	13.86	14.39
37 S	14.21	13.32	12.18	11.01	10.01	9.50	9.74	10.59	11.71	12.88	13.93	14.49
38 S	14.29	13.36	12.19	10.97	9.94	9.41	9.65	10.54	11.70	12.92	14.00	14.58
39 S	14.38	13.41	12.19	10.93	9.86	9.31	9.56	10.49	11.69	12.95	14.08	14.68
40 S	14.47	13.47	12.20	10.89	9.78	9.21	9.47	10.43	11.68	12.99	14.16	14.78
41 S	14.56	13.52	12.21	10.85	9.70	9.10	9.38	10.38	11.66	13.02	14.24	14.89
42 S	14.65	13.58	12.22	10.81	9.62	8.99	9.28	10.32	11.65	13.06	14.32	14.99
43 S	14.75	13.63	12.22	10.77	9.53	8.88	9.18	10.25	11.64	13.10	14.41	15.11
44 S	14.86	13.69	12.23	10.72	9.44	8.76	9.07	10.19	11.63	13.14	14.49	15.22
45 S	14.96	13.75	12.24	10.68	9.34	8.64	8.96	10.13	11.61	13.18	14.59	15.35
46 S	15.07	13.82	12.25	10.63	9.24	8.51	8.85	10.06	11.60	13.22	14.68	15.47
47 S	15.19	13.88	12.26	10.58	9.14	8.38	8.73	9.99	11.59	13.26	14.78	15.61
48 S	15.31	13.95	12.27	10.53	9.03	8.24	8.61	9.91	11.57	13.31	14.89	15.75
49 S	15.44	14.02	12.28	10.48	8.92	8.09	8.47	9.84	11.56	13.36	14.99	15.90
50 S	15.57	14.10	12.29	10.42	8.80	7.93	8.34	9.76	11.54	13.40	15.11	16.05
51 S	15.71	14.17	12.30	10.36	8.68	7.77	8.19	9.67	11.52	13.46	15.23	16.21
52 S	15.86	14.26	12.31	10.30	8.55	7.60	8.04	9.59	11.51	13.51	15.35	16.39
53 S	16.02	14.34	12.32	10.24	8.41	7.41	7.88	9.49	11.49	13.57	15.49	16.57
54 S	16.18	14.43	12.33	10.17	8.27	7.21	7.71	9.40	11.47	13.62	15.63	16.76
55 S	16.36	14.53	12.34	10.10	8.11	7.00	7.53	9.29	11.45	13.69	15.77	16.97
56 S	16.55	14.63	12.36	10.03	7.95	6.78	7.33	9.19	11.43	13.75	15.93	17.20
57 S	16.75	14.73	12.37	9.95	7.78	6.54	7.12	9.07	11.41	13.82	16.10	17.44
58 S	16.96	14.85	12.38	9.87	7.59	6.27	6.90	8.95	11.38	13.89	16.28	17.70
59 S	17.20	14.96	12.40	9.78	7.39	5.98	6.65	8.82	11.36	13.97	16.47	17.99
60 S	17.45	15.09	12.42	9.69	7.17	5.66	6.39	8.68	11.33	14.05	16.68	18.31
61 S	17.73	15.23	12.43	9.59	6.94	5.31	6.09	8.54	11.30	14.14	16.91	18.66
62 S	18.04	15.38	12.45	9.48	6.68	4.90	5.77	8.38	11.27	14.24	17.15	19.06
63 S	18.39	15.53	12.47	9.37	6.40	4.44	5.41	8.20	11.24	14.34	17.42	19.52
64 S	18.78	15.70	12.49	9.24	6.09	3.89	5.00	8.02	11.21	14.44	17.72	20.06
65 S	19.22	15.89	12.52	9.11	5.73	3.20	4.52	7.82	11.17	14.56	18.05	20.73
66 S	19.75	16.09	12.54	8.97	5.33	2.26	3.95	7.59	11.13	14.69	18.43	21.65
67 S	20.39	16.32	12.57	8.81	4.87	0.00	3.23	7.35	11.09	14.82	18.86	23.76
68 S	21.25	16.57	12.59	8.64	4.32	0.00	2.22	7.07	11.04	14.97	19.37	24.00
69 S	22.67	16.84	12.63	8.45	3.64	0.00	0.00	6.77	10.99	15.14	19.99	24.00
70 S	24.00	17.15	12.66	8.24	2.73	0.00	0.00	6.42	10.94	15.32	20.78	24.00
71 S	24.00	17.51	12.70	8.00	1.06	0.00	0.00	6.02	10.88	15.53	21.96	24.00
72 S	24.00	17.92	12.74	7.73	0.00	0.00	0.00	5.55	10.81	15.76	24.00	24.00
73 S	24.00	18.41	12.79	7.43	0.00	0.00	0.00	4.99	10.73	16.01	24.00	24.00
74 S	24.00	19.00	12.84	7.08	0.00	0.00	0.00	4.28	10.65	16.31	24.00	24.00
75 S	24.00	19.74	12.90	6.67	0.00	0.00	0.00	3.33	10.55	16.66	24.00	24.00

S_0 is calculated substituting 0° for 4° in the above expression,

i.e. as $S_0 = \frac{1}{7.5} \cos^{-1} (- \tan \phi \tan \delta)$.

(The inverse cosine expression in brackets must be evaluated in degrees. It represents half the daylength as an angle (90° = 6 hours). Thus doubling to obtain the daylength and dividing by 15° to obtain hours, the denominator becomes 7.5.)

Table 6.7(a) gives values of S_{04} for the representative dates in the month for monthly mean conditions for latitudes 26°N to latitude 75°N. It is thus simple to estimate σ_{4m} from S_m using latitudinal interpolation of the monthly values of S_{04}, the denominator. Table 6.7(b) gives the corresponding values of S_{04} for the region between 25°S and 25°N and Table 6.7(c) the values between 25°S and 55°S.

PROCEDURES CARRIED OUT AT AN HOURLY LEVEL FOR EACH MONTH

5. ESTABLISH COMPUTATIONAL TIMES OF DAY

5.1 Establish a set of times for hourly predictions. These are best set on each half hour between sunrise and sunset. For example, if the sun rises at 0815 hours and sets at 1545 hours, predictions are made at 0830, 0930, 1030, 1130, 1230, 1330, 1430 and 1530 hours.

5.2 Consider the symmetry of the solar geometry about solar noon. The predicted hourly irradiances on horizontal surfaces and on true north and south facing inclined surfaces are symmetrical about solar noon. For example the predictions at 1030 hours, which is one and a half hours before solar noon, will be equal to the predictions at 1330 hours, which is one and a half hours after solar noon. Thus hourly calculations need to be carried out for only half the day, the remaining half being an exact mirror image about noon. A similar principle applies to cases where predictions are made for two inclined surfaces which are arranged symmetrically about the north-south meridian. The hourly predictions for the two surfaces will be identical but will be mirrored about solar noon and will thus be in reverse time order. For example, the value calculated for an east facing surface at 1030 hours will be equal to the value calculated for a west surface at 1330 hours and so on.

Taking advantage of this symmetry can save large amounts of computation time.

6. ESTABLISH HOURLY SOLAR GEOMETRY

6.1 Calculate hourly solar altitude γ

$$\gamma = \sin^{-1}(\sin \phi \sin \delta + \cos \phi \cos \delta \cos(15°(t - 12)))$$

where ϕ is latitude (degrees)

δ is solar declination (degrees) (refer step 4.1)

t is local apparent time (hours).

Table 6.7a 4° sunrise/sunset daylength S_{04} (hours), latitudes 26° - 75° North.

Lati-tude	Jan 17	Feb 15	Mar 16	Apr 15	May 15	Jun 11	Jul 17	Aug 16	Sep 16	Oct 16	Nov 16	Dec 11
75 N	0.00	0.00	8.96	14.89	22.81	24.00	24.00	17.20	11.38	4.31	0.00	0.00
74 N	0.00	0.00	9.16	14.68	20.96	24.00	24.00	16.79	11.41	5.03	0.00	0.00
73 N	0.00	1.16	9.34	14.50	19.96	24.00	24.00	16.44	11.44	5.60	0.00	0.00
72 N	0.00	2.85	9.49	14.33	19.23	24.00	21.56	16.13	11.46	6.07	0.00	0.00
71 N	0.00	3.78	9.63	14.19	18.64	24.00	20.46	15.87	11.48	6.47	0.00	0.00
70 N	0.00	4.47	9.75	14.06	18.16	21.54	19.69	15.63	11.50	6.82	0.00	0.00
69 N	0.00	5.02	9.86	13.94	17.74	20.52	19.08	15.42	11.52	7.12	0.00	0.00
68 N	0.00	5.48	9.96	13.83	17.38	19.79	18.58	15.22	11.53	7.39	0.00	0.00
67 N	0.00	5.87	10.05	13.73	17.06	19.20	18.15	15.05	11.54	7.62	1.86	0.00
66 N	0.00	6.22	10.13	13.64	16.77	18.72	17.77	14.89	11.55	7.84	2.95	0.00
65 N	1.22	6.52	10.21	13.55	16.51	18.30	17.44	14.74	11.56	8.04	3.68	0.00
64 N	2.54	6.80	10.28	13.47	16.28	17.93	17.14	14.60	11.57	8.21	4.26	0.00
63 N	3.33	7.05	10.34	13.40	16.06	17.60	16.87	14.48	11.58	8.38	4.74	0.22
62 N	3.94	7.27	10.40	13.33	15.86	17.30	16.62	14.36	11.59	8.53	5.15	2.19
61 N	4.44	7.48	10.46	13.26	15.68	17.03	16.39	14.25	11.59	8.67	5.51	3.04
60 N	4.87	7.67	10.51	13.20	15.51	16.78	16.19	14.14	11.60	8.80	5.84	3.68
59 N	5.24	7.85	10.56	13.14	15.35	16.56	15.99	14.05	11.61	8.92	6.13	4.19
58 N	5.57	8.01	10.60	13.09	15.20	16.34	15.81	13.96	11.61	9.03	6.39	4.63
57 N	5.87	8.16	10.64	13.04	15.06	16.15	15.64	13.87	11.61	9.14	6.63	5.01
56 N	6.14	8.31	10.68	12.99	14.93	15.96	15.48	13.79	11.62	9.24	6.85	5.35
55 N	6.39	8.44	10.72	12.94	14.80	15.79	15.33	13.71	11.62	9.33	7.06	5.66
54 N	6.62	8.56	10.76	12.89	14.68	15.63	15.19	13.63	11.62	9.42	7.25	5.94
53 N	6.84	8.68	10.79	12.85	14.57	15.47	15.06	13.56	11.63	9.51	7.43	6.19
52 N	7.04	8.80	10.82	12.81	14.46	15.33	14.93	13.49	11.63	9.58	7.60	6.43
51 N	7.22	8.90	10.85	12.77	14.36	15.19	14.81	13.43	11.63	9.66	7.76	6.65
50 N	7.40	9.00	10.88	12.73	14.26	15.06	14.69	13.37	11.63	9.73	7.91	6.86
49 N	7.56	9.10	10.91	12.69	14.17	14.93	14.58	13.30	11.63	9.80	8.05	7.05
48 N	7.72	9.19	10.93	12.65	14.08	14.81	14.47	13.25	11.63	9.87	8.19	7.23
47 N	7.87	9.28	10.96	12.62	13.99	14.70	14.37	13.19	11.63	9.93	8.31	7.40
46 N	8.01	9.36	10.98	12.58	13.91	14.58	14.27	13.14	11.63	9.99	8.44	7.57
45 N	8.14	9.44	11.00	12.55	13.83	14.48	14.18	13.08	11.63	10.04	8.55	7.72
44 N	8.27	9.52	11.02	12.52	13.75	14.37	14.09	13.03	11.63	10.10	8.66	7.87
43 N	8.40	9.59	11.04	12.49	13.67	14.28	14.00	12.98	11.63	10.15	8.77	8.01
42 N	8.51	9.66	11.06	12.46	13.60	14.18	13.91	12.93	11.63	10.20	8.87	8.14
41 N	8.63	9.73	11.08	12.43	13.53	14.09	13.83	12.89	11.63	10.25	8.97	8.27
40 N	8.73	9.80	11.10	12.40	13.46	14.00	13.75	12.84	11.63	10.30	9.07	8.39
39 N	8.84	9.86	11.12	12.37	13.39	13.91	13.67	12.80	11.63	10.34	9.16	8.51
38 N	8.94	9.92	11.13	12.34	13.33	13.82	13.60	12.75	11.62	10.39	9.25	8.62
37 N	9.04	9.98	11.15	12.31	13.26	13.74	13.52	12.71	11.62	10.43	9.33	8.73
36 N	9.13	10.04	11.16	12.29	13.20	13.66	13.45	12.67	11.62	10.47	9.41	8.84
35 N	9.22	10.10	11.18	12.26	13.14	13.58	13.38	12.63	11.62	10.51	9.49	8.94
34 N	9.31	10.15	11.19	12.24	13.08	13.51	13.31	12.59	11.62	10.55	9.57	9.04
33 N	9.39	10.20	11.21	12.21	13.02	13.43	13.24	12.55	11.61	10.59	9.65	9.14
32 N	9.48	10.26	11.22	12.18	12.97	13.36	13.18	12.51	11.61	10.62	9.72	9.23
31 N	9.56	10.30	11.23	12.16	12.91	13.29	13.11	12.48	11.61	10.66	9.79	9.32
30 N	9.64	10.35	11.24	12.14	12.86	13.21	13.05	12.44	11.61	10.69	9.86	9.41
29 N	9.71	10.40	11.26	12.11	12.80	13.15	12.99	12.40	11.60	10.73	9.93	9.50
28 N	9.79	10.45	11.27	12.09	12.75	13.08	12.93	12.37	11.60	10.76	9.99	9.58
27 N	9.86	10.49	11.28	12.06	12.70	13.01	12.87	12.33	11.60	10.79	10.06	9.66
26 N	9.93	10.54	11.29	12.04	12.64	12.94	12.81	12.29	11.59	10.82	10.12	9.74

Table 6.7b 4° sunrise/sunset daylength S_{04} (hours), latitudes 25° North - 25° South.

Latitude	Jan 17	Feb 15	Mar 16	Apr 15	May 15	Jun 11	Jul 17	Aug 16	Sep 16	Oct 16	Nov 16	Dec 11
25 N	10.00	10.58	11.30	12.02	12.59	12.88	12.75	12.26	11.59	10.85	10.18	9.82
24 N	10.07	10.62	11.31	11.99	12.54	12.82	12.69	12.23	11.59	10.88	10.24	9.90
23 N	10.14	10.66	11.32	11.97	12.49	12.75	12.64	12.19	11.58	10.91	10.30	9.97
22 N	10.20	10.70	11.33	11.95	12.45	12.69	12.58	12.16	11.58	10.94	10.36	10.04
21 N	10.27	10.74	11.34	11.93	12.40	12.63	12.52	12.12	11.58	10.97	10.41	10.12
20 N	10.33	10.78	11.34	11.90	12.35	12.57	12.47	12.09	11.57	11.00	10.47	10.19
19 N	10.39	10.82	11.35	11.88	12.30	12.51	12.41	12.06	11.57	11.02	10.52	10.26
18 N	10.45	10.86	11.36	11.86	12.25	12.45	12.36	12.03	11.56	11.05	10.58	10.32
17 N	10.51	10.89	11.37	11.84	12.21	12.39	12.31	11.99	11.56	11.08	10.63	10.39
16 N	10.57	10.93	11.38	11.81	12.16	12.33	12.25	11.96	11.56	11.10	10.68	10.46
15 N	10.63	10.97	11.38	11.79	12.12	12.27	12.20	11.93	11.55	11.13	10.73	10.52
14 N	10.69	11.00	11.39	11.77	12.07	12.21	12.15	11.90	11.55	11.15	10.78	10.59
13 N	10.74	11.04	11.40	11.75	12.02	12.16	12.10	11.87	11.54	11.17	10.83	10.65
12 N	10.80	11.07	11.40	11.73	11.98	12.10	12.04	11.83	11.54	11.20	10.88	10.71
11 N	10.85	11.10	11.41	11.70	11.93	12.04	11.99	11.80	11.53	11.22	10.93	10.78
10 N	10.91	11.14	11.42	11.68	11.89	11.99	11.94	11.77	11.53	11.24	10.98	10.84
9 N	10.96	11.17	11.42	11.66	11.84	11.93	11.89	11.74	11.52	11.27	11.03	10.90
8 N	11.02	11.20	11.43	11.64	11.80	11.87	11.84	11.71	11.52	11.29	11.07	10.96
7 N	11.07	11.23	11.43	11.62	11.75	11.82	11.79	11.68	11.51	11.31	11.12	11.02
6 N	11.12	11.27	11.44	11.59	11.71	11.76	11.74	11.64	11.50	11.33	11.17	11.08
5 N	11.17	11.30	11.44	11.57	11.66	11.70	11.68	11.61	11.50	11.36	11.21	11.13
4 N	11.23	11.33	11.45	11.55	11.62	11.65	11.63	11.58	11.49	11.38	11.26	11.19
3 N	11.28	11.36	11.45	11.53	11.57	11.59	11.58	11.55	11.49	11.40	11.30	11.25
2 N	11.33	11.39	11.46	11.50	11.53	11.53	11.53	11.52	11.48	11.42	11.35	11.31
1 N	11.38	11.42	11.46	11.48	11.48	11.48	11.48	11.48	11.47	11.44	11.39	11.36
0	11.43	11.45	11.47	11.46	11.44	11.42	11.43	11.45	11.47	11.46	11.44	11.42
1 S	11.48	11.48	11.47	11.44	11.39	11.36	11.38	11.42	11.46	11.48	11.48	11.48
2 S	11.53	11.51	11.47	11.41	11.34	11.31	11.32	11.39	11.45	11.50	11.53	11.53
3 S	11.58	11.54	11.48	11.39	11.30	11.25	11.27	11.35	11.44	11.52	11.57	11.59
4 S	11.63	11.57	11.48	11.37	11.25	11.19	11.22	11.32	11.44	11.54	11.61	11.65
5 S	11.68	11.60	11.49	11.34	11.21	11.13	11.17	11.29	11.43	11.56	11.66	11.70
6 S	11.73	11.63	11.49	11.32	11.16	11.07	11.11	11.25	11.42	11.58	11.70	11.76
7 S	11.78	11.66	11.49	11.29	11.11	11.01	11.06	11.22	11.41	11.60	11.75	11.81
8 S	11.83	11.69	11.49	11.27	11.06	10.95	11.00	11.18	11.41	11.62	11.79	11.87
9 S	11.88	11.72	11.50	11.24	11.01	10.89	10.95	11.15	11.40	11.64	11.83	11.93
10 S	11.93	11.75	11.50	11.22	10.97	10.83	10.89	11.11	11.39	11.66	11.88	11.98
11 S	11.98	11.78	11.50	11.19	10.92	10.77	10.84	11.08	11.38	11.68	11.92	12.04
12 S	12.03	11.81	11.51	11.17	10.87	10.71	10.78	11.04	11.37	11.70	11.97	12.10
13 S	12.08	11.84	11.51	11.14	10.82	10.65	10.72	11.01	11.36	11.72	12.01	12.15
14 S	12.13	11.87	11.51	11.11	10.76	10.58	10.67	10.97	11.35	11.74	12.05	12.21
15 S	12.18	11.90	11.51	11.09	10.71	10.52	10.61	10.93	11.34	11.76	12.10	12.27
16 S	12.24	11.93	11.51	11.06	10.66	10.45	10.55	10.89	11.33	11.78	12.14	12.33
17 S	12.29	11.96	11.52	11.03	10.61	10.39	10.49	10.86	11.32	11.79	12.19	12.39
18 S	12.34	11.99	11.52	11.00	10.55	10.32	10.43	10.82	11.31	11.81	12.23	12.44
19 S	12.39	12.02	11.52	10.97	10.50	10.25	10.37	10.78	11.30	11.83	12.28	12.50
20 S	12.45	12.05	11.52	10.94	10.44	10.18	10.30	10.74	11.29	11.85	12.33	12.56
21 S	12.50	12.08	11.52	10.91	10.38	10.11	10.24	10.69	11.28	11.87	12.37	12.62
22 S	12.55	12.11	11.52	10.88	10.33	10.04	10.17	10.65	11.27	11.89	12.42	12.69
23 S	12.61	12.14	11.52	10.85	10.27	9.97	10.10	10.61	11.26	11.91	12.47	12.75
24 S	12.66	12.18	11.52	10.82	10.21	9.89	10.04	10.57	11.24	11.93	12.52	12.81
25 S	12.72	12.21	11.52	10.78	10.15	9.81	9.97	10.52	11.23	11.95	12.56	12.87

Table 6.7c 4° sunrise/sunset daylength S_{04} (hours), latitudes 26° - 75° South.

Latitude	Jan 17	Feb 15	Mar 16	Apr 15	May 15	Jun 11	Jul 17	Aug 16	Sep 16	Oct 16	Nov 16	Dec 11
26 S	12.78	12.24	11.52	10.75	10.08	9.74	9.90	10.47	11.22	11.97	12.61	12.94
27 S	12.83	12.27	11.52	10.72	10.02	9.66	9.82	10.43	11.20	11.99	12.66	13.00
28 S	12.89	12.30	11.52	10.68	9.95	9.57	9.75	10.38	11.19	12.01	12.71	13.07
29 S	12.95	12.34	11.52	10.65	9.89	9.49	9.67	10.33	11.17	12.03	12.77	13.14
30 S	13.01	12.37	11.52	10.61	9.82	9.40	9.59	10.28	11.16	12.05	12.82	13.21
31 S	13.07	12.41	11.52	10.57	9.75	9.31	9.51	10.23	11.14	12.08	12.87	13.28
32 S	13.14	12.44	11.52	10.53	9.67	9.22	9.43	10.18	11.13	12.10	12.93	13.35
33 S	13.20	12.48	11.52	10.49	9.60	9.13	9.35	10.12	11.11	12.12	12.98	13.42
34 S	13.27	12.51	11.52	10.45	9.52	9.03	9.26	10.07	11.09	12.14	13.04	13.50
35 S	13.33	12.55	11.52	10.41	9.44	8.93	9.17	10.01	11.08	12.16	13.09	13.57
36 S	13.40	12.59	11.52	10.36	9.36	8.83	9.07	9.95	11.06	12.18	13.15	13.65
37 S	13.47	12.63	11.51	10.32	9.27	8.72	8.98	9.89	11.04	12.21	13.21	13.73
38 S	13.54	12.66	11.51	10.27	9.19	8.61	8.88	9.83	11.02	12.23	13.28	13.81
39 S	13.62	12.70	11.51	10.22	9.10	8.50	8.78	9.76	11.00	12.26	13.34	13.90
40 S	13.69	12.75	11.51	10.17	9.00	8.38	8.67	9.69	10.98	12.28	13.40	13.99
41 S	13.77	12.79	11.50	10.12	8.91	8.26	8.56	9.62	10.95	12.30	13.47	14.08
42 S	13.85	12.83	11.50	10.07	8.80	8.13	8.44	9.55	10.93	12.33	13.54	14.17
43 S	13.93	12.87	11.49	10.01	8.70	7.99	8.32	9.47	10.91	12.36	13.61	14.26
44 S	14.02	12.92	11.49	9.95	8.59	7.85	8.19	9.40	10.88	12.38	13.68	14.36
45 S	14.11	12.97	11.49	9.89	8.47	7.70	8.06	9.31	10.86	12.41	13.76	14.47
46 S	14.20	13.02	11.48	9.83	8.36	7.55	7.92	9.23	10.83	12.44	13.84	14.57
47 S	14.30	13.06	11.48	9.76	8.23	7.39	7.78	9.14	10.80	12.47	13.92	14.68
48 S	14.40	13.12	11.47	9.70	8.10	7.21	7.63	9.05	10.77	12.50	14.00	14.80
49 S	14.50	13.17	11.46	9.62	7.96	7.03	7.47	8.95	10.74	12.53	14.09	14.92
50 S	14.61	13.23	11.46	9.55	7.81	6.84	7.30	8.85	10.70	12.56	14.18	15.04
51 S	14.72	13.28	11.45	9.47	7.66	6.63	7.11	8.74	10.67	12.59	14.27	15.17
52 S	14.84	13.34	11.44	9.39	7.49	6.41	6.92	8.63	10.63	12.62	14.37	15.31
53 S	14.96	13.40	11.43	9.30	7.32	6.17	6.71	8.51	10.59	12.66	14.48	15.46
54 S	15.09	13.47	11.42	9.21	7.13	5.91	6.49	8.38	10.55	12.70	14.59	15.61
55 S	15.23	13.54	11.41	9.11	6.93	5.63	6.25	8.24	10.51	12.73	14.70	15.77
56 S	15.37	13.61	11.40	9.01	6.72	5.32	5.99	8.10	10.46	12.77	14.82	15.94
57 S	15.53	13.69	11.39	8.90	6.49	4.98	5.71	7.95	10.42	12.82	14.95	16.13
58 S	15.69	13.76	11.38	8.78	6.24	4.59	5.40	7.79	10.36	12.86	15.08	16.32
59 S	15.86	13.85	11.36	8.65	5.96	4.15	5.05	7.61	10.31	12.91	15.23	16.53
60 S	16.05	13.94	11.35	8.52	5.66	3.63	4.66	7.42	10.25	12.95	15.38	16.76
61 S	16.25	14.03	11.33	8.38	5.32	2.98	4.20	7.22	10.19	13.00	15.54	17.00
62 S	16.47	14.13	11.31	8.22	4.94	2.10	3.66	6.99	10.12	13.06	15.72	17.27
63 S	16.71	14.24	11.30	8.05	4.50	0.00	2.99	6.75	10.05	13.12	15.91	17.57
64 S	16.96	14.35	11.27	7.87	3.98	0.00	2.05	6.48	9.97	13.18	16.12	17.89
65 S	17.25	14.48	11.25	7.67	3.34	0.00	0.00	6.18	9.88	13.24	16.34	18.26
66 S	17.57	14.61	11.23	7.45	2.50	0.00	0.00	5.84	9.79	13.31	16.59	18.67
67 S	17.92	14.75	11.20	7.21	0.97	0.00	0.00	5.45	9.69	13.39	16.86	19.16
68 S	18.33	14.91	11.17	6.94	0.00	0.00	0.00	5.01	9.58	13.47	17.16	19.73
69 S	18.80	15.08	11.14	6.63	0.00	0.00	0.00	4.48	9.46	13.56	17.51	20.45
70 S	19.35	15.27	11.10	6.29	0.00	0.00	0.00	3.83	9.33	13.66	17.90	21.43
71 S	20.04	15.49	11.06	5.89	0.00	0.00	0.00	2.96	9.18	13.76	18.35	23.74
72 S	20.97	15.72	11.01	5.42	0.00	0.00	0.00	1.54	9.01	13.88	18.89	24.00
73 S	22.53	15.99	10.96	4.85	0.00	0.00	0.00	0.00	8.82	14.01	19.55	24.00
74 S	24.00	16.30	10.90	4.13	0.00	0.00	0.00	0.00	8.61	14.16	20.41	24.00
75 S	24.00	16.65	10.83	3.15	0.00	0.00	0.00	0.00	8.36	14.33	21.71	24.00

Note: Computation time can be saved by calculating the values of $\sin \phi . \sin \delta$ and $\cos \phi . \cos \delta$ once only rather than repeating the same calculations for each hour.

6.2 Calculate hourly solar azimuth ψ

$$\cos \psi = (\sin \phi \sin \gamma - \sin \delta)/(\cos \phi \cos \gamma)$$

Before solar noon $\psi = -\cos^{-1}(\cos \psi)$ (degrees)

After solar noon $\psi = +\cos^{-1}(\cos \psi)$ (degrees)

7. ESTABLISH HOURLY RELATIVE AIR MASS, RAYLEIGH OPTICAL THICKNESS AND CLEAR SKY DIRECT BEAM IRRADIANCE AT NORMAL INCIDENCE

7.1 Calculate hourly relative air mass m

7.1.1 For solar altitude $\gamma \leqslant 10°$, using step 1 to obtain p/p_o.

$$m = \frac{p/p_o}{\sin \gamma + 0.15(\gamma + 3.885°)^{-1.253}} \quad \text{dimensionless}$$

7.1.2 For solar altitude $\gamma > 10°$, using step 1 to obtain p/p_o (dimensionless)

$$m = \frac{p}{p_o} \operatorname{cosec} \gamma \quad \text{dimensionless}$$

7.2 Calculate hourly Rayleigh optical thickness δ_R.

$$\delta_R = 1/(0.9\, m + 9.4) \quad \text{dimensionless}$$

7.3 Calculate Linke Turbidity Factor for actual solar altitude $T_L(\gamma)$

If air mass 2 Linke Turbidity Factor $T_L < 2.5$, then

$$T_L(\gamma) = T_L - (0.85 - 2.25 \sin \gamma + 1.11 \sin^2 \gamma)(T_L - 1)/1.5$$

dimensionless

otherwise

$$T_L(\gamma) = T_L - (0.85 - 2.25 \sin \gamma + 1.11 \sin^2 \gamma) \quad \text{dimensionless}$$

7.4 Calculate hourly clear sky direct beam irradiance at normal incidence I_c

$$I_c = I_{oj} . \exp(-T_L(\gamma) . \delta_R . m) \quad Wm^{-2}$$

where I_{oj} is extraterrestrial irradiance at normal incidence on day J Wm^{-2} (Refer Stage 4.3)

$T_L(\gamma)$ is Linke Turbidity Factor at solar altitude γ

7.5 Calculate hourly clear sky horizontal surface direct beam irradiance $I_c(0,0)$

$$I_c(0,0) = I_c \sin \gamma \qquad Wm^{-2}$$

8. ESTIMATE HOURLY CLEAR SKY HORIZONTAL SURFACE DIFFUSE IRRADIANCES FOR EACH MONTH D_c

8.1 Find the atmospheric transmittance coefficient q_a^m for absorption in a path length of air mass m for each hour using the value of T_L and solar altitude in Table 6.8. The table has been computed of the basis of Air Mass 2 Linke Turbidity Factor and already contains adjustments for the change of Linke Turbidity Factor with solar altitude. Use linear interpolation between values given in the Table.

8.2 Find the additional scattering correction f_1, using the value of T_L and solar altitude in Table 6.9 for each hour. The table has been computed on the basis of Air Mass 2 Linke Turbidity Factor and already contains adjustments for the change of Linke Turbidity Factor with solar altitude. Use linear interpolation between the values given in the table.

8.3 Calculate hourly clear sky horizontal surface diffuse irradiance D_c

$$D_c = 0.5\, f_1\, (I_{oj} \cdot q_a^m - I_c) \sin \gamma \qquad Wm^{-2}$$

where I_{oj} is the extraterrestrial irradiance at normal incidence (Wm^{-2}) (refer stage 4.3)

I_c is the clear sky direct beam irradiance at normal incidence (Wm^{-2}) (refer stage 7.4)

9. ESTIMATE HOURLY CLEAR SKY GLOBAL HORIZONTAL RADIATION

9.1 Obtain hourly global irradiance G_c (Wm^{-2}) by adding clear sky horizontal surface direct beam and diffuse components

$$G_c = I_c(0,0) + D_c \qquad Wm^{-2}$$

10. ESTIMATE HOURLY OVERCAST SKY HORIZONTAL SURFACE IRRADIANCES FOR EACH MONTH

10.1 Calculate hourly overcast sky horizontal surface diffuse irradiance D_b

$$D_b = K_d(2.6 + 182.6 \sin \gamma) \qquad Wm^{-2}$$

where γ is the solar altitude (degrees)

K_d is the correction to mean solar distance (dimensionless)

Note Under overcast skies the horizontal surface diffuse irradiance is equal to the horizontal surface global irradiance G_b.

$$G_b = D_b \qquad Wm^{-2}$$

Table 6.8 Transmittance through clean atmosphere allowing for absorption alone, q_a^m.

Solar Alt. (deg)	Air mass 2 Linke Turbidity Factor 1	2	3	4	5	6	7	8	9	10
0	0.641	0.635	0.625	0.611	0.597	0.583	0.569	0.555	0.541	0.527
2	0.664	0.657	0.646	0.632	0.618	0.603	0.589	0.574	0.560	0.545
4	0.686	0.678	0.667	0.652	0.637	0.622	0.607	0.592	0.577	0.562
6	0.707	0.698	0.686	0.670	0.655	0.639	0.624	0.609	0.593	0.578
8	0.726	0.716	0.703	0.687	0.671	0.656	0.640	0.624	0.608	0.592
10	0.744	0.733	0.719	0.703	0.687	0.671	0.655	0.638	0.622	0.606
12	0.761	0.749	0.735	0.718	0.701	0.685	0.668	0.652	0.635	0.619
14	0.776	0.764	0.749	0.732	0.715	0.698	0.681	0.664	0.647	0.630
16	0.791	0.777	0.762	0.745	0.727	0.710	0.693	0.676	0.659	0.641
18	0.805	0.790	0.774	0.757	0.739	0.722	0.704	0.686	0.669	0.651
20	0.817	0.802	0.785	0.768	0.750	0.732	0.714	0.696	0.679	0.661
22	0.829	0.813	0.796	0.778	0.760	0.742	0.724	0.706	0.688	0.670
24	0.840	0.823	0.806	0.788	0.769	0.751	0.733	0.714	0.696	0.678
26	0.851	0.833	0.815	0.796	0.778	0.759	0.741	0.722	0.704	0.685
28	0.860	0.842	0.823	0.805	0.786	0.767	0.749	0.730	0.711	0.692
30	0.869	0.850	0.831	0.812	0.793	0.775	0.756	0.737	0.718	0.699
32	0.877	0.858	0.839	0.820	0.800	0.781	0.762	0.743	0.724	0.705
34	0.885	0.865	0.845	0.826	0.807	0.788	0.768	0.749	0.730	0.710
36	0.892	0.872	0.852	0.832	0.813	0.793	0.774	0.755	0.735	0.716
38	0.899	0.878	0.858	0.838	0.818	0.799	0.779	0.760	0.740	0.720
40	0.905	0.884	0.863	0.843	0.824	0.804	0.784	0.764	0.745	0.725
42	0.911	0.889	0.868	0.848	0.828	0.808	0.789	0.769	0.749	0.729
44	0.916	0.894	0.873	0.853	0.833	0.813	0.793	0.773	0.753	0.733
46	0.921	0.898	0.877	0.857	0.837	0.817	0.797	0.777	0.756	0.736
48	0.925	0.902	0.881	0.861	0.840	0.820	0.800	0.780	0.760	0.740
50	0.929	0.906	0.884	0.864	0.844	0.824	0.803	0.783	0.763	0.743
52	0.933	0.910	0.888	0.867	0.847	0.827	0.806	0.786	0.766	0.745
54	0.937	0.913	0.891	0.870	0.850	0.830	0.809	0.789	0.768	0.748
56	0.940	0.916	0.894	0.873	0.853	0.832	0.812	0.791	0.771	0.750
58	0.943	0.919	0.896	0.876	0.855	0.835	0.814	0.794	0.773	0.752
60	0.945	0.921	0.899	0.878	0.857	0.837	0.816	0.796	0.775	0.754
62	0.947	0.923	0.901	0.880	0.859	0.839	0.818	0.797	0.777	0.756
64	0.950	0.925	0.902	0.882	0.861	0.840	0.820	0.799	0.778	0.758
66	0.951	0.927	0.904	0.883	0.863	0.842	0.821	0.801	0.780	0.759
68	0.953	0.928	0.906	0.885	0.864	0.843	0.823	0.802	0.781	0.760
70	0.954	0.930	0.907	0.886	0.865	0.845	0.824	0.803	0.782	0.761
72	0.956	0.931	0.908	0.887	0.866	0.846	0.825	0.804	0.783	0.762
74	0.957	0.932	0.909	0.888	0.867	0.846	0.826	0.805	0.784	0.763
76	0.958	0.933	0.910	0.889	0.868	0.847	0.826	0.806	0.785	0.764
78	0.958	0.933	0.910	0.890	0.869	0.848	0.827	0.806	0.785	0.764
80	0.959	0.934	0.911	0.890	0.869	0.848	0.827	0.807	0.786	0.765
82	0.959	0.934	0.911	0.890	0.870	0.849	0.828	0.807	0.786	0.765
84	0.959	0.934	0.912	0.891	0.870	0.849	0.828	0.807	0.786	0.765
86	0.960	0.935	0.912	0.891	0.870	0.849	0.828	0.807	0.786	0.765
88	0.960	0.935	0.912	0.891	0.870	0.849	0.828	0.807	0.786	0.765
90	0.960	0.935	0.912	0.891	0.870	0.849	0.828	0.807	0.786	0.765

Table 6.9 Additional scattering correction f_1 for estimation of clear sky diffuse irradiance.

Solar Alt. (deg)	Air mass 2 Linke Turbidity Factor									
	1	2	3	4	5	6	7	8	9	10
0	1.689	1.606	1.470	1.279	1.089	0.899	0.708	0.518	0.327	0.137
2	1.587	1.511	1.395	1.239	1.083	0.927	0.770	0.614	0.458	0.301
4	1.499	1.431	1.334	1.208	1.081	0.955	0.828	0.702	0.575	0.449
6	1.422	1.363	1.284	1.183	1.083	0.983	0.882	0.782	0.681	0.581
8	1.355	1.306	1.243	1.165	1.087	1.009	0.931	0.853	0.776	0.698
10	1.297	1.258	1.209	1.151	1.092	1.034	0.976	0.918	0.859	0.801
12	1.246	1.217	1.181	1.140	1.099	1.057	1.016	0.975	0.934	0.892
14	1.202	1.182	1.158	1.132	1.105	1.079	1.052	1.025	0.999	0.972
16	1.163	1.152	1.139	1.125	1.111	1.098	1.084	1.070	1.056	1.042
18	1.128	1.125	1.123	1.120	1.117	1.114	1.112	1.109	1.106	1.103
20	1.097	1.103	1.109	1.116	1.122	1.129	1.136	1.142	1.149	1.156
22	1.069	1.082	1.096	1.111	1.126	1.142	1.157	1.172	1.187	1.202
24	1.043	1.063	1.085	1.107	1.130	1.152	1.175	1.197	1.219	1.242
26	1.019	1.046	1.074	1.103	1.132	1.161	1.190	1.219	1.247	1.276
28	0.996	1.030	1.064	1.099	1.133	1.168	1.203	1.237	1.272	1.306
30	0.975	1.015	1.054	1.094	1.134	1.174	1.214	1.253	1.293	1.333
32	0.954	1.000	1.045	1.089	1.134	1.178	1.223	1.267	1.312	1.356
34	0.934	0.985	1.035	1.084	1.132	1.181	1.230	1.279	1.328	1.377
36	0.914	0.970	1.025	1.078	1.131	1.184	1.237	1.290	1.343	1.396
38	0.895	0.956	1.015	1.072	1.128	1.185	1.242	1.299	1.356	1.413
40	0.875	0.941	1.005	1.065	1.126	1.186	1.247	1.307	1.368	1.428
42	0.856	0.927	0.994	1.059	1.123	1.187	1.251	1.315	1.379	1.443
44	0.837	0.913	0.984	1.052	1.119	1.187	1.254	1.322	1.389	1.457
46	0.819	0.899	0.974	1.045	1.116	1.187	1.257	1.328	1.399	1.470
48	0.800	0.885	0.964	1.038	1.112	1.186	1.260	1.334	1.408	1.483
50	0.782	0.871	0.954	1.031	1.108	1.186	1.263	1.340	1.417	1.495
52	0.765	0.858	0.944	1.025	1.105	1.185	1.265	1.346	1.426	1.506
54	0.748	0.845	0.935	1.018	1.101	1.185	1.268	1.351	1.434	1.517
56	0.733	0.833	0.926	1.012	1.098	1.184	1.270	1.356	1.442	1.528
58	0.718	0.822	0.918	1.006	1.095	1.184	1.272	1.361	1.449	1.538
60	0.704	0.811	0.910	1.001	1.092	1.183	1.274	1.365	1.456	1.547
62	0.691	0.801	0.903	0.996	1.090	1.183	1.276	1.369	1.463	1.556
64	0.680	0.793	0.897	0.992	1.087	1.183	1.278	1.373	1.469	1.564
66	0.669	0.785	0.891	0.988	1.085	1.182	1.280	1.377	1.474	1.571
68	0.661	0.778	0.886	0.985	1.084	1.182	1.281	1.380	1.479	1.578
70	0.653	0.772	0.882	0.982	1.082	1.182	1.283	1.383	1.483	1.583
72	0.647	0.767	0.878	0.980	1.081	1.182	1.284	1.385	1.486	1.588
74	0.641	0.763	0.875	0.978	1.080	1.182	1.285	1.387	1.489	1.592
76	0.637	0.760	0.873	0.976	1.079	1.183	1.286	1.389	1.492	1.595
78	0.634	0.758	0.871	0.975	1.079	1.183	1.286	1.390	1.494	1.598
80	0.631	0.755	0.870	0.974	1.079	1.183	1.287	1.392	1.496	1.600
82	0.628	0.753	0.868	0.973	1.078	1.183	1.288	1.393	1.498	1.603
84	0.625	0.751	0.867	0.972	1.078	1.184	1.289	1.395	1.500	1.606
86	0.622	0.748	0.865	0.971	1.078	1.184	1.290	1.397	1.503	1.609
88	0.617	0.745	0.862	0.970	1.077	1.185	1.292	1.399	1.507	1.614
90	0.610	0.740	0.859	0.968	1.077	1.185	1.294	1.403	1.512	1.621

11. ESTIMATE HOURLY MONTHLY MEAN HORIZONTAL SURFACE DIRECT BEAM IRRADIANCES FOR EACH MONTH

11.1 Estimate additional correction function for estimating monthly mean direct beam irradiance f_5 either by tables using 11.1.1 or by direct computation using 11.1.2

11.1.1 From Table 6.10

Table 6.10 provides values of f_5 at 1° intervals of solar altitude for a range of values of σ_{4m}.

11.1.2 Or by direct computation

If solar altitude $\gamma < 45°$ degrees

$$f_5 = \sum_{j=0}^{3} \left(\sum_{i=0}^{2} (a_{ij} \ \gamma^i) \ \sigma_{4m}{}^j \right)$$

Otherwise

$$f_5 = \sum_{j=0}^{3} \left(\sum_{i=0}^{2} (a_{ij} \times 45^i) \ \sigma_{4m}{}^j \right)$$

where σ_{4m} is the 4 degree daily relative sunshine duration from Stage 4.7

The values of a_{ij} are given in Table 6.11

11.2 Calculate hourly average horizontal surface direct beam irradiance $I_m(0,0)$ using the value of f_5.

$$I_m(0,0) = f_5 \ \sigma_{4m} \ I_c(0,0) \qquad Wm^{-2}.$$

where $I_c(0,0)$ is clear sky horizontal surface direct beam irradiance (Wm^{-2}) (refer stage 7.5)

σ_{4m} is the monthly mean daily relative sunshine duration on 4° sunrise basis (refer stage 4.7)

12. ESTIMATE HOURLY MONTHLY MEAN HORIZONTAL SURFACE SKY DIFFUSE IRRADIANCES FOR EACH MONTH

12.1 The monthly mean hourly values of the diffuse irradiance are estimated by a relatively complex process evolved as the result of detailed studies at the hourly level of the diffuse radiation climatology of Western Europe. The method is based on the consequences of several observations concerning the nature of the basic diffuse irradiance observations.

Table 6.10 Additional correction f_5 for calculating monthly mean direct irradiance from clear sky direct irradiance.

Solar Altitude (deg)	Relative sunshine duration (based on 4° sunrise/sunset) 0.1	0.2	0.3	0.4	0.5	0.6	0.7	0.8	0.9	1.0
0	0.102	0.486	0.746	0.904	0.983	1.006	0.996	0.976	0.970	0.999
1	0.160	0.516	0.756	0.902	0.976	0.998	0.991	0.974	0.970	0.999
2	0.216	0.544	0.766	0.901	0.970	0.991	0.985	0.972	0.969	0.999
3	0.271	0.572	0.775	0.900	0.963	0.984	0.980	0.969	0.969	0.999
4	0.325	0.599	0.784	0.898	0.957	0.977	0.975	0.967	0.969	0.999
5	0.377	0.625	0.793	0.897	0.951	0.970	0.970	0.965	0.969	0.999
6	0.428	0.650	0.801	0.895	0.945	0.964	0.965	0.963	0.969	0.999
7	0.477	0.675	0.810	0.894	0.939	0.957	0.960	0.960	0.969	0.999
8	0.525	0.699	0.818	0.892	0.933	0.951	0.956	0.958	0.969	0.999
9	0.572	0.722	0.825	0.891	0.928	0.945	0.951	0.957	0.969	0.999
10	0.617	0.744	0.833	0.889	0.922	0.939	0.947	0.955	0.970	0.999
11	0.660	0.766	0.840	0.888	0.917	0.933	0.943	0.953	0.970	0.999
12	0.703	0.787	0.846	0.886	0.911	0.927	0.939	0.951	0.970	0.999
13	0.744	0.807	0.853	0.884	0.906	0.922	0.935	0.950	0.970	0.999
14	0.783	0.826	0.859	0.883	0.901	0.916	0.931	0.948	0.970	0.999
15	0.821	0.845	0.865	0.881	0.896	0.911	0.928	0.947	0.970	1.000
16	0.858	0.863	0.870	0.879	0.891	0.906	0.924	0.945	0.971	1.000
17	0.893	0.880	0.875	0.878	0.887	0.901	0.921	0.944	0.971	1.000
18	0.927	0.897	0.880	0.876	0.882	0.896	0.917	0.943	0.971	1.000
19	0.960	0.912	0.885	0.874	0.877	0.892	0.914	0.942	0.971	1.000
20	0.991	0.927	0.889	0.872	0.873	0.887	0.911	0.941	0.972	1.000
21	1.021	0.941	0.893	0.870	0.869	0.883	0.909	0.940	0.972	1.000
22	1.049	0.955	0.897	0.869	0.865	0.879	0.906	0.939	0.972	1.000
23	1.076	0.968	0.900	0.867	0.861	0.875	0.903	0.938	0.973	1.000
24	1.102	0.980	0.903	0.865	0.857	0.871	0.901	0.937	0.973	1.000
25	1.126	0.991	0.906	0.863	0.853	0.868	0.899	0.937	0.973	1.000
26	1.149	1.001	0.908	0.861	0.849	0.864	0.896	0.936	0.974	1.000
27	1.170	1.011	0.911	0.859	0.846	0.861	0.894	0.935	0.974	1.000
28	1.190	1.020	0.912	0.857	0.843	0.858	0.893	0.935	0.975	1.000
29	1.209	1.028	0.914	0.855	0.839	0.855	0.891	0.935	0.975	1.001
30	1.226	1.035	0.915	0.853	0.836	0.852	0.889	0.934	0.976	1.001
31	1.241	1.042	0.916	0.851	0.833	0.850	0.888	0.934	0.976	1.001
32	1.256	1.048	0.917	0.849	0.830	0.847	0.886	0.934	0.977	1.001
33	1.269	1.053	0.917	0.847	0.827	0.845	0.885	0.934	0.977	1.001
34	1.280	1.058	0.917	0.844	0.824	0.843	0.884	0.934	0.978	1.001
35	1.290	1.061	0.917	0.842	0.822	0.841	0.883	0.934	0.979	1.001
36	1.299	1.064	0.916	0.840	0.819	0.839	0.882	0.934	0.979	1.001
37	1.307	1.066	0.916	0.838	0.817	0.837	0.882	0.935	0.980	1.001
38	1.313	1.068	0.914	0.836	0.815	0.836	0.881	0.935	0.981	1.001
39	1.317	1.069	0.913	0.833	0.813	0.834	0.881	0.935	0.981	1.001
40	1.320	1.069	0.911	0.831	0.811	0.833	0.880	0.936	0.982	1.001
41	1.322	1.068	0.909	0.829	0.809	0.832	0.880	0.936	0.983	1.001
42	1.322	1.066	0.907	0.826	0.807	0.831	0.880	0.937	0.983	1.001
43	1.321	1.064	0.904	0.824	0.805	0.830	0.880	0.938	0.984	1.002
44	1.319	1.061	0.901	0.822	0.804	0.830	0.881	0.939	0.985	1.002
>45	1.315	1.057	0.898	0.819	0.802	0.829	0.881	0.939	0.986	1.002

1. The diffuse radiation under partially clouded conditions is higher than under either clear or overcast conditions. The maximum values occur when σ_{4m} is around 0.25 to 0.30. Linear interpolation between clear and overcast conditions is thus invalid.

2. For values of relative sunshine duration σ_{4m} below about 0.70, there is a strong linear correlation between the mean monthly mean hourly diffuse irradiance and the solar altitude at the mid point of each measurement hour.

3. For values of the relative sunshine duration σ_{4m} about 0.25, the influence of turbidity on the diffuse radiation received is relatively small.

4. For clear sky conditions the influence of turbidity is very marked, the diffuse radiation increasing strongly with increase in turbidity.

5. For dull and overcast conditions more turbid climates give less diffuse radiation than clearer climates.

6. At sunrise and sunset there is still some diffuse radiation.

The formulae devised and set out below attempt to take account of all these factors.

12.2 Computational process

The computation of the diffuse sky irradiance on horizontal surfaces proceeds in four stages:

a) Computation of the expected horizontal monthly mean diffuse irradiance for a value of σ_{4m} = 0.25 for the given solar altitude.

b) Linear interpolation of the monthly mean diffuse irradiance between the value at σ_{4m} = 0.25 and the clear day value D_c at σ_{4m} = 1.0, by inputting the actual value of the relative sunshine duration, σ_{4m}. to find the interpolated value.

c) Application of a pragmatic correction to allow for the deviations of the actual observed data from the simplified linear model predictions.

d) Application of an additional annual pragmatic correction to allow for the general clarity of atmosphere at a particular place.

These pragmatic corrections were based on hourly diffuse radiation data available for a number of European stations.

12.2.1 <u>Compute the monthly mean horizontal diffuse irradiance for a relative sunshine duration of</u> σ_{4m} = 0.25

The monthly mean diffuse irradiance for a monthly mean relative sunshine duration of 0.25, $D_{.25}$, is calculated for each solar altitude γ as

$$D_{.25} = (2 + 5.3\,\gamma)K_d \qquad Wm^{-2}$$

where K_d is the correction to mean solar distance. (Refer step 4.2).

12.2.2 <u>Interpolate to estimate uncorrected horizontal diffuse irradiance for actual observed relative sunshine duration</u> ${}_uD_m$.

The uncorrected value of the monthly mean horizontal diffuse irradiance ${}_uD_m$ at solar altitude γ, is given by

$${}_uD_m = \sigma_{4m}\,D_c + (1 - \sigma_{4m})(D_{.25} - 0.25\,\sigma_{4m}\,D_c)/.75 \quad Wm^{-2}$$

where D_c = clear day diffuse horizontal irradiance (Wm^{-2}) (Refer step 8.3.)

12.2.3 <u>Estimate pragmatic correction f_6 to apply to uncorrected value ${}_uD_m$ to obtain monthly mean hourly diffuse sky irradiance on a horizontal surface</u> D_m.

either by computation using 12.2.3.1 or using Tables, refer 12.2.3.2.

12.2.3.1 f_6 is computed from the given mean daily relative sunshine duration σ_{4m}. for each solar altitude using the following formula:

$$f_6 = \left(\sum_{i=0}^{3} b_i\,\sigma_{4m}^{\;i}\right)\left(\sum_{i=0}^{2} c_i\,\gamma^i\right)/\left(\sum_{i=0}^{2} d_i(\overline{a+b})^i\right)$$

where $(\overline{a+b})$ is the annual mean value of the Angstrom (a + b) and where b_i, c_i and d_i are given in Table 6.12.

It will be noted that the first part of the expression is constant over the whole day, while the value of the denominator is constant over the whole year. The middle expression is a function of solar altitude and its value changes from hour to hour. The denominator makes a correction for the effects of pollution and dust. The correction was based on global and diffuse radiation data from 15 European stations.

When monthly values of (a + b) are not available from which to obtain the annual mean value it is suggested the following values are adopted.

Very clear climates	0.80
Clear climates	0.76
Urban climates	0.72
Polluted areas/very large cities	0.68

12.2.3.2 Table 6.13 provides values of f_6 as a function of solar altitude and monthly mean daily relative sunshine duration computed for an annual mean value of $(\overline{a+b}) = .74$. At the base of the table, correction factors to apply for climates of differing clarity are included. The correction is made by multiplying together the two figures.

Table 6.11 Constants a_{ij} used in formula for calculating additional correction function f_5 for estimating monthly mean direct beam irradiance.

j \ i	0	1	2
0	-0.42907	0.098539	-0.0011429
1	6.1280	-0.46284	0.0050835
2	-8.5179	0.64546	-0.0067794
3	3.8176	-0.28110	0.0028390

Note: The practical limits of σ_{4m} as set by typical observed sunshine data are between about 0.1 and 0.7, but the function has been adjusted to converge on the clear sky values when $\sigma_{4m} = 1.0$.

Table 6.12 Constants used in formula for calculating additional correction functions f_6 for estimating monthly mean diffuse irradiance on horizontal surfaces.

i	b_i	c_i	d_i
0	0.5212	0.83202	7.12889
1	2.429	0.011619	-14.9747
2	-3.383	-1.8832×10^{-4}	9.10674
3	1.432	9.8559×10^{-7}	-

Table 6.13 Additional correction f_6 for calculating monthly mean diffuse irradiance .

Solar Altitude (deg)	Relative sunshine duration (based on 4° sunrise/sunset) 0.1	0.2	0.3	0.4	0.5	0.6	0.7	0.8	0.9
0	0.589	0.710	0.792	0.839	0.860	0.861	0.849	0.830	0.813
2	0.604	0.730	0.813	0.862	0.883	0.884	0.871	0.853	0.835
4	0.619	0.747	0.833	0.883	0.905	0.906	0.893	0.874	0.856
6	0.633	0.764	0.852	0.903	0.925	0.926	0.913	0.893	0.875
8	0.646	0.780	0.869	0.921	0.944	0.945	0.932	0.912	0.893
10	0.658	0.794	0.885	0.938	0.961	0.962	0.949	0.929	0.909
12	0.669	0.808	0.900	0.954	0.978	0.979	0.965	0.944	0.925
14	0.679	0.820	0.914	0.969	0.993	0.994	0.980	0.959	0.939
16	0.689	0.831	0.926	0.982	1.006	1.007	0.993	0.972	0.952
18	0.697	0.842	0.938	0.994	1.019	1.020	1.005	0.984	0.964
20	0.705	0.851	0.948	1.005	1.030	1.031	1.017	0.995	0.974
22	0.712	0.860	0.958	1.015	1.041	1.042	1.027	1.005	0.984
24	0.719	0.867	0.967	1.025	1.050	1.051	1.036	1.014	0.993
26	0.724	0.874	0.974	1.033	1.058	1.059	1.044	1.022	1.001
28	0.729	0.880	0.981	1.040	1.066	1.067	1.052	1.029	1.008
30	0.734	0.886	0.987	1.046	1.072	1.073	1.058	1.036	1.014
32	0.738	0.891	0.992	1.052	1.078	1.079	1.064	1.041	1.020
34	0.741	0.895	0.997	1.057	1.083	1.084	1.069	1.046	1.024
36	0.744	0.898	1.001	1.061	1.087	1.088	1.073	1.050	1.028
38	0.747	0.901	1.004	1.065	1.091	1.092	1.077	1.054	1.032
40	0.749	0.904	1.007	1.067	1.094	1.095	1.080	1.057	1.035
42	0.750	0.906	1.009	1.070	1.096	1.097	1.082	1.059	1.037
44	0.752	0.907	1.011	1.072	1.098	1.099	1.084	1.061	1.039
46	0.753	0.908	1.012	1.073	1.099	1.101	1.085	1.062	1.040
48	0.753	0.909	1.013	1.074	1.100	1.101	1.086	1.063	1.041
50	0.754	0.910	1.013	1.074	1.101	1.102	1.087	1.063	1.041
52	0.754	0.910	1.014	1.075	1.101	1.102	1.087	1.064	1.041
54	0.754	0.910	1.014	1.074	1.101	1.102	1.087	1.063	1.041
56	0.753	0.909	1.013	1.074	1.101	1.102	1.086	1.063	1.041
58	0.753	0.909	1.013	1.074	1.100	1.101	1.086	1.063	1.041
60	0.753	0.908	1.012	1.073	1.100	1.101	1.085	1.062	1.040
62	0.752	0.908	1.012	1.072	1.099	1.100	1.085	1.061	1.039
64	0.752	0.907	1.011	1.072	1.098	1.099	1.084	1.061	1.039
66	0.751	0.907	1.010	1.071	1.097	1.098	1.083	1.060	1.038
68	0.751	0.906	1.010	1.070	1.097	1.098	1.082	1.059	1.037
70	0.750	0.905	1.009	1.070	1.096	1.097	1.082	1.059	1.037
72	0.750	0.905	1.009	1.069	1.096	1.097	1.081	1.058	1.036
74	0.750	0.905	1.008	1.069	1.095	1.096	1.081	1.058	1.036
76	0.750	0.905	1.008	1.069	1.095	1.096	1.081	1.058	1.036
78	0.750	0.905	1.009	1.069	1.096	1.097	1.081	1.058	1.036
80	0.750	0.906	1.009	1.070	1.096	1.097	1.082	1.059	1.037
82	0.751	0.907	1.010	1.071	1.097	1.098	1.083	1.060	1.038
84	0.752	0.908	1.012	1.072	1.099	1.100	1.085	1.061	1.039
86	0.754	0.909	1.013	1.074	1.101	1.102	1.086	1.063	1.041
88	0.755	0.912	1.016	1.077	1.103	1.105	1.089	1.066	1.044
90	0.757	0.914	1.019	1.080	1.107	1.108	1.092	1.069	1.047

Additional corrections to diffuse, based on annual Angstrom (a+b)
(multiply corrections from table above by these values)

(a+b)	correction	(a+b)	correction
0.66	0.853	0.76	1.026
0.68	0.894	0.78	1.046
0.70	0.933	0.80	1.058
0.72	0.969	0.82	1.063
0.74	1.000	0.84	1.060

12.4 Correct the value of the uncorrected sky diffuse on a horizontal surface to obtain the required corrected value of the monthly mean sky irradiance falling on a horizontal surface.

$$D_m = f_6 \, {}_uD_m \qquad Wm^{-2}$$

13. ESTIMATE MONTHLY MEAN HOURLY GLOBAL IRRADIANCES AND DAILY GLOBAL IRRADIATION

13.1 Obtain hourly monthly mean horizontal surface global irradiance G_m. by adding mean direct beam and mean diffuse irradiance

$$G_m = I_m(0,0) + D_m \qquad Wm^{-2}$$

14. NORMALIZE, IF HORIZONTAL OBSERVATIONS ARE AVAILABLE, TO TABULATED OR MAPPED VALUES OF GLOBAL SOLAR RADIATION

14.1 The inclined plane tables for monthly mean conditions which appear in the CEC Inclined Surface Radiation Atlas have been prepared by normalizing the horizontal predicted values so that the daily totals of the hourly global irradiances were set equal to the observed values of the monthly mean daily global irradiation on a horizontal surface at each site. This process allows for any inaccuracies in the estimation of the atmospheric properties. It assumes that the reported radiation values for horizontal surfaces are reliable i.e. they are made by properly calibrated instruments of adequate field performance. In preparing the tables the reported observed data were normally accepted as correct which may not always be the case. An important exception was Italy where the observed global irradiation values were obtained with Robitsch Actinographs. Here the observed horizontal data had been adjusted downwards to allow for instrumental shortcomings before incorporation in the CEC European Solar Radiation Atlas, Vol. I, 2nd Edition.

14.2 The method of normalization used is to simply compare the predicted daily global totals with the corresponding observed daily global irradiation for the site. Each predicted hourly value of direct beam and diffuse irradiance is multiplied by the ratio (daily observed/daily predicted horizontal surface global radiation), to obtain an adjusted set of hourly values of the direct beam and diffuse horizontal surface irradiance. On each table in Chapter 4 of the Inclined Surface Atlas, the normalization factor that had to be applied to the raw computations to match the observed data is given. In general these normalization factors were relatively low, i.e. less than 5%. A large normalization factor above 10% may imply that either

i) the site is relatively heavily obstructed by surrounding hills, etc.

or ii) the instrumentation was poorly calibrated and adjusted.

or iii) the instrument used was unsuitable for making accurate radiation measurements of the required standard.

or iv) the ground is continuously snow covered in that month.

Stations known to be of high observational reliability rarely revealed big estimation errors when computing the horizontal irradiances. Refer Appendix 8.

14.3 If the normalization process is to be adopted, it must be applied after computation of the horizontal irradiances and before the computation of the inclined surface irradiances.

14.4 The normalization process for any site in Europe may be carried out using any convenient horizontal surface global radiation maps, for example those which have been published by the Commission of the European Communities in the European Solar Radiation Atlas, Vol. 1, 2nd Edition (1). Alternatively other published maps and tables such as appear in the UK CIBS Guide A2, Weather and Solar Data, 1982 (2) or the Atlas Solaire Francais (3) can be used for the process of normalization. In general it is assumed, however, that most users of this publication would use the new horizontal surface global radiation maps published in the CEC European Solar Radiation Atlas, Vol. 1, 2nd edition for normalization purposes.

14.5 If separate values for the monthly mean global and diffuse daily irradiation on horizontal surfaces are available, the normalization process can be applied to the two components separately, so the sum of the hourly values of the direct beam irradiance on the horizontal surface is made equal to the difference between the daily global and diffuse monthly mean daily horizontal surface irradiation, and the sum of the hourly sky diffuse irradiances on a horizontal surface is made equal to the observed horizontal surface monthly mean diffuse irradiation. However, accurate results will only be obtained if the diffuse radiation observations are properly corrected for shading ring obstruction effects before being used in the computational process. Details about improved shading ring corrections for the European regions may be found in Solar Energy R&D in the European Community, Series F, Volume 2, Solar Radiation Data, D. Reidel, 1983 (4), especially under Action 4.1 reported on pages 200-248 in that publication.

14.6 At this stage all the processes described in Figure 6.1(a) are complete. These horizontal computations are used for all subsequent slope calculations.

COMPUTATIONS FOR INCLINED PLANES

Figure 6.2 shows the structure of this part of the computational process.

15. ESTABLISH HOURLY INCLINED SURFACE GEOMETRY FOR EACH SURFACE AND FOR EACH MONTH

15.1 Calculate wall solar azimuth angle α_S (degrees) for each hour

$$\alpha_S = \alpha - \psi \quad \text{degrees}$$

where α is the surface azimuth angle (degrees)

and ψ is the solar azimuth angle (degrees)

NB In order to ensure that α_S lies in the range $-180° < \alpha_S < 180°$, one of the following adjustments may be required.

If $\alpha_S < -180°$ then $\alpha_S \rightarrow \alpha_S + 360°$

If $\alpha_S > 180°$ then $\alpha_S \rightarrow \alpha_S - 360°$

15.2 Calculate cosine of the angle of incidence of the solar beam on the surface $\cos \nu$ for each hour

$$\cos \nu = \sin \beta \cos \gamma \cos \alpha_S + \cos \beta \sin \gamma$$

If $\cos \nu < 0$ then the sun lies behind the surface. To avoid negative values, which would be incorrect, the following restriction should be applied:

If $\cos \nu < 0$ then $\cos \nu = 0$

16. ESTIMATE THE HOURLY CLEAR SKY INCLINED SURFACE IRRADIANCES FOR EACH MONTH AND FOR EACH SURFACE

16.1 Calculate clear sky inclined surface direct beam irradiance $I_c(\beta,\alpha)$ at each hour

$$I_c(\beta,\alpha) = I_c \cos \nu \quad \text{Wm}^{-2}$$

16.2 Calculate the clear sky diffuse sky irradiance on the plane $D_c(\beta,\alpha)$

16.2.1 Find the ratio f_2 for each hour using Table 6.14 a-j, where f_2 is the ratio of clear sky inclined surface to horizontal surface diffuse irradiance. The values must be obtained by interpolation in Tables 6.14 a-j according to the value of T_L. It should be noted that the tables have been computed to use the air mass 2 Linke Turbidity Factors as inputs. The necessary adjustments for solar altitude are already incorporated and values of $T_L(\gamma)$ should not be used with the table. High accuracy cannot be expected at low angles of solar elevation. Below a solar elevation of 2.5° it is suggested the values for 2.5 should be used.

Table 6.14(a) Ratio of inclined surface clear sky diffuse irradiance to clear sky horizontal diffuse irradiance as a percentage

Air Mass 2 Linke Turbidity Factor = 1.5

Angle of Slope	Solar altitude (degrees)	Wall solar azimuth angle (degrees) 0	15	30	45	60	75	90	105	120	135	150	165	180
15	2.5	155	153	147	138	127	114	100	96	98	99	100	100	100
	5	130	129	126	121	114	107	100	97	97	98	98	99	99
	15	123	122	120	117	113	107	100	100	98	96	94	93	93
	30	119	119	117	115	112	107	99	100	99	98	96	95	95
	45	115	115	114	112	109	105	99	100	100	99	98	97	96
	60	111	111	110	109	107	103	99	100	101	101	100	99	99
	75	108	108	107	107	105	103	100	101	102	103	102	102	102
	90	106	106	106	106	106	106	106	106	106	106	106	106	106
30	2.5	202	199	188	171	149	125	100	99	103	104	104	105	105
	5	154	151	146	136	124	111	100	98	101	102	103	103	104
	15	141	139	136	130	122	111	99	97	94	95	96	96	96
	30	133	132	129	125	119	110	98	96	95	92	89	87	86
	45	125	124	122	119	114	106	97	96	96	94	92	90	89
	60	117	117	115	113	109	103	96	97	97	97	95	94	93
	75	110	110	109	108	105	102	97	99	100	101	100	99	98
	90	107	107	107	107	107	107	107	107	107	107	107	107	107
45	2.5	239	234	219	194	163	132	99	99	103	104	105	106	106
	5	170	167	159	145	129	112	99	98	102	103	104	104	105
	15	152	150	144	136	125	112	97	92	94	95	95	95	95
	30	141	139	135	129	121	109	95	91	87	85	85	85	85
	45	129	128	125	120	114	104	92	90	88	85	82	79	78
	60	118	117	115	112	107	100	91	91	90	89	86	84	83
	75	108	107	106	104	101	97	92	93	94	94	93	91	91
	90	103	103	103	103	103	103	103	103	103	103	103	103	103
60	2.5	263	256	238	208	170	134	98	97	100	101	102	103	103
	5	178	174	164	148	127	110	97	96	99	99	100	101	102
	15	155	153	146	136	123	108	94	89	90	91	91	91	91
	30	142	140	135	127	117	104	90	83	81	81	81	80	80
	45	127	125	122	116	108	98	86	82	77	76	75	74	73
	60	112	111	109	105	99	92	85	82	80	77	74	72	71
	75	100	99	98	96	92	89	84	84	84	83	81	80	79
	90	93	93	93	93	93	93	93	93	93	93	93	93	93
75	2.5	271	264	243	210	169	131	94	93	93	93	94	95	96
	5	177	173	162	143	121	103	93	92	92	92	93	94	95
	15	151	149	141	130	115	100	89	84	83	83	83	84	84
	30	135	133	128	119	108	96	83	76	74	74	73	73	72
	45	118	117	113	106	98	88	78	71	69	69	68	67	66
	60	102	101	98	93	88	82	76	71	67	67	66	65	64
	75	87	87	85	83	80	77	74	72	70	69	67	65	64
	90	80	80	80	80	80	80	80	80	80	80	80	80	80
90	2.5	265	257	236	202	159	123	89	85	84	83	84	85	86
	5	167	163	151	132	109	93	87	84	82	82	83	84	85
	15	140	137	130	118	102	89	82	77	74	73	73	74	74
	30	123	121	115	106	95	84	74	69	66	64	64	63	63
	45	104	103	99	92	84	75	68	64	61	60	59	58	57
	60	87	86	83	78	73	68	65	62	60	59	58	57	56
	75	71	70	69	66	64	63	62	60	59	58	58	57	57
	90	63	63	63	63	63	63	63	63	63	63	63	63	63

Table 6.14(b) Ratio of inclined surface clear sky diffuse irradiance to clear sky horizontal diffuse irradiance as a percentage

Air Mass 2 Linke Turbidity Factor = 2.0

Angle of Slope	Solar altitude (degrees)	Wall solar azimuth angle (degrees)												
		0	15	30	45	60	75	90	105	120	135	150	165	180
15	2.5	156	154	148	139	127	114	100	95	97	98	98	98	98
	5	130	129	126	121	114	107	100	95	96	96	97	97	97
	15	123	123	121	117	113	107	100	98	96	94	92	90	90
	30	119	119	117	115	111	106	99	98	97	95	94	93	92
	45	115	114	113	111	109	104	99	98	98	97	95	94	94
	60	110	110	109	108	106	103	99	99	99	99	98	97	96
	75	107	107	106	105	104	102	99	100	101	101	100	100	100
	90	104	104	104	104	104	104	104	104	104	104	104	104	104
30	2.5	204	200	189	171	149	125	99	97	100	101	102	102	103
	5	154	152	146	136	124	111	99	95	99	100	100	101	101
	15	141	140	136	129	121	110	98	94	91	91	91	92	92
	30	133	132	129	124	118	109	97	94	91	87	84	82	81
	45	124	123	121	118	112	105	96	94	92	89	87	85	84
	60	116	115	114	111	107	102	95	95	94	93	91	89	89
	75	108	108	107	106	103	100	96	96	97	97	96	95	95
	90	103	103	103	103	103	103	103	103	103	103	103	103	103
45	2.5	241	235	220	195	163	132	98	97	100	101	102	102	103
	5	170	168	159	145	128	111	98	95	98	99	100	101	101
	15	152	150	145	136	124	110	96	88	89	90	90	90	90
	30	141	139	135	128	119	107	93	87	81	79	79	78	78
	45	128	127	123	118	111	102	90	86	82	78	74	72	70
	60	115	115	112	109	104	97	89	87	85	83	80	78	77
	75	104	104	103	101	98	94	89	89	89	88	87	86	85
	90	97	97	97	97	97	97	97	97	97	97	97	97	97
60	2.5	265	258	239	209	170	134	96	95	96	97	98	99	99
	5	179	175	164	148	126	108	95	93	95	95	96	97	97
	15	156	154	147	136	121	106	92	84	85	85	85	85	85
	30	141	139	134	126	115	101	87	77	74	74	73	73	72
	45	125	123	119	113	105	95	83	76	69	67	67	66	65
	60	109	108	106	101	96	89	81	77	73	69	66	64	62
	75	96	95	94	91	88	84	80	79	77	76	74	73	72
	90	87	87	87	87	87	87	87	87	87	87	87	87	87
75	2.5	274	266	245	211	169	131	92	90	89	89	90	91	92
	5	177	173	162	143	120	101	91	88	88	88	88	89	90
	15	152	149	141	129	113	98	86	79	78	77	77	77	78
	30	135	133	127	117	105	92	79	70	67	66	66	65	65
	45	116	114	110	103	94	84	73	63	61	60	59	59	58
	60	98	97	94	89	83	77	71	64	59	58	57	57	56
	75	82	82	80	78	75	72	69	66	63	60	58	57	56
	90	72	72	72	72	72	72	72	72	72	72	72	72	72
90	2.5	267	260	238	202	159	122	86	82	80	79	80	81	82
	5	167	163	151	132	108	91	84	80	78	77	78	79	80
	15	140	137	129	117	101	86	78	72	68	67	67	67	67
	30	122	119	113	104	92	79	69	63	58	57	56	56	55
	45	101	100	95	88	79	70	62	57	53	52	51	50	49
	60	82	81	78	73	68	63	58	54	52	50	49	49	48
	75	66	65	63	61	58	56	55	53	51	50	50	49	49
	90	55	55	55	55	55	55	55	55	55	55	55	55	55

Table 6.14(c) Ratio of inclined surface clear sky diffuse irradiance to clear sky horizontal diffuse irradiance as a percentage

Air Mass 2 Linke Turbidity Factor = 3.0

Angle of Slope	Solar altitude (degrees)	Wall solar azimuth angle (degrees) 0	15	30	45	60	75	90	105	120	135	150	165	180
15	2.5	157	155	149	139	127	114	100	91	93	94	94	94	94
	5	130	129	126	120	114	106	100	92	91	92	92	92	92
	15	124	123	121	117	112	106	99	96	93	89	86	85	84
	30	120	119	117	114	111	105	99	96	94	92	90	88	88
	45	114	114	113	111	108	104	99	97	95	93	92	91	90
	60	110	109	108	107	105	102	98	98	97	96	95	94	94
	75	105	105	105	104	103	101	99	99	99	98	98	97	97
	90	102	102	102	102	102	102	102	102	102	102	102	102	102
30	2.5	206	202	191	172	149	124	98	92	94	95	95	96	96
	5	154	152	146	135	122	110	98	90	92	93	93	94	94
	15	143	141	137	129	120	109	97	89	83	84	84	84	84
	30	133	132	129	123	116	107	95	90	84	80	75	73	72
	45	123	122	120	116	111	103	94	90	86	82	79	77	76
	60	114	113	112	109	105	100	94	91	89	87	85	83	83
	75	106	105	104	103	101	98	94	93	93	92	91	90	90
	90	99	99	99	99	99	99	99	99	99	99	99	99	99
45	2.5	244	239	222	196	163	131	96	91	92	93	94	94	95
	5	171	168	159	144	126	109	95	89	90	91	91	92	92
	15	154	152	146	135	122	108	93	81	80	80	80	80	80
	30	141	139	134	127	117	104	90	80	72	69	68	68	67
	45	126	125	121	116	108	99	88	80	74	68	64	61	60
	60	112	112	109	105	100	94	86	82	78	74	71	69	68
	75	100	100	99	97	94	90	86	84	83	81	80	79	78
	90	91	91	91	91	91	91	91	91	91	91	91	91	91
60	2.5	269	262	242	210	170	133	92	87	88	88	88	89	90
	5	179	175	164	146	124	105	91	85	85	86	86	87	87
	15	158	155	147	135	119	103	87	76	75	74	74	74	74
	30	141	139	133	124	111	97	82	69	64	63	62	62	61
	45	122	121	117	110	101	90	78	68	59	55	55	54	54
	60	105	104	101	97	91	84	76	69	63	58	55	52	51
	75	90	90	88	86	83	79	75	72	69	67	65	64	63
	90	78	78	78	78	78	78	78	78	78	78	78	78	78
75	2.5	278	271	248	213	168	129	87	81	80	79	80	81	82
	5	178	174	161	141	117	98	85	79	78	77	78	79	79
	15	153	151	142	128	111	94	79	70	67	66	66	66	66
	30	134	132	125	115	101	87	72	60	56	55	54	54	53
	45	113	111	106	98	89	78	67	53	50	48	48	47	47
	60	93	92	89	84	77	70	63	54	47	46	46	45	45
	75	76	75	74	71	68	64	61	57	53	50	48	46	45
	90	62	62	62	62	62	62	62	62	62	62	62	62	62
90	2.5	272	264	241	204	158	120	79	73	70	69	70	70	71
	5	167	163	150	130	104	87	77	71	68	67	67	68	69
	15	142	139	130	115	97	81	70	62	58	56	56	56	56
	30	120	118	111	100	87	73	60	53	48	46	45	45	44
	45	97	96	91	83	73	63	53	47	42	40	39	39	38
	60	76	75	72	67	60	54	49	44	41	39	38	38	37
	75	58	58	56	53	50	48	46	43	41	39	39	39	39
	90	44	44	44	44	44	44	44	44	44	44	44	44	44

Table 6.14(d) Ratio of inclined surface clear sky diffuse irradiance to clear sky horizontal diffuse irradiance as a percentage

Air Mass 2 Linke Turbidity Factor = 4.0

Angle of Slope	Solar altitude (degrees)	Wall solar azimuth angle (degrees) 0	15	30	45	60	75	90	105	120	135	150	165	180
15	2.5	158	156	150	140	127	114	99	88	89	89	89	89	90
	5	130	129	126	120	113	106	99	89	87	87	87	88	88
	15	125	124	122	118	112	106	99	94	89	85	82	80	79
	30	120	119	117	114	110	105	99	95	92	89	86	85	84
	45	114	114	113	110	107	103	98	96	93	91	89	88	88
	60	109	109	108	107	104	102	98	97	95	94	93	92	92
	75	105	105	104	103	102	100	98	98	98	97	96	96	96
	90	101	101	101	101	101	101	101	101	101	101	101	101	101
30	2.5	208	204	192	173	148	124	97	87	88	89	89	89	90
	5	154	152	145	135	121	109	97	85	86	87	87	87	87
	15	144	142	138	130	120	108	96	85	77	77	77	77	77
	30	134	133	129	123	116	106	95	87	80	74	69	66	65
	45	123	122	119	115	109	102	94	88	83	78	75	73	72
	60	113	112	111	108	104	99	93	89	86	84	81	80	79
	75	104	104	103	101	99	96	93	92	90	89	88	87	87
	90	97	97	97	97	97	97	97	97	97	97	97	97	97
45	2.5	247	241	224	197	163	130	94	85	85	85	86	86	86
	5	171	168	158	143	124	108	93	83	83	83	83	84	84
	15	156	154	147	136	122	106	91	74	73	73	73	73	73
	30	141	140	134	126	115	102	88	76	65	62	61	61	60
	45	125	124	120	114	106	97	86	77	69	62	57	54	53
	60	111	110	107	104	98	92	84	79	73	69	66	64	63
	75	98	98	96	94	91	88	84	81	79	77	75	74	74
	90	87	87	87	87	87	87	87	87	87	87	87	87	87
60	2.5	272	265	244	211	169	131	88	80	79	79	79	80	80
	5	179	175	163	145	122	103	87	78	77	77	77	78	78
	15	160	157	149	135	118	100	83	69	67	66	66	66	66
	30	141	139	133	123	110	94	79	62	56	55	55	54	53
	45	121	120	115	108	98	87	75	63	52	48	48	47	47
	60	103	102	99	94	88	80	73	65	58	52	48	45	44
	75	87	87	85	83	79	76	71	68	64	61	59	58	57
	90	74	74	74	74	74	74	74	74	74	74	74	74	74
75	2.5	281	273	250	213	167	127	81	73	71	70	70	71	71
	5	177	173	160	139	114	94	80	71	69	68	68	69	69
	15	156	153	143	128	109	90	74	63	59	58	58	57	57
	30	134	132	125	113	99	83	67	53	49	47	47	46	45
	45	111	109	104	96	85	74	62	47	43	41	41	40	39
	60	90	89	86	80	73	66	59	49	41	39	39	38	38
	75	72	71	70	67	64	60	57	51	47	43	41	40	39
	90	57	57	57	57	57	57	57	57	57	57	57	57	57
90	2.5	275	266	242	204	156	118	72	65	61	59	60	60	61
	5	167	162	149	127	101	83	70	63	59	58	58	59	59
	15	143	140	130	115	95	77	63	55	50	48	48	48	48
	30	120	118	110	98	84	68	54	46	41	39	38	37	37
	45	95	93	88	79	69	58	47	41	36	34	33	32	32
	60	73	72	68	63	56	49	43	38	34	32	32	31	31
	75	54	53	51	48	45	42	40	37	35	33	33	33	32
	90	38	38	38	38	38	38	38	38	38	38	38	38	38

Table 6.14(e) Ratio of inclined surface clear sky diffuse irradiance to clear sky horizontal diffuse irradiance as a percentage

Air Mass 2 Linke Turbidity Factor = 5.0

Angle of Slope	Solar altitude (degrees)	Wall solar azimuth angle (degrees) 0	15	30	45	60	75	90	105	120	135	150	165	180
15	2.5	159	157	150	140	127	114	99	85	86	86	86	86	86
	5	130	129	126	120	113	106	99	87	84	84	84	84	84
	15	126	125	122	118	112	106	99	93	87	82	78	76	75
	30	120	120	118	114	110	105	98	94	90	87	84	82	82
	45	115	114	113	110	107	103	98	95	92	90	88	87	86
	60	109	109	108	106	104	101	98	96	95	93	92	91	90
	75	104	104	104	103	102	100	98	98	97	96	96	95	95
	90	100	100	100	100	100	100	100	100	100	100	100	100	100
30	2.5	209	205	193	173	148	124	96	84	84	84	84	85	85
	5	155	152	145	134	121	108	96	82	82	82	82	82	83
	15	146	144	139	130	120	108	95	82	72	72	72	72	72
	30	135	134	130	123	115	105	94	85	77	70	65	61	60
	45	123	122	119	115	109	101	93	86	80	76	72	69	68
	60	112	112	110	107	103	98	92	88	85	81	79	77	76
	75	103	103	102	100	98	95	92	91	89	88	86	85	85
	90	95	95	95	95	95	95	95	95	95	95	95	95	95
45	2.5	249	243	225	198	162	130	92	80	80	80	80	80	81
	5	171	168	158	143	123	107	91	78	78	78	78	78	78
	15	158	156	148	137	122	105	89	70	68	67	67	67	67
	30	142	140	135	126	115	101	86	73	60	56	56	55	55
	45	125	124	120	114	105	95	84	74	65	58	53	49	47
	60	110	109	106	102	97	90	83	77	71	66	62	60	59
	75	97	96	95	93	90	86	82	80	77	75	73	71	71
	90	85	85	85	85	85	85	85	85	85	85	85	85	85
60	2.5	274	267	245	211	169	131	86	75	73	73	73	73	74
	5	179	175	163	144	120	101	85	73	71	71	71	71	72
	15	162	159	150	136	118	99	81	64	61	60	60	60	60
	30	142	140	133	122	108	93	76	58	51	50	49	48	48
	45	121	119	114	107	96	85	73	59	48	43	43	42	41
	60	102	101	97	92	86	78	71	62	54	48	43	40	39
	75	85	84	83	80	77	73	69	65	61	58	56	54	53
	90	71	71	71	71	71	71	71	71	71	71	71	71	71
75	2.5	284	275	252	214	166	126	77	68	65	63	64	64	65
	5	177	173	159	138	112	92	76	66	63	62	62	62	63
	15	158	155	145	129	108	88	71	58	53	52	52	51	51
	30	135	132	125	113	97	81	64	48	44	42	41	41	40
	45	110	108	103	94	83	71	59	43	38	37	36	35	34
	60	88	87	83	78	71	63	56	45	36	35	34	34	33
	75	69	69	67	64	61	57	54	48	43	39	37	35	34
	90	54	54	54	54	54	54	54	54	54	54	54	54	54
90	2.5	277	268	244	204	155	116	67	59	54	53	53	54	54
	5	166	162	148	126	99	80	66	58	53	51	52	52	52
	15	146	142	132	115	94	74	59	50	44	42	42	42	42
	30	120	118	110	97	81	65	50	42	36	34	33	32	32
	45	94	92	86	77	66	54	43	37	32	29	28	28	27
	60	70	69	65	60	52	46	39	34	30	28	28	27	26
	75	50	50	48	45	42	39	36	33	31	29	29	29	28
	90	34	34	34	34	34	34	34	34	34	34	34	34	34

Table 6.14(f) Ratio of inclined surface clear sky diffuse irradiance to clear sky horizontal diffuse irradiance as a percentage

Air Mass 2 Linke Turbidity Factor = 6.0

Angle of Slope	Solar altitude (degrees)	Wall solar azimuth angle (degrees) 0	15	30	45	60	75	90	105	120	135	150	165	180
15	2.5	159	157	150	140	127	114	99	83	83	83	83	83	83
	5	130	129	126	120	112	106	99	84	81	81	81	81	81
	15	127	126	123	118	112	106	99	92	85	79	75	72	71
	30	121	120	118	115	110	105	98	93	89	85	82	80	80
	45	115	114	113	110	107	103	98	95	91	89	87	85	85
	60	109	109	108	106	104	101	98	96	94	92	91	90	89
	75	104	104	103	103	101	100	98	97	96	96	95	94	94
	90	100	100	100	100	100	100	100	100	100	100	100	100	100
30	2.5	210	206	194	174	148	124	96	81	81	81	81	81	81
	5	155	152	145	134	120	108	96	79	79	78	79	79	79
	15	147	145	140	131	120	107	95	80	69	69	68	68	68
	30	136	134	130	124	115	105	93	83	74	67	61	57	56
	45	123	122	119	115	108	101	93	85	79	73	69	66	65
	60	112	111	110	107	102	97	92	87	83	80	77	75	74
	75	103	102	101	100	98	95	92	90	88	87	85	84	83
	90	94	94	94	94	94	94	94	94	94	94	94	94	94
45	2.5	250	244	226	198	162	130	91	77	76	76	76	76	76
	5	171	168	158	142	122	106	90	75	74	73	74	74	74
	15	160	157	150	137	122	105	88	66	64	63	63	63	63
	30	143	141	135	126	114	100	86	70	57	52	52	51	50
	45	125	124	120	113	104	94	83	72	63	55	49	45	43
	60	109	108	106	101	96	89	82	75	69	64	60	57	56
	75	96	95	94	92	89	85	82	78	75	73	71	69	69
	90	84	84	84	84	84	84	84	84	84	84	84	84	84
60	2.5	276	268	246	212	168	131	84	71	69	68	68	69	69
	5	179	175	162	143	119	100	83	69	67	66	66	67	67
	15	164	161	152	137	117	98	79	60	57	56	56	56	55
	30	143	141	134	122	108	91	75	55	47	46	45	44	44
	45	121	119	114	106	95	83	71	57	44	40	39	38	37
	60	101	99	96	91	84	77	69	60	51	45	40	37	35
	75	84	83	81	79	75	72	68	63	59	56	53	51	50
	90	69	69	69	69	69	69	69	69	69	69	69	69	69
75	2.5	285	277	252	214	165	126	75	64	60	59	59	59	60
	5	177	172	159	137	110	91	74	62	58	57	57	58	58
	15	160	156	146	129	108	87	68	54	49	48	47	47	47
	30	135	133	125	112	96	79	62	45	40	38	37	37	36
	45	110	108	102	93	81	69	57	40	35	33	32	31	30
	60	87	85	82	76	68	61	54	42	33	32	31	30	29
	75	67	67	65	62	58	55	52	46	41	36	34	32	30
	90	52	52	52	52	52	52	52	52	52	52	52	52	52
90	2.5	278	269	244	204	154	116	64	55	50	48	49	49	49
	5	166	161	147	125	97	78	62	54	49	47	47	47	48
	15	147	144	133	115	93	73	56	46	40	38	38	38	37
	30	121	118	110	96	80	63	47	38	33	30	29	29	28
	45	93	91	85	75	64	52	41	34	29	26	25	24	23
	60	68	67	63	57	50	43	36	31	28	25	25	24	23
	75	48	47	45	42	39	36	34	31	28	27	26	25	25
	90	32	32	32	32	32	32	32	32	32	32	32	32	32

Table 6.14(g) Ratio of inclined surface clear sky diffuse irradiance to clear sky horizontal diffuse irradiance as a percentage

Air Mass 2 Linke Turbidity Factor = 7.0

Angle of Slope	Solar altitude (degrees)	Wall solar azimuth angle (degrees)												
		0	15	30	45	60	75	90	105	120	135	150	165	180
15	2.5	159	157	150	140	127	114	99	81	81	81	81	81	81
	5	130	129	125	120	112	105	99	83	79	79	79	79	79
	15	127	126	123	119	113	106	99	91	84	77	73	70	69
	30	121	120	118	115	110	104	98	93	88	84	81	79	78
	45	115	114	113	110	107	103	98	94	91	88	86	84	83
	60	109	109	107	106	104	101	98	96	94	92	90	89	89
	75	104	104	103	102	101	100	98	97	96	95	95	94	94
	90	100	100	100	100	100	100	100	100	100	100	100	100	100
30	2.5	210	206	193	173	148	124	96	79	78	78	78	78	78
	5	154	152	145	134	119	107	95	76	76	76	76	76	76
	15	148	146	140	131	120	107	94	78	66	66	65	65	65
	30	136	135	130	124	115	104	93	82	72	64	58	54	53
	45	123	122	119	114	108	100	92	84	77	72	67	64	63
	60	112	111	109	106	102	97	92	87	82	79	76	74	73
	75	102	102	101	99	97	94	92	89	87	86	84	83	82
	90	94	94	94	94	94	94	94	94	94	94	94	94	94
45	2.5	250	244	226	198	162	130	90	74	73	72	73	73	73
	5	170	167	157	141	121	105	90	72	71	70	70	71	71
	15	161	158	150	138	121	104	88	64	61	60	60	59	59
	30	144	142	136	126	114	100	85	68	54	49	48	48	47
	45	125	124	119	113	104	94	83	71	61	52	46	42	40
	60	109	108	105	101	95	88	81	74	67	62	58	55	53
	75	95	94	93	91	88	84	81	77	74	72	69	68	67
	90	83	83	83	83	83	83	83	83	83	83	83	83	83
60	2.5	275	268	246	211	167	130	83	69	66	65	65	65	65
	5	178	174	161	142	118	99	82	67	64	63	63	63	63
	15	166	162	153	137	117	97	78	58	54	53	52	52	52
	30	143	141	134	122	107	90	74	52	44	42	42	41	40
	45	120	118	113	105	94	82	70	55	41	37	36	35	34
	60	100	99	95	90	83	76	68	58	49	42	37	34	32
	75	82	82	80	77	74	70	67	62	58	54	51	49	48
	90	69	69	69	69	69	69	69	69	69	69	69	69	69
75	2.5	285	276	252	213	164	125	73	61	57	55	56	56	56
	5	176	171	157	136	109	89	72	59	55	54	54	54	54
	15	161	158	147	129	107	86	67	51	46	44	44	44	43
	30	136	133	125	111	95	77	60	42	37	35	34	33	33
	45	109	107	101	92	80	67	55	37	32	30	29	28	27
	60	85	84	80	74	67	59	52	40	31	29	28	27	26
	75	66	65	63	60	57	53	50	44	39	34	31	29	28
	90	50	50	50	50	50	50	50	50	50	50	50	50	50
90	2.5	277	269	243	203	153	115	62	53	47	45	45	46	46
	5	164	159	145	123	95	77	60	51	45	44	44	44	44
	15	148	145	133	115	92	71	54	44	38	35	35	35	34
	30	120	118	109	95	78	61	45	36	30	27	27	26	25
	45	92	90	84	74	62	50	39	31	26	24	23	22	21
	60	67	65	62	55	48	41	34	29	25	23	22	21	20
	75	46	46	44	40	37	34	32	29	26	25	24	23	22
	90	30	30	30	30	30	30	30	30	30	30	30	30	30

Table 6.14(h) Ratio of inclined surface clear sky diffuse irradiance to clear sky horizontal diffuse irradiance as a percentage

Air Mass 2 Linke Turbidity Factor = 8.0

Angle of Slope	Solar altitude (degrees)	Wall solar azimuth angle (degrees) 0	15	30	45	60	75	90	105	120	135	150	165	180
15	2.5	159	157	150	140	126	114	99	79	79	79	79	79	79
	5	130	129	125	119	112	105	99	81	77	77	77	77	77
	15	127	126	123	119	112	106	99	90	82	76	71	68	66
	30	121	120	118	115	110	104	98	92	87	83	80	78	77
	45	115	114	112	110	107	103	98	94	90	87	85	83	83
	60	109	108	107	106	104	101	98	95	93	91	90	89	88
	75	104	104	103	102	101	100	98	97	96	95	94	94	93
	90	100	100	100	100	100	100	100	100	100	100	100	100	100
30	2.5	209	205	193	172	147	124	95	77	76	76	76	76	76
	5	153	151	144	133	119	107	95	74	74	73	73	73	73
	15	148	146	141	131	120	107	94	77	64	63	63	63	62
	30	136	135	130	123	114	104	93	81	71	62	56	52	50
	45	123	122	119	114	108	100	92	84	76	70	66	63	61
	60	112	111	109	106	102	97	91	86	82	78	75	72	71
	75	102	101	100	99	97	94	91	89	87	85	83	82	81
	90	94	94	94	94	94	94	94	94	94	94	94	94	94
45	2.5	249	242	225	196	160	130	90	72	71	70	70	70	70
	5	169	166	156	140	120	104	89	70	68	68	68	68	68
	15	162	159	151	138	121	104	87	62	58	57	57	57	56
	30	144	142	135	126	113	99	84	67	51	46	46	45	44
	45	125	123	119	112	103	93	82	70	59	50	44	39	37
	60	108	107	104	100	94	88	81	73	66	60	56	53	51
	75	94	94	92	90	87	84	81	77	73	71	68	66	65
	90	83	83	83	83	83	83	83	83	83	83	83	83	83
60	2.5	273	266	244	209	165	130	82	66	63	62	62	62	62
	5	176	172	160	140	116	98	81	64	61	60	60	60	60
	15	166	163	153	137	117	96	77	55	51	50	50	49	49
	30	143	141	133	121	106	89	73	50	41	40	39	38	38
	45	120	118	113	104	93	81	69	53	39	34	33	32	31
	60	99	98	94	89	82	75	67	57	48	41	35	31	29
	75	82	81	79	76	73	70	66	61	56	53	50	48	46
	90	68	68	68	68	68	68	68	68	68	68	68	68	68
75	2.5	282	274	250	211	162	124	72	59	54	53	53	53	53
	5	173	169	155	134	107	88	71	57	52	51	51	51	51
	15	162	158	147	129	106	85	66	49	44	42	41	41	40
	30	135	132	124	111	94	76	59	40	35	33	32	31	30
	45	108	106	100	91	79	66	54	35	30	28	27	26	25
	60	84	83	79	73	66	58	51	38	29	27	26	25	24
	75	65	64	62	59	55	52	49	43	37	33	29	27	25
	90	50	50	50	50	50	50	50	50	50	50	50	50	50
90	2.5	275	266	241	201	151	114	61	50	44	42	42	43	43
	5	162	157	143	121	93	75	59	49	43	41	41	41	41
	15	149	145	133	115	92	70	52	42	35	33	32	32	31
	30	120	117	108	95	77	60	44	34	28	25	24	23	22
	45	91	89	82	73	60	48	37	30	24	22	21	20	19
	60	66	64	60	54	46	39	33	28	24	21	20	19	18
	75	45	44	42	39	35	32	30	27	24	23	22	21	20
	90	29	29	29	29	29	29	29	29	29	29	29	29	29

Table 6.14(i) Ratio of inclined surface clear sky diffuse irradiance to clear sky horizontal diffuse irradiance as a percentage

Air Mass 2 Linke Turbidity Factor = 9.0

Angle of Slope	Solar altitude (degrees)	Wall solar azimuth angle (degrees) 0	15	30	45	60	75	90	105	120	135	150	165	180
15	2.5	158	156	149	139	126	114	99	78	78	78	78	78	78
	5	129	128	124	119	111	105	99	80	75	75	75	75	75
	15	127	126	123	119	112	106	98	89	81	75	69	66	65
	30	121	120	118	115	110	104	98	92	87	82	79	77	76
	45	115	114	112	110	107	102	98	94	90	87	84	83	82
	60	109	108	107	106	103	101	98	95	93	91	89	88	87
	75	104	103	103	102	101	100	98	97	96	95	94	93	93
	90	99	99	99	99	99	99	99	99	99	99	99	99	99
30	2.5	208	204	191	171	146	124	95	75	74	74	74	74	74
	5	152	150	143	132	118	106	95	73	72	71	71	71	71
	15	148	146	141	131	119	107	94	76	62	61	61	61	60
	30	136	135	130	123	114	104	93	80	69	61	54	50	48
	45	123	122	119	114	107	100	92	83	76	69	65	61	60
	60	111	111	109	106	101	96	91	86	81	77	74	72	70
	75	102	101	100	99	96	94	91	89	86	85	83	81	81
	90	94	94	94	94	94	94	94	94	94	94	94	94	94
45	2.5	246	240	223	194	159	129	89	71	68	68	68	68	68
	5	167	164	154	138	119	103	89	68	66	65	65	65	65
	15	162	159	151	138	121	103	86	60	56	55	55	54	54
	30	144	142	135	125	113	99	84	66	50	44	44	43	42
	45	125	123	119	112	103	93	82	69	58	49	42	38	35
	60	108	107	104	100	94	87	81	72	65	59	55	52	50
	75	94	93	92	90	87	83	80	76	73	70	67	65	64
	90	82	82	82	82	82	82	82	82	82	82	82	82	82
60	2.5	271	263	241	207	163	129	81	64	61	60	60	60	60
	5	174	170	157	138	114	97	80	62	59	58	58	58	57
	15	166	163	153	137	116	95	77	53	49	48	47	47	46
	30	143	141	133	121	105	89	72	49	39	38	37	36	35
	45	119	118	112	104	93	81	69	52	38	33	32	31	30
	60	99	97	94	88	81	74	66	56	46	39	33	30	28
	75	81	80	79	76	72	69	66	60	55	52	48	46	45
	90	67	67	67	67	67	67	67	67	67	67	67	67	67
75	2.5	279	271	247	208	160	123	71	57	52	50	50	50	50
	5	171	166	153	131	105	86	70	55	50	48	48	48	48
	15	162	158	147	129	106	84	64	47	41	39	39	38	38
	30	135	132	124	110	93	75	58	38	33	31	30	29	28
	45	108	106	100	90	78	65	53	34	29	27	25	24	23
	60	84	82	79	72	65	57	50	37	27	25	24	23	22
	75	64	63	61	58	55	51	48	42	36	31	28	25	24
	90	49	49	49	49	49	49	49	49	49	49	49	49	49
90	2.5	271	263	238	198	148	112	59	48	42	40	40	40	40
	5	159	154	140	118	91	73	57	47	41	39	39	38	38
	15	148	144	133	114	91	69	50	40	33	31	30	29	29
	30	119	116	108	94	76	58	42	32	26	23	22	21	20
	45	90	88	82	72	59	47	36	28	23	20	19	18	17
	60	65	63	59	53	45	38	31	26	22	20	19	18	17
	75	44	43	41	38	34	31	29	26	23	22	21	20	19
	90	28	28	28	28	28	28	28	28	28	28	28	28	28

Table 6.14(j) Ratio of inclined surface clear sky diffuse irradiance to clear sky horizontal diffuse irradiance as a percentage

Air Mass 2 Linke Turbidity Factor = 10.0

Angle of Slope	Solar altitude (degrees)	Wall solar azimuth angle (degrees) 0	15	30	45	60	75	90	105	120	135	150	165	180
15	2.5	158	156	149	139	126	114	99	77	76	76	76	76	76
	5	129	128	124	118	111	105	99	79	74	74	74	74	74
	15	128	127	124	119	112	105	98	89	81	74	68	65	64
	30	121	121	118	115	110	104	98	92	86	82	78	76	75
	45	115	114	113	110	107	102	98	94	90	87	84	82	82
	60	109	108	107	106	104	101	98	95	93	91	89	88	87
	75	104	103	103	102	101	99	98	97	96	95	94	93	93
	90	99	99	99	99	99	99	99	99	99	99	99	99	99
30	2.5	207	203	191	171	145	123	95	74	73	72	72	72	72
	5	152	149	142	131	117	106	95	71	70	70	70	70	70
	15	149	147	141	132	120	106	94	75	60	60	59	59	59
	30	137	135	131	124	114	104	92	80	69	60	53	49	47
	45	123	122	119	114	108	100	92	83	75	69	64	61	59
	60	112	111	109	106	101	96	91	85	81	77	73	71	70
	75	102	101	100	99	96	94	91	88	86	84	82	81	80
	90	93	93	93	93	93	93	93	93	93	93	93	93	93
45	2.5	245	239	222	194	158	128	89	69	67	66	66	66	65
	5	167	163	153	138	118	103	88	67	64	63	63	63	63
	15	162	159	151	138	121	103	86	58	55	53	53	53	52
	30	144	142	136	126	113	98	83	65	48	43	42	42	41
	45	125	124	119	112	103	92	81	68	57	48	41	36	34
	60	108	107	104	100	94	87	80	72	65	59	54	51	49
	75	94	93	92	90	87	83	80	76	72	69	67	65	64
	90	82	82	82	82	82	82	82	82	82	82	82	82	82
60	2.5	270	262	240	206	162	127	80	63	59	57	57	57	57
	5	173	169	157	138	113	96	79	60	57	55	55	55	55
	15	167	163	153	137	116	95	76	52	47	46	45	45	44
	30	144	141	133	121	106	88	71	47	38	36	35	35	34
	45	120	118	113	104	93	80	68	51	37	31	30	29	28
	60	99	98	94	89	81	74	66	55	46	38	32	28	26
	75	81	80	79	76	72	68	65	59	55	51	48	45	44
	90	66	66	66	66	66	66	66	66	66	66	66	66	66
75	2.5	278	270	245	207	159	121	70	55	50	48	48	48	48
	5	170	165	152	130	104	85	68	53	48	46	46	46	46
	15	162	158	147	129	106	83	63	45	40	37	37	36	36
	30	136	133	124	110	93	74	57	37	31	29	28	27	26
	45	108	106	100	90	78	65	52	32	27	25	24	23	22
	60	84	83	79	73	65	57	49	36	26	24	23	22	21
	75	64	63	61	58	54	50	47	41	35	30	27	24	23
	90	47	47	47	47	47	47	47	47	47	47	47	47	47
90	2.5	270	261	236	197	147	110	57	46	40	38	38	38	37
	5	158	153	139	117	89	71	55	45	39	36	36	36	36
	15	149	145	133	114	90	68	49	38	31	29	28	28	27
	30	120	117	108	94	76	57	40	31	24	22	21	20	19
	45	91	88	82	72	59	46	34	27	22	19	18	17	16
	60	65	64	60	53	45	37	30	25	21	19	18	17	16
	75	44	43	41	38	34	30	28	25	22	21	20	19	18
	90	26	26	26	26	26	26	26	26	26	26	26	26	26

16.2.2 Calculate hourly clear sky inclined surface diffuse irradiance $D_c(\beta, \alpha)$

$$D_{cs}(\beta, \alpha) = f_2 \, D_c \qquad Wm^{-2}$$

16.3 Calculate hourly clear sky inclined surface ground reflected diffuse irradiance $R_c(\beta, \alpha)$

$$R_c(\beta, \alpha) = f_3 \, G_c \qquad Wm^{-2}$$

where f_3 is the ratio of inclined surface ground reflected diffuse irradiance to horizontal surface global irradiance (see step 3)

16.4 Obtain hourly clear sky inclined surface total diffuse irradiance $D_c(\beta, \alpha)$ by adding ground reflected and sky diffuse components

$$D_c(\beta,\alpha) = D_{cs}(\beta,\alpha) + R_c(\beta,\alpha) \qquad Wm^{-2}$$

16.5 Obtain hourly clear sky inclined surface global irradiance $G_c(\beta,\alpha)$ by adding direct beam and total diffuse

$$G_c(\beta,\alpha) = I_c(\beta,\alpha) + D_c(\beta,\alpha) \qquad Wm^{-2}$$

17. ESTIMATE HOURLY OVERCAST SKY INCLINED SURFACE IRRADIANCES FOR EACH ANGLE OF SLOPE β FOR EACH MONTH

Surfaces having the same angle of slope but different azimuths will produce identical estimates of the overcast sky irradiances so calculations need not be repeated for planes of the same inclination.

17.1 Calculate overcast sky inclined surface diffuse irradiance $D_{bs}(\beta,\alpha)$ at each hour

$$D_{bs}(\beta,\alpha) = f_4 \,.\, D_b \qquad Wm^{-2}$$

where f_4 is the ratio of inclined surface diffuse to horizontal surface diffuse under an overcast sky. (Refer step 2)

D_b is the overcast day horizontal diffuse irradiance (Wm^{-2})

17.2 Calculate hourly overcast sky inclined surface ground reflected irradiance $R_b(\beta,\alpha)$

$$R_b(\beta,\alpha) = f_3 \, G_b \qquad Wm^{-2}$$

where f_3 is the ratio of ground reflected diffuse irradiance on an inclined surface to global irradiance on a horizontal surface, refer step 3.

17.3 Obtain overcast sky inclined surface total diffuse irradiance $D_b(\beta , \alpha)$ hour by hour by adding overcast sky diffuse and ground reflected diffuse components

$$D_b(\beta,\alpha) = D_{bs}(\beta,\alpha) + R_b(\beta,\alpha) \qquad Wm^{-2}$$

Note Under overcast skies the inclined surface total diffuse irradiance is equal to the inclined surface global irradiance $G_b(\beta,\alpha)$

$$G_b(\beta,\alpha) = D_b(\beta,\alpha) \qquad Wm^{-2}$$

18. ESTIMATE HOURLY MONTHLY MEAN INCLINED SURFACE IRRADIANCES FOR EACH MONTH AND FOR EACH SURFACE

18.1 Calculate hourly monthly mean inclined surface direct beam irradiance $I_m(\beta,\alpha)$

$$I_m(\beta,\alpha) = f_5\ \sigma_{4m}\ I_c(\beta,\alpha)\ Wm^{-2}$$

where $I_c(\beta,\alpha)$ is clear sky inclined surface direct beam irradiance (Wm^{-2}) (refer step 16.1)

σ_{4m} is the monthly mean daily relative sunshine duration on 4 degree sunrise/sunset basis (refer step 4.7)

f_5 is the monthly mean direct beam correction (refer step 11.1)

18.2 Calculate hourly monthly mean inclined plane sky diffuse irradiance $D_m(\beta,\alpha)$

18.2.1 The computation first requires that the horizontal component of the monthly mean diffuse irradiance at any solar altitude is split into two components, a clear sky component, and an overcast sky/cloudy sky component. This split is achieved using the monthly mean relative sunshine duration σ_{4m}. The two separate horizontal components for clear and overcast/partially cloudy skies are then converted into inclined surface values using the appropriate slope algorithms outlined earlier. It is assumed that the contribution from the clouds in the case of the partially clouded sky can be estimated on the basis that the mean relative radiance distribution is adequately described by the Moon and Spencer overcast sky luminance distribution function.

The two components are then added together to estimate the monthly mean diffuse irradiance reaching the slope from the sky. Finally a pragmatic correction to increase the diffuse sky irradiation on surfaces lying between 45° and 135° to the sun's direction is applied. This correction was derived by comparing observation with prediction.

18.2.2 Estimate monthly mean hourly diffuse component attributed to clear sky on the surface $_{uc}D_m(\beta,\alpha)$ before final correction

$$_{uc}D_m(\beta,\alpha) = \sigma_{4m} D_{cs}(\beta,\alpha) \qquad Wm^{-2}$$

where $D_{cs}(\beta,\alpha)$ is the clear sky inclined surface diffuse irradiance (Wm^{-2}) (Refer step 16.2)

18.2.3 Estimate monthly mean hourly overcast/partially clouded sky component on horizontal surface $_{pc}D_m$ for given daily relative sunshine duration σ_{4m}

$$_{pc}D_m = D_m - \sigma_{4m} D_c \qquad Wm^{-2}$$

where D_m is the monthly mean hourly diffuse irradiance on a horizontal surface. Wm^{-2} Refer step 12.4

and D_c is the clear sky hourly diffuse irradiance on a horizontal surface. Wm^{-2} Refer step 8.3

18.2.4 Convert value overcast/partially clouded diffuse to slope using Moon and Spencer relationship before final correction

$$_{upc}D_m(\beta,\alpha) = f_4 \, _{pc}D_m \qquad Wm^{-2}$$

where f_4 is the ratio of inclined surface diffuse to horizontal surface diffuse under an overcast sky. (Refer step 2)

18.2.5 Add hourly sky diffuse components to obtain uncorrected monthly mean sky diffuse on plane $_uD_m(\beta,\alpha)$

$$_uD_m(\beta,\alpha) = {}_{uc}D_m(\beta,\alpha) + {}_{pc}D_m(\beta,\alpha) \qquad Wm^{-2}$$

18.2.6 Apply, if modulus of wall solar azimuth angle α_S lies between 45° and 135°, additional correction as follows to obtain monthly mean sky irradiance on a plane $D_{ms}(\beta,\alpha)$

$$D_{ms}(\beta,\alpha) = {}_uD_m(\beta,\alpha)(1 + \sin\beta \sin 2(\alpha_S - 45) \times (.19 - .14 \sin\gamma)) \; Wm^{-2}$$

where β is slope tilt (degrees)

α_S is the modulus of the wall solar azimuth angle in degrees,

and γ is the solar altitude (degrees)

Otherwise if $-45 < \alpha_S < 45$ or $\alpha_S > 135$ or $\alpha_S < -135$

$$D_{ms}(\beta,\alpha) = {}_uD_{ms}(\beta,\alpha) \qquad Wm^{-2}$$

18.3 Calculate hourly monthly mean inclined surface ground reflected diffuse irradiance $R_m(\beta,\alpha)$

$$R_m(\beta,\alpha) = f_3 \, G_m \qquad Wm^{-2}$$

where f_3 is the ratio of inclined surface ground reflected diffuse irradiance to horizontal surface global irradiance (refer step 3)

G_m is the monthly mean hourly horizontal surface global irradiance (Wm^{-2}) (refer step 13)

18.4 Obtain hourly monthly mean inclined plane total diffuse irradiance $D_m(\beta, \alpha)$ by adding average sky diffuse and ground reflected diffuse components

$$D_m(\beta,\alpha) = D_{ms}(\beta,\alpha) + R_m(\beta,\alpha) \quad Wm^{-2}$$

18.5 Obtain hourly monthly mean inclined surface global irradiance $G_m(\beta, \alpha)$ by adding mean direct beam and mean diffuse irradiance (sky and ground)

$$G_m(\beta,\alpha) = I_m(\beta,\alpha) + D_m(\beta,\alpha) \quad Wm^{-2}$$

19. DAILY INTEGRATION PROCEDURES, SURFACE BY SURFACE, MONTH BY MONTH

19.1 Obtain daily irradiation totals for each surface by simply adding together the corresponding hourly irradiances for each component for the day ($Wm^{-2} \times h\ d^{-1}$). (This conversion process from irradiance to irradiation depends on using exact hourly spaced integration steps). The sum of these component daily totals will be the daily global irradiation in $Whm^{-2}d^{-1}$.

The symmetry of the hourly irradiance predictions may be used to reduce the amount of effort required at this stage. This symmetry is discussed in detail in section 5.2.

On horizontal surfaces and on north and south facing inclined surfaces the daily total can be calculated by adding together either the morning (up to 11.30) or afternoon (12.30 onwards) hourly irradiances and then multiplying the result by a factor of two.

On any pair of equally inclined surfaces which are arranged symmetrically about the north-south meridian the daily irradiation totals will be identical.

Correct application of these methods can reduce the number of daily integrations required by up to one half.

19.2 Repeat month by month as required to get complete hourly and daily sets of required data for each required surface.

20. COMPUTATION OF MONTHLY MEAN MAXIMUM VALUES OF SOLAR RADIATION

20.1 The computational procedure outlined in this section is the method used to produce the tables giving the monthly mean maximum daily values of solar radiation on inclined planes included in the CEC European Solar Radiation Atlas, Vol. II, Inclined Surfaces. In general all the computational procedures are identical with the procedures already described with one or two important exceptions. The most important is, of course, that, for clear days, only those calculations relevant to the clear day part of the problem are performed. The other key difference is that different computational dates and hence different declinations are used. The computational dates for days of maximum horizontal surface radiation were selected to allow for the fact that days of maximum solar radiation on horizontal surfaces tend to occur towards the end of the month when the declination is increasing and towards the beginning of the month when the declination is decreasing. In June the maximum values tend to be clustered around the solstice on June 21st. The design dates used were selected from a statistical study of the days in the month during which maximum solar radiation is most likely to occur in Northern Europe.

20.2 Monthly mean maximum radiation on horizontal surfaces for latitudes above 25°

The following computational steps for obtaining clear day maximum predictions are different from those used for monthly mean values:

Inputs: The value of S_{max}, the monthly mean maximum daily sunshine duration, is used in place of S_m.

Step 4.1 Select the representative values of the solar declination using the monthly mean maximum day columns in Table 6.2 for the appropriate hemisphere.

Step 4.2 Tabulate monthly values of the correction to mean solar distance K_d using the column for monthly mean maxima in Table 6.2.

Step 4.3 Tabulate monthly values of I_{oj} (Wm^{-2}) using the clear day values in Table 6.3 for the appropriate hemisphere.

Step 4.4.1 Proceed exactly as indicated in 4.4.1, but use Table 6.15 instead of Table 6.4 to estimate the air mass 2 Linke Turbidity, when using procedures 4.4.1.1 or 4.4.1.2.

Step 4.5 Use the table of Astronomical Daylengths calculated for the clear day dates, Table 6.16 (a) or (b) in place of Table 6.6 to calculate sunrise and sunset.

Step 4.6 Calculate the 4 degree daylength for clear days S_{04max}. using Table 6.17 (a) or (b) which is based on the computational dates for clear days, in place of Table 6.7. Substitute the monthly mean maximum value of

Table 6.15 Monthly values of the constant f_c used in the estimation of representative air mass 2 Linke Turbidity Factors T_L for clear days using monthly mean values of Angstrom (a + b) derived from daily observations on a month by month basis

$$T_L = f_c - f_c(a + b)$$

based on observed global and diffuse radiation data for 13 European stations.

Northern Hemisphere above 25°N	Jan	Feb	Mar	Apr	May	Jun	Jul	Aug	Sep	Oct	Nov	Dec
f_c*	13.5	13.9	13.4	15.4	15.4	15.4	14.9	16.3	14.6	13.2	12.8	11.6
Southern Hemisphere below 25°S	Jul	Aug	Sep	Oct	Nov	Dec	Jan	Feb	Mar	Apr	May	Jun

In the light of present knowledge a value of 15.0 appears appropriate for the region between 25°N and 25°S.

* Values for the Southern Hemisphere are tentative.

Table 6.16a Astronomical daylength S_0 (hours) on days of monthly maximum global irradiation, latitudes 26°-75° North.

Latitude	Jan 29	Feb 26	Mar 29	Apr 28	May 29	Jun 21	Jul 4	Aug 4	Sep 4	Oct 4	Nov 4	Dec 4
75 N	0.00	7.14	13.72	21.42	24.00	24.00	24.00	24.00	15.72	9.78	0.00	0.00
74 N	0.00	7.51	13.61	20.25	24.00	24.00	24.00	24.00	15.46	9.93	2.12	0.00
73 N	0.00	7.82	13.51	19.44	24.00	24.00	24.00	24.00	15.23	10.06	3.41	0.00
72 N	0.00	8.09	13.42	18.81	24.00	24.00	24.00	21.67	15.03	10.18	4.25	0.00
71 N	2.36	8.33	13.34	18.30	24.00	24.00	24.00	20.56	14.85	10.28	4.90	0.00
70 N	3.42	8.54	13.26	17.87	24.00	24.00	24.00	19.78	14.69	10.38	5.43	0.00
69 N	4.17	8.74	13.20	17.49	24.00	24.00	24.00	19.18	14.54	10.46	5.88	0.00
68 N	4.76	8.91	13.14	17.17	22.55	24.00	24.00	18.68	14.41	10.54	6.26	0.00
67 N	5.25	9.06	13.08	16.87	21.22	24.00	23.14	18.25	14.29	10.61	6.60	2.06
66 N	5.67	9.21	13.03	16.61	20.40	22.23	21.51	17.88	14.18	10.68	6.90	3.10
65 N	6.04	9.34	12.98	16.38	19.77	21.11	20.64	17.55	14.08	10.74	7.16	3.82
64 N	6.37	9.46	12.94	16.16	19.26	20.36	19.98	17.26	13.99	10.79	7.41	4.40
63 N	6.66	9.57	12.90	15.97	18.82	19.77	19.45	17.00	13.90	10.84	7.63	4.88
62 N	6.92	9.68	12.86	15.79	18.43	19.28	19.00	16.75	13.82	10.89	7.83	5.29
61 N	7.16	9.78	12.83	15.62	18.09	18.86	18.61	16.53	13.74	10.94	8.02	5.66
60 N	7.38	9.87	12.79	15.46	17.79	18.49	18.26	16.33	13.67	10.98	8.19	5.98
59 N	7.59	9.95	12.76	15.32	17.51	18.15	17.94	16.14	13.61	11.02	8.36	6.28
58 N	7.78	10.03	12.73	15.18	17.26	17.85	17.66	15.96	13.54	11.06	8.51	6.55
57 N	7.96	10.11	12.71	15.06	17.02	17.58	17.40	15.80	13.48	11.09	8.65	6.79
56 N	8.12	10.18	12.68	14.94	16.80	17.33	17.16	15.65	13.43	11.13	8.78	7.02
55 N	8.28	10.25	12.65	14.82	16.60	17.10	16.94	15.50	13.38	11.16	8.91	7.23
54 N	8.42	10.31	12.63	14.72	16.41	16.88	16.73	15.37	13.33	11.19	9.03	7.43
53 N	8.56	10.37	12.61	14.61	16.24	16.68	16.54	15.24	13.28	11.22	9.14	7.62
52 N	8.69	10.43	12.59	14.52	16.07	16.49	16.36	15.12	13.23	11.25	9.24	7.79
51 N	8.81	10.49	12.57	14.43	15.91	16.31	16.19	15.00	13.19	11.27	9.34	7.95
50 N	8.93	10.54	12.55	14.34	15.76	16.15	16.02	14.89	13.15	11.30	9.44	8.11
49 N	9.04	10.59	12.53	14.25	15.62	15.99	15.87	14.79	13.11	11.32	9.53	8.26
48 N	9.15	10.64	12.51	14.17	15.49	15.84	15.72	14.68	13.07	11.35	9.62	8.40
47 N	9.25	10.69	12.49	14.10	15.36	15.69	15.59	14.59	13.03	11.37	9.71	8.53
46 N	9.35	10.74	12.47	14.02	15.23	15.56	15.45	14.50	13.00	11.39	9.79	8.66
45 N	9.45	10.78	12.46	13.95	15.12	15.42	15.33	14.41	12.96	11.41	9.87	8.78
44 N	9.54	10.82	12.44	13.88	15.00	15.30	15.20	14.32	12.93	11.43	9.94	8.90
43 N	9.63	10.86	12.43	13.82	14.90	15.18	15.09	14.24	12.90	11.45	10.01	9.01
42 N	9.71	10.90	12.41	13.75	14.79	15.06	14.98	14.16	12.86	11.47	10.08	9.12
41 N	9.79	10.94	12.40	13.69	14.69	14.95	14.87	14.08	12.83	11.49	10.15	9.22
40 N	9.87	10.98	12.38	13.63	14.59	14.84	14.76	14.01	12.81	11.51	10.22	9.32
39 N	9.95	11.01	12.37	13.58	14.50	14.74	14.66	13.94	12.78	11.52	10.28	9.42
38 N	10.02	11.05	12.36	13.52	14.41	14.64	14.57	13.87	12.75	11.54	10.34	9.51
37 N	10.09	11.08	12.35	13.46	14.32	14.54	14.47	13.80	12.72	11.56	10.40	9.61
36 N	10.16	11.11	12.33	13.41	14.23	14.45	14.38	13.74	12.70	11.57	10.46	9.69
35 N	10.23	11.15	12.32	13.36	14.15	14.36	14.29	13.67	12.67	11.59	10.52	9.78
34 N	10.30	11.18	12.31	13.31	14.07	14.27	14.20	13.61	12.65	11.60	10.57	9.86
33 N	10.36	11.21	12.30	13.26	13.99	14.18	14.12	13.55	12.62	11.62	10.62	9.95
32 N	10.42	11.24	12.29	13.21	13.91	14.09	14.04	13.49	12.60	11.63	10.68	10.03
31 N	10.48	11.27	12.28	13.17	13.84	14.01	13.96	13.43	12.58	11.65	10.73	10.10
30 N	10.54	11.30	12.26	13.12	13.77	13.93	13.88	13.37	12.55	11.66	10.78	10.18
29 N	10.60	11.33	12.25	13.07	13.69	13.85	13.80	13.32	12.53	11.67	10.83	10.25
28 N	10.66	11.35	12.24	13.03	13.62	13.78	13.73	13.26	12.51	11.69	10.88	10.32
27 N	10.72	11.38	12.23	12.99	13.55	13.70	13.65	13.21	12.49	11.70	10.92	10.40
26 N	10.77	11.41	12.22	12.94	13.49	13.63	13.58	13.16	12.47	11.71	10.97	10.47

Table 6.16b Astronomical daylength S_0 (hours) on days of monthly maximum global irradiation, latitudes 26°-75° South.

Lati-tude	Date Jan 29	Feb 26	Mar 29	Apr 28	May 29	Jun 21	Jul 4	Aug 4	Sep 4	Oct 4	Nov 4	Dec 4
26 S	13.58	13.11	12.42	11.63	10.93	10.45	10.73	11.36	12.13	12.90	13.46	13.63
27 S	13.65	13.16	12.44	11.61	10.88	10.38	10.67	11.33	12.14	12.94	13.53	13.70
28 S	13.72	13.21	12.46	11.59	10.83	10.31	10.62	11.30	12.15	12.98	13.60	13.78
29 S	13.80	13.26	12.47	11.57	10.78	10.23	10.56	11.27	12.15	13.02	13.67	13.86
30 S	13.87	13.31	12.49	11.56	10.73	10.16	10.49	11.24	12.16	13.06	13.74	13.93
31 S	13.95	13.37	12.51	11.54	10.68	10.08	10.43	11.21	12.17	13.11	13.81	14.02
32 S	14.03	13.42	12.54	11.52	10.62	10.00	10.37	11.18	12.17	13.15	13.88	14.10
33 S	14.11	13.48	12.56	11.50	10.57	9.92	10.30	11.15	12.18	13.20	13.96	14.18
34 S	14.20	13.53	12.58	11.48	10.51	9.84	10.24	11.11	12.19	13.24	14.04	14.27
35 S	14.28	13.59	12.60	11.46	10.46	9.76	10.17	11.08	12.19	13.29	14.11	14.36
36 S	14.37	13.65	12.62	11.44	10.40	9.67	10.10	11.05	12.20	13.34	14.20	14.45
37 S	14.46	13.72	12.65	11.42	10.34	9.58	10.03	11.01	12.21	13.39	14.28	14.55
38 S	14.56	13.78	12.67	11.40	10.27	9.49	9.95	10.97	12.21	13.44	14.37	14.64
39 S	14.65	13.85	12.69	11.38	10.21	9.39	9.88	10.94	12.22	13.49	14.46	14.74
40 S	14.75	13.92	12.72	11.35	10.14	9.30	9.80	10.90	12.23	13.55	14.55	14.85
41 S	14.86	13.99	12.74	11.33	10.08	9.19	9.71	10.86	12.24	13.61	14.64	14.96
42 S	14.97	14.06	12.77	11.31	10.00	9.09	9.63	10.82	12.25	13.66	14.74	15.07
43 S	15.08	14.14	12.80	11.28	9.93	8.98	9.54	10.77	12.26	13.72	14.85	15.18
44 S	15.19	14.21	12.83	11.26	9.86	8.86	9.45	10.73	12.27	13.79	14.95	15.30
45 S	15.31	14.29	12.86	11.23	9.78	8.75	9.36	10.68	12.27	13.85	15.06	15.43
46 S	15.44	14.38	12.89	11.20	9.70	8.62	9.26	10.64	12.28	13.92	15.18	15.56
47 S	15.57	14.47	12.92	11.17	9.61	8.49	9.16	10.59	12.29	13.99	15.30	15.70
48 S	15.71	14.56	12.95	11.15	9.52	8.36	9.05	10.54	12.31	14.06	15.42	15.84
49 S	15.86	14.65	12.99	11.11	9.43	8.22	8.94	10.48	12.32	14.14	15.56	15.99
50 S	16.01	14.75	13.02	11.08	9.34	8.07	8.82	10.43	12.33	14.22	15.70	16.15
51 S	16.17	14.86	13.06	11.05	9.23	7.91	8.70	10.37	12.34	14.30	15.84	16.32
52 S	16.34	14.97	13.10	11.01	9.13	7.74	8.57	10.31	12.35	14.39	16.00	16.50
53 S	16.52	15.08	13.14	10.98	9.02	7.57	8.43	10.25	12.36	14.48	16.16	16.69
54 S	16.71	15.20	13.18	10.94	8.90	7.38	8.29	10.18	12.38	14.57	16.33	16.89
55 S	16.92	15.33	13.23	10.90	8.78	7.18	8.14	10.11	12.39	14.67	16.52	17.11
56 S	17.14	15.47	13.27	10.86	8.65	6.96	7.98	10.04	12.41	14.78	16.71	17.34
57 S	17.38	15.61	13.32	10.81	8.51	6.73	7.80	9.96	12.42	14.89	16.93	17.59
58 S	17.64	15.77	13.38	10.77	8.36	6.48	7.62	9.88	12.44	15.01	17.15	17.87
59 S	17.92	15.93	13.43	10.72	8.20	6.21	7.42	9.79	12.46	15.14	17.40	18.17
60 S	18.23	16.11	13.49	10.66	8.03	5.91	7.20	9.70	12.48	15.28	17.67	18.50
61 S	18.58	16.30	13.55	10.61	7.85	5.57	6.97	9.60	12.50	15.42	17.97	18.87
62 S	18.97	16.51	13.62	10.55	7.65	5.20	6.72	9.49	12.52	15.58	18.30	19.29
63 S	19.41	16.73	13.69	10.48	7.43	4.78	6.44	9.38	12.54	15.75	18.67	19.79
64 S	19.94	16.98	13.77	10.41	7.20	4.28	6.13	9.26	12.56	15.93	19.09	20.38
65 S	20.58	17.25	13.85	10.34	6.94	3.68	5.78	9.12	12.59	16.13	19.58	21.14
66 S	21.43	17.55	13.94	10.26	6.65	2.92	5.39	8.98	12.62	16.35	20.16	22.28
67 S	22.94	17.89	14.04	10.17	6.33	1.77	4.93	8.82	12.65	16.59	20.91	24.00
68 S	24.00	18.28	14.15	10.08	5.97	0.00	4.39	8.65	12.68	16.86	22.00	24.00
69 S	24.00	18.72	14.26	9.98	5.56	0.00	3.73	8.46	12.72	17.16	24.00	24.00
70 S	24.00	19.25	14.39	9.86	5.07	0.00	2.84	8.25	12.76	17.50	24.00	24.00
71 S	24.00	19.90	14.53	9.74	4.48	0.00	1.34	8.02	12.80	17.89	24.00	24.00
72 S	24.00	20.74	14.69	9.60	3.74	0.00	0.00	7.75	12.85	18.35	24.00	24.00
73 S	24.00	22.05	14.87	9.44	2.71	0.00	0.00	7.45	12.90	18.90	24.00	24.00
74 S	24.00	24.00	15.07	9.26	0.01	0.00	0.00	7.10	12.96	19.58	24.00	24.00
75 S	24.00	24.00	15.29	9.06	0.00	0.00	0.00	6.70	13.03	20.48	24.00	24.00

Table 6.17a 4° sunrise/sunset daylength S_{04} (hours) on days of monthly maximum global irradiation, latitudes 26°-75° North.

Lati-tude	Jan 29	Feb 26	Mar 29	Apr 28	May 29	Jun 21	Jul 4	Aug 4	Sep 4	Oct 4	Nov 4	Dec 4
75 N	0.00	3.99	11.65	17.56	24.00	24.00	24.00	20.13	13.51	7.49	0.00	0.00
74 N	0.00	4.77	11.66	17.12	24.00	24.00	24.00	19.29	13.40	7.81	0.00	0.00
73 N	0.00	5.38	11.67	16.74	24.00	24.00	24.00	18.64	13.30	8.08	0.00	0.00
72 N	0.00	5.88	11.68	16.41	22.40	24.00	24.00	18.11	13.22	8.33	0.00	0.00
71 N	0.00	6.30	11.69	16.12	20.96	24.00	23.06	17.67	13.14	8.54	0.00	0.00
70 N	0.00	6.66	11.70	15.86	20.07	22.07	21.29	17.29	13.07	8.73	1.90	0.00
69 N	0.00	6.97	11.70	15.63	19.40	20.86	20.34	16.95	13.00	8.90	3.07	0.00
68 N	0.00	7.25	11.71	15.43	18.85	20.06	19.65	16.65	12.94	9.05	3.85	0.00
67 N	2.15	7.50	11.71	15.24	18.39	19.43	19.08	16.39	12.89	9.19	4.45	0.00
66 N	3.13	7.72	11.72	15.07	17.99	18.91	18.61	16.15	12.83	9.31	4.95	0.00
65 N	3.83	7.93	11.72	14.91	17.64	18.47	18.20	15.93	12.79	9.43	5.37	0.00
64 N	4.38	8.11	11.72	14.77	17.33	18.09	17.84	15.73	12.74	9.54	5.74	0.00
63 N	4.85	8.28	11.72	14.63	17.04	17.74	17.52	15.54	12.70	9.63	6.07	1.91
62 N	5.25	8.44	11.72	14.51	16.79	17.44	17.23	15.37	12.66	9.73	6.36	2.88
61 N	5.60	8.58	11.72	14.39	16.55	17.16	16.96	15.21	12.62	9.81	6.63	3.56
60 N	5.91	8.72	11.73	14.28	16.33	16.90	16.72	15.07	12.59	9.89	6.87	4.10
59 N	6.20	8.84	11.73	14.18	16.13	16.67	16.49	14.93	12.55	9.96	7.09	4.56
58 N	6.46	8.96	11.73	14.08	15.94	16.45	16.29	14.80	12.52	10.03	7.29	4.95
57 N	6.70	9.07	11.72	13.99	15.77	16.25	16.09	14.68	12.49	10.10	7.48	5.31
56 N	6.91	9.17	11.72	13.90	15.60	16.06	15.91	14.56	12.46	10.16	7.66	5.62
55 N	7.12	9.27	11.72	13.82	15.44	15.88	15.74	14.45	12.43	10.21	7.82	5.91
54 N	7.31	9.36	11.72	13.74	15.30	15.71	15.58	14.35	12.41	10.27	7.98	6.17
53 N	7.48	9.44	11.72	13.66	15.16	15.56	15.43	14.25	12.38	10.32	8.12	6.41
52 N	7.65	9.52	11.72	13.59	15.03	15.41	15.29	14.15	12.35	10.37	8.26	6.63
51 N	7.81	9.60	11.72	13.52	14.90	15.26	15.15	14.06	12.33	10.42	8.39	6.84
50 N	7.95	9.68	11.72	13.46	14.78	15.13	15.02	13.98	12.31	10.46	8.51	7.04
49 N	8.09	9.75	11.71	13.39	14.67	15.00	14.89	13.89	12.28	10.50	8.62	7.22
48 N	8.23	9.81	11.71	13.33	14.56	14.88	14.77	13.81	12.26	10.54	8.74	7.39
47 N	8.35	9.88	11.71	13.27	14.45	14.76	14.66	13.74	12.24	10.58	8.84	7.56
46 N	8.47	9.94	11.71	13.21	14.35	14.64	14.55	13.66	12.22	10.62	8.94	7.71
45 N	8.59	10.00	11.70	13.16	14.25	14.54	14.45	13.59	12.20	10.65	9.04	7.86
44 N	8.70	10.06	11.70	13.11	14.16	14.43	14.34	13.52	12.18	10.68	9.13	8.00
43 N	8.80	10.11	11.70	13.05	14.07	14.33	14.25	13.45	12.16	10.72	9.22	8.14
42 N	8.91	10.16	11.69	13.00	13.98	14.23	14.15	13.39	12.14	10.75	9.30	8.26
41 N	9.00	10.21	11.69	12.95	13.89	14.14	14.06	13.33	12.12	10.78	9.39	8.39
40 N	9.10	10.26	11.69	12.91	13.81	14.04	13.97	13.26	12.10	10.81	9.46	8.51
39 N	9.19	10.31	11.68	12.86	13.73	13.95	13.88	13.20	12.08	10.83	9.54	8.62
38 N	9.27	10.35	11.68	12.81	13.65	13.87	13.80	13.15	12.07	10.86	9.61	8.73
37 N	9.36	10.40	11.68	12.77	13.58	13.78	13.72	13.09	12.05	10.88	9.68	8.83
36 N	9.44	10.44	11.67	12.73	13.50	13.70	13.64	13.03	12.03	10.91	9.75	8.94
35 N	9.52	10.48	11.67	12.68	13.43	13.62	13.56	12.98	12.02	10.93	9.82	9.03
34 N	9.59	10.52	11.66	12.64	13.36	13.54	13.49	12.93	12.00	10.96	9.89	9.13
33 N	9.67	10.56	11.66	12.60	13.29	13.47	13.41	12.87	11.98	10.98	9.95	9.22
32 N	9.74	10.60	11.66	12.56	13.22	13.39	13.34	12.82	11.97	11.00	10.01	9.31
31 N	9.81	10.63	11.65	12.52	13.16	13.32	13.27	12.77	11.95	11.02	10.07	9.40
30 N	9.88	10.67	11.65	12.48	13.09	13.25	13.20	12.72	11.93	11.04	10.13	9.49
29 N	9.95	10.70	11.64	12.44	13.03	13.18	13.13	12.68	11.92	11.06	10.18	9.57
28 N	10.01	10.74	11.64	12.40	12.96	13.11	13.06	12.63	11.90	11.08	10.24	9.65
27 N	10.07	10.77	11.63	12.37	12.90	13.04	13.00	12.58	11.89	11.10	10.29	9.73
26 N	10.14	10.80	11.63	12.33	12.84	12.97	12.93	12.53	11.87	11.12	10.34	9.80

Table 6.17b 4° sunrise/sunset daylength S_{04} (hours) on days of monthly maximum global irradiation, latitudes 26°-75° South.

Lati-tude	Jan 4	Feb 4	Mar 4	Apr 4	May 4	Jun 4	Jul 29	Aug 29	Sep 28	Oct 29	Nov 28	Dec 22
26 S	12.93	12.48	11.82	11.03	10.30	9.79	10.09	10.75	11.54	12.28	12.82	12.97
27 S	12.99	12.53	11.83	11.00	10.25	9.71	10.03	10.72	11.54	12.32	12.88	13.04
28 S	13.06	12.57	11.85	10.98	10.19	9.63	9.96	10.68	11.54	12.35	12.94	13.11
29 S	13.12	12.62	11.86	10.96	10.13	9.55	9.89	10.65	11.54	12.39	13.00	13.18
30 S	13.19	12.66	11.87	10.93	10.07	9.47	9.83	10.61	11.54	12.43	13.06	13.25
31 S	13.26	12.71	11.89	10.91	10.01	9.38	9.75	10.57	11.54	12.46	13.13	13.32
32 S	13.33	12.76	11.90	10.88	9.95	9.29	9.68	10.53	11.54	12.50	13.19	13.39
33 S	13.40	12.81	11.92	10.86	9.89	9.20	9.61	10.49	11.54	12.54	13.26	13.47
34 S	13.48	12.86	11.93	10.83	9.82	9.11	9.53	10.45	11.54	12.58	13.33	13.55
35 S	13.55	12.91	11.94	10.80	9.76	9.01	9.45	10.41	11.54	12.62	13.40	13.62
36 S	13.63	12.96	11.96	10.77	9.69	8.91	9.37	10.37	11.54	12.66	13.47	13.70
37 S	13.71	13.01	11.97	10.75	9.62	8.81	9.29	10.32	11.54	12.70	13.54	13.79
38 S	13.79	13.06	11.99	10.71	9.54	8.70	9.20	10.28	11.54	12.74	13.62	13.87
39 S	13.88	13.12	12.00	10.68	9.47	8.59	9.11	10.23	11.54	12.78	13.69	13.96
40 S	13.96	13.18	12.02	10.65	9.39	8.47	9.02	10.18	11.53	12.83	13.77	14.05
41 S	14.05	13.23	12.03	10.62	9.31	8.36	8.92	10.13	11.53	12.87	13.85	14.14
42 S	14.14	13.29	12.05	10.58	9.22	8.23	8.82	10.07	11.53	12.92	13.94	14.24
43 S	14.24	13.35	12.06	10.54	9.13	8.10	8.71	10.02	11.53	12.96	14.02	14.33
44 S	14.33	13.42	12.08	10.50	9.04	7.97	8.61	9.96	11.52	13.01	14.11	14.44
45 S	14.43	13.48	12.10	10.46	8.94	7.82	8.49	9.90	11.52	13.06	14.20	14.54
46 S	14.54	13.55	12.11	10.42	8.84	7.67	8.37	9.84	11.52	13.11	14.30	14.65
47 S	14.65	13.62	12.13	10.38	8.74	7.52	8.25	9.77	11.51	13.17	14.40	14.76
48 S	14.76	13.69	12.15	10.33	8.63	7.35	8.12	9.70	11.51	13.22	14.50	14.88
49 S	14.88	13.77	12.17	10.29	8.52	7.18	7.98	9.63	11.50	13.28	14.61	15.01
50 S	15.00	13.85	12.19	10.24	8.39	6.99	7.83	9.56	11.50	13.34	14.72	15.13
51 S	15.13	13.93	12.20	10.18	8.27	6.79	7.68	9.48	11.49	13.40	14.84	15.27
52 S	15.27	14.01	12.22	10.13	8.13	6.58	7.52	9.39	11.49	13.47	14.96	15.41
53 S	15.41	14.10	12.24	10.07	7.99	6.35	7.34	9.31	11.48	13.53	15.09	15.56
54 S	15.56	14.20	12.27	10.01	7.84	6.11	7.16	9.22	11.47	13.60	15.23	15.72
55 S	15.72	14.29	12.29	9.94	7.68	5.84	6.96	9.12	11.46	13.68	15.37	15.89
56 S	15.89	14.40	12.31	9.88	7.51	5.55	6.75	9.02	11.45	13.75	15.52	16.07
57 S	16.07	14.50	12.33	9.80	7.32	5.23	6.52	8.91	11.44	13.84	15.68	16.26
58 S	16.27	14.62	12.36	9.73	7.12	4.87	6.27	8.79	11.43	13.92	15.85	16.46
59 S	16.47	14.74	12.38	9.64	6.91	4.46	6.00	8.67	11.42	14.01	16.04	16.68
60 S	16.70	14.87	12.41	9.56	6.68	3.99	5.70	8.53	11.41	14.11	16.23	16.91
61 S	16.94	15.01	12.44	9.46	6.42	3.43	5.36	8.39	11.40	14.21	16.45	17.17
62 S	17.20	15.16	12.47	9.36	6.14	2.71	4.98	8.23	11.38	14.32	16.68	17.45
63 S	17.49	15.31	12.50	9.25	5.83	1.63	4.55	8.07	11.36	14.43	16.93	17.76
64 S	17.81	15.49	12.53	9.14	5.48	0.00	4.04	7.88	11.35	14.56	17.20	18.10
65 S	18.17	15.67	12.57	9.01	5.08	0.00	3.42	7.69	11.33	14.69	17.50	18.49
66 S	18.57	15.87	12.60	8.87	4.62	0.00	2.60	7.47	11.31	14.84	17.84	18.93
67 S	19.04	16.10	12.64	8.72	4.07	0.00	1.22	7.23	11.28	14.99	18.23	19.45
68 S	19.60	16.34	12.69	8.55	3.38	0.00	0.00	6.96	11.26	15.17	18.67	20.08
69 S	20.28	16.61	12.73	8.37	2.44	0.00	0.00	6.65	11.23	15.36	19.18	20.89
70 S	21.20	16.92	12.78	8.17	0.01	0.00	0.00	6.31	11.19	15.56	19.81	22.12
71 S	22.84	17.26	12.84	7.94	0.00	0.00	0.00	5.91	11.16	15.80	20.62	24.00
72 S	24.00	17.66	12.90	7.68	0.00	0.00	0.00	5.44	11.12	16.06	21.80	24.00
73 S	24.00	18.13	12.97	7.38	0.00	0.00	0.00	4.88	11.07	16.36	24.00	24.00
74 S	24.00	18.69	13.04	7.04	0.00	0.00	0.00	4.17	11.02	16.70	24.00	24.00
75 S	24.00	19.38	13.13	6.63	0.00	0.00	0.00	3.20	10.96	17.10	24.00	24.00

observed bright sunshine S_{max} from the tables in the European Solar Radiation Atlas, Vol. I, 2nd Edition. If the station is not in Atlas, if possible attempt to obtain monthly mean maximum values of S_{max} from records. Compute the daily relative sunshine duration for days of high radiation as S_{max}/S_{04max}. If values of S_{max} cannot be obtained assume the ratio to be 0.90 for clear days. Alternatively adopt the method given in Appendix 3.

Step 5.1 Set the computational times to match the daylengths in Stage 4.5.

Steps 5.2 to 7.5 Then proceed through to step 7.5 using the new values relating to the solar geometry and the atmospheric turbidity.

New Step 7.6 Finally, introduce a new step 7.6 to allow for the effects of slight amounts of cloud on the direct beam availability.

Set $I_c(0,0) = I_c(0,0) \times S_{max}/S_{04max}$.

Then compute clear day sky diffuse on a horizontal plane as before. Stop horizontal calculations on completion of step 9. Then proceed to normalization if values of G_{max} the observed monthly mean maximum daily radiation on a horizontal surface for a specific month are available. These values are available in the systematic tables published in the European Solar Radiation Atlas, Vol. 1, Second Edition. The individual normalization process is carried out as before by multiplying the hourly computed values of the global, the direct beam and the diffuse irradiance by the ratio (daily observed clear day mean maximum daily global radiation/predicted clear day mean maximum daily global radiation). Obviously the normalization process can only be carried out if observed values of G_{max} are available. All the clear day tables in the CEC Solar Radiation Inclined Surface Atlas were normalized against G_{max} with the exception of Berlin where no observed values were available for G_{max} and S_{max}.

20.3 The method proposed above is not suitable for latitudes between 23° N and 23° S, as the times of year when the sun is highest will depend on position in relation to the equator. In this band it is suggested that the monthly mean values of declination etc. be used. It follows that the only additional calculation for S_{max} values is to complete the new step 7.6 having carried out the basic calculations for average days.

21. SLOPE IRRADIATION ASSOCIATED WITH MONTHLY MEAN MAXIMUM HORIZONTAL IRRADIATION

The final steps are to proceed to inclined plane calculations using the methods in Sections 15 - 16. Passing over steps 17 and 18, one proceeds to step 19 to obtain the inclined plane monthly irradiation values associated with the days of monthly mean maximum

PROFORMA 1: MONTHLY MEAN PREDICTIONS – HORIZONTAL SURFACE CLEAR DAY SOLAR RADIATION VALUES – EC METHOD

Station name: PARIS/TRAPPES **Site clarity:** URBAN **Month:** FEB

Latitude ϕ: deg 48 min 46 N+,S− + **Longitude** λ: deg 02 min 01 E+,W− + **Declination** δ **from Table 6.2** $\pm$ − 12.81 Degrees

Station height (m): 168

AH2T$_L$ Table 6.4: 03.31

Sunrise hrs 07 min 00 $\pm$ − $\sin\delta \sin\phi$ .1667

Sunset hrs 17 min 00 $\pm$ + $\cos\delta \cos\phi$ .6427 $\pm$ + $\sin\delta$ .9751 p/p_0 .9832

I_{oj} Table 6.3: 1401 Wm^{-2}

Angstrom coefficients
Monthly values Angstrom (a + b) .75
Annual mean Angstrom (a + b) .75

[1] Local apparent time t a.m.	[1] p.m.	[2] Cosine of hour angle ω	[3] $\sin\gamma = (\cos\delta\cos\phi)\cos\omega + (\sin\delta\sin\phi)$	[4] Solar altitude γ	[5] $\cos\gamma$	[6] Solar azimuth ψ (− a.m., + p.m.) $\cos^{-1}(\sin\phi\sin\gamma - \sin\delta)/\cos\phi\cos\gamma$	[7] Optical air mass. If $\gamma < 10°$ then: $m = p/p_0/(\sin\gamma + 0\cdot15(\gamma + 3\cdot885)^{-1\cdot25}$ otherwise: $m = p/p_0/\sin\gamma$	[8] Rayleigh optical thickness $\delta_R = 1/(0\cdot9m + 9\cdot4)$	[9] Linke turbidity for actual solar altitude $T_L(\gamma)$	[10] Direct beam irradiance normal to beam $I_c = I_{oj}\exp(-m\delta_R T_L(\gamma))$	[11] Direct beam irradiance on a horizontal surface $I_c(0,0)$ [[3] × [10]]	[12] Atmospheric transmittance coefficient absorption alone. q_a^m. Table 6.8	[13] Additional scattering factor f_1 Table 6.9	[14] Diffuse irradiance on horizontal surface D_c $\frac{1}{2}[I_{oj} \times [12] - [10]] \times [13] \times [3]$	[15] Global irradiance on a horizontal surface G_c [11] + [14]
				Deg.		Deg.				Wm^{-2}	Wm^{-2}			Wm^{-2}	Wm^{-2}
0330	2030	−0.6088	-	-	-	-	-	-	-	-	-	-	-	-	-
0430	1930	−0.3827	-	-	-	-	-	-	-	-	-	-	-	-	-
0530	1830	−0.1305	-	-	-	-	-	-	-	-	-	-	-	-	-
0630	1730	0.1305	-	-	-	-	-	-	-	-	-	-	-	-	-
0730	1630	0.3827	0.0793	4.55	0.9968	±64.65	10.96	0.0519	2.63	314	25	0.667	1.283	32	57
0830	1530	0.6088	0.2246	12.98	0.9744	±52.54	4.38	0.0749	2.91	539	121	0.736	1.159	64	185
0930	1430	0.7934	0.3432	20.07	0.9393	±39.19	2.86	0.0835	3.10	668	229	0.780	1.111	81	310
1030	1330	0.9239	0.4271	25.28	0.9042	±24.37	2.30	0.0872	3.22	735	314	0.806	1.086	91	405
1130	1230	0.9914	0.4705	28.07	0.8824	± 8.32	2.09	0.0886	3.27	765	360	0.818	1.075	96	456
							Half day totals Whm^{-2}			3021	1049	Half day totals Whm^{-2}		364	1413
							Daily totals Whm^{-2}			6042	2098	Daily totals Whm^{-2}		728	2826

horizontal surface irradiation. It should be noted that, for certain surfaces in certain months, the days of monthly mean maximum inclined surface radiation may not coincide with the days of monthly mean maximum horizontal surface radiation. In Europe this phenomenon is particularly encountered on south facing vertical surfaces between March and September, especially at lower latitudes. It is the consequence of the impact of the changing solar geometry across the course of the month on the value of the cosine of angle of incidence. It is often helpful to plot the inclined plane predictions against the day of year on a graph to perceive the detailed course of the irradiation fluxes on specific inclined planes across the course of each month.

22. WORKED EXAMPLE - COMPUTATION OF MONTHLY MEAN HOURLY AND DAILY VALUES OF SOLAR RADIATION ON AN INCLINED SURFACE

22.1 Accuracy required

This worked example is provided to demonstrate the method in detail, and to provide a basis for checking where computer programs are developed.

It is suggested that the calculations should be carried out using 4 significant figures for trigonometrical quantities and expressing the solar altitude and azimuth in the decimal form in degrees to two decimal places. The Air Mass 2 Linke Turbidity Factor should be computed to 2 decimal places, e.g. 3.31. The air mass should be expressed to 2 decimal places, and the Rayleigh optical thickness to 4 decimal places, e.g. .0749. The irradiance values should be rounded to the nearest Wm^{-2}.

When data are drawn from the associated Tables, the most accurate results will be achieved using graphical interpolation, as the functions are not linear. Linear interpolation will produce slight differences in some of the more widely spaced tables, but the errors will usually be less than ½%. It is suggested that interpolated values drawn from Tables should be expressed to the same number of significant figures as the basic table from which they are drawn. The boxes at the head of the proforma described below are designed to limit input accuracy to the appropriate level, i.e. the 4 degree daylength is expressed to 2 decimal places, and σ_{4m} to 3 decimal places and so on.

22.2 Arrangement of the proformas for systematic computation

Four working proformas are provided on which to carry out the calculations. The calculations should proceed in sequence from Proforma 1 to Proforma 4.

The first task is to enter the data into the various boxes at the head of each Table, using the supporting Tables in the text where appropriate. The boxes are, as far as possible, arranged to be vertically above the computational stage in which the number in them is actually used in the computational process at the hourly level. It is most important to enter the signs correctly. Once the

PROFORMA 2: HORIZONTAL SURFACE MONTHLY MEAN SOLAR RADIATION PREDICTIONS – EC METHOD

Station name: PARIS/TRAPPES **Site type:** URBAN **AM2 T_L** 03.31 **Month:** FEB

Monthly mean duration of bright sunshine S_m 3.10 hours | Year **Period from** 1966 **to** Year 1975 | **Angstrom coefficients Month** .75

4° degree SR/SS daylength from Table 6.7 9.12 hours | σ_{4m} 0.340 | **Annual mean** .75

[17] From Form 1 Col. [3]; [18] K_d from Table 6.2: 1.025; [19] and [20] From Form 1 Col. [4] Col. [11]; [23] K_d from Table 6.2: 1.025; [24] From Form 1 Col. [14]

[16] Local apparent time t		[17] Sin γ	[18] Overcast day diffuse D_b (= global G_b on hor.) $K_d(2.6 + 182.6\sin\gamma)$	[19] Solar altitude γ	[20] Clear day direct beam horizontal irradiance $I_c(0,0)$	[21] Direct beam correction function for monthly means Table 6.10	[22] Monthly mean direct beam irradiance $I_m(0,0)$ $[[20] \times [21] \times \sigma_{4m}]$	[23] Monthly mean horizontal diffuse irradiance $\sigma_{4m} = 0.25D_{0.25}$ $= K_d(2 + 5.3\gamma)$	[24] Clear day diffuse irradiance D_c	[25] Monthly mean clear day contributions $\sigma_{4m} \times [24]$	[26] $0.25 \times D_c$ i.e. $0.25 \times [24]$	[27] Uncorrected overcast day contribution $(1 - \sigma_{4m})(D_{0.25} - 0.25D_c)/0.75$	[28] Diffuse correction function f_6 from Table 6.13 i.e. correct for annual (a + b)	[29] Monthly mean horizontal diffuse irradiance D_m $[[25] + [27]] \times [28]$	[30] Monthly mean global horizontal irradiance G_m $[22] + [29]$
a.m.	p.m.		Wm^{-2}	Deg.	Wm^{-2}		Wm^{-2}	Wm^{-2}	Wm^{-2}	Wm^{-2}	Wm^{-2}	Wm^{-2}		Wm^{-2}	Wm^{-2}
0330	2030	-	-	-	-	-	-	-	-	-	-	-	-	-	-
0430	1930	-	-	-	-	-	-	-	-	-	-	-	-	-	-
0530	1830	-	-	-	-	-	-	-	-	-	-	-	-	-	-
0630	1730	-	-	-	-	-	-	-	-	-	-	-	-	-	-
0730	1630	0.0793	17.5	4.55	25	0.839	7	27	32	11	8	17	0.874	24	31
0830	1530	0.2246	44.7	12.98	121	0.867	36	73	64	22	16	50	0.946	68	104
0930	1430	0.3432	66.9	20.07	229	0.880	69	111	81	27	20	80	0.989	106	175
1030	1330	0.4271	82.6	25.28	314	0.884	94	139	91	31	23	102	1.013	135	229
1130	1230	0.4705	90.7	28.07	360	0.885	108	155	96	33	24	115	1.023	151	259

Totals	Overcast [18]	Mean beam [22]	Diffuse [29]	Global [30]
Half day totals Whm^{-2}	302.4	314	484	798
Daily totals Whm^{-2}	605	628	968	1596

Half day totals Whm^{-2} overcast: 302.4; Daily totals Whm^{-2} overcast: 605

Half day totals Whm^{-2} mean beam: 314; Daily day totals Whm^{-2} mean beam: 628

Half day totals Whm^{-2} Diffuse and global: 484, 798; Daily totals Whm^{-2} Diffuse and global: 968, 1596

daylength has been established, it becomes possible to mark off the hours for which the calculations actually have to be performed. The form has been designed to allow calculations for any month of the year up to latitude 60°N and 60°S. Except in mid-summer, at high latitudes, only some of the lines will need to be used.

Each column in each table is numbered. The column numbers progressively increase from Table 1 to Table 4, so each column is specifically numbered with a unique reference number. Mathematical operations are often indicated as a mathematical operation on the columns, i.e. Proforma 4, Column 59 the statement = (57) + (58) means Columns 57 and Column 58 are added together to obtain the values for Column 59. When using Proformas 2 to 4, frequently numbers are transferred from the hourly columns on one form to another. The instructions for such transfers are given at the head of the hourly columns. On the completion of Proforma 2, a decision has to be made whether the calculations are to be normalized against observed data or not. Normalization has no effect on the values in Proforma 3, nor does it have any effect on any clear day data used in Proforma 4, but it does affect the values of $I_m(0,0)$ in Column 51, D_m in Column 53, and $R_m(\beta,\alpha)$ in Column 62 in proforma 4. The normalization process should be carried out before this transfer of numbers is done.

If calculations are to be carried out for differing amounts of sunshine in different years, Proformas 1 and 3 will be unaltered, and only the computations on Proformas 2 and 4 will need to be performed.

22.3 Example worked for Paris/Trappes

22.3.1 Inputs

Station: Paris/Trappes   Latitude: 48°46'N   Longitude: 02°01'E
Height: 168m.

Month: February

Observed monthly mean hours of bright sunshine: 3.1 hours
CEC Solar Radiation Atlas, Vol. I, 2nd edition

Observed global radiation on a horizontal surface: 1590 $Whm^{-2}\ d^{-1}$ (WRR)
CEC Solar Radiation Atlas, Vol. I, 2nd edition

Derived value of Angstrom (a + b) for February: 0.75
CEC Solar Radiation Atlas, Vol. I, 2nd Edition.

Annual mean value of Angstrom (a + b): 0.75
CEC Solar Radiation Atlas, Vol. I, 2nd Edition.

Slope of surface for which estimates are required: 45°

Orientation of surface, measured from due south: -45°
(West positive, East negative)
The bearing of the normal to the surface is thus 135°. It is measured from true north, not magnetic north.

PROFORMA 3: INCLINED SURFACE CLEAR DAY AND OVERCAST PREDICTIONS OF SOLAR RADIATION – EC METHOD

Station name: PARIS/TRAPPES

Ground albedo ρ_s .20

AM2 T_L 03.31

Month: FEB

Surface orientation α with respect to due S in N hemis., due N in S hemis. (-E, +W) − 45.00

Ground reflectance factor f_3 .029

Overcast sky slope conversion factor f_4 .798

Surface slope degrees β 45.00

	Enter from Proforma 1 Col. [4]	Col. [6]		Surface slope degrees β 45.00	From Form 1 Col. [10]	Clear Day Beam on Slope	From Form 1 Col. [14]	Clear Day Diffuse on slope				Clear Day Global on Slope	From Form 2 Col. [18]	Overcast Day Diffuse and Global on Slope		
[31]	[32]	[33]	[34]	[35]	[36]	[37]	[38]	[39]	[40]	[41]	[42]	[43]	[44]	[45]	[46]	[47]
Local apparent time	Solar altitude γ	Solar azimuth ψ −ve East +ve West	Wall solar azimuth angle α_{wF} $\alpha_w = (\psi - \alpha)$ *Refer ro footnote	Cosine of angle of incidence $\cos \nu = \cos \gamma \sin \beta \cos \alpha_w + \sin \gamma \cos \beta$ If $\cos \nu < 0$ then $\cos \nu = 0$	Direct beam normal irradiance I_c	Direct beam slope irradiance $I_c(\beta, \alpha)$ [35] × [36]	Clear sky diffuse horizontal D_c	f_2 interpolate from Tables 5.14 A-J	Clear sky diffuse on slope $D_{cs}(\beta, \alpha)$ [38] × [39]	Ground reflected slope diffuse, $R_c(\beta, \alpha)$ Proforma 1 [15] × f_3	Total diffuse on slope, $D_c(\beta, \alpha)$ [40] + [41]	Global on slope, $G_c(\beta, \alpha)$ [37] + [42]	Overcast sky horizontal irradiance D_b	Overcast sky diffuse on slope $D_{bs}(\beta, \alpha)$ [44] × f_4	Overcast sky ground reflected slope Form 2 [18] × f_3 (f_3 is above Col. 41)	Overcast sky global on slope $D_b(\beta \cdot \alpha)$ [45] + [46]
Hours	Deg.	Deg.*	Deg.		Wm-2	Wm-2	Wm-2		Wm-2	Wm-2	Wm-2	Wm-2	Wm-2	Wm-2	Wm-2	Wm-2
0330	-	– -	-	-	-	-	-	-	-	-	-	-	-	-	-	-
0430	-	– -	-	-	-	-	-	-	-	-	-	-	-	-	-	-
0530	-	– -	-	-	-	-	-	-	-	-	-	-	-	-	-	-
0630	-	– -	-	-	-	-	-	-	-	-	-	-	-	-	-	-
0730	4.55	−64.65	−19.65	0.7199	314	226	32	1.726	55	1.7	57	283	17.5	14	0	14
0830	12.98	−52.54	− 7.54	0.8419	539	454	64	1.554	99	5.4	104	558	44.7	36	1	37
0930	20.07	−39.19	5.81	0.9034	668	603	81	1.503	122	9.1	131	734	66.9	53	2	55
1030	25.28	−24.37	20.63	0.9004	735	662	91	1.423	129	11.9	141	803	82.6	66	2	68
1130	28.07	− 8.32	36.68	0.8331	765	637	96	1.327	127	13.4	140	777	90.7	72	3	75
1230	28.07	+ 8.32	53.32	0.7054	765	540	96	1.222	117	13.4	130	670	90.7	72	3	75
1330	25.28	+24.37	69.37	0.5272	735	387	91	1.104	100	11.9	112	499	82.6	66	2	68
1430	20.07	+39.19	84.19	0.3099	668	207	81	0.970	79	9.1	88	295	66.9	53	2	55
1530	12.98	+52.54	97.54	0.0684	539	37	64	0.851	54	5.4	59	96	44.7	36	1	37
1630	4.55	+64.65	109.65	0.0000	314	0	32	0.874	28	1.7	30	30	17.5	14	0	14
1730	-	+ -	-	-	-	-	-	-	-	-	-	-	-	-	-	-
1830	-	+ -	-	-	-	-	-	-	-	-	-	-	-	-	-	-
1930	-	+ -	-	-	-	-	-	-	-	-	-	-	-	-	-	-
2030	-	+ -	-	-	-	-	-	-	-	-	-	-	-	-	-	-
					Totals Whm-2	3753		Totals Whm-2	910	83	992	4745	Totals Whm-2	482	16	498

* Values must be between −180° and 180°. Adjust using Stage 15.1.

22.3.2 Detailed workings

The detailed workings are given on Proformas 1 to 4. The predicted value of the horizontal surface monthly mean global irradiance from Proforma 2 is 1595 Whm^{-2} d^{-1}. The normalization factor is thus 1590/1596 = .9962, which in this particular case is very small.

Typically the normalization factor is a few percent. In the Tables in the Atlas, the normalization factors are rounded to the nearest percent, and hence in the tables for Paris/Trappes, it will be noted a value of 1.00 is entered for the normalization factor for monthly mean estimates.

Blank proformas for photocopying are provided in Appendix 10.

23. MONTHLY MEAN MAXIMUM SOLAR RADIATION COMPUTATIONS

Proformas 1 and 3 can be used as the basis of such calculations, but the appropriate declination for days of maximum global radiation must be used. The correct date in the month from Table 6.2 should be entered on the Proforma. The astronomical daylength S_{0max} and the 4 degree sunrise daylength S_{04max} must be selected from Tables 6.16 and 6.17. Values of the monthly mean maximum hours of bright sunshine S_{max} are given in the station tables in the CEC Solar Radiation Atlas, Vol. I, Second Edition. These values should be used in place of the monthly mean values. Proformas 2 and 4 are not used, but, before adding the direct irradiance in Column 11 to the diffuse irradiance in Column 14 the values in Column 11 should be multiplied by S_{max}/S_{04max}. The data on Proforma 1 can be normalized to match the observed values of G_{max} if available. The corrected values of $I_c(0,0)$ and D_c are then entered into Columns 36, 38 and 41 of Proforma 3 before completing the inclined plane calculation. The Tables computed in the CEC Atlas for clear days include the normalization factors that had to be adopted to match the predicted G_{max} horizontal data to the observed values of G_{max} reported in the horizontal surface Atlas for example. The normalization factor for G_{max} for Paris/Trappes for February was found to be 1.06. This was the highest normalization value for G_{max} values for any month in the year at Paris/Trappes where the observed value of G_{max} was 3302 Whm^{-2} d^{-1} WRR, and the observed value of S_{max} was 8.8 hours in February.

24. AUTOMATION OF THE PROCESS ON MICROCOMPUTERS

The above processes of computation are tedious because of the large number of operations that have to be completed. An interactive microcomputer program for carrying out the computation rapidly has been developed in the Department of Building Science, University of Sheffield. This runs on an Apple IIE with 48K of memory. Its use requires no programming experience. Enquiries concerning the availability and costs of the microcomputer program should be addressed to: Head of the Department of Building Science, University of Sheffield, Western Bank, Sheffield, S10 2TN, UK. Telephone: (0742) 78555 ext. 4708 Telex: 547216 UG SHEF G.

PROFORMA 4: INCLINED SURFACE MONTHLY MEAN SOLAR RADIATION PREDICTIONS – EC METHOD

Station name: PARIS/TRAPPES   Normalisation factor N.F. 0.9962   AM2 T_L 03.31   Month: FEB   Albedo .20

Surface inclination 45.00 degrees   Surface azimuth (-E, +W) – 45.00 degrees   Rel. 4 degree sunshine duration σ_{4m} 0.340   Ground reflection factor f_3 .029

Overcast sky conversion factor f_4 .798

[48] Local apparent time	[49] Sine of solar altitude Sin γ (From Form 1 Col. [3])	[50] Cosine of angle of incidence Cos ν (From Form 3 Col. [35])	[51] Mean hor. direct beam irradiance, $I_m(0,0)$ Normalised values [22] × N.F. (From Form 2 col [22]*)	[52] Monthly mean direct beam slope irradiance, $I_m(\beta,\alpha)$ [51] × [50]/[49]	[53] Monthly mean hor. diffuse irradiance D_m Normalised values [29] × N.F. (From Form 2 col [29]*)	[54] Clear sky horizontal diffuse irradiance D_c (From Form 1 Col. [14])	[55] Monthly mean clear sky horizontal irradiance contribution $\sigma_{4m} D_c$ (From Form 2 Col. [25])	[56] Monthly mean overcast partially clouded horizontal sky irradiance contribution [53] – [55]	[57] Monthly mean overcast/partially clouded slope sky irradiance contribution $_{pc}D_m$ [56] × f_4	[58] Monthly clear sky slope irradiance contribution $_cD_m$ = [39] × [54] × σ_{4m}	[59] Uncorrected monthly mean sky diffuse slope irradiance [57] + [58]	[60] If 45° < |α_w| < 135°, compute additional correction factor $f_8 = (1 + \sin\beta \sin 2(\alpha_s - 45°) \times (0.19 - 1.4 \sin\gamma))$. Else $f_8 = 1$	[61] Corrected monthly mean sky diffuse irradiance on slope $D_m(\beta,\alpha)$ [60] × [59]	[62] Reflected mean diffuse on slope $R_m(\beta,\alpha)$ using normalised values of G_m f_3 × [30] × N.F.	[63] Monthly mean global irradiance on slope, $G_m(\beta,\alpha)$ [52] + [61] + [62]
Hours			Wm^{-2}	Wm^{-2}	Wm^{-2}	Wm^{-2}	Wm^{-2}	Wm^{-2}	Wm^{-2}	Wm^{-2}	Wm^{-2}		Wm^{-2}	Wm^{-2}	Wm^{-2}
0330	-	-	-	-	-	-	-	-	-	-	-	-	-	-	-
0430	-	-	-	-	-	-	-	-	-	-	-	-	-	-	-
0530	-	-	-	-	-	-	-	-	-	-	-	-	-	-	-
0630	-	-	-	-	-	-	-	-	-	-	-	-	-	-	-
0730	0.0793	0.7199	7	64	24	32	11	13	10	19	29	1.000	29	1	94
0830	0.2246	0.8419	36	135	67	64	22	45	36	34	70	1.000	70	3	208
0930	0.3432	0.9034	68	179	106	81	27	79	63	41	104	1.000	104	5	288
1030	0.4271	0.9004	94	198	135	91	31	104	83	44	127	1.000	127	7	332
1130	0.4705	0.8331	108	191	150	96	33	117	93	43	136	1.000	136	8	335
1230	0.4705	0.7054	108	162	150	96	33	117	93	40	133	1.025	136	8	306
1330	0.4271	0.5272	94	116	135	91	31	104	83	34	117	1.069	125	7	248
1430	0.3432	0.3099	68	61	106	81	27	79	63	27	90	1.098	99	5	165
1530	0.2246	0.0684	36	11	67	64	22	45	36	19	55	1.108	61	3	75
1630	0.0793	0.0000	7	0	24	32	11	13	10	10	20	1.098	22	1	23
1730	-	-	-	-	-	-	-	-	-	-	-	-	-	-	-
1830	-	-	-	-	-	-	-	-	-	-	-	-	-	-	-
1930	-	-	-	-	-	-	-	-	-	-	-	-	-	-	-
2030	-	-	-	-	-	-	-	-	-	-	-	-	-	-	-
			Daily totals Whm^{-2}	1117								Daily totals Whm^{-2}	909	48	2074

* After normalisation

CHAPTER 6 - References

1. Commission of the European Communities (1984), European Solar Radiation Atlas, Vol. I, Global radiation on horizontal surfaces, 2nd Edition, Ed. W. Palz, Verlag TUV, Rheinland.

2. Chartered Institution of Building Services (1982), CIBS Guide A2, Weather and Solar Data, 1982, Chartered Institution of Building Services, Delta House, 222 Balham High Road, London, SW12 9BS.

3. Claux, P., Gilles, R., Pesso, A. & Raoust, M. (1982), Atlas Solaire Francais, PYC Edition, 254 rue de Vaugirard, 75740 Paris Cedex 15.

4. Palz, W. (1983), Solar Energy R&D in the European Community, Series F, Volume 2, Solar Radiation Data, D. Reidel Publishing Company.

CHAPTER 7

OTHER APPLICATIONS OF THE EUROPEAN COMMUNITY SOLAR RADIATION MODEL

7.1 Introduction

The EC diffuse sky irradiance model has the important advantage of being based on a combination of two well researched radiance models, one model describing clear skies, and the other model describing skies associated with overcast/partially cloudy conditions. The overcast radiance model is the well established Moon and Spencer formula, long adopted as the CIE standard for overcast sky daylighting design.

It follows, as a consequence of this theoretical formulation, that the radiance of any patch of sky may be determined directly from the EC solar irradiation model. This makes it possible to study both the effects of obstructions on diffuse sky radiation availability, and also to study the performance of low concentration collectors, which harness some of the radiant energy from the sky, as well as the direct beam.

This Chapter first explains how to estimate, at hourly intervals, the radiance of any determined patch of sky for clear and overcast conditions. It then explains how to estimate the monthly mean hourly radiance for any sky patch.

This Chapter also explains the use of the EC model for daylighting studies. As discussed in Chapter 3, there are close links between the luminance patterns of the sky and the radiance patterns of the sky. The link between the irradiance model and the illuminance model can be achieved by using the concept of the luminous efficiency,defined separately for direct sunlight and for skylight. These are the conversion factors needed to convert beam and diffuse irradiance values into the corresponding illuminance values. Once this conversion has been achieved the two luminance pattern models can be used to estimate values of the luminance of the sky for clear sky, overcast sky and monthly mean conditions for any defined point in the sky.

These modifications make it possible to use the EC solar radiation model for daylighting studies, which are coming to play an increasingly important role in the study of passive solar building design because the energy economics of buildings depend both on the economics of electrical energy consumption for lighting, as well as on the economics of energy consumption for space heating.

7.2 Radiance pattern - clear skies

It is necessary first of all to establish the solar geometry. For this purpose, it is best to use the techniques already described in Chapter 6. The output data on the proformas described in Chapter 6 can be used as the inputs in the subsequent radiance calculations.

The essential steps are:

i) Establish Linke Turbidity Factor at air mass 2 from Angstrom (a + b), and convert this for the actual solar altitude for the hour under study, $T_L(\gamma)$.

ii) Establish the clear sky diffuse irradiance, D_c, using the standard EC methodology.

iii) Using the polynomial described in Appendix 7, estimate the clear sky zenith radiance L_{cz} from D_c. The ratio L_{cz}/D_c is a function of Linke Turbidity Factor $T_L(\gamma)$ and solar altitude γ.

iv) Find the sky patch - sun azimuth angle for patch (Θ, α) from its patch azimuth angle from due south α and the solar azimuth angle ψ , $\alpha_s = \psi - \alpha$.

v) Using Equation 3.14 with the constants given in Equation 3.21 in Chapter 3, estimate the radiance of the selected patches relative to the zenith radiance, $L_c(\Theta,\alpha)/L_{cz}$.

vi) The required clear sky radiance $L_c(\Theta, \alpha)$ may be found by applying the ratio from v) to the value of L_{cz} found under iii). The units will be watts per square metre per steradian.

7.3 Radiance pattern - overcast skies

The mean radiance of the overcast sky at any solar altitude may be estimated as follows:

i) Estimate the overcast sky mean diffuse irradiance for that solar altitude, D_b using the standard procedures of Chapter 4.

ii) Use the standard Moon and Spencer conversion ratio to find the overcast sky zenith radiance, L_{bz}, $L_{bz} = (9/7\pi)D_b = 0.409\ D_b$ watts/metre^2/steradian.

iii) For the selected patch elevation Θ, estimate the value of the ratio $L_b(\Theta)/L_{bz}$ from the Moon and Spencer formula

$$L_b(\Theta) = L_{bz}\frac{(1 + 2 \sin \Theta)}{3}$$

iv) Combine (ii) and (iii) to find $L_b(\Theta)$.

7.4 Monthly mean radiance patterns

The monthly mean radiance pattern for any patch (Θ,α) at any hour is established as follows:

i) Estimate the two components of the diffuse sky horizontal irradiance on a horizontal surface ${}_{pc}D_m$ and $\sigma_{4m}\ D_c$ using the standard methodology described in Chapter 6, Section 18.23.

ii) Use Appendix 7 to convert the clear sky diffuse value $\sigma_{4m}\ D_c$ to the corresponding zenith radiance as a function of $T_L(\gamma)$ and solar altitude γ.

iii) Find the sky patch - sun azimuth angle for patch (Θ,α) from its patch azimuth angle from due south α and the solar azimuth angle ψ, $\alpha_s = \psi - \alpha$.

iv) Using Equation 3.14, Chapter 3, with the associated constants in Equation 3.21, compute ratio ${}_{c}L_{m}(\Theta,\alpha)/L_{c}$.

v) Combining (ii) and (iv) estimate clear sky radiance contribution for patch $L_{c}(\Theta,\alpha)$.

vi) Estimate the overcast sky/partially clouded sky contribution to the zenith radiance using the Moon and Spencer relationship, i.e. ${}_{pc}L_{mz} = 0.409\ {}_{pc}D_{m}$.

vii) Convert overcast sky partially clouded sky zenith luminance L_{bz} to slope value ${}_{pc}L_{m}(\Theta\ ,\ \alpha)$ for elevation value of Θ of patch, using Moon and Spencer formula as in Section 7.3. ${}_{pc}L_{m}(\Theta,\alpha) = {}_{pc}L_{mz}(\ 1 + 2 \sin \Theta)/3$.

viii) Add the two radiance components to obtain the hourly value of monthly mean diffuse radiance for patch, $L_{m}(\Theta,\alpha)$.

ix) Repeat above at appropriate intervals of time of day and year.

Once the monthly mean radiance patterns are established, it becomes possible to examine the effect of obstructions on the monthly mean sky irradiation falling on inclined surfaces, either at the hourly level, or the daily level by summing the hourly radiance contributions for each patch over the day. In view of the large number of calculations involved, it is best to do these calculations using appropriate computing techniques. Radiance files can be very easily bred from irradiance files using the techniques described above. A standard computing package for doing this has been developed in the Department of Building Science, University of Sheffield.

7.5 Terminology for prediction of illuminance and luminance

This section of Chapter 7 deals with the estimation of illuminance and luminance. In order to distinguish between illuminance and irradiance the same basic symbols have been retained but the illuminance symbols are prefixed with the symbol E (Eclairage), thus $(ED)_{b}$ represents the horizontal surface diffuse illuminance on an overcast day, $EI_{c}(0,\ 0)$ represents the beam illuminance on a horizontal surface on clear days, $(EL)_{cz}$ represents the clear sky zenith luminance, and so on.

7.6 Luminous efficiency of the direct beam - cloudless day conditions

The illuminance normal to the direct beam for clear sky conditions may be estimated from the Linke Turbidity Factor using Equations 7.1 and 7.2 below. The relationship between the atmospheric turbidity factor for light, T_{v}, and Linke Turbidity Factor, T_{L}, was derived using the spectroradiometric measurements of Dogniaux at Uccle (1). These measurements, which were commissioned as part of the CEC Solar Radiation Data Programme, gave details of the measured direct beam spectrum, waveband by waveband, for a number of cloudless days in Belgium. Associated measurements of Linke Turbidity Factor and air mass were also made at the same time. By weighting the observed spectral energy, waveband by waveband, by the CIE luminosity curve for the light adapted eye, the illuminance normal to the direct beam could be calculated from the spectral observations. As the CIE weighting curve for luminosity peaks very sharply around 555 nm, illumination can be treated as approximately monochromatic radiation as far as illuminance prediction is concerned, but obviously not so as far as colour rendering is concerned. Knowing the extraterrestrial irradiance and the air mass, it was thus a simple matter, adopting Beer's law, to calculate the overall transmittance of the atmosphere for light. Using fundamental theory, the light transmittance through a perfectly clean Rayleigh sky was determined, assuming an appropriate associated ozone content. By comparing the actual illuminance transmittance with the perfectly clean sky illuminance transmittance, a turbidity factor for light was determined, T_v, following precisely the same methods as are used to derive the Linke Turbidity Factor. While the Illumination Turbidity Factor T_v is mathematically defined in the same way as the Linke Turbidity Factor, the Rayleigh optical depth for illumination δ_v will be different from the Rayleigh optical depths δ_R used in irradiance calculations. The optical depth for illumination may be considered, for practical purposes, as being independent of air mass, i.e. the radiation is treated as monochromatic, so Beer's Law applies.

The derived set of values of T_v from Belgium were then related to the associated observed Linke Turbidity Factors, measurement by measurement, and the results correlated using polynomial regression (Equation 7.1). A full account of the derivation of this algorithm may be found in Page and Thompson (2). Once the luminous efficiency of the direct beam $K_{SB}(\gamma)$ is obtained, it is then a simple matter to convert the direct beam irradiance calculated as explained in Chapter 6 into an illuminance by multiplying by the beam luminous efficiency.

The Illumination Turbidity Factor for the CIE weighted visible spectrum $T_V(\gamma)$ is estimated from the Linke Turbidity Factor $T_L(\gamma)$ using equation 7.1.

$$T_v(\gamma) = 0.7868 + 0.12652\ T_L(\gamma) + 0.08666\ T_L(\gamma)^2 \quad \text{dimensionless} \qquad (7.1)$$

Then the direct beam normal illuminance for cloudless conditions may be estimated, calculating the air mass m in the standard way, making allowance for height effects as:

$$(EI)_c = (EI)_o \exp(-\ m\ \delta_v\ T_v(\gamma)) \quad \text{lux} \qquad (7.2)$$

where $(EI)_o$ = Extraterrestrial illuminance at normal incidence corrected to mean solar distance (132.3 x K_d), klux

$T_L(\gamma)$ = Linke Turbidity Factor at solar altitude γ, dimensionless

m = Optical air mass, dimensionless

δ_v = Rayleigh optical depth for centre of visual spectrum

The value of δ_v may be taken to be 0.1265.

Combining 7.1 and 7.2 one obtains

$$(EI)_c = (EI)_o \exp(-0.1265\ m(0.7868 + 0.12652\ T_L(\gamma) + 0.08666\ T_L(\gamma)^2)) \quad \text{klux} \qquad (7.3)$$

The luminous efficiency of the direct beam may be obtained by dividing equation 7.3 by the beam irradiance:

$$K_{sB}(\gamma) = (EI)_o/I_o \frac{\exp(-0.1265\ m(0.7868 + 0.12652\ T_L(\gamma) + 0.08666\ T_L(\gamma)^2))}{\exp(-\ m\ \delta_R\ T_L(\gamma))} \quad lm.w^{-1} \qquad (7.4)$$

The solar constant $I_O = 1367\ Wm^{-2}$ and the extraterrestrial direct beam illuminance at mean solar distance $(EI)_o = 132.3$ klux, so it follows that

$$(EI)_o/I_o = \frac{132.3 \times 1000}{1367} = 96.78\ lm.w^{-1}$$

Therefore equation 7.4 becomes

$$K_{sB}(\gamma) = 96.78 \times \frac{\exp(-0.1265\ m(0.7868 + 0.12652\ T_L(\gamma) + 0.08666\ T_L(\gamma)^2))}{\exp(-\ m\ \delta_R\ T_L(\gamma))} \quad lm.w^{-1} \qquad (7.5)$$

δ_R is a function of air mass, whereas δ_V is not. The clear sky direct beam illuminance values on any slope $(EI)_C(\ ,\alpha)$ are obtained by multiplying the values of $I_C(\beta,\alpha)$ by $K_{sB}(\gamma)$. The values found for $K_{sB}(\gamma)$ for a range of solar altitudes and Linke Turbidity Factors $T_L(\gamma)$ are given in Table 7.1.

7.7 Luminous efficiency of light from the clear sky, K_{sd}

The luminous efficiency of light from the clear sky K_{sd} was also estimated from the Dogniaux spectroradiometric observations using the CIE standard luminosity curve for the light adapted eye, using the spectral observations for the sky alone. It was assumed that 90% of the energy lay in the range 300 to 1000 nm. The values for K_{sB} for a range of Linke Turbidity Factors are given in Table 7.1. K_{sB} was found to be equal to $97.55 + 13.01\ T_L(\gamma) - 0.8283\ T_L(\gamma)^2$ lumens/watt where T_L is the Linke Turbidity Factor. No dependence on solar altitude was detected. The clear sky horizontal diffuse sky illuminance is thus simply calculated by multiplying the computed clear sky horizontal diffuse irradiance in watts/metre² by K_{sd} to obtain the horizontal sky illuminance in lux, i.e. $K_{sd}\ D_c$.

7.8 Global illuminance and global luminous efficiency horizontal surfaces - clear skies

The global illuminance on a horizontal surface is $K_{sB}(\gamma) I_c \sin\gamma + K_{sd}\ D_c$ where I_c is the clear sky direct beam irradiance, Wm^{-2}, γ is the solar altitude and D_c is the clear sky diffuse horizontal irradiance. The luminous efficiency for global radiation on a horizontal surface is thus $(K_{sB}(\gamma)\ I_c \sin\gamma + K_{sd}\ D_c)/(I_c \sin\gamma + D_c)$ lumens/watt. Values computed with this model are given in Table 7.1.

Table 7.1 Luminous efficiencies predicted from model proposed in this Chapter
Units: Lumens/watt

Solar Altitude		Clear sky Linke Turbidity Factor $T_L(\gamma)$ 2	4	6	8	10	Overcast sky All $T_L(\gamma)$
5	Beam	49.8	38.1	12.2	1.6	0.1	-
	Global (Hor)	69.9	99.4	122.6	137.4	141.1	114.5
10	Beam	78.5	77.0	47.7	18.3	4.4	-
	Global (Hor)	86.2	105.0	115.8	126.4	132.4	117.1
15	Beam	89.2	91.7	68.4	36.7	14.1	-
	Global (Hor)	93.6	108.9	114.7	120.4	125.0	119.3
20	Beam	94.3	98.0	79.8	50.5	24.9	-
	Global (Hor)	97.4	110.6	114.3	116.7	119.0	121.3
30	Beam	98.1	101.9	89.1	65.7	40.7	-
	Global (Hor)	100.3	111.0	113.2	112.5	111.3	124.4
40	Beam	99.3	102.6	92.5	73.0	50.4	-
	Global (Hor)	101.2	110.2	111.8	110.0	107.0	126.4
50	Beam	99.7	102.6	94.1	77.2	56.6	-
	Global (Hor)	101.3	109.3	110.6	108.4	104.5	127.2
60	Beam	99.8	102.5	95.0	79.8	60.7	-
	Global (Hor)	101.2	108.5	109.7	107.4	103.1	126.8
70	Beam	99.8	102.3	95.5	81.4	63.3	-
	Global (Hor)	101.1	107.9	109.1	106.8	102.3	125.4
Clear sky All altitudes	Sky diffuse	127.5	136.9	144.0	148.7	151.1	

7.9 Diffuse illuminance on slopes - clear days

a) Sky diffuse illuminance

Conversion to slopes for sky diffuse illuminance is achieved using the same algorithms as are used for inclined surface radiation predictions. This involves some approximation because the formula for the luminance or brightness distribution of the clear sky is not identical with that used to describe the radiance distribution of the clear sky. However, the luminous efficiency of the clear sky is not known with great precision, and the accuracy achievable does not warrant the introduction of new algorithms that give nearly the same, but not identical predictions for slopes. The prediction methods of Chapter 6 are used. Thus the slope sky illuminance is given by K_{sd} $D_c \times f_2$ lux where D_c is the clear sky diffuse irradiance on the horizontal surface and f_2 is the conversion factor for the given slope, given in Chapter 6, Tables 6.14a to 6.14b.

b) Reflected light from the ground - clear days

The light reflected from the ground falling on any slope is estimated in the same way as the radiation reflected from the ground, using the global horizontal illuminance together with the ground visual reflectance for light. However, the albedo and the visual reflectance may have different values for the same surface and care must be exercised in the choice of ground reflectance. In particular, natural vegetation which absorbs heavily in the red and blue parts of the spectrum, has a visual reflectance of around 10% compared with an albedo of 20%. Plants are highly reflective in the near infra red as photographs taken in the near infra red show, but not in the visible, where a considerable amount of energy is absorbed for photosynthesis.

7.10 Prediction of the illuminance on inclined surfaces on overcast days

The luminous efficiency on overcast days in Europe may be related to solar altitude by the following formula:

$$K_{sb}(\gamma) = 111.72 + 0.5939\,\gamma - 0.0057\,\gamma^2 \quad \text{lumens/watt} \qquad (7.6)$$

The estimation of the horizontal surface diffuse illuminance from the sky is simply obtained by multiplying the overcast day horizontal irradiance G_b (=D_b) in Wm^{-2} by $K_{sb}(\gamma)$ to obtain the illuminance in lux. The diffuse illuminance on a surface (θ, α) is given by $K_{sb}(\gamma) \times D_b \times f_4$. f_4 is obtained from Table 6.1 in Chapter 6. As $D_b = G_b$ the ground reflected component is obtained in the standard way as $K_{sb}(\gamma) \times D_b \times f_3$ lux, setting $f_3 = 0.5 \; \rho'_g(1.0 - \cos\beta)$ where ρ'_g is now the visual reflectance of the ground surface and β is the slope tilt.

7.11 Monthly mean illuminances on inclined planes

The direct beam illuminance on any surface $EI_m(\beta,\alpha)$ is simply calculated as $K_{sd}(\gamma)\; I_m(\beta,\alpha)$. The diffuse sky irradiance contribution $D_m(\beta,\alpha)$ is split into its two components as described in Chapter 6, Section 18.2.3 - 18.2.6. The clear sky slope component is multiplied by K_{sd} to obtain the clear sky contribution to the slope illuminance. The overcast sky/partially clouded sky component is multiplied by $K_{sb}(\gamma)$ to obtain the overcast sky component. The ground reflected component is estimated from the monthly mean global horizontal illuminance,

$(EG)_m = (K_{sB}(\gamma)\ I_m(0,0) + K_{sd}\ D_m)$, and the ground reflectance ρ'_g using the standard formula:

$$ER_m(\beta, \alpha) = 0.5(EG)_m\ \rho'_g(1.0 - \cos\beta)\ \text{lux} \qquad (7.7)$$

The monthly mean illuminance on the slope $EG_m(\beta,\alpha)$ will be the sum of these four components.

The monthly mean luminous efficiency for the slope can be obtained by dividing $EG_m(\beta,\alpha)$ by $G_m(\beta,\alpha)$. It must be noted that the monthly mean global luminous efficiency for sloping surfaces may be very different to the global luminous efficiency for horizontal surfaces, as the luminous efficiencies of the different components are not the same (Refer Table 7.1) and the relative weight of the various components changes from slope to slope according to time of day and year.

7.12 Estimation of the luminance of the sky

Estimates of the luminance of the sky are important in daylighting design. The basic processes to estimate the luminance of any sky patch are the same as used for radiance estimates, but the Liebelt sky luminance distribution function, equation 7.8, which adopts the clear sky angular notation of Chapter 3, is used in place of the sky radiance distribution function to estimate the clear sky luminance contribution. The scientific treatment of the overcast sky contribution is identical to the treatment of radiance prediction, as both the radiance and the luminance distribution functions are the same. The Liebelt luminance formula is:

$$\frac{(EL)_c(\Theta, \alpha_s)}{(EL)_{cz}} =$$

$$\frac{(1 - e^{-0.088m\ T_L(\gamma)})(1 + X_1(e^{-3\eta\pi/180} - 0.009) + X_2\cos^2\eta)}{(1 - e^{-0.088\ T_L(\gamma)})(1 + X_2(e^{-3(90-\gamma)\pi/180} - 0.009) + X_2\cos^2(90-\gamma))} \qquad (7.8)$$

where γ = solar altitude, degrees

η = angle between sun's direction, and sky patch (Θ,α), degrees

$X_1 = 0.6155 + 1.9687\ T_L(\gamma) \quad 2 < T_L < 8$

$X_2 = 0.60 - 0.038\ T_L(\gamma) \quad 2 < T_L < 8$

One additional algorithm is needed in order to calculate $(EL)_{cz}$ from the diffuse horizontal illuminance. The required equation was developed from the Liebelt formulation, Equation 7.8. For luminance calculations, the radiance formula in Appendix 7 should be replaced by the following formula:

$$(ED)_c/(EL)_{cz} = \sum_{i=0}^{3}\sum_{j=0}^{3} a_{ij}\ T_L(\gamma)^j\ \gamma^i \quad \text{steradians} \qquad (7.9)$$

where $T_L(\gamma)$ = Linke Turbidity Factor at solar altitude γ, dimensionless

γ = solar altitude, degrees

and where the values of a_{ij} are given by

i \ j	0	1	2	3
0	7.3124	-0.24678	0.01666	0.0
1	-0.02211	0.03108	-3.1814E-3	1.0656E-4
2	-0.6398E-3	-0.8826E-3	0.083375E-3	-2.7709E-6
3	0.2636E-5	0.5826E-5	-0.051411E-5	1.6459E-8

The units of luminance are candelas/metre². Obviously one works with the values of $(ED)_c$, $(ED)_b$, and $(ED)_m$ as the starting point for the luminance calculations in place of the irradiances D_c, D_b, and D_m.

Proceeding in this way, following the detailed procedures of sections 7.2 - 7.4 above, a detailed analysis can be prepared of the clear sky, mean overcast sky and monthly mean sky luminance patterns. Such data form the fundamental basis of daylighting energy economy studies, and are readily derived from these modifications of the solar radiation model.

References

1. Dogniaux, R., (1981), Distribution spectrale du rayonnement solaire à Uccle, Institut Royal Météorologique de Belgique, Misc. Series B, No. 52, IRMB, Brussels.

2. Page, J.K. & Thompson, J.L., (1982), Modelling daylight availability, Proceedings of CIBS Lighting Division Conference, Warwick, 1982, CIBS, London.

APPENDICES

APPENDIX 1

BASIC ALGORITHMS NEEDED TO PREDICT THE MOTION OF THE SUN, DAYLENGTH, AND THE EXTRATERRESTRIAL IRRADIANCE

In the following algorithms the extraterrestrial irradiance (algorithm 11) and solar declination (algorithm 3) are taken to be constant over the course of a day. Although both these quantities do in fact vary continuously, the use of a single daily value is sufficiently accurate for most practical purposes.

1. DAY NUMBER J

The number of the day in the year measured from noon on the 31st December, i.e. noon 1st Jan, J = 1, 1st Feb = 31 + 1 = 32. In leap years the number of days is 366. Normally a 365 day year is used for solar calculations to avoid the complications of leap years. Hence 1st March = 31 + 28 + 1 = 60.

2. DAY ANGLE J' degrees

The day angle expresses the day number J as an angle from 1200 hours on the 31st December. A year length of 365.25 days is used to take account of the four year leap year effect. A very precise estimate of solar position requires that the effects of the annual discontinuities introduced by leap years be properly considered, but such accuracy is not necessary in most solar energy applications.

INPUTS

Day Number J

ALGORITHM 2

$$J' = 360° \times J/365.25 \quad \text{degrees} \qquad (A1.2)$$

3. SOLAR DECLINATION δ degrees

Solar declination is the angle between the sun's rays and the equatorial plane. The declination has a positive value when the sun is north of the equator and a negative value when it is south of the equator. The declination becomes positive at the vernal equinox (21st March) and negative at the autumnal equinox (23rd September). The maximum value of the declination is +23°27' and the minimum value is -23°27'.

INPUTS

Day Angle J' degrees

ALGORITHM 3

$$\delta = \sin^{-1}(0.3978 \sin(J' - 80.2° + 1.92° \sin(J' - 2.80°))) \quad \text{degrees} \qquad (A1.3)$$

Note: Chapter 6, Table 6.2 contains monthly values of the declination recommended for use in the standard EC computation procedure. The values are based on a slightly more accurate basic algorithm than Algorithm A1.3 which were then used in a daily integration procedure to obtain the correct weighted value for each month to allow for the variation in declination across that month. Different data are used for days of maximum global radiation on a horizontal surface G_{max} to those used for monthly mean estimates.

4. ASTRONOMICAL DAYLENGTH S_o hours

Astronomical daylength is the computed time during which the centre of the solar disc is above an altitude of zero degrees (without allowance for atmospheric refraction).

INPUTS

Latitude of site (N + ve, S - ve)	ϕ	degrees
Solar declination	δ	degrees

ALGORITHM 4

$$S_o = \frac{1}{7.5} \cos^{-1}(- \tan \delta \tan \phi) \quad \text{hours} \qquad (A1.4.1)$$

N.B. When located within the Arctic and Antarctic circles

If $(- \tan \delta \tan \phi) > 1$, then $S_o = 0$ hours

or if $(- \tan \delta \tan \phi) < -1$, then $S_o = 24$ hours

The value of S_o can be used to calculate the times of sunrise t_r and sunset t_s in local apparent (solar) time.

$$t_r = 12 - S_o/2 \quad \text{hours} \qquad (A1.4.2)$$

$$t_r = 12 + S_o/2 \quad \text{hours} \qquad (A1.4.3)$$

Note: Table 6.6 a), b) and c) contains values of the astronomical daylength for latitudes between 75°S and 75°N calculated for the recommended values of solar declination given in Table 6.2. The times of sunrise and sunset in local apparent time can easily be derived using equations A1.4.2 and A1.4.3.

5. THE CONVERSION OF LOCAL STANDARD TIME TO LOCAL APPARENT TIME

The calculation methods adopted in this book use local apparent time (solar time) as the time system. This simplifies and shortens the computation. However, it may sometimes be necessary to produce hourly irradiance predictions using local standard time (clock time). This requires the conversion from local standard time into local apparent time.

The difference between local standard time and local apparent time varies with both time of year and the longitude of the site in relation to the reference longitude of the particular time zone. The basic annual variation is first calculated by the equation of time E.T. (A1.5.1). This result is then used in the actual conversion from local mean time to local apparent time (A1.5.2).

INPUTS

Day angle	J'	degrees
Longitude of site (positive to the east of Greenwich)	λ	degrees
Reference longitude of the time zone (positive to the east of Greenwich)	λ_{ST}	degrees
Correction for summer time, if used (usually +1 in summer, zero in winter. The dates of introduction and termination of summer vary from country to country)	c	hours

ALGORITHM 5

$$E.T. = -0.128 \sin(J' - 2.80°) - 0.165 \sin(2J' + 19.7°) \text{ hours} \quad (A1.5.1)$$

$$L.A.T. = L.M.T. + \frac{\lambda - \lambda_{ST}}{15} + E.T. - c \quad \text{hours} \quad (A1.5.2)$$

where E.T. is the equation of time, hours

L.A.T. is local apparent time, hours

L.M.T. is local standard time, hours

Note:

The latitude of the time reference zones for different European countries are

0° Iceland, Ireland, Portugal, UK.

+15° Albania, Austria, Belgium, Czechoslovakia, Denmark, France, German Democratic Republic, Federal Republic of Germany, Gibraltar, Hungary, Italy, Luxembourg, Malta, Netherlands, Norway, Poland, Spain, Sweden, Switzerland, Yugoslavia.

+30° Cyprus, Greece, Roumania.

6. SOLAR HOUR ANGLE ω degrees

The solar hour angle expresses the time of day in terms of the angle of rotation of the earth about its axis from its solar noon position at a specific place. As the earth rotates 360° about its axis in 24 hours, in one hour the rotation is 15°. By convention the hour angle is negative before noon and positive after noon, i.e. 0900 LAT represents an hour angle of -45° and 1500 LAT represents an hour angle of +45°.

INPUTS

Local Apparent Time (L.A.T.)	t	hours

ALGORITHM 6

$$\omega = 15(t - 12) \text{ degrees} \qquad (A1.6)$$

7. SOLAR ALTITUDE γ degrees

Solar altitude is the angle between the centre of the solar disc and the horizontal plane.

INPUTS

Latitude of site	ϕ	degrees
Solar declination	δ	degrees
Solar hour angle	ω	degrees

ALGORITHM 7

$$\gamma = \sin^{-1}(\sin\phi \sin\delta + \cos\omega \cos\phi \cos\delta) \qquad (A1.7)$$

Notes: For computation at the hourly level it is economical to calculate daily values of $(\sin\phi \sin\phi)$ and $(\cos\phi \cos\phi)$ first because the values remain constant for the day, and then to retain the values for consequent calculation hour by hour at different values of ω.

8. SOLAR AZIMUTH ψ degrees

In the Northern hemisphere the solar azimuth angle is the angle between the vertical plane containing the direction of the sun, and the vertical plane running true north-south measured from south. The azimuth angle has a positive value when the sun is to the west of south, i.e. during the afternoon in solar time. It has a negative value when the sun is east of south. For the Southern hemisphere the reference direction is true north. The sun's position is often described as a bearing from true north, and this is sometimes incorrectly referred to as the solar azimuth angle. It is important to adopt the correct definition in using the algorithms below:

INPUTS

Latitude of site (N + ve, S - ve)	λ	degrees
Solar declination (Algorithm 3)	δ	degrees

Solar hour angle (Algorithm 6) ω degrees

Solar altitude (Algorithm 7) γ degrees

ALGORITHM 8

$$\cos \psi = (\sin \phi \sin \gamma - \sin \delta)/\cos \phi \cos \gamma \quad (A1.8.1)$$

$$\sin \psi = \cos \delta \sin \omega/\cos \gamma \quad (A1.8.2)$$

If $\sin \psi < 0 \quad = -\cos^{-1}(\cos \psi)$

If $\sin \psi > 0 \quad = \cos^{-1}(\cos \psi)$

Note: In the southern hemisphere: $\cos \psi = -(\sin \phi \sin \gamma - \sin \delta)/\cos \phi \cos \gamma$

9. WALL SOLAR AZIMUTH ANGLE α_S degrees

The wall solar azimuth angle is the angle between the vertical plane containing the normal to the surface and the vertical plane passing through the centre of the solar disc, i.e. it is the resolved angle on the horizontal plane between the direction of the sun and the normal to the surface. The values lie between -180° and +180°.

Note: In the Northern hemisphere the surface azimuth angle is the angle between a vertical plane containing the normal to the surface and south. In the Southern hemisphere it is expressed in relation to north.

INPUTS

Solar azimuth angle measured from due south in Northern hemisphere and from due north in Southern hemisphere (easterly values - ve) (Algorithm 8) ψ degrees

Surface azimuth angle measured from due south in Northern hemisphere and from due north in Southern hemisphere (easterly values - ve) α degrees

ALGORITHM 9

$$\alpha_S = \psi - \alpha \quad \text{degrees} \quad (A1.9.1)$$

If $\alpha_S > 180$ then $\alpha_S = \alpha_S - 360°$. degrees (A1.9.2)

If $\alpha_S < -180$ then $\alpha_S = \alpha_S + 360°$ degrees (A1.9.3)

Sign convention: Sun anti-clockwise from normal on plan: sign negative

Sun clockwise from normal on plan: sign positive

10. COSINE OF THE ANGLE OF INCIDENCE OF THE DIRECT BEAM ON A SURFACE $\cos \nu$

INPUTS

Solar altitude	γ	degrees
Solar azimuth	ψ	degrees
Surface inclination	δ	degrees
Wall solar azimuth angle	α_S	degrees

ALGORITHM 10

$$\cos \nu = \cos \gamma \sin \beta \cos \alpha_S + \sin \gamma \cos \beta \qquad (A1.10.1)$$

N.B. If $\cos \nu$ is negative, the sun lies behind the surface
i.e. for $\cos \nu > 0.0$ the sun is on the surface.

For horizontal surfaces the above algorithm reduces to the simpler form:

$$\cos \nu = \sin \gamma \qquad (A1.10.2)$$

A simpler form may also be used on vertical surfaces

$$\cos \nu = \cos \gamma \cos \alpha_S \qquad (A1.10.3)$$

Note: The algorithm is based on the centre of the solar disc without allowance for atmospheric refraction.

11. EXTRATERRESTRIAL IRRADIANCE AT NORMAL INCIDENCE I_{oj} Wm^{-2}.

INPUTS

Day Angle	J'	degrees

ALGORITHM 11

$$I_{oj} = I_o \times (1.0 + 0.03344 \cos(J' - 2.80°)) \quad Wm^{-2} \qquad (A1.11)$$

where I_o is the solar constant: 1367 Wm^{-2}.

and (1.0 + 0.03344 cos(J' - 2.80°)) is the correction K_d to allow for the variations in sun-earth distance from its mean value.

Note: Table 6.3 in Chapter 6 provides values of I_{oj} for the middle day of each month computed using this algorithm. These values are, of course, independent of latitude.

12. 4 DEGREE SUNRISE/SUNSET DAYLENGTH

4 degree sunrise/sunset daylength is the length of time during which the centre of the solar disc is above an altitude of 4 degrees, which is the solar altitude at which a sunshine recorder typically starts to burn.

INPUTS

Latitude of site (N + ve, S - ve)	λ	degrees
Solar declination	δ	degrees

ALGORITHM 12

$$S_{04} = \frac{1}{7.5} \cos^{-1}(\sin 4^\circ - \sin \phi \sin \gamma)/\cos \phi \cos \gamma) \quad \text{hours} \qquad (A1.12.1)$$

If $((\sin 4^\circ - \sin \phi \sin \gamma)/(\cos \phi \cos \gamma)) > 1$, then
the sun never rises above 4° during day and $S_{04} = 0$ hours

If $((\sin 4^\circ - \sin \phi \sin \gamma)/(\cos \phi \cos \gamma)) < -1$,
then the sun never falls below 4° during the day and $S_{04} = 24$ hours.

APPENDIX 2

A REVIEW OF METHODS FOR ASSESSING THE LINKE TURBIDITY FACTOR IN THE ABSENCE OF MEASUREMENTS OF SOLAR DIRECT BEAM INTENSITY

Summary

This appendix first gives details of various methods available for assessing the Linke Turbidity Factor for Europe in the absence of direct measurements. It gives particular attention to the problem of the variation of Linke Turbidity Factor with solar altitude. The various methods available to estimate the Linke Turbidity Factor are set out first, followed by a discussion of the inter-relationships between them. It is based on a paper of M.R. Dogniaux (1).

Methods for estimating the Linke Turbidity Factor

Method 1

The relationship of Dogniaux

Dogniaux, based on a statistical analysis of the Linke Turbidity Factor at Uccle (2), adopted the following relationship in 1975 (3)

$$T_L(\gamma) = \left(\frac{85 + \gamma}{39.5e^{-w} + 47.4} + 0.1\right) + (16 + 0.22\, w)\ \beta_A \qquad \text{(A2.1)}$$

where γ is the elevation of the sun, degrees

w is the depth of precipitable water vapour, cm

β_A is the Angstrom turbidity coefficient, dimensionless.

In the absence of available observational data for w and β_A, Dogniaux suggested the adoption of the following classification of the impact of different types of climate on $T_L(\gamma)$.

Precipitable water vapour

Polar regions or deserts - dry air	w = 0.5 to 1 cm
Temperate climates	w = 2 to 4 cm
Tropical climates - humid air	w > 5 cm

(Additional information on precipitable water vapour may be found in Chapter 2, Table 2.1 and Table 2.2 in the main text).

Angstrom Turbidity coefficient β_A

Rural site	$\beta_A = 0.05$

Urban site $\beta_A = 0.10$

Industrial site $\beta_A = 0.20$

This formula can be applied to any site.

Method 2

The relationship formula of Dogniaux - Sneyers

This specific relationship for Uccle established following a statistical study of the stability of the atmospheric turbidity at Uccle over the period 1951-1970 (4) is of the form

$$T_L(\gamma) = a_0 + a_1\gamma + a_2 \cos (30.MN) \quad \text{(A2.2)}$$

where γ is the solar altitude in degrees

MN is the month number, Jan MN = 1, Feb MN = 2, ... Dec MN = 12.

a_0, a_1 and a_2 are constants.

The coefficients a_0, a_1 and a_2 were estimated by a selective harmonic analysis of monthly means of the Linke Turbidity Factor for the period 1951-1980. The expression takes account of the hourly variation through the term $(a_1\gamma)$ and of the daily variation through the term $(a_2 \cos(30.MN))$, so providing an annual account of the transmission characteristics of the atmosphere in the absence of clouds, without the need to input the precipitable water vapour. It predicts the peak value in June. According to whether one wishes to refer to the monthly mean Linke Turbidity $\bar{T}_L(\gamma)$, when values of a_0, a_1 and a_2 derived from the mean values of the cloudless day direct beam irradiance I_c are used, or to conditions of minimum monthly Linke Turbidity $T_L(\gamma)_{MIN}$, when values of a_0, a_1 and a_2 corresponding to mean monthly minimum values of the Turbidity Factor are used, one adopts one or other of the following formulae:

Mean conditions for month

$$\bar{T}_L(\gamma) = 3.372 + 0.053\ \gamma + 0.296 \cos(30.MN) \quad \text{(A2.3)}$$

Minimum Linke Turbidity Factor for month

$$T_L(\gamma)_{MIN} = 2.730 + 0.027\ \gamma - 0.198 \cos(30.MN) \quad \text{(A2.4)}$$

These equations, however, do not lend themselves to generalisations for other sites, as the coefficients a_0, a_1 and a_2 are site specific. The relationship function must be drawn up for each site on the basis of measurements at that specific site.

Method 3

The relationship function of Dogniaux - Lemoine

A more general approach is provided in the study "Classification of radiation sites in terms of different indices of atmospheric transparency" (5)(6). The authors analysed the impacts of both variations in latitude and the sum of the mean monthly coefficients a and b in the Angstrom regression formula on the air mass 2 Linke Turbidity Factor T_L.

The Angstrom regression formula is

$$G_d/G_{od} = a + b\, S_d/S_{od} \qquad \text{(A2.5)}$$

where G_d is the daily global irradiation on the horizontal plane at a given site on a specific day

G_{od} is the corresponding extraterrestrial daily irradiation on a horizontal plane outside the atmosphere on that day. Here estimated with a solar constant of 1367 Wm^{-2}, subsequently corrected to mean solar distance.

S_d is the daily observed bright sunshine in hours

S_{od} is the computed astronomical daylength in hours on that day at the site in question.

The authors working on data for the Northern Hemisphere defined an Index of Atmospheric Transparency (in French, indice de transparence atmosphérique) which they abbreviated to ATI which was related to the sum of the monthly mean values of Angstrom regression coefficients a and b. The expression they developed for the air mass 2 Linke Turbidity Factor was

$$T_L = 22.76 + 0.0536\ \phi - 27.78\ \text{ATI} \qquad \text{(A2.6)}$$

where ϕ is the latitude in degrees

ATI is the atmospheric transparency index.

The ATI, in the absence of a precise determination, can be estimated from an assessment of site atmospheric conditions and the general nature of the site. The basic values of ATI were developed for latitude 45°N and are given in Table A2.1, together with the method of correcting the value of ATI for other latitudes, using the correction factor which is included in Table A2.1.

Once the air mass 2 Linke Turbidity Factor has been derived from Equation A2.6, Dogniaux has proposed that the solar altitude correction of Page, based on a parametisation of the WMO correction method should be adopted to correct the value to the actual air mass. This correction is discussed in the main body of Chapter 2. The significance of this approach lies in the fact that values of the Angstrom a and b can be derived relatively simply for any station where daily observations of G_d and S_d are available, and where they are not available a systematic methodology for estimating the values is provided. As already mentioned, the CEC European Solar Radiation Atlas, Vol. I, Horizontal Surfaces, 2nd Edition, contains systematic values of (a + b) for 100 European sites.

Method 4

Relationship functions of Perrin de Brichambaut

In the book "Le gisement solaire: évaluation de la ressource énergétique" (7), Perrin de Brichambaut and Ch. Vauge, prepared a relationship of the type:

$$T_L = a_o + a_1 \beta_A + a_2 \ln w \qquad (A2.7)$$

where β_A is the Angstrom Turbidity Coefficient

w is the precipitable water vapour in cm

a_o, a_1, a_2 are constants adjustable to local conditions.

Based on experience, they adopted the following mean values for a_o, a_1 and a_2:

$$a_o = 2.5 \quad a_1 = 1.6 \quad a_2 = 0.5 \qquad (A2.8)$$

Taking account of the normal humidity burden found in the atmosphere they suggested the following relationships

For temperate regions

$$T_L = 2.6 + 16\ \beta_A (\pm 0.3) \qquad (A2.9)$$

For regions in the inter-tropical humid zone

$$T_L = 3.2 + 17\ \beta_A (\pm 0.4) \qquad (A2.10)$$

Using equation A2.7 with the coefficients in A2.8, they prepared estimates given in Table A2.2.

The authors mention an effect of station elevation implying a correction $T_1 = -0.35$ z where z is the site elevation in km. They also mention a correction T_2 to allow for solar elevation effects on T_L of 0.2, but, for simplification, suggested this correction could be omitted.

Table A2.1 Estimation of the atmospheric transparency index ATI as a function of site based on Dogniaux and Lemoine (5)
$ATI = 0.69051 + 0.00193\ \phi + R$
where ϕ is the latitude in degrees and R allows for the effects of local turbidity as indicated below.

Turbidity class	Atmospheric clarity	Typical nature of site	Atmospheric parameters β_A	Atmospheric parameters w cm	Factor R	ATI at ϕ=45
1	Very clear	Dry air mountain climates polar climates some deserts*	<0.05	<2	0.04 ± 0.02	>0.79
2	Clear	Rural areas in temperate climates	0.05 -0.10	2 - 4	0.00 ± 0.02	0.79 -0.76
3	Slightly polluted	Urban areas in temperate climates	0.10 -0.20	2 - 4	-0.04 ± 0.02	0.76 -0.72
4	Very polluted or very water laden atmosphere	Industrial climates and humid tropical climates	>0.20 >0.20	2 - 4 >5	<-0.06	0.72

* While the value of w in desert climates is invariably low, the value of β_A depends on the dust burden which may be quite high. There is, therefore, a danger of exaggerating the energy available in desert areas. The dust burden often depends on season (9).

Table A2.2 Linke Turbidity Factors suggested by Perrin de Brichambaut and Vauge (7).

Colour of the sky	Visibility at surface Km	Turbidity coefficient A	Vapour pressure at surface e mb			
			3-5	5-8	9-16	18-30
Deep blue	>100	0.01-0.02	2.0	2.3	2.6	2.9
Pure blue	60-100	0.03-0.06	2.6	2.9	3.2	3.5
Pale blue	30- 50	0.08-0.15	3.4	3.7	4.0	4.3
Milky blue to whitish	12- 25	0.20-0.40	4.9	5.2	5.6	5.8
			0.5	1	2	4
			Precipitable H_2O in cm			

Method 5

Method for estimating air mass 2 Linke Turbidity Factor based on methodology proposed in World Meteorological Organisation Technical Note No. 172, p.121-124

The method proposed by W.M.O. (8) for estimating the Linke Turbidity Factor has the great advantage that no prior knowledge is required of the turbidity characteristics of the region in which the method is to be applied. It is related to Method 4.

The daily or monthly representative inputs needed are:

(1) Site height z, m
(2) Water vapour pressure in millibars
(3) Colour of the sky.

The method starts by adopting an air mass 2 Linke Turbidity Factor of 3.3, and adjusing this number, which is an average sea level value to allow for the effects of site height, water vapour absorption, aerosol and solar elevation. The effects of aerosols are assessed through a judgement of the colour of the sky.

Correction factor T_1.

The adjustment factor for height suggested by WMO is $T_1 = -0.35$ z/1000 where z is the station height in metres. This allows for the normal decrease of turbidity and humidity with height.

T_1 varies from 0 (sea level) to -0.7 (z = 2000 m).

Correction factor T_2.

The adjustment for humidity is

$$T_2 = 0.5 \ln e - 1.1 \qquad \text{(A2.11)}$$

where e is the water vapour pressure at ground level in millibars.

This adjustment allows for the influence of the amount of precipitable water which is statistically linked to the water vapour pressure at ground level

T_2 varies from -0.4 (e = 4 mb) to + 0.6 (e = 30 mb)

Correction factor T_3.

The adjustment for turbidity is assessed as a function of the blueness of the sky, and takes account of the quantity of atmospherical aerosols, but not their specific spectral effects. The values applied are given in Table A2.3.

The air mass 2 Linke Turbidity Factor is found by adding the corrections

$$T_L = 3.3 + T_1 + T_2 + T_3 \qquad \text{(A2.12)}$$

Table A2.3 Correction factors T_3 for method 5 to allow for effects of aerosol through colour of the sky. Source: WMO (8).

Colour of sky	Deep blue	Pure blue	Medium blue	Pale blue	Milky blue	Whitish
Estimated coefficient of turbidity	0.02	0.04	0.06	0.10	0.20	> 0.35
T_3	-0.8	-0.3	0	+0.7	+2.1	>+4.0

Table A2.4 Correction Factor T_4 for method 5 to allow for the effects of solar elevation on the Linke Turbidity Factor. Source: WMO (8).

Elevation angle of sun	8°	15°	30°	45°	>60°
sin γ	0.14	0.25	0.50	0.71	>0.85
T_4	-0.6	-0.3	0	+0.1	+0.3

Correction factor T_4

The air mass 2 Linke Turbidity Factor then has to be adjusted for solar elevation. This further correction T_4 is found from the Table A2.4. Thus we have

$$T_L(\gamma) = T_L + T_4 \qquad (A2.13)$$

Solar altitude dependence of $T_L(\gamma)$ implicit in the various formulae

Formula A2.1 predicts the slope of $T_L(\gamma)$ as a function of solar altitude γ as $1/(39.5\ e^{-w} + 47.4)$, i.e. only dependent on water vapour content.

Formula A2.2 predicts the coresponding function as:
0.053 γ for mean conditions
0.027 γ for conditions of minimum turbidity.

Formula A2.6 is corrected by Page's algorithm based on the WMO Tables, refer Chapter 2, paragraph 19.

Formula A2.10 is uncorrected for solar altitude but mentions a variation of ±0.2.

Formula A2.13 is the tabular formulation used by Page to derive his algorithm, as discussed in Chapter 2, paragraph 19.

Using a sea level solar altitude of 30° (Air mass 2) as the reference it is simple to compute the consequent correction factors as a function of γ. The Dogniaux relationships are linear with solar altitude and the WMO relationships are not. The resulting corrections are plotted in Figure A2.1 for w = 1.5 cm for a value of $T_L = 4.0$. Given the fact that the attenuation is a function of m, it appears that the WMO curvilinear form is more likely to be representative of the underlying physical facts than the linear relationships. The slope of the Dogniaux Sneyers line is greater than the other two, because, in effect, it contains two factors, the summer winter variation in clarity and the summer winter change in solar altitude.

Comparison of the various prediction methods for T_L with observed air mass 2 Linke Turbidity Factors at Uccle

Dogniaux has compared the predictions of the various formulae with the observed values of the air mass 2 T_L, for a very long series of carefully supervised pyrheliometric measurements at Uccle. Figure A2.2 shows the comparison between observation and prediction to Methods 1, 2, 3 and 5. The WMO method appears the best and it is also fairly simple to use. The sharp increase in atmospheric turbidity found in northern Europe in July and August however is not reflected properly in any of the modelling techniques. In order to make these comparative estimates, Dogniaux had to make certain subjective judgements. $_A$ was set constant and equal to 0.17 to obtain estimates of monthly mean values for Uccle, and at 0.09 to obtain estimates for conditions of minimum turbidity. The precipitable water vapour was derived from Hahn's formula $w = 0.17\ \bar{e}$ using the observed monthly mean vapour pressure at Uccle. The ATI was set at $0.75 - 0.0012\ \bar{e}$

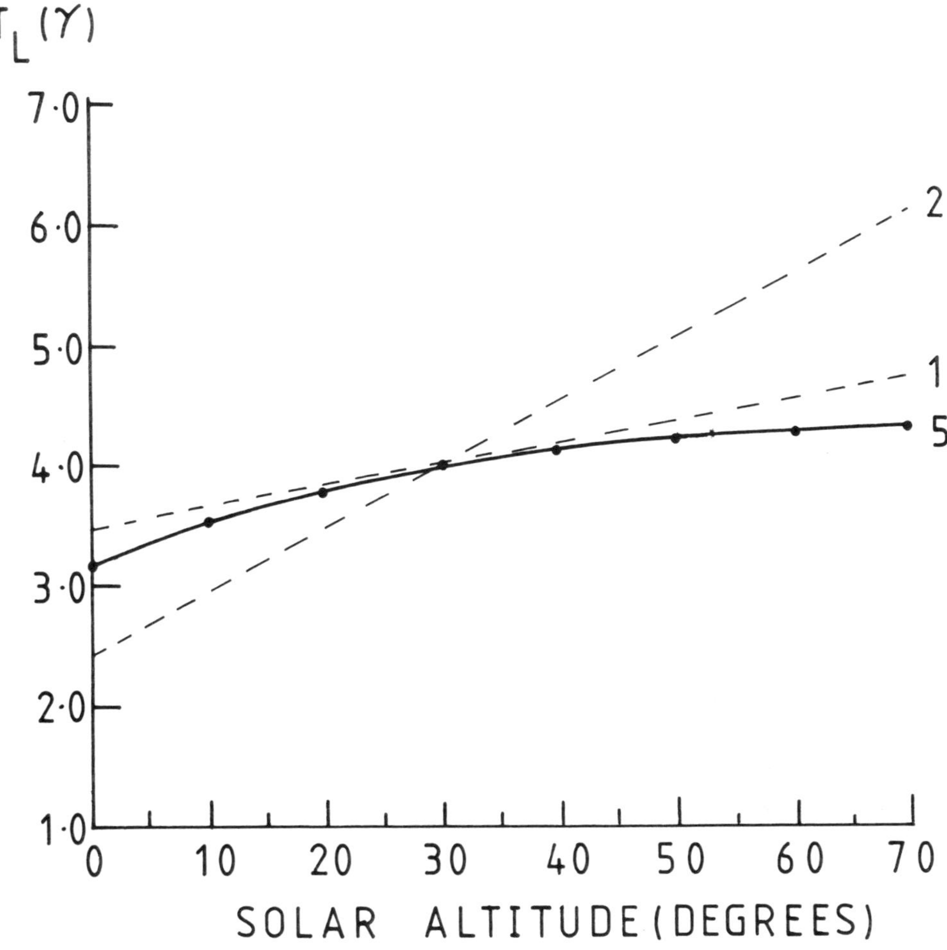

Figure A2.1 The values of $T_L(\gamma)$ for a given solar altitude for an air mass 2 T_L value of 4, line 1, predicted by method 1, line 2, predicted by method 2, curve 5, predicted by method 5, as modified by Page.

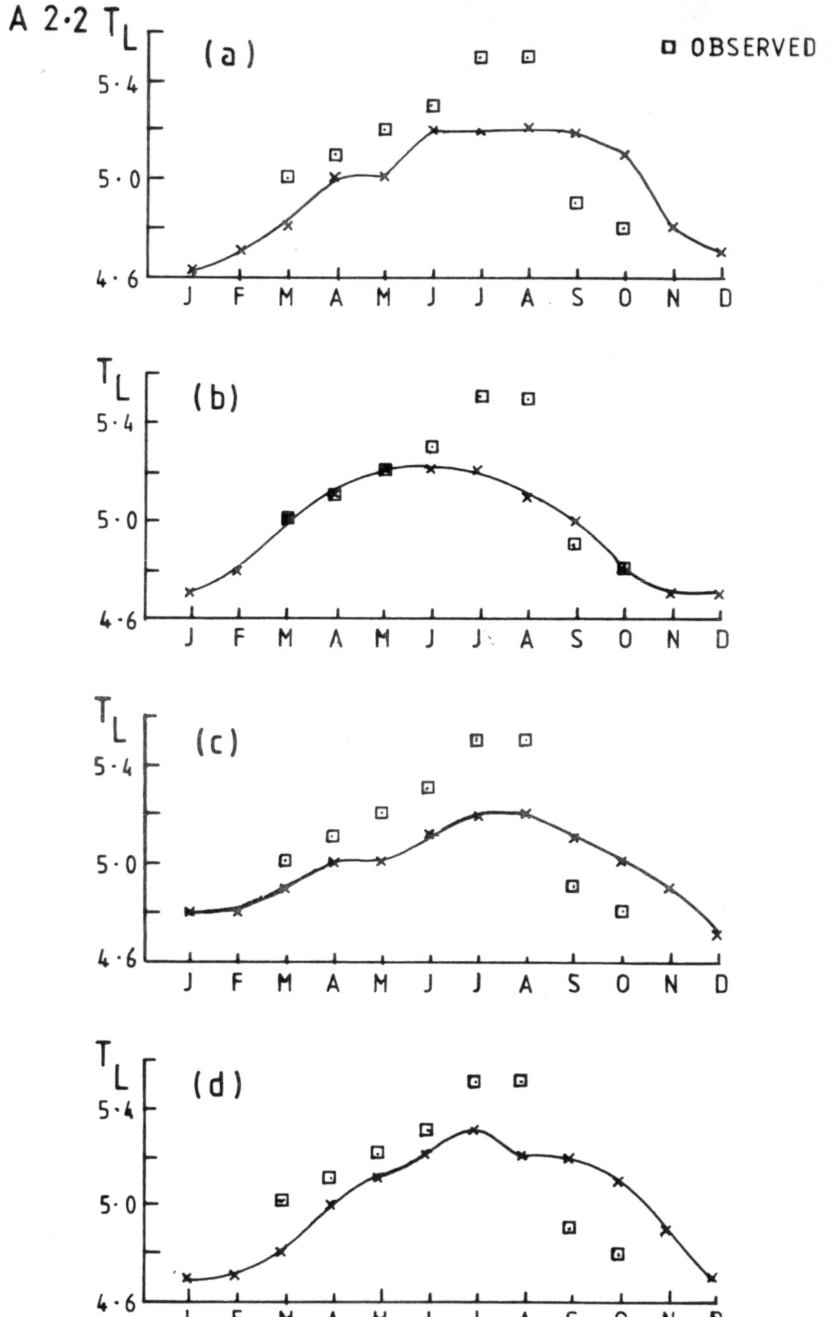

Figure A2.2 The relationship between predicted and observed values of T_L at Uccle using different methods (a) Method 1, (b) Method 2, (c) Method 3 and (d) Method 5. Observed values in squares. None of the methods predict the July-August peak correctly.

for monthly mean conditions and at 0.80 - 0.0012 $\bar{e}$ for days of minimum turbidity. The introduction of a second harmonic term into the Dogniaux formula to reproduce the July-August peak could produce a significant increase in its prediction accuracy.

Conclusions

The estimation of Linke Turbidity Factor at any specific site presents difficulties due to the subjective judgement that has to be exercised concerning the aerosol content of the atmosphere in the absence of site specific measurements of T_L. Until more observed values of T_L are available for the European area, it will remain difficult to be sure that appropriate values of T_L are being adopted at any specific site. There is a need to better order the existing European data on turbidity, so a better general scientific picture is established of the clarity of the atmosphere across the whole region at different times of year. In practical terms, it appears solar altitude is the key factor influencing the Linke Turbidity Factor at any specific site.

Appendix 2 - References

1. Dogniaux, R., (1984), De l'influence de l'estimation du facteur total de trouble atmosphérique sur l'évaluation du rayonnement solaire direct par ciel clair - Application aux données radiométriques de l'IRM à Uccle, IRM, Misc. Serie C, No. 20.

2. Dogniaux, R. and Doyen, P., (1968), Analyse statistique du trouble atmospherique à Uccle à partir d'observations radiométriques - Période de référence 1951-1965, IRM Publ. Série A, No. 65.

3. Dogniaux, R. and Lemoine, M., (1976), Programme de calcul des éclairements solaires énergétiques et lumineux de surfaces orientées et inclinées, Ciel serein et ciel couvert, IRM, Misc. Série C, No. 14.

4. Dogniaux, R. and Sneyers, R., (1972), Sur la stabilité du trouble atmosphérique à Uccle au cours de la période 1951-1970, IRM Publ. Série A, No. 75.

5. Dogniaux, R. and Lemoine, M., (1983), Classification of radiation sites in terms of different indices of atmospheric transparency, Proceedings of the 1983 International Daylighting Conference, Phoenix, Arizona, USA, Thomas Vonder, AIA-Editor.

6. Dogniaux, R. and Lemoine, M., (1983), Classification of radiation sites in terms of different indices of atmospheric transparency, in Solar Energy R&D in the European Community, Series F, Vol. 2, Solar Radiation Data, (Ed. W. Palz), D. Reidel Publishing Company, p.94-107.

7. Perrin de Brichambaut, Ch. et Vauge, Ch., (1982), Le gisement solaire: évaluation de la ressource énergétique, TEC et DOC Lavoisier, 75008 Paris.

8. WMO, (1981), Meteorological aspects of solar radiation as an energy source, WMO Technical Note No. 172, WMO, No. 557.

9. Nimmo, B. and Said, S.A.M., (1982), Direct and total radiation measurements for Dhahran, Saudi Arabia, in Proc. Solar World Forum, Brighton, Ed. Hall, D.O. and Morton, J., Pergamon Press, p.2343-2348.

APPENDIX 3

THE RELATIONSHIP BETWEEN MONTHLY MEAN MAXIMUM DAILY SUNSHINE, S_{max} AND MONTHLY MEAN DAILY SUNSHINE

It became evident during the course of the development of the clear day model that there were relationships between S_{max}, the monthly mean maximum daily sunshine, and S_m, the monthly mean value of daily sunshine, published in the CEC European Solar Radiation Atlas, Vol. I, 2nd edition, Global radiation on horizontal surfaces. Places with low relative values of S_m tended to have low values of S_{max}.

As values of daily mean sunshine are much more widely available than values of S_{max} which is needed to estimate the clear day solar radiation on slopes, if the relationship can be quantified, then it becomes possible to make estimates from daily mean sunshine alone. The relationships were developed on a dimensionless basis to take account of the effects of daylength. Data for a hundred stations were used.

The ratios S_{max}/S_{04max} and S_m/S_o were first computed where S_{max}/S_{04max} is the ratio of observed S_{max} to the daylength calculated on a 4 degree sunrise basis, using the declinations appropriate to G_{max} days (Refer Chapter 6, Table 6.2) and where S_m/S_o is the mean monthly relative duration of daily sunshine calculated on the basis of astronomical sunrise. Values of S_m/S_o appear in the station tables in the CEC Solar Radiation Atlas, Vol. I. (For other stations, S_o. should be calculated using the mid-month values of the solar declination, which is a different date to that used to calculate S_{04max} (Refer Chapter 6, Table 6.2 and Tables 6.6(a) to 6.6(c)).

The data were then split into latitude bands of 35-40, 40-45, 45-50, 50-55, 55-60 and above 60 degrees north. Linear regressions were sought relating S_{max}/S_{04max} to S_m/S_o of the form

$$S_{max}/S_{04max} = a_o + a_1 S_m/S_o \tag{A3.1}$$

Table A3.1 gives the results of these linear regression studies. The strength of the correlation varied. In general the correlation was quite acceptable but sometimes it was not statistically significant. Table A3.1 does, however, provide a method for estimating S_{max} from S_m in the absence of available observations of S_{max}, so makes the methodology described for clear day calculations available to any European station where values of S_m and (a + b) are available. Where values of a + b are not available, they can be first estimated using the method of Dogniaux and Lemoine described in Appendix 1 under method 3.

Table A3.1 Regression of S_{max}/S_{40max} against S_m/S_o $\sigma_{4max} = a_o + a_1 S_m/S_o$.

Band 35-40 degrees

Sites: Athens, Cagliari, Crotone, Ustica, Messina, Trapani, Gela, Pantelleria

	Jan	Feb	Mar	Apr	May	Jun	Jul	Aug	Sep	Oct	Nov	Dec
a_o	.970	.518	.554	.327	.312	.338	.224	.247	.485	.393	.612	.614
a_1	-.096	.823	.638	1.024	.913	.833	.883	.879	.618	.828	.598	.730
Correlation	.108	.723	.706	.906	.881	.877	.950	.936	.700	.852	.646	.692

Band 40-45 degrees

Sites: Millau, Carpentras, Nice, Bologna, Genova, Capo Mele, Pisa, Ancona, Pianosa, Pescara, Vigna di Valle, Roma, Amendola, Olbia, Napoli, Brindisi, Alghero, Capo Palinuro, Banja Luka, Beograd, Negotin, Sarajevo, Zlatibor, Split, Pristina, Bar, Bitola

	Jan	Feb	Mar	Apr	May	Jun	Jul	Aug	Sep	Oct	Nov	Dec
a_o	.727	.737	.872	.882	.868	.902	.895	.848	.774	.726	.783	.715
a_1	.554	.458	.071	.080	.108	.052	.057	.129	.233	.354	.377	.634
Correlation	.772	.743	.153	.149	.167	.085	.010	.278	.445	.683	.563	.800

Band 45-50 degrees

Sites: Salzburg, Innsbruck, Sonnblick, Klagenfurt, Bratislava, Trappes, Nancy, Macon, Limoges, Wuerzburg, Trier, Weihenstephan, Hohenpeissenberg, Bolzano, Udine, Trieste, Venezia, Milano, Torino, Zakopane, Jersey, Zurich, Davos, Locarno Monti, Ljubljana, Zagreb, Wien

	Jan	Feb	Mar	Apr	May	Jun	Jul	Aug	Sep	Oct	Nov	Dec
a_o	.699	.724	.821	.908	.865	.949	.986	.907	.746	.733	.763	.845
a_1	.653	.525	.168	.006	.122	-.040	-.102	.028	.317	.356	.461	.267
Correlation	.679	.446	.171	0.000	.134	.045	.149	.036	.279	.445	.487	.322

Table A3.1 Regression of S_{max}/S_{40max} against S_m/S_o $\sigma_{4max} = a_o + a_1 S_m/S_o$.
continued

Band 50-55 degrees

Sites: Oostende, Melle, Uccle, Saint Hubert, Hradec Kralove, Jyndevad, Norderney, Hamburg, Braunschweig, Braunlage, Birr, Kilkenny, Valentia, Groningen, Den Helder, De Bilt, Vlissingen, Kolobrzeg, Suwalki, Warszawa, Aldergrove, Cawood, S. Bonnington, Cambridge, Aberporth, London, Kew, Bracknell, Maastricht

	Jan	Feb	Mar	Apr	May	Jun	Jul	Aug	Sep	Oct	Nov	Dec
a_o	.605	.704	.819	.709	.602	.799	.717	.770	.718	.555	.749	.449
a_1	1.402	.672	.133	.545	.778	.342	.512	.334	.406	1.023	.524	2.754
Correlation	.465	.451	.157	.493	.680	.587	.795	.746	.529	.707	.390	.703

Band 55-60 degrees

Sites: Hojbakkegard, Karlstad, Stockholm, Visby, Svalov, Torslanda, Eskdalemuir

	Jan	Feb	Mar	Apr	May	Jun	Jul	Aug	Sep	Oct	Nov	Dec
a_o	.474	.762	.726	.690	.748	.804	.729	.803	.806	.587	.700	.564
a_1	2.984	.668	.471	.601	.424	.293	.432	.287	.265	1.037	1.117	2.828
Correlation	.925	.701	.762	.934	.986	.917	.936	.856	.572	.907	.381	.915

Band 60-65 degrees

Sites: Bergen, Lerwick (Not satisfactory for practical application)

	Jan	Feb	Mar	Apr	May	Jun	Jul	Aug	Sep	Oct	Nov	Dec
a_o	-1.076	.981	.577	.134	.690	.813	.497	.690	-.290	N.A.	.098	-.093
a_1	19.7	-.500	.850	2.080	.575	.280	1.360	.556	4.600	N.A.	4.100	12.600

Correlation Too few stations to achieve workable correlations

APPENDIX 4

THE USE OF SWISS SKY RADIANCE MEASUREMENTS TO ESTIMATE DIFFUSE IRRADIANCE FROM THE SKY

prepared by P. Valko, Schweizerische Meteorologische Anstalt, Zurich, Switzerland

The methodology used to estimate the sky diffuse irradiance from radiance patterns is of some general interest in solar studies, so is reported in detail here.

In order to split the diffuse irradiance into a sky diffuse component and a ground reflected component, it is possible to envisage a network of pyranometers of perfect cosine response arranged at different tilts and orientations, fitted with an artifical horizon, exposed alongside other pyranometers operating simultaneously without such a screening at the same tilts and orientations to provide a direct approach to this splitting process. Apart from difficulties (errors by overheating and cosine errors among others), such a system would not provide information about the radiance distribution which is also in its own right most useful information for solar energy research. Efforts to compute radiance distributions exclusively from multipyranometer measurements have been made by Slåen (1). The theoretical model developed, however, yields results of limited validity compared with detailed scientific observations of radiance.

The alternative is to compute values of the diffuse irradiance components on any inclined surface from observations of the radiance. It would, in principle, be sufficient to measure, by scanning, the sky radiance and reflected radiance with an instrument of suitable practicable angular resolution - presupposing the site is of homogeneous reflectivity. This is generally not the case. Radiance scanners furthermore are unsuitable to operate in routine networks.

Linking observed horizontal surface irradiance empirically with radiance distributions is of evident advantage, as pyranometers with horizontal exposure are in wide use. However, when considering the results discussed in Chapter 3, part I, one has to take account of the fact that the silicon diode scanners cover only the spectral range up to 1100 nm, while the thermopiles of the pyranometers cover the range up to about 3000 nm. Due to this spectral gap, changing water vapour content of the atmosphere would, for instance, affect any probable close regressions between L_{CZ} and D_c. This regression is particularly important, since it can give immediate access to the absolute values of L_{CZ} and so leading through to the graphical solution of equation 3.2, to establish the values of the radiance at any point in the sky, and on the ground.

To reduce errors due to the spectral differences between the two instruments, an adjustment for different sensor properties had first to be made. Absolute radiances in Wm^{-2} steradian^{-1} could be determined from the known spectral response and geometrical characteristics of the radiance meter (Heimo (2)). Denoting the measured radiances by $L'(\Theta,\alpha)$ and their integral for the horizontal surface by D'_c

$$D'_c = \int_0^{\frac{\pi}{2}} \int_0^{2\pi} L'(\Theta, \alpha) \sin\Theta \cos\Theta \, d\Theta \, d\alpha \qquad (A4.1)$$

The adjusting factor C is defined as

$$C = \frac{D_c}{D'_c} \qquad (A4.2)$$

whereby D_c means the "true" horizontal surface diffuse irradiance derived from measurements with the horizontal pyranometer of G_c and the absolute radiometer of I_c as

$$D_c = G_c - I_c \sin\gamma \qquad (A4.3)$$

Equation A4.1 has to be solved numerically using the 121 individual measured radiance values. Values of radiance were taken at 18° spacing of elevation Θ, and 15° spacing of azimuth α.

Considering the angular grid for scanning the sky (Chapter 3.1), each measured value $L'_c(\Theta, \alpha_s)$ was assigned to an angular area element on a unit sphere of $\Delta\Theta \cdot \cos\Theta \cdot \Delta\alpha$, i.e. (18°) . (cos Θ) . (15°), or expressed in steradians ($\pi^2/180^2$) . (18°) . (cos Θ) . (15°). The radiance is assumed uniform over each area element along the elevation circles Θ = 36°, 54° and 72° as well as around the zenith value at Θ = 90°. For the lowest sky belt, the radiances measured at Θ = 18° have been taken as representative of the whole range between the horizon at Θ = 0° and the elevation circle Θ = 27°. Radiances measured around the horizontal circle itself (Θ = 0°) have been omitted to avoid possible site specific effects on the adjusting procedure. The numerical integration thus considers only 97 radiances instead of the observed 121 values. In fact often less than 97 points were available, since, taking account of possible uncertainties in positioning the instruments, observations in the vicinity of the sun, i.e. readings for all points where η, the incidence angle between the sun's position and the respective point in the sky, is < 7.5°, have been left out.

Accordingly, as a numerical approximation, D'_c may be evaluated as follows:

$$D'_c = \sum_{i=1}^{4} \sum_{j=0}^{23} L'_c(\Theta_i, \alpha_j) \int_{\Theta_i - \Delta\Theta'}^{\Theta_i + \Delta\Theta} \int_{\alpha_j - \Delta\alpha}^{\alpha_j + \Delta\Theta} \sin\Theta \cos\Theta \, d\Theta \, d\alpha$$

$$+ \pi L'_{cz}[1 - \sin^2(\Theta_4 + \Delta\Theta)] \ \mathrm{Wm}^{-2} \qquad (A.4.4)$$

where

Θ_i = 18° i . π/180° radians, i = 1 → 4

α_j = 15° j . π/180° radians, j = 0 → 23

$\Delta\alpha$ = 7.5° . π/180° radians

$\Delta\Theta$ = 9° . π/180° radians

$\Delta\Theta' = \Delta\Theta$ radians for $i \geq 2$

$\Delta\Theta' = \Theta_1$ radians for $i = 1$. (This allows for the bottom zone running down to the horizon)

Considering that

$$\int_{\alpha_j - \Delta\alpha}^{\alpha_j + \Delta\alpha} d\alpha = 2\Delta\alpha$$

and

$$\int_{\Theta_i - \Delta\Theta'}^{\Theta_i + \Delta\Theta} \sin\Theta \cos\Theta \, d\Theta = \tfrac{1}{2}[\sin^2(\Theta_i + \Delta\Theta) - \sin^2(\Theta_i - \Delta\Theta')]$$

It follows

$$D'_c = \Delta\alpha \sum_{i=1}^{4} \sum_{j=0}^{23} L'_c(\Theta_i, \alpha_j)[\sin^2(\Theta_i + \Delta\Theta_i) - \sin^2(\Theta_i - \Delta\Theta'_i)]$$

$$+ \pi L'_{cz} [1 - \sin^2 (\Theta_4 + \Delta\Theta)]$$

or inserting the angular values

$$D'_c \approx 7.5° \; \pi/180° \; [\sin^2 27° - 0°] \sum_{j=0}^{23} L'_c(18°, \alpha_j)$$

$$+ [\sin^2 45° - \sin^2 27°] \sum_{j=0}^{23} L'_c(36°, \alpha_j)$$

$$+ [\sin^2 63° - \sin^2 45°] \sum_{j=0}^{23} L'_c(54°, \alpha_j)$$

$$+ [\sin^2 81° - \sin^2 63°] \sum_{j=0}^{23} L'_c(72°, \alpha_j)$$

$$+ L'_{cz}[1 - \sin^2 81°] \qquad \text{(A4.5)}$$

The last term is, of course, the contribution from the small area around the zenith.

By using equation A4.5 for each set of measurements to evaluate D'_c, the measured radiances $L'_c(\Theta,\alpha)$ could individually be calibrated for the whole spectral range of the pyranometers by

$$C \, . \, L'_c(\Theta,\alpha) = L_c(\Theta,\alpha) \qquad \text{(A4.6)}$$

The mean of 267 values of C was 1.103. 95% of the values lay between 0.950 and 1.300. Taking account of the measurement errors and the limitations of the numerical integration, this variation is acceptable. A 10% mean loss of sky radiance between 1100 and 3000 nm at an average water vapour content of the air is realistic.

The relationship between L_{CZ} and D_C could then be investigated using equation A4.2. The ratio $(\pi\ C\ L'_{CZ})/D_C = \pi L_{CZ}/D_C$ relates the zenith radiance to the irradiance. For an isotropically radiating sky $L_c(\Theta,\alpha) = L_{cz}$ and the value of $\pi L_{cz}/D_c$ is 1.

The departure of the value of the ratio from 1 is therefore an intuitive measure for demonstrating to what extent the irradiances induced by real skies deviate from those from an isotropic sky of uniform radiance L_{cz}. Computed values of the ratio $\pi L_{CZ}/D_C$ for the five examples given in Chapter 3, Table 3.1 are entered in row 22 of that Table.

Inclined surfaces integration

The use of equation $L_{CZ} = 1/\pi\ D\ \ f(\gamma)$, which defines the mean trend discussed in Chapter 3, Section 8, especially Figure 3.17, and the radiance observations illustrated by Figure 3.6c, permits the sky component $D_{CS}(\beta, \alpha_S)$ alone of the total diffuse irradiance $D_c(\beta, \alpha)$ on any tilted surface to be estimated by numerical integration.

The geometric weight of the single radiances $L_c(\beta, \alpha)$ for the general case is cos ν. Thus sin Θ in equation A4.1 which is the cosine angle of incidence on a horizontal surface has to be replaced by the cosine of the angle of incidence of the element measured from the centre on the inclined surface which is given by:

$$\cos \nu = \sin \Theta \cdot \cos \beta + \cos \Theta \cdot \cos \alpha_s \cdot \sin \beta \qquad \text{(A4.7)}$$

The radiance integral in its general form and its numerical solution for several specific cases is discussed in detail in a recent report of Bener (3). Using a quadratic interpolation procedure, the radiance field was computed first for the grid $\Delta\Theta = 2°$, $\Delta\alpha = 6°$ and, further on, linearly for every $\Delta\Theta = \Delta\alpha = 1°$. Radiances were thus immediately available at altogether 32,401 points in each hemisphere before the Bener-algorithm was adapted to each measured distribution. This was done to have enough freedom both to vary the number of sub-divisions used and to choose the precise configuration of area-elements when solving the double integral for specific cases. Generally values of $D_{CS}(\beta,\alpha_S)$ were computed, but the algorithm can, of course, be applied both for reflected radiances or for the sky for surfaces of either positive or negative inclination.

The fit of the numerical approximation depends on the number of area-elements into which the sky is sub-divided for reproducing the irradiance of the horizontal surface. For this purpose the sky was sub-divided in 360 different ways by shifting the chosen grid of area elements in steps of 1° around through all azimuths. The analytical solution, which is, of course, independent of this rotation, can be approximated within a band width of about 0.5% variation by choosing a fine meshed grid of 72 x 72 = 5,184 elements. Since this choice is also a question of computer costs, and as the accuracy of measurements is anyhow limited, 36 x 36 = 1,296 sub-divisions have been chosen for routine computations.

Reflected irradiances and the problem of spectral corrections

The ground reflected irradiance field $R_C(\beta, \alpha_s)$ was found by subtracting the $D_{CS}(\beta, \alpha_S)$ field shown in Figure 3.6d from the $D_C(\beta, \alpha_S)$ field shown in Figure 3.6b. The same result can also be achieved by numerical integration of $L_c(-\Theta, \alpha_s)$ with respect to the 77 pyranometer surfaces. The results of this latter integration were used for correcting evidently erroneous entries, given in parenthesis in Figure 3.6h, which result from the limitations of the instrumentation, the numerical integration process and also the spectral adjustment process, i.e. values of $D_{cs} > D_c$ occurred for $\beta > 0$. It has to be noted for inclined planes that the spectral irradiance is dependent on a combination of sky spectral properties of a specific part of the sky and ground reflectance properties at a specific point on the ground. The spectral corrections for the components of the radiance of a given angle depend on the azimuth angle with respect to the sun α_S and the solar elevation γ and so could not be computed using equation A4.2. Studies were carried out introducing a second spectral adjustment for downward facing surfaces C_R defined as

$$C_R = \frac{R_c}{R'_c} \tag{A4.8}$$

Using the pyranometer value R_c for the inverted horizontal surface and the analogous integral R'_c were not successful for the EMPA-site but may yield improvements for homogeneous snow cover. A refinement of the form

$$C^* = \frac{D_c(\beta, \alpha_s)}{CD_{cs}(\beta, \alpha_s) + C_R\, R_c(\beta, \alpha_s)} \tag{A4.9}$$

where two separate correction coefficients are applied to allow for the different spectral characteristics of sky and ground may perhaps reduce the errors in Figure 3.6h.

Appendix 4 - References

1. Slåen, T., (1984), A method for computing the angular distribution of solar radiation from multipyranometer observations, Proceedings, Solar World Congress, Perth, 1983, Vol.4, pp.2189-2193.

2. Heimo, A., (1979), Determination of the calibration factor of sky radiometers (unpublished communication).

3. Bener, P., (1986), Berechnung der Irradiance auf eine Beliebig orientierte Flache aus der Verteilung der Radiance uber den Himmel. Part of the Final Report to Chapter 3, Reference 15, p.44 + 20 Figures.

APPENDIX 5

DEVELOPMENT OF SUITABLE SIMPLIFIED ALGORITHMS FOR ESTIMATING THE CLEAR SKY DIFFUSE RADIATION ON SLOPES

A5.1 Introduction

Appendix 4 has already explained the principles, whereby working from a defined radiance distribution of the clear sky, the diffuse irradiance on any slope may be estimated. The radiance distribution adopted for the clear sky computations of slope irradiance in the CEC Inclined Surface Atlas is given by Equation 3.14 in the main text. These processes of numerical integration are, however, relatively slow, and the computing time involved in performing such numerical integrations is quite considerable. To save computing time, a simplified model was required.

The first stage in preparing a simplified model was to perform numerical integrations to determine the ratio of the slope sky diffuse irradiance to the horizontal diffuse irradiance for a range of Linke turbidities and solar azimuth angles α_s for a range of solar altitudes for a range of slopes. The sky was divided into zonal elements of 5° in elevation and 5° in azimuth. Performing the integrations for horizontal planes as well as for vertical planes, enabled the ratio $D_{cs}(\beta,\alpha_s)/D_c$ to be computed from the radiance distribution described by Equation 3.14.

The next stage was to pragmatically develop a simpler model which fitted the results of the numerical integration, but which was much faster in computational time. The process of simplification, of course, implied some loss of accuracy, but it was impractical to have a station specific slope radiation model that required long numerical integrations over the whole sky hemisphere, every time a computation had to be performed for a specific time of day in a specific month. The algorithm that was developed using Equation 3.14 is described in the following pages. It was used throughout the Sheffield University studies on this CEC project.

Now Valko's work in Switzerland described in Chapter 3, Part I, has reached a mature stage of development, it should be possible to produce an improved simplified formula describing the clear sky radiance pattern, and consequently to develop improved rapid algorithms for predicting the cloudless day slope irradiance from the predicted cloudless day diffuse irradiance on a horizontal surface.

A detailed description of the algorithm developed for the CEC Inclined Surface Solar Radiation project and used by the University of Sheffield team throughout this project, including the production of the CEC European Solar Radiation Atlas, Vol. II, Inclined Surfaces (1), follows.

A5.2 Augmented direct beam sky diffuse model

The method first uses the concept of splitting the diffuse flux from the cloudless sky into two components, a component associated with the position of the sun which can be spatially associated with the direct beam and a background component assumed to come from a sky of uniform radiance. This enables considerable simplification to be achieved. The split adopted was determined by numerical integration using the radiance distribution function to determine the aureole component within an angle of 30° around the solar disc. This aureole component was then assumed to be spatially located at the centre of the solar disc. Its magnitude varied with the Linke Turbidity Factor $T_L(\gamma)$ and with solar altitude. The remaining diffuse energy ${}_bD_c$ was then assumed to be background, and to come from a clear sky of uniform radiance. Thus, on a horizontal plane,

$$D_c = {}_bD_c + {}_cD_c \sin \gamma \qquad \text{(A5.1)}$$

where ${}_bD_c$ is the background diffuse irradiance on a horizontal surface, Wm^{-2}.

${}_cD_c \sin \gamma$ is the component associated with the direction of the sun resolved into the horizontal plane, Wm^{-2}.

By numerically integrating over a wide range of solar altitudes between 0 and 90° and a wide range of Linke Turbidity Factors between 1.5 and 10.0, a systematic set of values of the ratio ${}_bD_c/D_c$ were computed, and then, using standard polynomial regression analysis techniques, the relationship was stated algorithmically. The resulting algorithm was:

$${}_bD_c/D_c = \sum_{i=0}^{2} \left(\sum_{j=0}^{2} a_{ij} \, T_L(\gamma))^j \right)^i \quad \text{dimensionless} \qquad \text{(A5.2)}$$

where a_{ij} is a set of constants given in the table below.

Values of a_{ij} used in equation A5.2

j \ i	0	1	2
0	9.502×10^{-1}	-1.340×10^{-3}	-1.907×10^{-5}.
1	-2.458×10^{-2}	-1.817×10^{-3}	1.673×10^{-5}.
2	9.574×10^{-4}	9.282×10^{-5}	-8.634×10^{-7}.

This ratio may be used directly to calculate the horizontal surface background irradiance from the total horizontal surface diffuse irradiance D_c. The horizontal surface circumsolar component may then be obtained as the difference

$$_cD_c \sin\gamma = (D_c - {_bD_c}) \quad Wm^{-2} \qquad (A5.3)$$

where $_cD_c \sin\gamma$ is the diffuse irradiance on a horizontal surface from the circumsolar region in a cloudless sky.

A5.3 Estimating the slope irradiance due to diffuse radiation from the clear sky using simple augmented direct model

With two additional inputs, the angle of tilt of the plane β and the angle of incidence of sun on the surface ν, it is then possible to estimate the diffuse sky irradiance on any slope.

The background diffuse irradiance $_bD_c(\beta,\alpha)$ on a surface of slope β may be estimated from

$$_bD_c(\beta,\alpha) = {_bD_c}\ 0.5(1 + \cos\beta)\ Wm^{-2} \qquad (A5.4)$$

The circumsolar diffuse irradiance on a surface of slope β and orientation α having angle of solar incidence ν, $_cD_c(\beta,\alpha)$ is given by

$$_cD_c(\beta,\alpha) = {_cD_c} \cos\nu \quad Wm^{-2} \qquad (A5.5)$$

By rearranging equations A5.4 and A5.5, also using A5.3, one may estimate f_2*, the ratio of the clear sky inclined surface diffuse irradiance to horizontal surface diffuse irradiance, using the simple two component model, for surfaces facing the sun, $\cos\nu$ +ve

$$f_2* = 0.5({_bD_c}/D_c)(1 + \cos\beta) + (1 - {_bD_c}/D_c)\cos\nu/\sin\gamma \qquad (A5.6a)$$

otherwise, for $\cos\nu$ -ve

$$f* = 0.5({_bD_c}/D_c)(1 + \cos\beta) \qquad (A5.6b)$$

This two component model is not particularly accurate as it cannot properly handle the complex impact of the actual sky radiance distribution with the horizon brightening effect and the radiance minimum at 90° to the sun's position. The next section describes a pragmatic algorithmic modification which makes this augmented model substantially more accurate.

A5.4 Pragmatic correction to the simplified augmented direct model for cloudless skies

Principles

Any simple augmented model for predicting clear sky diffuse radiation on slopes suffers from important defects because factors like finite aureole size and the natural sky brightening around the clear sky horizon cannot be handled properly. The errors in augmented models tend to be greatest when the sun is just going off a surface, or has just moved behind a surface and also when the sun is low, or very high, which, for vertical surfaces, is a special case of the sun being nearly off the surface. There is a very large body of experience that indicates the errors depend mainly on the wall solar azimuth angle. Atmospheric turbidity is also an important factor. The horizon brightening effect is proportionally greatest when the turbidity is low. With very turbid conditions, the effect is reversed and the clear sky horizon radiance is lower than the radiance nearer the zenith.

If an accurate, turbidity dependent sky radiance distribution model is available, it is possible to obtain values of the clear sky irradiance on slopes by numerical integration, but this is very expensive in computing time. In order to confer greater accuracy than the simple augmented model can provide, combined with reasonable speed in computation, the results of detailed numerical integration were used to develop a pragmatic correction procedure to the augmented model that requires vastly less computing time than full numerical double integration. Nevertheless the correction procedures are quite complex, so they are best mounted on a suitable main frame, mini or micro computer to provide rapid calculation. The results can also be systematically tabulated (Refer Chapter 6, Tables 6.14a to 6.14g) to provide data for manual calculation. The procedures below developed can handle surfaces of any slope and orientation.

The additional correction is calculated as a function of solar altitude, Linke Turbidity Factor, slope tilt and wall solar azimuth angle, interpolating between two out of five basic wall solar azimuth angle directions, 0°, 45°, 90°, 135° and 180°, according to the wall solar azimuth angle. (0° is facing the sun and 180° is facing away from the sun). The process calculates the additional correction factor for the two orientations lying either side of the actual orientation. The mathematical interpolation procedures, mainly of a non linear kind, are used to derive the corrections for intermediate wall solar azimuth angles. Only two of the five basic values need to be calculated for any specific orientation. The basic interpolative procedures are illustrated in Figure A.5.1. A special procedure had to be adopted for solar altitudes above 70°. Due to convergence problems, a constant value for the correction ratio is adopted for solar altitudes below 5°.

A5.5 Detailed algorithms for finding correction factor to simplified two component augmented model

INPUTS

Linke Turbidity Factor	$T_L(\gamma)$	Dimensionless
Solar altitude	γ	Degrees
Wall solar azimuth angle, absolute value in degrees, i.e. if $\alpha_S < 0°$ then $\alpha_S = -\alpha_S$	α_S	Degrees
Surface inclination from horizontal	β	Degrees

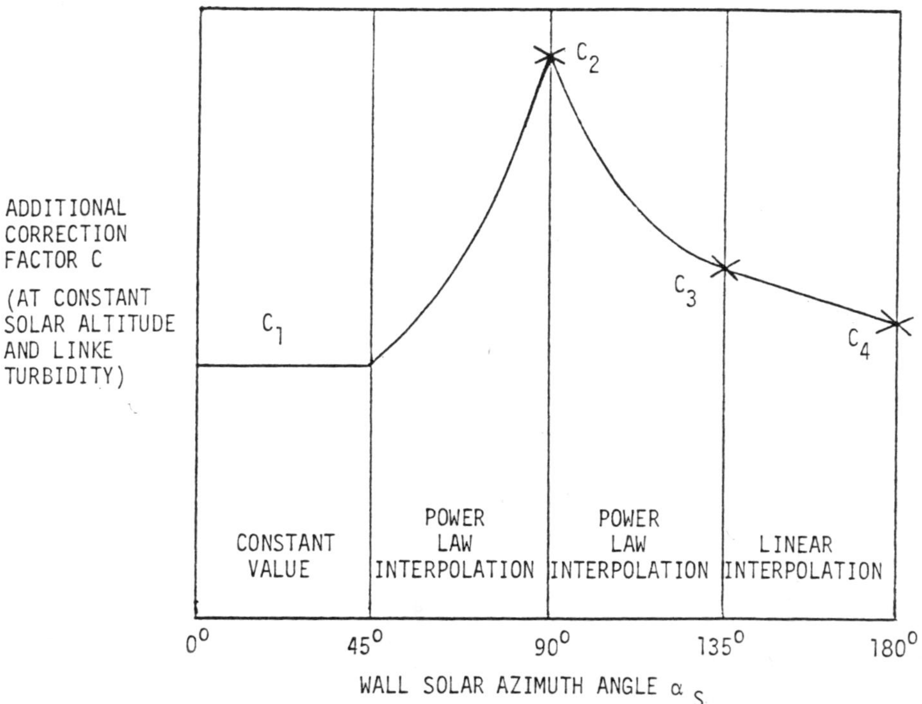

Figure A5.1 Interpolative procedures used for the calculation of the additional correction factor in the prediction of clear sky inclined surface diffuse irradiance

The correction factor f is found using the following calculation procedure. This begins either by calculating a single value of f or by calculating two values of f and interpolating between them, according to wall solar azimuth angle. If the solar altitude is greater than 70° a further modification is required in Stage 9. Finally, in Stage 10, the value of f is used to correct the ratio f_2* which was calculated using the simple two component model using Equation A.7.5.

STAGE 1 STARTING LOGIC

1.1 If $\gamma \leqslant 70°$ go to stage 1.3

1.2 If solar altitude γ is above 70°, then temporarily set solar altitude to a value of 70° to use in the calculations until the last stage. Save actual solar altitude value as γ_s for final interpolation between the 70° value and the zenith value in Stage 9.
i.e. if $\gamma > 70°$ then first set $\gamma_s = \gamma$
then set $\gamma = 70°$ and go to Stage 1.4

1.3 If solar altitude γ is less than 5°, set γ equal to 5°, saving the actual value as γ_s
i.e. if $0° > \gamma > 5°$, then set $\gamma_s = \gamma, \gamma = 5°$

1.4 If wall solar azimuth angle is greater than 135°, go to Stage 5.

1.5 If wall solar azimuth is greater than 90°, go to Stage 3.

STAGE 2 CORRECTION FACTOR C_1 FOR $\alpha_s < 45°$

2.1 If the wall solar azimuth angle α_s is less than 90° then first calculate C_1, the correction factor at wall solar azimuth angles less than or equal to 45°, using the following formula:

$$C_1 = (f_1(T_L(\gamma)) + f_2(\gamma)) \times f_3(T_L(\gamma))$$

where:

i) $f_1(T_L(\gamma)) = \sum_{i=0}^{4} a_i (T_L(\gamma))^i$

setting

$a_o = 2.036$; $a_1 = -0.3236$; $a_2 = 0.05578$; $a_3 = -5.094 \times 10^{-3}$;

$a_4 = 1.856 \times 10^{-4}$.

ii) $f_2(\gamma) = \sum_{i=0}^{5} a_i \gamma^i$.

setting

$a_o = -1.0440$; $a_1 = 0.092434$; $a_2 = -3.2014 \times 10^{-3}$;

$a_3 = 5.6317 \times 10^{-5}$; $a_4 = -5.0829 \times 10^{-7}$; $a_5 = 1.8944 \times 10^{-9}$.

iii) $f_3(T_L(\gamma), \gamma)$ is a function of Linke Turbidity Factor and solar altitude

(a) If $\gamma < 30°$ then

$$f_3(T_L(\gamma),\gamma) = 1.024 - 0.001\,\gamma$$

otherwise

(b) $f_3(T_L(\gamma),\gamma) = 0.994 + 0.000375(\gamma - 30)(0.3\,T_L(\gamma) - 0.8)$

2.2 For wall solar azimuth angle $\alpha_S < 45°$, no further calculations are required until Stage 9 is reached.

If $\alpha_S < 45°$ then $f = C_1$ and go to Stage 9.

STAGE 3 CORRECTION FACTOR C_2 AT $\alpha_S = 90°$

For wall solar azimuth angles greater than 45° (except precisely 90°, 135° and 180°), the correction factor must be obtained by interpolation between two adjacent computed values of f.

3.1 If $45° < \alpha_S < 135°$ then calculate C_2, the correction factor at a wall solar azimuth angle of 90° using the following formula, where β is the surface inclination, in degrees

$$C_2 = (f_4(T_L(\gamma)) + f_5(T_L(\gamma),\gamma) - 1.0)\,\frac{\beta}{90.0} + 1.0$$

where:

i) $f_4(T_L(\gamma)) = \sum_{i=0}^{3} a_i(T_L(\gamma))^i.$

setting:

$a_0 = 2.182;\ a_1 = -02738;\ a_2 = 0.03868;\ a_3 = -1.838 \times 10^{-3}$

ii) $f_5(T_L(\gamma), \gamma)$ is given by:

a) If $\gamma < 40°$

$$f_5 T_L(\gamma)\gamma) = 0.0004\,(30.0 - \gamma)(6.0 - T_L(\gamma))$$

b) If $\gamma > 40°$ and if $T_L(\gamma) > 4.0$

$$f_5(T_L(\gamma), \gamma) = 0.004\,(T_L(\gamma) - 8.0)$$

c) If $\gamma > 40°$ and if $T_L(\gamma) < 4.0$

$$f_5(T_L(\gamma), \gamma) = -0.02 + 0.0014(4.0 - T_L(\gamma))(\gamma - 45)$$

If wall solar azimuth angle = 90° then no further calculations are required until Stage 9

If $\alpha_S = 90°$ then $f = C_2$ and go to Stage 9

If $\alpha_S > 90°$ then go to Stage 5.

STAGE 4 INTERPOLATION SECTOR $45° < \alpha_S < 90°$

4.1 For wall solar azimuth angles $45° < \alpha_S < 90°$ interpolate between C_1 and C_2, using the following procedures to find the value of C.

If $45.0 < \alpha_S < 90.0$ then

$$f = C_1 + (C_2 - C_1)\left(\frac{\alpha_S - 45°}{45°}\right)^{\alpha_1}$$

where the exponent α_1 is given by:

$$\alpha_1 = 3.8 + 0.1\ T_L(\gamma) - 0.03$$

Then go to Stage 9

STAGE 5 CORRECTION FACTOR C_3 AT $\alpha_S = 135°$

5.1 If the wall solar azimuth angle is greater than 90°, then calculate C_3, the value of the correction factor at a wall solar azimuth angle of 135°, as follows.

$$C_3 = (f_6(T_L(\gamma)) + f_7(\gamma))\ f_8(T_L(\gamma),\gamma)$$

where:

i) $f_6(T_L(\gamma)) = \sum_{i=0}^{4} a_i (T_L(\gamma))^i.$

setting:

$a_0 = 2.183$; $a_1 = -0.5058$; $a_2 = 0.08540$; $a_3 = -7.091 \times 10^{-3}$; $a_4 = 2.217 \times 10^{-4}$.

ii) $f_7(\gamma) = \sum_{i=0}^{2} a_i\ \gamma^i.$

setting:

$a_0 = 0.2008$; $a_1 - -0.01164$; $a_2 = 1.662 \times 10^{-4}$.

iii) $f_8(T_L(\gamma), \gamma)$ is given by

a) If $\gamma < 40°$ then

$$f_8(T_L(\gamma),\gamma) = 1.0$$

b) If $T_L(\gamma) > 4.0$ and if $\gamma > 40°$ then:

$$f_8(T_L(\gamma),\gamma) = 0.990 + 0.00027\ (\gamma - 10°)(T_L(\gamma) - 5.0)$$

c) If $T_L(\gamma) < 4.0$ and if $\gamma > 40°$ then

$$f_8(T_L(\gamma),\gamma) = 1.041 - 0.018\ T_L(\gamma)$$

5.2 If wall solar azimuth angle $\alpha_S = 135°$, no further calculation is required until step 9

If $\alpha_S = 135°$ then $f = C_3$ and go to Stage 9

If $\alpha_S > 135°$ then go to Stage 7

STAGE 6 INTERPOLATION SECTOR 90° - 135°

6.1 If the wall solar azimuth angle lies between 90° and 135°, interpolate between C_2 and C_3 to find correction as follows

If $90° < \alpha_S < 135°$ then

$$f = C_3 + (C_2 - C_3)\left(\frac{135° - \alpha_s}{45°}\right)^{\alpha_2}.$$

where the power exponent α_2 is a function of solar altitude

$$\alpha_2 = 2.05 - 0.009\ \gamma$$

Then go to Stage 9.

STAGE 7 CORRECTION FACTOR C_4 AT $\alpha_S = 180°$

7.1 For wall solar azimuth angles between 135° and 180° calculate C_4, the correction at the 180° wall solar azimuth position, using the following formula

$$C_4 = f_9(T_L(\gamma)) + f_{10}(\gamma)$$

where

i) $$f_9 = \sum_{i=0}^{4} a_i(T_L(\gamma))^i.$$

setting

$a_o = 2.210$; $a_1 = -0.5508$; $a_2 = 0.09598$; $a_3 = -8.368 \times 10^{-3}$;

$a_4 = 2.768 \times 10^{-4}$

ii) $$f_{10} = \sum_{i=0}^{2} a_i\ \gamma^i.$$

setting

$a_o = 0.3112$; $a_1 = -0.01734$; $a_2 = 2.227 \times 10^{-4}$.

7.2 For wall solar azimuth angle 180° no further calculations are needed

If $\alpha_S = 180°$ then $f = C_4$ and go to Stage 9.

STAGE 8 INTERPOLATION SECTOR 135° - 180°

8.1 If the wall solar azimuth angle lies between 135° and 180°, then interpolate between C_3 and C_4 to find the value of f, using linear interpolation

If $135° < \alpha_S < 180°$, then

$$f = C_4 + (C_3 - C_4)\left(\frac{180.0 - \alpha_S}{45.0}\right)$$

STAGE 9 INTERPOLATION PROCEDURES NEEDED ONLY FOR SOLAR ALTITUDES ABOVE 70°

9.1 If solar altitude $\gamma < 70°$, then go to Stage 10

9.2 Reset γ to actual value i.e. $\gamma = \gamma_s$

9.3 For solar altitude $\gamma > 70°$, correct value of f obtained using $\gamma = 70°$ by interpolation with the zenith value, using the actual value of the solar altitude γ.

Thus if $\gamma > 70.0$ then

$$f = f + \left(\frac{\gamma - 70°}{20°}\right)^{1.7} \times (f_{11}(T_L(\gamma)) - f)$$

where $f_{11} = \sum_{i=0}^{3} a_i (T_L(\gamma))^i$.

setting:

$$a_o = 2.810;\ a_1 = -0.5621;\ a_2 = 0.07694;\ a_3 = -3.469 \times 10^{-3}$$

STAGE 10 DETERMINATION OF CLEAR SKY INCLINED SURFACE TO HORIZONTAL SURFACE DIFFUSE IRRADIANCE RATIO f_2

10.1 If the solar altitude was set equal to 5° in stage 1 then reset it equal to the actual value

i.e. if $\gamma = 5°$ then $\gamma = \gamma_s$.

10.2 Calculate the uncorrected ratio $f_2{*}(90,\alpha_S)$ for a vertical surface of the same azimuth using equation A5.6.

either a) If $\alpha_S < 90°$ then

$$f_2{*}(90,\alpha_S) = 0.5({}_bD_c/D_c) + (1 - {}_bD_c/D_c) \cos\gamma \cos\alpha_S/\sin\gamma$$

or b) If $\alpha_S > 90°$ then

$$f_2{*}(90,\alpha_S) = 0.5({}_bD_c/D_c)$$

where

$f_2{*}(90,\alpha_S)$ is the clear sky inclined surface to horizontal surface diffuse irradiance ratio for a vertical surface of the same azimuth as the surface for which the prediction is required.

${}_bD_c/D_c$ is the ratio of background to total diffuse on a horizontal surface calculated according to equation A5.2.

10.3 Apply the correction factor f to the uncorrected ratio $f_2{*}$ from equation A5.6 to obtain the corrected ratio f_2 as follows

$$f_2 = f_2{*} + (f - 1.0) \sin\beta \ .\ f_2{*}(90,\alpha_S)$$

Chapter 6 contains systematic tables of f_2 for different values of the air mass 2 Linke Turbidity Factor. The values of f_2 in these Tables were computed by first of all establishing the value of $T_L(\gamma)$ for the given solar altitude, using equations 2.23 and 2.24. Then the expression above was evaluated numerically. If computing facilities are available it is easier to mount the above algorithms on the computer than to carry out interpolations between the various tables in Chapter 6 Tables 6.14a to 6.14g.

APPENDIX 6

THE ESTIMATION OF MEAN MONTHLY HOURLY VALUES OF BRIGHT SUNSHINE FROM MONTHLY MEAN DAILY BRIGHT SUNSHINE

The monthly mean daily relative sunshine duration, σ_d, sometimes called percentage possible sunshine, is a familiar concept in applied climatology. It is simply defined as the ratio of the monthly mean daily hours of bright sunshine to the monthly mean astronomical daylength. σ_d does not unfortunately describe the distribution of sunshine across the day, which is an important defect when attempting to estimate monthly means of the irradiance at specific times of day, especially on slopes facing east and west where the peak energies occur relatively early and late in the day. If hourly sunshine observations are available, one may also calculate a monthly mean hourly relative sunshine duration σ_h for a given time of day in a specific month. σ_h is the mean monthly hourly sunshine recorded during the period lying half an hour before or after the specified hour during a particular month divided by the number of observations at that hour in that month. Unfortunately, few stations extract values of observed sunshine on an hourly basis, so a technique for estimating σ_h from σ_d is sometimes useful in applied solar energy studies.

Figure A6.1 shows the values of the ratio σ_h/σ_d for 13 West German stations as reported by the Institut für Lichttechnik (A6.1)(A6.2). It will be noted that the values of the ratio are well above one around noon, whereas, close to sunrise and sunset, they are well below 0.5. The monthly mean hourly relative sunshine duration may thus be correlated with solar altitude. There are two reasons for this strong correlation with solar altitude. Firstly, it has to be remembered that broken clouds may have an appreciable vertical depth which will produce a more substantial solar cut off at low solar altitudes compared with conditions when the sun is higher. Therefore, the lower the sun, the greater the probability for a given broken cloud cover that some clouds will stand in the way of the direct beam. Secondly the lower the sun, the greater the atmospheric attenuation of the direct beam, so the sunshine recorder card burn threshold which is typically around 150 - 200 Wm^{-2} has a greater probability of not being reached when the sun is low than when it is high, especially on more turbid days. The consequence is the observed monthly mean hourly relative sunshine duration becomes less as the solar altitude decreases. This factor particularly influences the direct beam radiation availability early and late in the day.

The sunshine studies carried out on hourly sunshine data for Germany showed that the observed hourly data for the whole of the country could be collapsed into a dimensionless form using a concept which the Institut für Lichttechnik called relative sunshine probability. The relative sunshine probability is defined as the ratio of the actual monthly mean hourly relative sunshine duration at a given solar altitude to the predicted relative sunshine duration value when the sun is vertically overhead. Theoretical studies showed the correlation was not linear. A pragmatic mathematical function for relative sunshine probability as a function of

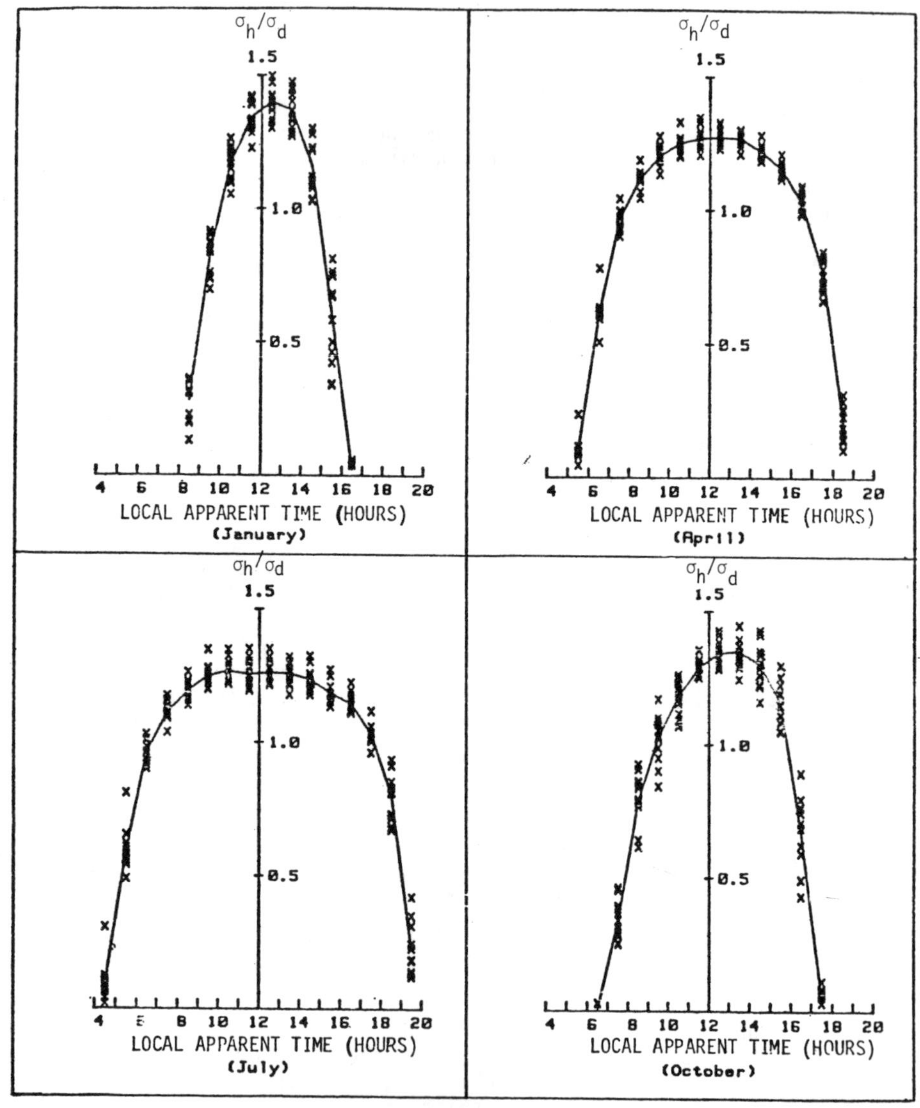

Figure A6.1 Daily variation of monthly mean hourly relative sunshine duration σ_h expressed as a fraction of monthly mean daily relative sunshine duration $\sigma_m = S_m/S_o$ for 13 West German stations (1951-70). Source: Krochmann, Rattunde and Aydinli.

solar altitude was found to cover adequately all German data. This function also gave good results using observations at Locarno-Monti. Observed UK hourly sunshine data was also shown to conform to a relatively similar pattern. It should be noted, however, that this approach produces a symmetrical sunshine function because it does not differentiate between morning and afternoon sunshine amounts.

The pragmatic relationship between relative sunshine probability and solar altitude is shown graphically in Figure A6.2. This function can be expressed mathematically by:

$$\sigma_{rel} = 2.5 \tan \gamma \text{ for } \gamma < 10° \qquad \text{dimensionless}$$

and

$$\sigma_{rel} = 1.0 - 0.1/\tan \gamma \text{ for } \gamma > 10° \qquad \text{dimensionless}$$

If one integrates the relative sunshine probability function over the daylength, i.e.

$$\int_{t_r}^{t_s} \sigma_{rel} \, dt$$

where t_r is the time of sunrise and t_s is the time of sunset, one obtains the integrated relative daylength S_{orel} in hours, i.e. the daylength on a relative sunshine weighted basis. The integration of σ_{rel} in the above equation was carried out numerically. In the Berlin study, a six minute time step was used. The integral is a function of latitude and month. The value of S_{orel}, the weighted daylength, is, of course, shorter than the astronomical daylength. Thus, defining σ_r as the monthly mean daily sunshine duration on a weighted sunshine probability basis, σ_r may be calculated from the weighted daylength S_{orel} as

$$\sigma_r = S_m/S_{orel} \qquad \text{dimensionless}$$

where S_m is the monthly mean daily duration of bright sunshine (hours).

Table A6.1 gives monthly values of S_{orel} for latitudes 30°-75°N.

The value of σ_h is higher, of course, than the value of the average daily relative sunshine duration σ_r because of the influence of the solar elevation weighting on relative sunshine probability on S_{orel}.

One may then obtain the estimated mean monthly hourly relative sunshine duration σ_h for any given hour in a particular month by multiplying σ_r by the relative sunshine probability σ_{rel} for the actual solar altitude at that time of day.

$$\sigma_h = \sigma_r \cdot \sigma_{rel} \qquad \text{dimensionless}$$

The effect of using the relative sunshine probability concept in the modelling procedure compared with using the classical definition of monthly mean daily relative sunshine duration σ_m is to raise the hourly relative sunshine duration around noon and to decrease it sharply when the sun is very low. The method, however, does not allow asymmetries of sunshine between morning and afternoon to be considered, which remains one weakness of the approach.

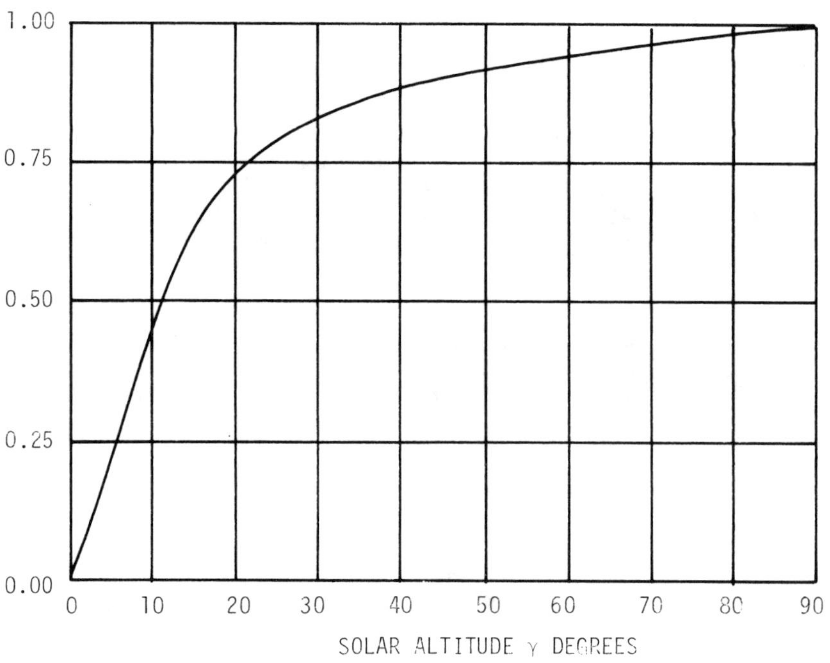

Figure A6.2 Relative sunshine probability as a function of solar altitude. Source: Krochmann, Rattunde and Aydinli.

Table A6.1 Relative sunshine weighted daylength (hours), S_{orel}, at different latitudes based on declination values from Table 6.2.

LATITUDE	MONTH											
	J	F	M	A	M	J	J	A	S	O	N	D
30	7.14	7.99	9.01	9.96	10.66	10.98	10.84	10.26	9.41	8.39	7.41	6.87
31	7.04	7.92	8.97	9.96	10.69	11.03	10.88	10.27	9.38	8.33	7.31	6.75
32	6.93	7.84	8.94	9.96	10.72	11.08	10.92	10.29	9.36	8.27	7.21	6.64
33	6.82	7.76	8.90	9.96	10.75	11.12	10.96	10.30	9.34	8.21	7.11	6.52
34	6.70	7.68	8.86	9.96	10.78	11.17	11.00	10.31	9.31	8.14	7.01	6.39
35	6.59	7.60	8.82	9.96	10.81	11.22	11.03	10.32	9.29	8.08	6.91	6.26
36	6.47	7.52	8.77	9.95	10.84	11.26	11.07	10.34	9.26	8.01	6.80	6.13
37	6.34	7.43	8.73	9.95	10.87	11.31	11.11	10.35	9.23	7.93	6.68	6.00
38	6.21	7.34	8.68	9.95	10.91	11.36	11.15	10.36	9.21	7.86	6.57	5.86
39	6.08	7.25	8.64	9.94	10.93	11.41	11.19	10.37	9.18	7.79	6.45	5.71
40	5.95	7.15	8.59	9.94	10.97	11.46	11.24	10.38	9.14	7.71	6.33	5.56
41	5.80	7.05	8.53	9.93	11.00	11.51	11.28	10.39	9.11	7.63	6.20	5.40
42	5.66	6.95	8.48	9.92	11.03	11.56	11.32	10.40	9.08	7.54	6.06	5.24
43	5.51	6.84	8.42	9.92	11.06	11.61	11.36	10.41	9.04	7.46	5.93	5.07
44	5.35	6.73	8.37	9.91	11.09	11.67	11.41	10.42	9.00	7.37	5.78	4.90
45	5.18	6.62	8.31	9.90	11.13	11.72	11.45	10.42	8.97	7.28	5.64	4.72
46	5.01	6.50	8.24	9.89	11.16	11.78	11.50	10.43	8.93	7.18	5.48	4.53
47	4.83	6.37	8.18	9.88	11.20	11.84	11.54	10.44	8.88	7.08	5.32	4.34
48	4.65	6.24	8.11	9.87	11.23	11.90	11.59	10.45	8.84	6.97	5.15	4.13
49	4.46	6.11	8.04	9.86	11.27	11.96	11.64	10.46	8.79	6.86	4.98	3.92
50	4.26	5.97	7.97	9.85	11.30	12.02	11.69	10.47	8.74	6.75	4.80	3.69
51	4.04	5.82	7.89	9.83	11.34	12.09	11.75	10.47	8.69	6.63	4.61	3.46
52	3.82	5.67	7.81	9.82	11.38	12.16	11.80	10.48	8.64	6.51	4.41	3.22
53	3.59	5.51	7.73	9.80	11.43	12.23	11.86	10.49	8.59	6.38	4.20	2.97
54	3.35	5.34	7.64	9.79	11.47	12.30	11.92	10.50	8.53	6.25	3.98	2.71
55	3.10	5.17	7.55	9.77	11.51	12.38	11.98	10.51	8.47	6.10	3.76	2.44
56	2.84	4.98	7.46	9.76	11.56	12.47	12.05	10.52	8.41	5.96	3.52	2.16
57	2.57	4.79	7.35	9.74	11.61	12.56	12.12	10.53	8.34	5.80	3.27	1.90
58	2.29	4.59	7.25	9.72	11.67	12.65	12.20	10.54	8.27	5.64	3.00	1.64
59	2.01	4.38	7.14	9.70	11.72	12.75	12.28	10.55	8.20	5.47	2.73	1.39
60	1.75	4.15	7.02	9.68	11.78	12.86	12.36	10.56	8.12	5.28	2.45	1.15
61	1.50	3.91	6.90	9.66	11.85	12.98	12.46	10.57	8.04	5.09	2.17	0.93
62	1.25	3.66	6.77	9.63	11.92	13.11	12.56	10.59	7.95	4.89	1.90	0.72
63	1.02	3.40	6.64	9.61	11.99	13.25	12.66	10.61	7.86	4.68	1.64	0.52
64	0.80	3.12	6.49	9.59	12.08	13.41	12.79	10.62	7.77	4.45	1.38	0.34
65	0.59	2.84	6.34	9.56	12.17	13.59	12.92	10.64	7.67	4.21	1.14	0.19
66	0.40	2.54	6.17	9.54	12.27	13.80	13.07	10.67	7.56	3.96	0.91	0.07
67	0.24	2.24	6.00	9.51	12.39	14.06	13.24	10.69	7.45	3.69	0.69	0.00
68	0.10	1.96	5.82	9.48	12.51	14.38	13.45	10.72	7.33	3.41	0.49	0.00
69	0.01	1.69	5.62	9.46	12.66	14.70	13.71	10.76	7.20	3.11	0.31	0.00
70	0.00	1.42	5.41	9.43	12.84	15.01	14.02	10.80	7.07	2.80	0.15	0.00
71	0.00	1.16	5.18	9.40	13.07	15.32	14.34	10.85	6.92	2.48	0.04	0.00
72	0.00	0.91	4.94	9.37	13.37	15.62	14.64	10.91	6.77	2.19	0.00	0.00
73	0.00	0.68	4.68	9.35	13.67	15.91	14.94	10.98	6.60	1.89	0.00	0.00
74	0.00	0.46	4.41	9.32	13.97	16.20	15.24	11.07	6.42	1.60	0.00	0.00
75	0.00	0.27	4.11	9.30	14.27	16.48	15.53	11.19	6.23	1.32	0.00	0.00

However, detailed studies carried out in the University of Sheffield using observed monthly mean hourly values of global and diffuse radiation for several stations in Europe, for which hourly sunshine values were also available, showed that the use of hourly sunshine duration did not improve the accuracy of prediction of the monthly mean hourly components of solar radiation either on the horizontal plane, or on inclined surfaces. In fact, it sometimes made these estimates worse for inclined surfaces, because of the important influence of the threshold burn level of sunshine recorders on the methodology.

Because $\sigma_{rel} = 0$ when the solar altitude $\gamma = 0$, the Berlin method predicts the total extinction of the monthly mean direct beam at sunrise and sunset, a result not in accord with observation. The direct beam is thus improperly accounted for at low solar altitudes. This makes little impact on the horizontal values of the irradiance because cos ν at low solar altitudes is so small, but for surfaces facing the sun, the errors become significantly large. As there is no way of determining the actual relative hourly sunshine duration at low solar altitude from imperfect sunshine recorders the Berlin methodology was abandoned as a method for producing the CEC tables as it was found that the use of relative sunshine duration produced systematic errors of considerable magnitude for the direct beam intensity at the hourly level. An alternative more accurate, but more complex approach for assessing the monthly mean beam irradiance at the hourly level is described in the main text of Chapter 5. At the daily level, the overestimates at one period balance the underestimates at another period of the day, and so the daily predictions of the Berlin model are surprisingly reliable in spite of the defect of the approach at the hourly level.

Appendix 6 - References

A6.1 Krochmann, J. (1974), Über die Sonnenscheinwahrscheinlichkeit in der Bundesrepublik Deutschland, Lichttechnik 25, 10, 11.

A6.2 Aydinli, S. (1981), Über die Berechnung der zur Verfugung Stehenden Solarenergie und des Tageslichtes. Dissertation TU Berlin, Fortschrittsberichte der VDI Zeitschriften-Reihe 6, No.79.

APPENDIX 7

DETERMINATION OF THE RATIO OF THE CLOUDLESS SKY DIFFUSE SKY IRRADIANCE/ILLUMINANCE ON A HORIZONTAL SURFACE TO THE ZENITH RADIANCE/LUMINANCE OF THAT SKY

Predicted ratio (horizontal diffuse irradiance/zenith radiance) - clear skies

The ratio D_c/L_{cz}, where D_c is the horizontal surface clear sky diffuse irradiance, and L_{cz} is the corresponding clear sky zenith radiance, is important for the rapid computation of the relationships between diffuse sky irradiance and the radiance of any part of the sky. The value of D_c/L_{cz} for a sky of uniform radiance is π (= 3.142). Table A7.1 gives the values of this ratio computed for Linke Turbidity Factors of 1.5, 2, 4, 6, 8, and 10 using the different theoretical models discussed in Chapter 3 for the prediction of the luminance or radiance of the cloudless sky.

Five radiance/luminance distribution formulae were used to produce Table A7.1.

1. Modified Libelt equation: Equation 3.13.

2. Formula used to produce CEC clear sky slope irradiance tables: Equation 3.14.

3. Formula modified to match broadly with Valko's results: Equation 3.21.

4. Kittler's formula adopted as standard by the Commission Internationale de l'éclairage (CIE): Equation 3.11.

5. Gusev's formula: Equation 3.12.

Valko has presented in Figure 3.17 observed values of the ratio $\pi L_{cz}/D_c$, i.e. the reciprocal of the function in Table A7.1 multiplied by π. For a uniform sky this ratio would be 1. In practice, when the sun is low, the observed value of the ratio is around 0.4. The zenith is close to the position of minimum sky radiance at 90° on the sun's meridian, so the radiance of the zenith is low relative to that of the horizon zone, when the sun is very low. With the high sun, the ratio exceeds one, as the zenith radiance is then associated with the aureole around the sun and so is high relative to the rest of the sky. When the sun has an elevation of about 60°, the value of the ratio is about 1.

The values of $\pi L_{cz}/D_c$ predicted using the new radiance formula equation 3.21 have been plotted onto Valko's data in Figure 3.30. It will be seen that the predicted curves overlay the observed data very well, except the values for the mountain top site of the Weissfluhjoch, where the observed values of the ratio are somewhat lower than predicted. The model was not, however, developed to deal with sites of such great altitude.

Using the data in Table A7.1 a suitable algorithm has been devised to enable the zenith radiance to be predicted from the diffuse sky irradiance on a horizontal surface for any solar altitude and where relevant, any Linke Turbidity Factor $T_L(\gamma)$. This is given in Table A7.2. This eliminates the need to use full numerical integration methods to establish the relationship $\pi L_{cz}/D_c$. Once L_{cz} has been established it becomes straightforward to estimate $L_c(\Theta,\alpha_s)$ for any element of the sky using the radiance/luminance distribution formula, Equation 3.21.

Table A7.1 Computed ratio of D_c/L_{cz} at stated solar elevations

$T_L(\gamma)$		D_c/L_{cz} Solar altitude									
		1	10	20	30	40	50	60	70	80	90
2.0	Modified Liebelt		6.98	6.74	6.21	5.46	4.61	3.73	2.87	2.09	1.49
	CEC formula		6.87	6.73	6.36	5.76	4.88	3.76	2.57	1.57	0.878
	New formula	7.38	7.47	7.16	6.46	5.54	4.54	3.55	2.64	1.87	1.260
4.0	Modified Liebelt		6.84	6.69	6.13	5.26	4.24	3.22	2.31	1.58	1.02
	CEC formula		6.84	6.79	6.35	5.53	4.40	3.16	2.06	1.24	0.706
	New formula	7.41	7.67	7.42	6.60	5.45	4.20	3.05	2.10	1.38	0.875
6.0	Modified Liebelt		6.80	6.70	6.10	5.14	4.02	2.94	2.03	1.34	0.846
	CEC formula		6.84	6.82	6.32	5.36	4.12	2.88	1.85	1.11	0.642
	New formula	7.86	8.18	7.83	6.78	5.38	3.97	2.76	1.83	1.17	0.724
8.0	Modified Liebelt		6.81	6.73	6.09	5.05	3.86	2.76	1.86	1.20	0.748
	CEC formula		6.84	6.82	6.26	5.21	3.93	2.70	1.72	1.04	0.607
	New formula	8.56	8.85	8.28	6.96	5.34	3.82	2.58	1.68	1.05	0.643
10.0	Modified Liebelt		6.84	6.77	6.08	4.97	3.74	2.63	1.75	1.11	0.686
	CEC formula		6.84	6.81	6.19	5.09	3.79	2.59	1.65	0.998	0.585
	New formula	11.97	11.02	8.93	6.78	4.96	3.52	2.42	1.61	1.037	0.646
	Kittler CIE	7.01	7.24	7.06	6.40	5.41	4.29	3.20	2.25	1.51	0.970
	Gusev		8.10	7.89	7.00	5.70	4.29	3.03	2.03	1.30	0.812

Table A7.2 Algorithm to predict the ratio of the horizontal irradiance cloudless days/zenith radiance.

$$D_c/L_{cz} = \sum_{i=0}^{4} \left(\sum_{j=0}^{3} a_{ij}(T_L(\gamma))^j \right) \gamma^i$$

(Refer equation 3.21 for basis of numerical integration) Applicable for values of $T_L(\gamma)$ between 2 and 8 and solar altitudes 70° and below, the following constants gives the relevant values, with an accuracy usually better than 1%, except at γ = 70° when the error may rise to 3.8% at high turbidities.

i \ j	0	1	2
0	7.670	-2.415×10^{-1}	4.375×10^{-2}
1	-1.352×10^{-2}	3.202×10^{-2}	-2.006×10^{-3}
2	-1.283×10^{-3}	-1.003×10^{-3}	1.437×10^{-5}
3	1.133×10^{-5}	7.637×10^{-6}	5.188×10^{-7}
4	-3.488×10^{-8}	-8.237×10^{-9}	-4.890×10^{-9}

APPENDIX 8

ACCURACY OF PREDICTION OF DAILY GLOBAL RADIATION ON A HORIZONTAL SURFACE ACHIEVED BY THE EUROPEAN COMMUNITY SOLAR RADIATION MODEL

The model proposed estimates the daily global irradiation by summing the hourly values of the irradiation. The key inputs in addition to station latitude and height are

i) monthly mean values of the observed bright sunshine S and monthly mean maximum values of the observed bright sunshine S_{max}.

ii) monthly values of the sum of the derived Angstrom regression coefficients (a + b) for estimating daily global irradiation from daily bright sunshine.

The air mass 2 Linke Turbidity Factors used are derived variables obtained from input (ii). Complete input data for 96 of the 102 sites was available together with observed data of monthly mean daily global irradiation on a horizontal surface, G, and also of monthly mean maximum daily global irradiation, G_{max}.

The results of prediction can be compared with the observations made of G and G_{max} at each site. The error is reported here as the normalisation factor i.e. observed monthly mean daily horizontal global solar radiation divided by predicted monthly mean horizontal daily global solar radiation. All the hourly horizontal predicted irradiance values were multiplied by the normalisation factor, before computation of the inclined surface irradiance values proceeded.

The normalisation factors for each site are given at the base of each of the two station tables. There are different sets of normalisation factors, one for monthly mean values and the other for monthly mean maximum values. The clear days are normalized against G_{max} and the monthly means against G.

Table 5.15 in the main text summarises statistically the prediction errors of the basic monthly mean model. Statistically the model slightly underestimates when considered against the whole data set. However one has to consider that some of the observed data has been obtained using Robitzsch pyranometers which tend to overestimate the solar radiation especially on clear days

Table A8.1 contains a more detailed discussion for all cases where the difference between prediction and observation for monthly means exceeds 10%. Table A8.2 gives the corresponding analysis for clear days for all observations where the error exceeded $\pm$ 10%.

In general the models perform well, and in some cases where they do not perform well, there are good reasons for raising doubts about the accuracy or suitability of the input data either on grounds of instrumental suitability, or calibration shortcomings. There is also the problem of assessing the effects of obstructions like mountains. As detailed data on the obstruction characteristics of sites was not generally available, it was not possible to consider this problem scientifically. When obstruction occurs, the monthly mean bright sunshine is normally reduced proportionately more than the observed global radiation on horizontal surfaces, as the contribution of the low sun to horizontal irradiation is small. Thus severe obstruction is likely to produce an underestimate of global radiation, i.e. normalisation factor well above 1.00, because the relative daily sunshine duration input, i.e. S/S_{04}, will be too low.

The main weakness of the model is that it does not take account of changes in the inter-reflection of diffuse energy between sky and ground, thus the diffuse energy from the sky is likely to be underestimated when the ground is snow covered. The model, it appears, would produce improved estimates if the diffuse radiation were increased by about 10% when the ground was snow covered. As no systematic information was available about snow cover duration, corrections for snow cover could not be applied in producing the CEC European Solar Radiation Atlas, Vol. II. If snow cover data could be systematised on a European basis, modelling accuracy could be improved.

Table A8.1 Analysis of modelling errors in the prediction of mean horizontal surface daily irradiation exceeding ±10% in Tables prepared for CEC Inclined Surface Radiation Atlas

Errors are reported as the normalisation factor given in the tables, i.e. observed global radiation/predicted global radiation.

Stations with 4 or more months showing horizontal surface global radiation modelling prediction errors of greater than ±10%

SONNBLICK, 3106m — Jan 1.22, Feb 1.27, Mar 1.30, Apr 1.30, May 1.19, Oct 1.14, Nov 1.16, Dec 1.19.

Comment: Model not suitable for such high altitudes, especially as effects of reflected radiation from snow not modelled.

FICHTELBERG, 1214m — Jan 0.81, Feb 0.79, Mar 0.82, Apr 0.86, May 0.84, Jun 0.80, Jul 0.83, Aug 0.89, Sep 0.89, Oct 0.83, Nov 0.79, Dec 0.78.

Comment: Reason for relatively low observed values in relation to prediction not understood. The values of the Angstrom a + b were estimated by comparison with the Bundesrepublik Deutschland. The estimates appear to have ascribed too clear an atmosphere.

LOGRANO, 353m — Mar 1.11, Apr 1.14, May 1.11, Sep 1.31

Comment: CEC Atlas values estimated from sunshine.

ZARAGOZA, 258m — Jan 1.17, Feb 1.17, Mar 1.17, Apr 1.36, May 1.22, Jun 1.23, Jul 1.22, Aug 1.26, Sep 1.26, Sep 1.35, Oct 1.28, Nov 1.29.

Comment: CEC Atlas values estimated from sunshine.

TORTOSA, 44m — Jan 1.25, Feb 1.26, Mar 1.25, Apr 1.14, May 1.17, Jun 1.11, Nov 0.89.

Comment: CEC Atlas values estimated from sunshine.

MADRID, 669m — May 0.88, Jun 0.89, Jul 0.87, Oct 0.88, Nov. 0.85.

Comment: Values of Angstrom a + b seem exceptionally high for an urban site March - August leading on to overestimates in modelling.

BERGEN, 45m — Jan 1.17, Feb 1.13, Nov 1.11, Dec 1.15.

Comment: Very high latitude with considerable winter snow. Model does not allow for reflection from snow enhancing diffuse radiation. Probably some mountain obstruction of sunshine as well.

Prediction errors greater than ±10% in March - September with less than 4 months in year with prediction errors above 10%

PHILADELPHIA Mar 1.13.

Comment: Short record 2 years only

PORTO May 1.11, Jun 1.11, Jul 1.11

Comment: Instrument not known possibly Robitzsch pyranometer

COIMBRA Mar 1.12, Apr 1.11.

Comment: Instrument not known possibly Robitzsch pyranometer

ZAKOPANE, 857m May 0.88, Jun 0.85, Jul 0.85

Comment: Summer values seem low in relation to sunshine. Station height might indicate some obstruction.

Prediction errors of greater than ±10% in winter October - February with less than 4 months a year with prediction errors above ± 10%

December only

KILKENNY 1.18, LERWICK 1.13, JERSEY 0.89, ST. HUBERT 0.87, BERLIN 1.11, TRIER0.88, SVALOV 1.16, PRISTINA 1.11.

Comment: Difficult to make accurate predictions and estimates with low sun. Influence of snow at some sites.

January only

WEIHENSTEPHAN 1.12, SVALOV 1.11, STOCKHOLM 1.14, ZURICH 1.11.

Comment: Influence of snow likely.

February only

INNSBRUCK 1.11, KARLSTAD 1.13.

Comment: Influence of snow likely.

POTSDAM 0.88.

Comment: Urban pollution possibly underestimated.

Two months in winter

KLAGENFORT Jan 1.12, Feb 1.13.

Comment: Influence of snow likely.

ALIARTOS Dec 0.84, Jan 0.86.

Comment: Reasons for apparently low observed values not understood.

VISBY Dec 1.12, Jan 1.12.

Comment: Influence of snow likely.

LJUBLJANA Nov 0.84, Dec 0.88.

Comment: Derived values of Angstrom (a + b) in winter generally appear to be unreasonably high in Yugoslavia.

DAVOS Dec 1.12, Jan 1.19.

Comment: Influence of snow and obstruction.

Three months in winter

CAWOOD Nov 1.16, Dec 1.19, Jan 1.14.

Comment: Station of modest calibration status

BRATISLAVIA Nov 0.87, Dec 0.86, Jan 0.85.

Comment: Atmospheric clarity probably overestimated.

WARSAW Oct 0.89, Nov 0.86, Dec 0.88.

Comment: Atmospheric clarity probably overestimated.

Table A8.2 Analysis of modelling errors exceeding ±10% in the prediction of monthly mean maximum values of global irradiation on a horizontal surface G_{max} in the Tables prepared for the CEC Inclined Surface Radiation Atlas

The figures given against each station are the normalisation values for the tables, i.e. observed global radiation/predicted global radiation

Stations with 3 months or more with prediction errors greater than ±10%

DAVOS, 1590m — Oct 1.15, Nov 1.19, Dec 1.20, Jan 1.24.

Comment: Influence of mountains shortens sunlight availability in mid-winter, hence input S_{max}/S_{max40} probably too low. Model does not allow for the effects of inter-reflection between snow and sky.

FICHTELBERG, 1214m — Jan 0.86, Feb 0.88, Oct 0.83, Dec 0.89.

Comment: A similar feature is observed in the monthly mean predictions - the reasons for the low values are not understood however. Values of Angstrom a + b were estimated by comparison with stations in Bundesrepublik Deutschland. The turbidities appear higher in the DDR.

LOGRANO, 353m — Feb 1.19, Jun 1.17, Jul 1.16, Sep 1.14, Oct 1.17, Dec 1.22.

Comment: CEC Atlas values estimated from sunshine.

ZARAGOZA, 258m — Mar 1.14, Apr 1.24, May 1.19, Jun 1.27, Jul 1.31, Aug 1.32, Sep 1.20, Nov 1.16, Dec 1.21.

Comment: CEC Atlas values estimated from sunshine.

SALAMANCA, 793m — Jan 0.85, Feb 0.87, Mar 0.89, Apr 0.89.

Comment: CEC Atlas values estimated from sunshine.

JYNDEVAD, 15m — Feb 1.13, Oct 1.12, Dec 1.18.

Comment: Short record and relatively low sun.

PHILADELPHIA, 138m — Jan 0.81, Jul 0.87, Dec 0.88.

Comment: Reasons not understood, but only two year record.

BOLZANO, 241m — Feb 1.11, Sep 1.14, Dec 1.15.

Comment: Italian data from Robitzsch pyranometer. Arbitrary downward adjustments possibly too low. Sunshine record appears low, possibly mountain obstruction.

Stations with less than 3 months with prediction errors greater than ±10%

PIANOSA Jan 1.12, Oct 1.11.

PANTELLERIA Jan 1.13.

TRAPANI Nov 1.11.

Comment: Italian data was based on data observed on Robitzsch pyranometers. A pragmatic reduction factor was applied to allow for over-estimates. The pragmatic reduction was possibly too low for certain months.

HEILIGENDAMM Jan 0.82, Feb 0.88.

POTSDAM Jan 0.87, Dec 0.80.

DRESDEN Jan 0.81, Feb 0.87.

Comment: Values of Angstrom a + b had to be estimated by comparison with stations in the Bundesrepublik Deutschland. Insufficient allowance seems to have been made for the turbidity characteristics of the atmosphere in mid-winter in the Deutsches Demokratische Republik.

LOCARNO-MONTI Dec 1.11

Comment: Mountains lead to obstruction of sunshine, S/S_0 probably too low. Model does not allow for effects of inter-reflections from snow.

COIMBRA Jun 1.11.

Comment: Instrument not known, possibly Robitzsch Actinograph. Predictions generally low.

BERGEN Dec 0.84, Jan 0.89.

Comment: Very high latitude with low sun, obstruction by mountains distinct possibility.

KARLSTAD Dec 1.29.

Comment: Very high latitude, high measurement/prediction accuracy difficult to achieve with such a low sun.

STOCKHOLM Oct 1.15.

Comment: Reason for this odd high value not understood.

SVALOV Dec 1.26.

Comment: Relatively high latitude with low sun, high measurement/prediction accuracy difficult to achieve.

NEGOTIN Oct 1.11, Dec 1.17.

SARAJEVO Dec 1.17.

ZLATIBOR	Dec 1.15.
Comment:	The values of the Angstrom a + b appear to be exceptionally high in winter for most of the Yugoslav stations. Instrument type not known, nor calibration status of network.
MADRID	Nov 0.89.
Comment:	The reported radiation at Madrid is generally less than expected for the sunshine throughout the year.
ZAKOPANE	Jan 1.14, Dec 1.13.
Comment:	Reason not known, possibly influence of snow.

APPENDIX 9

COMPARISON OF OBSERVED RADIANCE RATIOS FOR LOW ELEVATIONS IN THE SKY WITH PREDICTED VALUES

The observed data concerning low elevation sky patches arrived too late to be reviewed in the main body of the book. This Appendix compares the results of the Swiss observations at Le Locle and EMPA given in Figures 3.16a to 3.16f with the theoretical radiance model proposed for clear skies in this book, Equation 3.14, with the constants given in Equation 3.21.

The computed values of the ratio L_c/L_{cz} were first estimated at 5° intervals of solar altitude for the range of solar azimuths given in Figure 3.16a to Figure 3.16e for air mass 2 Linke Turbidity Factors of 2 and 4. These values were then plotted logarithmically and the observed data from Figures 3.16a to 3.16e superimposed.

The following conclusions were drawn:

Le Locle - clear site - ground typically snow covered

Azimuth angle 5°, sky patch altitude 18°

The observed data are about twice the predicted values close to the sun's direction, for solar elevations up to 30°. The difference then falls steadily until, at 50° solar altitude, the difference is negligible. It was assumed an air mass 2 Turbidity was representative.

Azimuth angle 5°, sky patch altitude 12°

The results were similar to the 18° patch but the peak differences were smaller (about 1.75 compared with 2).

Azimuth angle 5°, sky patch altitude 8°

The difference between prediction and observation is better than for the 12° and 18° patches. The pattern is the same, but the peak differences were about 1.50 expressed as a ratio.

Azimuth angle 5°, sky patch altitude 5°

The observed curve shows more curvature at low solar altitudes than the theoretical model. As a result, prediction and theory are very close around 10° solar altitude. At 30° solar altitude, the observed values are about 1.5 times the predicted values. At 50° solar elevation, the theory and observation again coincide.

Azimuth angle 60°, sky patch altitude 18°

The observations are slightly higher than prediction. For $T_L = 2$, for solar altitudes between 10° and 40°, the difference is about 10%. At 50 degrees solar altitude, the results coincide.

Azimuth angle 60°, sky patch altitude 12°

The observations are very close to prediction for $T_L = 2$. They are slightly above for low solar elevations, and slightly below for high solar elevations.

Azimuth angle 60°, sky patch altitude 8°

The observations lie between the values predicted for $T_L = 2$ and $T_L = 4$. At 10° solar elevation they correspond with the $T_L = 4$ predicted values and at 30°-40° with the $T_L = 2$ predicted values. At 50° solar elevation they correspond to the values for $T_L = 3$.

Azimuth angle 60°, sky patch altitude 5°

The observed curve has a strong curvature which is not predicted by theory. Between 20° and 30° solar elevation the observed values lie within the band of predictions for $T_L = 2$, and $T_L = 4$, being close to the $T_L = 4$ values at 20° solar elevation, the $T_L = 2$ values at 30°-40° and $T_L = 3$ values at 50° degrees.

Azimuth angle 90°, sky patch altitude 18° and patch 12°

The observed values are well predicted by the model, lying within the band of predicted values for $T_L = 2$ and $T_L = 4$, and fairly close to the $T_L = 2$ values.

Azimuth angle 90°, sky patch altitude 8°

The observations all lie within the range of values predicted for $T_L = 2$ and $T_L = 4$. However, greater curvature is apparent. At 10°, the observations correspond to predicted values for $T_L = 4$ and between 30°-40° solar elevation, to values for $T_L = 2.5$.

Azimuth angle 90°, sky patch altitude 5°

The curvature of the observations is very strong compared with prediction. At 20° solar altitude, the observations are close to the $T_L = 4$ predictions, in the region 30°-40° close to the $T_L = 2.2$ predictions.

Azimuth angle 180°, sky patch altitude 18°

The observed values are close to the predicted values for $T_L = 3.5$. The observed azimuth 180° values exceed the 90° azimuth values below 30°. The theoretical model predicts a crossover at a similar solar altitude.

Azimuth angle 180°, sky patch altitude 12°

There is greater curvature than the theoretical model predicts. The observed values lie within the range of predicted values for $T_L = 2$ and $T_L = 4$ except around solar elevations of 10°. The observed crossover point with the 90° observation occurs at about 36°. Theory predicts about 45° degrees.

Azimuth angle 180°, sky patch altitude 8°

The observed curvature is stronger than for the 8 degrees patch. However observed values lie within the predicted range for $T_L = 2$ and $T_L = 4$.

Azimuth angle 180°, sky patch altitude 5°

The observed curves show very strong curvature not predicted by theory. However, the values for solar elevations above 20° lie between the predicted values for $T_L = 2$ and $T_L = 4$.

Conclusions for Le Locle

1. The model underpredicts the observed radiances close to the sun's direction. However, the model was not developed for sites with snow cover.

2. Some observations show greater curvature than the model predicts. This curvature is not evident in the EMPA results, and the scientific reasons for it are not clear.

3. In general there is good agreement between the model and observation from low sky patch altitudes, considering the fact the range of radiance observations is two orders of magnitude.

Observations for EMPA

Azimuth angle 15°, sky patch altitude 9°

Assuming a representative turbidity of 4, the observations lie above the predicted values by a factor of 25-50%.

Azimuth angle 15°, sky patch altitude 3°

The observations are matched fairly closely, if a T_L value of 3.5 is assumed in the theoretical model.

Azimuth angle 60°, sky patch altitude 9°

The observed values are about 30% above the predicted $T_L = 2$ values at 15° solar altitude. They coincide with the $T_L = 2$ values at 30° solar altitude. For the range 40°-55°, they coincide with the $T_L = 2.8$ theoretical values.

Azimuth angle 60°, sky patch altitude 3°

The observed values are close to the values predicted for $T_L = 3.5$.

Azimuth angle 90°, sky patch altitude 9°

The 90° degree values lie above the 180° degree values, except at large solar altitudes. This is the opposite result to that predicted by theory. It was not a result found at Le Locle. The slope of the curve is greater than predicted by the theory. The observed values, however, lie in the range predicted for $T_L = 2$ and $T_L = 4$.

Azimuth angle 90°, sky patch altitude 3°

The 90 degree values always lie above the 180 degree value, a result not predicted by theory. The observed 90 degree values are close to the values predicted for $T_L = 3.5$.

Azimuth angle 180°, sky patch altitude 9°

The observed values are close to the predicted values for $T_L = 3.5$, but the observed slope is shallower.

Azimuth angle 180°, sky patch altitude 3°

The observed values, which are always below the 90 degree values, are low compared with the theory. They are about 20% below the predicted T_L = 4 values.

Conclusions for EMPA

1. The model underpredicts the radiance slightly for small solar azimuth angles.

2. The reason why the 90° azimuth values are above the 180° azimuth values is not clear.

3. In general the model predictions are satisfactory.

General conclusions

1. The general validity of the theoretical radiance model for low elevation sky patches has been confirmed.

2. Its weakness in predicting the observed radiance close to the sun remains. However, it has not been possible to examine whether the instrumentation can adequately resolve the radiance field in this region, in the presence of very strong nearby radiance fields, which would augment the primary signal.

3. The role of back-reflected sky diffuse radiation from snow covered ground has not been examined in the modelling process, so less accurate prediction results are likely in presence of snow.

APPENDIX 10

BLANK PROFORMAS FOR RADIATION CALCULATIONS USING EC METHODOLOGY, DESK TOP METHOD

It is suggested for desk top calculations that the blank forms that follow should be copied, using an enlarging photocopier, so that adequate space is available in which to enter the figures. Chapter 6 provides an example of the four Proformas with figures already entered.

PROFORMA 1: MONTHLY MEAN PREDICTIONS – HORIZONTAL SURFACE CLEAR DAY SOLAR RADIATION VALUES – EC METHOD

Station name: ________________ **Site clarity:** ________________ **Month:** ___

Latitude ϕ deg ___ min ___ N+,S– ___ **Longitude λ** deg ___ min ___ E+,W– ___ **Declination δ** from Table 6.2 ± ___ Degrees ___

Sunrise hrs ___ min ___ ± ___ $\sin\delta \sin\phi$. ___

Station height (m): ___

AH2T$_L$ Table 6.4 ___

Angstrom coefficients
Monthly values
Angstrom (a + b) . ___

Sunset hrs ___ min ___ ± ___ $\cos\delta \cos\phi$. ___ ± ___ $\sin\delta$. ___ p/p_0 . ___

I_{oj} Table 6.3 ___ Wm^{-2}

Annual mean
Angstrom (a + b) . ___

[1] Local apparent time t		[2] Cosine of hour angle ω	[3] $\sin\gamma = (\cos\delta \cos\phi) \cos\omega + (\sin\delta \sin\phi)$	[4] Solar altitude γ	[5] $\cos\gamma$	[6] Solar azimuth ψ (– a.m., + p.m.) $\cos^{-1}(\sin\phi \sin\gamma - \sin\delta)/\cos\phi \cos\gamma$
a.m.	p.m.			Deg.		Deg.
0330	2030	–0.6088				±
0430	1930	–0.3827				±
0530	1830	–0.1305				±
0630	1730	0.1305				±
0730	1630	0.3827				±
0830	1530	0.6088				±
0930	1430	0.7934				±
1030	1330	0.9239				±
1130	1230	0.9914				±

[7] Optical air mass. If $\gamma < 10°$ then: $m = p/p_0/(\sin\gamma + 0\cdot15\,(\gamma + 3\cdot885)^{-1\cdot25}$ otherwise: $m = p/p_0/\sin\gamma$	[8] Rayleigh optical thickness $\delta_R = 1/(0\cdot9m + 9\cdot4)$	[9] Linke turbidity for actual solar altitude $T_L(\gamma)$	[10] Direct beam irradiance normal to beam $I_c = I_{oj}\exp(-m\delta_R T_L(\gamma))$	[11] Direct beam irradiance on a horizontal surface $I_c(0,0)$ [[3] × [10]]
			Wm^{-2}	Wm^{-2}
Half day totals Whm^{-2}				
Daily totals Whm^{-2}				

[12] Atmospheric transmittance coefficient absorption alone. $q_a{}^m$. Table 6.8	[13] Additional scattering factor f_1 Table 6.9	[14] Diffuse irradiance on horizontal surface D_c $\frac{1}{2}[I_{oj} \times [12] - [10]] \times [13] \times [3]]$
		Wm^{-2}
Half day totals Whm^{-2}		
Daily totals Whm^{-2}		

[15] Global irradiance on a horizontal surface G_c [11] + [14]
Wm^{-2}

PROFORMA 2: HORIZONTAL SURFACE MONTHLY MEAN SOLAR RADIATION PREDICTIONS – EC METHOD

Station name: ☐☐☐☐☐☐☐☐☐☐☐☐☐☐☐☐☐☐ **Site type:** ☐☐☐☐☐☐☐☐☐☐☐☐☐☐ **AM2 T_L** ☐☐.☐☐ **Month:** ☐☐☐

Monthly mean duration of bright sunshine S_m ☐☐.☐☐ **hours**   **Period from** Year ☐☐☐☐ **to** Year ☐☐☐☐   **Angstrom coefficients Month** ☐☐☐

4° degree SR/SS daylength from Table 6.7 ☐☐.☐☐ **hours** σ_{4m} ☐☐.☐☐   **Annual mean** ☐☐☐

[16] Local apparent time t		[17] Sin γ	[18] Overcast day diffuse D_b (= global G_b on hor.) $K_d(2.6 + 182.6\sin\gamma)$	[19] Solar altitude γ	[20] Clear day direct beam horizontal irradiance $I_c(0,0)$	[21] Direct beam correction function for monthly means Table 6.10	[22] Monthly mean direct beam irradiance $I_m(0,0)$ [[20] × [21] × σ_{4m}]
		From Form 1 Col. [3]	K_d from Table 6.2 ☐☐.☐☐	From Form 1 Col. [4]	From Form 1 Col. [11]		
a.m.	p.m.		Wm^{-2}	Deg.	Wm^{-2}		Wm^{-2}
0330	2030						
0430	1930						
0530	1830						
0630	1730						
0730	1630						
0830	1530						
0930	1430						
1030	1330						
1130	1230						
		Half day totals Whm^{-2} overcast				Half day totals Whm^{-2} mean beam	
		Daily totals Whm^{-2} overcast				Daily day totals Whm^{-2} mean beam	

[23] Monthly mean horizontal diffuse irradiance $\sigma_{4m} = 0.25D_{0.25} = K_d(2 + 5.3\gamma)$	[24] Clear day diffuse irradiance D_c	[25] Monthly mean clear day contributions $\sigma_{4m} \times$ [24]	[26] $0.25 \times D_c$ i.e., $0.25 \times$ [24]	[27] Uncorrected overcast day contribution $(1 - \sigma_{4m})(D_{0.25} - 0.25D_c)/0.75$	[28] Diffuse correction function f_6 from Table 6.13 i.e. correct for annual (a + b)	[29] Monthly mean horizontal diffuse irradiance D_m [[25] + [27]] × [28]	[30] Monthly mean global horizontal irradiance G_m [22] + [29]
K_d from Table 6.2 ☐☐.☐☐	From Form 1 Col. [14]						
Wm^{-2}	Wm^{-2}	Wm^{-2}	Wm^{-2}	Wm^{-2}		Wm^{-2}	Wm^{-2}
					Half day totals Whm^{-2} Diffuse and global		
					Daily totals Whm^{-2} Diffuse and global		

PROFORMA 3: INCLINED SURFACE CLEAR DAY AND OVERCAST PREDICTIONS OF SOLAR RADIATION – EC METHOD

Station name: ☐☐☐☐☐☐☐☐☐☐☐☐☐☐☐ **Ground albedo** ρ_g ☐.☐☐ **AM2** T_L ☐☐.☐☐ **Month:** ☐☐☐

Surface orientation α **with respect to due S in N hemis., due N in S hemis.** -E, +W ☐ ☐☐.☐☐ **Ground reflectance factor** f_3 ☐☐☐☐ **Overcast sky slope conversion factor** f_4 ☐.☐☐☐

Surface slope degrees β ☐☐.☐☐

	Enter from Proforma 1 Col. [4]	Col. [6]		
[31]	[32]	[33]	[34]	[35]
Local apparent time	Solar altitude γ	Solar azimuth ψ –ve East +ve West	$\alpha_w = (\psi - \alpha)$ *Refer ro footnote	Cosine of angle of incidence $\cos\nu = \cos\gamma\sin\beta\cos\alpha_w + \sin\gamma\cos\beta$ If $\cos\nu < 0$ then $\cos\nu = 0$
Hours	Deg.	Deg.*	Deg.	
0330 0430 0530				
0630 0730 0830				
0930 1030 1130				
1230 1330 1430				
1530 1630 1730				
1830 1930 2030				

* Values must be between –180° and 180°. Adjust using Stage 15.1.

From Form 1 Col. [10]	Clear Day Beam on Slope	From Form 1 Col. [14]	Clear Day Diffuse on slope				Clear Day Global on Slope
[36]	[37]	[38]	[39]	[40]	[41]	[42]	[43]
Direct beam normal irradiance I_c	Direct beam slope irradiance $I_c(\beta,\alpha)$ [35] × [36]	Clear sky diffuse horizontal D_c	f_2 interpolate from Tables 5.14 A-J	Clear sky diffuse on slope $D_{cs}(\beta,\alpha)$ [38] × [39]	Ground reflected slope diffuse, $R_c(\beta,\alpha)$ Proforma 1 [15] × f_3	Total diffuse on slope, $D_c(\beta,\alpha)$ [40] + [41]	Global on slope, $G_c(\beta,\alpha)$ [37] + [42]
Wm-2	Wm-2	Wm-2		Wm-2	Wm-2	Wm-2	Wm-2
Totals Whm-2			Totals Whm-2				

From Form 2 Col. [18]	Overcast Day Diffuse and Global on Slope		
[44]	[45]	[46]	[47]
Overcast sky horizontal irradiance D_b	Overcast sky diffuse on slope $D_{bs}(\beta,\alpha)$ [44] × f_4	Overcast sky ground reflected slope Form 2 [18] × f_3 (f_3 is above Col. 41)	Overcast sky global on slope $D_b(\beta \cdot \alpha)$ [45] + [46]
Wm-2	Wm-2	Wm-2	Wm-2
Totals Whm-2			

PROFORMA 4: INCLINED SURFACE MONTHLY MEAN SOLAR RADIATION PREDICTIONS – EC METHOD

Station name: ☐☐☐☐☐☐☐☐☐☐☐☐☐☐☐☐☐☐ **Normalisation factor N.F.** ☐.☐☐☐☐☐ **AM2 T_L** ☐☐.☐☐ **Month:** ☐☐☐

Albedo .☐☐

Surface inclination ☐☐☐.☐☐ degrees **Surface azimuth** -E, +W ☐ ☐☐☐.☐☐ degrees **Rel. 4 degree sunshine duration σ_{4m}** ☐☐☐.☐☐ **Ground reflection factor f_3** .☐☐☐

Overcast sky conversion factor f_4 .☐☐☐

[48]	[49]	[50]	[51]	[52]	[53]	[54]	[55]	[56]	[57]	[58]	[59]	[60]	[61]	[62]	[63]
	From Form 1 Col. [3]	From Form 3 Col. [35]	From Form 2 col [22]*		From Form 2 col [29]*	From Form 1 Col. [14]	From Form 2 Col. [25]								
Local apparent time	Sine of solar altitude Sin γ	Cosine of angle of incidence Cos ν	Mean hor. direct beam irradiance, $I_m(0,0)$ Normalised values [22] × N.F.	Monthly mean direct beam slope irradiance, $I_m(\beta,\alpha)$ [51] × [50]/[49]	Monthly mean hor. diffuse irradiance D_m Normalised values [29] × N.F.	Clear sky horizontal diffuse irradiance D_c	Monthly mean clear sky horizontal irradiance contribution $\sigma_{4m}D_c$	Monthly mean overcast partially clouded horizontal sky irradiance contribution [53] – [55]	Monthly mean overcast/partially clouded slope sky irradiance contribution $_{pc}D_m$ [56] × f_4	Monthly clear sky slope irradiance contribution $_cD_m$ = [39] × [54] × σ_{4m}	Uncorrected monthly mean sky diffuse slope irradiance [57] + [58]	If $45° < \lvert\alpha_w\rvert < 135°$, compute additional correction factor $f_8 = (1 + \sin\beta \sin 2(\alpha_s - 45°) \times (0.19 - 1.4 \sin\gamma))$. Else $f_8 = 1$	Corrected monthly mean sky diffuse irradiance on slope $D_m(\beta,\alpha)$ [60] × [59]	Reflected mean diffuse on slope $R_m(\beta,\alpha)$ using normalised values of G_m f_3 × [30] × N.F.	Monthly mean global irradiance on slope, $G_m(\beta,\alpha)$ [52] + [61] + [62]
Hours			Wm^{-2}	Wm^{-2}	Wm^{-2}	Wm^{-2}	Wm^{-2}	Wm^{-2}	Wm^{-2}	Wm^{-2}	Wm^{-2}		Wm^{-2}	Wm^{-2}	Wm^{-2}
0330 0430 0530															
0630 0730 0830															
0930 1030 1130															
1230 1330 1430															
1530 1630 1730															
1830 1930 2030															
			Daily totals Whm^{-2}								Daily totals Whm^{-2}				

* After normalisation

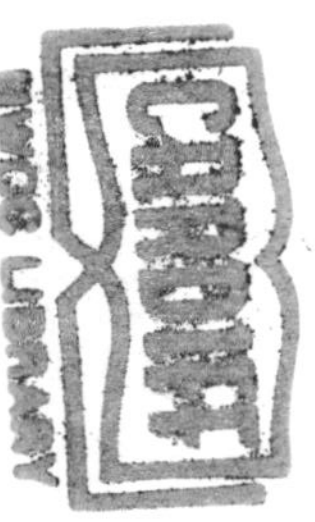